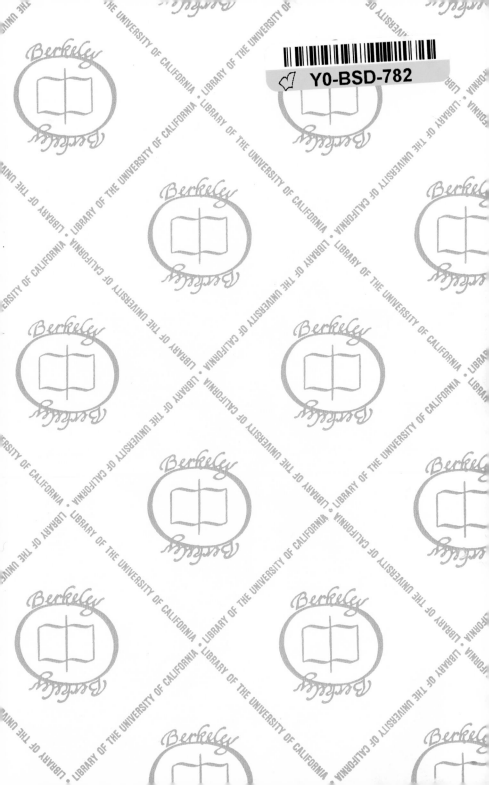

NATIONAL NUCLEAR ENERGY SERIES

Manhattan Project Technical Section

Division VIII — Volume 5

THE CHEMISTRY
OF URANIUM

The Element, Its Binary and Related Compounds

THE CHEMISTRY OF URANIUM

The Element, Its Binary and Related Compounds

BY

JOSEPH J. KATZ

Chemistry Division, Argonne National Laboratory

AND

EUGENE RABINOWITCH

Research Professor, University of Illinois

DOVER PUBLICATIONS, INC.
NEW YORK

This new Dover edition, first published in 1961, is an unabridged and unaltered republication of the first edition published by the McGraw-Hill Book Company, Inc. in 1951 under the title *The Chemistry of Uranium. Part I: The Element, Its Binary and Related Compounds*. The words "Part I" have been deleted from the title of this edition because the work is complete in itself and plans for a second part have been indefinitely suspended.

Manufactured in the United States of America

Dover Publications, Inc.
180 Varick Street
New York 14, N. Y.

FOREWORD

The wartime project for development of atomic energy was a remarkable feat of cooperation and accomplishment by government, science, industry, labor, and the military services aimed exclusively at the military application of atomic energy. Our present national atomic energy program, expanding upon the previous developments, is directed not only toward the assurance of national security but also toward the realization of the immense potential benefits atomic energy holds for our civilization. The record of progress and the results of extensive scientific investigations and engineering development are contained in the National Nuclear Energy Series. This knowledge, which offers the basis of world-wide benefits from nuclear science, is being published in the established scientific tradition, not solely to meet the precise needs of science but also in support of the high goals of the American people set forth in the Atomic Energy Act. The work reported in this series is a tribute to all the scientists engaged in both the Manhattan Project and the postwar Atomic Energy Commission program.

Gordon Dean, Chairman
U. S. Atomic Energy Commission

ACKNOWLEDGMENT

The Manhattan Project Technical Section of the National Nuclear Energy Series embodies results of work done in the nation's wartime atomic energy program by numerous contractors, including Columbia University. The arrangements for publication of the series volumes were effected by Columbia University, under a contract with the United States Atomic Energy Commission. The Commission, for itself and for the other contractors who contributed to this series, wishes to record here its appreciation of this service of Columbia University in support of the national nuclear energy program.

PREFACE

This volume is one of a series which has been prepared as a record of the research work done under the Manhattan Project and the Atomic Energy Commission. The name Manhattan Project was assigned by the Corps of Engineers, War Department, to the far-flung scientific and engineering activities which had as their objective the utilization of atomic energy for military purposes. In the attainment of this objective, there were many developments in scientific and technical fields which are of general interest. The National Nuclear Energy Series (Manhattan Project Technical Section) is a record of these scientific and technical contributions, as well as of the developments in these fields which are being sponsored by the Atomic Energy Commission.

The declassified portion of the National Nuclear Energy Series, when completed, is expected to consist of some 60 volumes. These will be grouped into eight divisions, as follows:

Division   I — Electromagnetic Separation Project
Division   II — Gaseous Diffusion Project
Division III — Special Separations Project
Division IV — Plutonium Project
Division   V — Los Alamos Project
Division VI — University of Rochester Project
Division VII — Materials Procurement Project
Division VIII — Manhattan Project

Soon after the close of the war the Manhattan Project was able to give its attention to the preparation of a complete record of the research work accomplished under Project contracts. Writing programs were authorized at all laboratories, with the object of obtaining complete coverage of Project results. Each major installation was requested to designate one or more representatives to make up a committee, which was first called the Manhattan Project Editorial Advisory Board, and later, after the sponsorship of the Series was assumed by the Atomic Energy Commission, the Project Editorial Advisory Board. This group made plans to coordinate the writing programs at all the installations, and acted as an advisory group in all matters affecting the Project-wide writing program. Its last meeting was held on Feb. 9, 1948, when it recommended the publisher for the Series.

The names of the Board members and of the installations which they represented are given below.

Atomic Energy Commission	
Public and Technical Information Service	Alberto F. Thompson
Technical Information Division, Oak Ridge Extension	Brewer F. Boardman
Office of New York Operations	Charles Slesser, J. H. Hayner, W. M. Hearon *
Brookhaven National Laboratory	Richard W. Dodson
Carbide & Carbon Chemicals Corporation (K-25)	R. B. Korsmeyer, W. L. Harwell, D. E. Hull, Ezra Staple
Carbide & Carbon Chemicals Corporation (Y-12) †	Russell Baldock
Clinton Laboratories ‡	J. R. Coe
General Electric Company, Hanford	T. W. Hauff
General Electric Company, Knolls Atomic Power Laboratory	John P. Howe
Kellex Corporation	John F. Hogerton, Jerome Simson, M. Benedict
Los Alamos	R. R. Davis, Ralph Carlisle Smith
National Bureau of Standards	C. J. Rodden
Plutonium Project	
Argonne National Laboratory	R. S. Mulliken, H. D. Young
Iowa State College	F. H. Spedding
Medical Group	R. E. Zirkle
SAM Laboratories §	G. M. Murphy
Stone & Webster Engineering Corporation	B. W. Whitehurst
University of California	R. K. Wakerling, A. Guthrie
University of Rochester	D. R. Charles, M. J. Wantman

* Represented Madison Square Area of the Manhattan District.

† The Y-12 plant at Oak Ridge was operated by Tennessee Eastman Corporation until May 4, 1947, at which time operations were taken over by Carbide & Carbon Chemicals Corporation.

‡ Clinton Laboratories was the former name of the Oak Ridge National Laboratory.

§ SAM (Substitute Alloy Materials) was the code name for the laboratories operated by Columbia University in New York under the direction of Dr. H. C. Urey, where much of the experimental work on isotope separation was done. On Feb. 1, 1945, the administration of these laboratories became the responsibility of Carbide & Carbon Chemicals Corporation. Research in progress there was transferred to the K-25 plant at Oak Ridge in June, 1946, and the New York laboratories were then closed.

Many difficulties were encountered in preparing a unified account of Atomic Energy Project work. For example, the Project Editorial Advisory Board was the first committee ever organized with representatives from every major installation of the Atomic Energy Project. Compartmentation for security was so rigorous during the war that it had been considered necessary to allow a certain amount of duplication of effort rather than to permit unrestricted circulation of research information between certain installations. As a result, the writing programs of different installations inevitably overlap markedly in many scientific fields. The Editorial Advisory Board has exerted itself to reduce duplication in so far as possible and to eliminate discrepancies in factual data included in the volumes of the NNES. In particular, unified Project-wide volumes have been prepared on Uranium Chemistry and on the Analysis of Project Materials. Nevertheless, the reader will find many instances of differences in results or conclusions on similar subject matter prepared by different authors. This has not seemed wholly undesirable for several reasons. First of all, such divergencies are not unnatural and stimulate investigation. Second, promptness of publication has seemed more important than the removal of all discrepancies. Finally, many Project scientists completed their contributions some time ago and have become engrossed in other activities so that their time has not been available for a detailed review of their work in relation to similar work done at other installations.

The completion of the various individual volumes of the Series has also been beset with difficulties. Many of the key authors and editors have had important responsibilities in planning the future of atomic energy research. Under these circumstances, the completion of this technical series has been delayed longer than its editors wished. The volumes are being released in their present form in the interest of presenting the material as promptly as possible to those who can make use of it.

The Editorial Advisory Board

The Manhattan Project Technical Section of the National Nuclear
Energy Series is intended to be a comprehensive account of the scientific and technical achievements of the United States program for
the development of atomic energy. It is not intended to be a detailed
documentary record of the making of any inventions that happen to be
mentioned in it. Therefore, the dates used in the Series should be
regarded as a general temporal frame of reference, rather than as
establishing dates of conception of inventions, of their reduction to
practice, or of occasions of first use. While a reasonable effort has
been made to assign credit fairly in the NNES volumes, this may, in
many cases, be given to a group identified by the name of its leader
rather than to an individual who was an actual inventor.

AUTHORS' PREFACE

Chemistry has played an important role in the development of nuclear energy even though the more fundamental and spectacular achievements — the chain-reacting pile and the atomic bomb — were primarily the work of physicists.

In the preparation of materials to be used for the separation of uranium isotopes and in the conversion of uranium to plutonium, as well as in the separation of plutonium from uranium, a knowledge of the chemical properties of uranium and its compounds was required which could not be derived from the perusal of prewar publications in this undeveloped field. A program of experimental research in uranium chemistry had to be inaugurated at the very inception of the Atomic Energy Project.

When new chemical information began to accumulate and much of the older literature data was found to be incorrect or inexact, a need for a reliable compilation of important facts in this field became acutely felt in various branches of the Atomic Energy Project. Accordingly, the Information Division of the Metallurgical Project at the University of Chicago was asked in 1943 to undertake the preparation of a "Uranium Chemistry Handbook."

Shortly before the outbreak of the war in Europe, the "Uranium" volume of the well-known "Gmelins Handbuch der anorganischen Chemie" was published in Germany. The first plan of the Information Division was to supplement this publication by a compilation of the newer results published in what since has become known as "open literature," together with the data scattered in hundreds of Project reports. Preparation of this supplement was begun by several members of the Information Division, particularly Mrs. Marylin Howe. Soon, however, a more ambitious plan was conceived. The well-known Gmelin Handbook is a complete, but rather uncritical, and, particularly in its physicochemical parts, almost mechanical assembly of good and bad data. It was felt that the research in the field would be assisted by a well-organized critical review of the whole field of uranium chemistry, including both the older work and the results obtained in Project laboratories.

Work in this direction was started late in 1944, with the Project research workers benefiting continuously by the growing collection of materials: first a bibliographic file, then a file of abstracts from published articles and Project reports, and finally drafts of various chapters which were circulated to the groups particularly interested in their contents.

After the war, the preparation of the monograph slowed down for a variety of reasons. However, the need for its completion in the interest of continued progress of nuclear energy research was obvious. The Manhattan District "declassification code" determined that the results of fundamental research in the field of uranium chemistry could and should be published openly to stimulate the broad development of chemical science, while specific chemical-technological procedures used in the laboratories and plants of the Atomic Energy Project remained undisclosed. Since only a part of the information on basic uranium chemistry contained in the vast agglomeration of Manhattan District wartime research reports could be expected to find its way into regular publication channels, and thus become available to the investigators who needed it in their research, the task of collecting this information, separating it from the technological descriptions and calculations with which it often was interwoven, and integrating it with the pre-Project knowledge of the chemistry of uranium and its compounds remained an important task. Consequently, after leaving the Information Division of the Manhattan Project in 1946, the authors continued to prepare this work on a part-time basis, and the first part of this treatise is now presented to the public. It deals with the element uranium—its occurrence, preparation, and physical and chemical properties—and simple uranium compounds whose chemistry has been largely developed in the laboratories of the Atomic Energy Project—hydrides, oxides, halides, and similar compounds. Roughly, it is the field described by the vague term "dry uranium chemistry."

The second half of the treatise, now in active albeit necessarily slow preparation, will deal with the properties of uranium ions in various states and their reactions in solution. It is hoped that, with the completion of this part, workers in the field of uranium chemistry will have at their disposal an adequate presentation of the chemical properties of this previously but little known element. It is also intended to publish, as a companion volume to this treatise, a collection of original research papers in the field of uranium chemistry originating in Project laboratories. These papers will contain a more detailed description of the methods used and results obtained than can be included in a review volume such as this.

The Project material discussed in this volume originated in a large number of laboratories. The Manhattan Project Laboratory at

Ames, Iowa (now the Iowa State College Institute for Atomic Research), under the leadership of Prof. F. H. Spedding was a leading contributor to the study of uranium metal and its simple compounds. A large part of this volume is based on the work of this group. Considerable work in uranium chemistry was carried out at Brown University under Prof. C. A. Kraus; at Columbia University (later the SAM Laboratories) under Profs. H. C. Urey, R. H. Crist, and W. F. Libby; and in the Chemistry Department of the University of California at Berkeley under Prof. W. H. Latimer. The Radiation Laboratory of the University of California at Berkeley was an active contributor to this work, as was the Y-12 Laboratory (Electromagnetic Separation Plant) under the supervision of Drs. C. E. Larson and H. S. Young. Numerous other laboratories active during the war years also made significant contributions. It is the sincere hope of the authors, in view of the rather unconventional manner in which much of this work is now being presented to the scientific public, that there has been a proper attribution of credit to individual scientists for their contributions. This is a difficult and at times all but impossible task in view of the tremendous volume of reports, the rapidly fluctuating research teams, and the not inconsiderable degree of anonymity which characterized much of the wartime research of the Manhattan Project.

Sincere thanks are due Prof. Robert S. Mulliken, formerly head of the Information Division of the Metallurgical Project, who outlined the task and assisted in every way. It is also a pleasure to express thanks to Dr. Hoylande D. Young, Director of the Information Division at the Argonne National Laboratory, for her unfailing assistance and cooperation. Several members of the staff of the Information Division, particularly Mr. Paul A. Schulze and Mrs. Carolyn Baer, contributed valuable technical assistance in the preparation of the manuscript.

The authors also wish to thank Profs. Harrison S. Brown and Clyde A. Hutchison, Jr., and in particular Profs. Warren C. Johnson and Willard F. Libby, all of the University of Chicago, for the time they patiently spent in reviewing the whole manuscript for publication. Valuable suggestions were received from Drs. L. I. Katzin, H. R. Hoekstra, J. C. Hindman, and Sherman Fried of the Argonne National Laboratory; from Dr. J. R. McNally, Jr., of the Y-12 Laboratory of the Carbide and Carbon Chemicals Division, Union Carbide and Carbon Corporation; from Dr. Michael Fleisher of the U. S. Geological Survey; from Profs. D. Jerome Fisher and W. H. Zachariasen of the University of Chicago; and from Prof. R. E. Rundle of the Iowa

State College of Agriculture and Mechanic Arts. The authors are also indebted for criticism and comments to a large number of scientists in various laboratories of the Atomic Energy Project who read the manuscript while in preparation.

Joseph J. Katz
Eugene Rabinowitch

November, 1950

CONTENTS

Part 1

THE ELEMENT URANIUM

Chapter 1

ISOTOPIC COMPOSITION AND ATOMIC WEIGHT OF NATURAL URANIUM

Chapter 2

PROPERTIES OF THE URANIUM ATOM

Chapter 3

URANIUM IN NATURE

Chapter 7

INTERMETALLIC COMPOUNDS AND ALLOY
SYSTEMS OF URANIUM

Part 3

BINARY COMPOUNDS OF URANIUM OTHER THAN HALIDES

Chapter 8

THE URANIUM-HYDROGEN SYSTEM

Chapter 9

URANIUM BORIDES, CARBIDES, AND SILICIDES

Chapter 10

URANIUM COMPOUNDS WITH ELEMENTS OF GROUP Va

Chapter 11

URANIUM OXIDES, SULFIDES, SELENIDES, AND TELLURIDES

Part A. Uranium Oxides and Hydroxides

Part 4

URANIUM HALIDES AND RELATED COMPOUNDS

Chapter 12

NONVOLATILE FLUORIDES OF URANIUM

Chapter 13

URANIUM HEXAFLUORIDE

Chapter 14

URANIUM-CHLORINE COMPOUNDS

Chapter 15

BROMIDES, IODIDES, AND PSEUDOHALIDES OF URANIUM

Chapter 16

URANIUM OXYHALIDES

Part 1

THE ELEMENT URANIUM

Chapter 1

ISOTOPIC COMPOSITION AND ATOMIC WEIGHT
OF NATURAL URANIUM

In this chapter a brief account is given of the isotopic composition of natural uranium. A full discussion of the nuclear properties of all the natural and synthetic isotopes is beyond the scope of this volume, these being treated in detail in Division IV, Volume 17 A, of the National Nuclear Energy Series. Here attention is directed to a few fundamental data pertaining to the isotopic composition, atomic weight, and the interrelations of the three naturally occurring isotopes.

1. ISOTOPIC COMPOSITION

Natural uranium contains the three isotopes U^{238} (UI), U^{234} (UII), and U^{235} (actinouranium, AcU). Several comparatively short-lived isotopes have been obtained artificially by various nuclear reactions involving actinium, thorium, protactinium, or stable uranium isotopes as initial materials. The radioactive constants of the naturally occurring and artificially prepared isotopes are given in Table 1.1. The isotopes U^{238} (UI) and U^{234} (UII) are members of the same radioactive decay series, the so-called "(4n + 2)" series,

$$\text{UI} \xrightarrow{\alpha} \text{UX}_1 \xrightarrow{\beta} \text{UX}_2 \xrightarrow{\beta} \text{UII} \xrightarrow{\alpha}$$

They must therefore be present in a constant ratio, equal to the ratio of their half-lives, in all uranium sources in which radioactive equilibrium has been established.

The equilibrium ratio of the isotopes U^{235} and U^{234} has been found by mass-spectrographic analysis to be 141.5 (Chamberlain, Williams, and Yuster, 1946). A value of 138.0 for the U^{238}/U^{235} ratio has been found at Columbia (SAM Columbia 1). From these two values there is calculated a value of 19,530 for the U^{238}/U^{234} ratio, equivalent to 0.0051 per cent for the U^{234} content.

3

The result of the isotopic analysis of uranium by Nier (1939a) is illustrated in Fig. 1.1. The U^{234} peak is so low that only an approximate determination of the ratio U^{238}/U^{234} is possible. A series of comparisons gave values between 13.0×10^3 and 20.8×10^3 with an average of $17,000 \pm 10$ per cent, which is in fair agreement with the above value.

Table 1.1 — Radioactive Constants of Natural and Artificial Uranium Isotopes*

Mass no.	Type	Radiation Energy, mev	Half-life	Amount in natural uranium, %
		Natural isotopes		
234 (UII)	α	4.76	2.35×10^5 years	0.005
235 (AcU)	α	4.52	8.91×10^8 years	0.71
			7.07×10^8 years	
			8.52×10^8 years	
238 (UI)	α	4.21	4.51×10^9 years	99.28
		Artificial isotopes		
228	α (80%) K (20%)	6.72	9.3 min	
229	α (~20%) K (~80%)	6.42	58 min	
230	α	5.85	20.8 days	
231	K		4.2 days	
232	α	5.3	70 years	
233	α, γ, e^-	4.8	1.6×10^5 years	
237	β^-, γ, e^-		6.8 days	
239	β^-, γ, e^-		23.5 min	

*Seaborg and Perlman (1948).

Actinouranium, U^{235}, is not a member of the (4n + 2) series, and its origin in natural uranium is open to speculation. Measurements show that its concentration is remarkably constant.

The first mass-spectrographic proof of the presence of U^{235} in natural uranium was given by Dempster (1935), who estimated the concentration as less than 1 per cent. An exact determination was first made by Nier (1939a); Table 1.2 shows the results obtained by Nier with different uranium minerals. The average of these values is 139 ± 1. Table 1.2 shows no noticeable effect of the age of the mineral.

The ratio U^{238}/U^{235} has been redetermined by Columbia University workers (SAM Columbia 1). They used UF_5^+ and UF_6^+ ions from UF_6

in a standard source and U^+ ions in a surface ionization source. The results are shown in Table 1.3. A comparison of two African ores, one Canadian ore, and a Colorado carnotite showed differences of no more than 0.03 per cent.

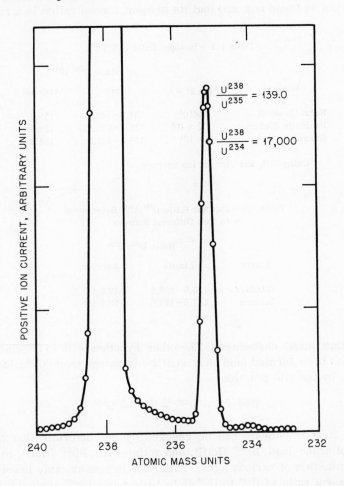

Fig. 1.1—Isotopes of uranium. The circles are experimentally determined points.

The values obtained with the two sources agree within statistical error, and the Columbia value, 138.0 ± 0.3, is used in this discussion. The early value of Nier (1939a) of 139 ± 1 is within statistical agreement. Perhaps an isotopic discrimination effect occurs in surface ionization because of different volatility of the two isotopes.

The constant content of U^{235} in natural uranium minerals of varying age can be explained in two ways: One hypothesis is to assume that U^{235} was present in the natural isotope mixture existing at the time of formation of the earth in a considerably larger proportion (relative to U^{238}) than is found now and that its present concentration is a residue

Table 1.2 — Isotopic Ratio U^{238}/U^{235}

Mineral	Age, years	Ratio U^{238}/U^{235}[*]	
		Limits	Average
Kolm (Sweden)	4×10^8	137.5 – 140.7	139.0
Uraninite (Ontario)	1.0×10^9	135.3 – 140.8	138.9
Dakeite	1×10^3	136.6 – 141.3	138.8

[*]Using UCl_4 and UBr_4 as ion sources.

Table 1.3 — Isotopic Ratio U^{238}/U^{235} Determined with Two Different Sources

Source	Ratio U^{238}/U^{235}	
	Limits	Average
Standard	135.0 – 138.6	137.0 ± 0.7
Surface	137.9 – 139.0	138.0 ± 0.3

from this initial endowment. The other hypothesis is to assume that U^{235} has been formed (and may still be forming) from U^{238}, for example, by the pile reaction

$$U^{238} + n \rightarrow U^{239} \overset{2\beta}{\rightarrow} Pu^{239} \overset{\alpha}{\rightarrow} U^{235}$$

The first hypothesis can be checked by the determination of the ratio of urano-lead, Pb^{206} (RaG), and actino-lead, Pb^{207} (AcD), in uranium minerals of various ages. This ratio is considerably lower than the present ratio of U^{238} to U^{235} (5 to 10 per cent Pb^{207} against 0.7 per cent U^{235}) and increases with the age of the mineral. The values found by Nier (1939b) for minerals of age 2×10^8 to 13×10^8 years could be fitted on a theoretical curve constructed with the values 0.046 for the present ratio of activities of the actinium series and the uranium series and 139 for the present ratio of concentration of U^{238} and U^{235}. Using these constants, Nier calculated the values that the U^{238}/U^{235} ratio must have had in past geological periods (Table 1.4).

The values in the last column of Table 1.4 are not so large as those obtained in some earlier estimates based on a higher value for the present ratio U^{238}/U^{235} and a lower value for the relative activity of the actinium series. This means that the importance of the actinium series in the thermal history of the earth is, according to Nier, not so great as it was formerly supposed to be.

Table 1.4 — U^{238}/U^{235} Ratio in the Past*

No. of years ago	U^{238}/U^{235}	Activity ratio, Ac series/U series
0×10^8	139.0	0.046
4×10^8	100.2	0.065
8×10^8	72.2	0.089
12×10^8	52.1	0.123
16×10^8	37.5	0.171
20×10^8	26.9	0.238

*After Nier (1939b).

In order to account as closely as possible for the Pb^{207}/Pb^{206} ratios found in minerals of various ages, Nier (1939b) had to choose the above-mentioned value of 0.046 for the ratio of activities of the actinium series and the uranium series at the present time (Fig. 1.2). Combining this value with Nier's concentration ratio, $U^{235}/U^{238} = 1/139$, and using a half-life of 4.51×10^9 years for U^{238}, a half-life of about 7×10^8 years is obtained for U^{235}. Direct measurements of the activity ratio of the two series, on the other hand, gave values close to 0.040, leading to a half-life of 8.1×10^8 years for U^{235}. As shown in Fig. 1.2, the observed values of the percentage ratio Pb^{207}/Pb^{206} are about 20 per cent higher than those calculated for an activity ratio of 0.040. The reason for this discrepancy, which can be interpreted either as an excess of actino-lead or as a deficiency of urano-lead, is as yet unclear. Nier's suggestion in 1939 was that the experimental value of the activity ratio might be too low. More recent work (British 1) indicates, however, that the U^{235} half-life is probably 8.9×10^8 years, which would make the activity ratio even lower than 0.040.

Since, on the whole, the results of the mass-spectrographic analysis of uranium and lead agree with the hypothesis of an initial U^{235} endowment, the extent of additional U^{235} formation during the lifetime of the earth must remain an open question. This formation is inevitable whenever U^{238} atoms are exposed to neutrons, and neutrons are always present in the earth, originating in cosmic-ray absorption, spontaneous fission, and (α,n) reactions. There seems to be no reason why the extent of U^{235} formation caused by these neutrons should be

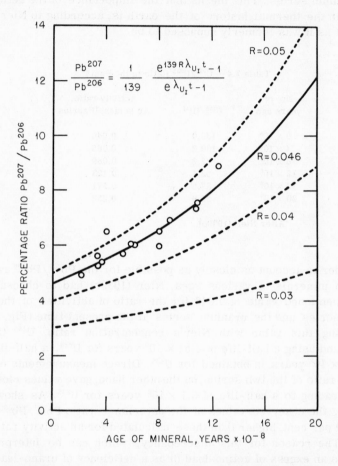

Fig. 1.2—Ratio of percentage of Pb^{207} to Pb^{206} (corrected for common lead impurities) as a function of the age of the mineral from which the lead was extracted. The curves are plotted according to the formula in the diagram for various assumed values of R (the ratio of activities of the actinium and uranium series). The circles are experimentally determined points (Nier, 1939b).

the same for uranium from various sources. Therefore evidence of the production of U^{235} from U^{238} after the solidification of the earth's crust can be sought in small variations of the U^{235} content of various minerals rather than in the presence of an approximately constant amount of about 0.7 per cent in all of them.

2. ATOMIC WEIGHT

Precision measurements of the atomic weight of uranium have been made by Hönigschmid and coworkers (1936). These measurements were based on the determination of the ratios UX_4/Ag and

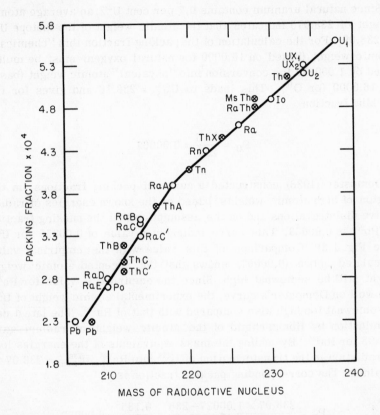

Fig. 1.3—Theoretical packing-fraction curve for the radioactive elements, computed from the energy of disintegration and the mass of helium 4.0039, assuming a value for lead equal to 2.3×10^{-4} (Dempster, 1938) . ○ represents the uranium disintegration series, and ⊗ represents the thorium disintegration series.

UX_4/4AgX, with X being either Br or Cl. The earlier determinations by Hönigschmid (1915, 1916, 1928) gave, for the atomic weight of uranium, values between 238.13 and 238.18. A value of 238.14 was adopted as the most probable one by the International Committee for Atomic Weights in 1928. This value, however, appeared too high in comparison with the independently known atomic weight of radium, and the newer determination of Hönigschmid and Witner (1936) confirmed that it was erroneous. From 18 determinations of the ratio $UCl_4/4Ag$ and 20 determinations of the ratio $UBr_4/4Ag$ (obtained by nephelometric titration), the two average values U = 238.073 and U = 238.076 were calculated, and a value of 238.07 was therefore generally adopted in 1936.

Since natural uranium contains 0.7 per cent U^{235}, an average atomic weight of 238.075 indicates that the atomic weight of the isotope U^{238} is 238.09_5. For the calculation of the packing fraction this "chemical" atomic weight (based on 16.0000 for natural oxygen) must be multiplied by 1.000272 for conversion into "physical" atomic weight (based on 16.0000 for O^{16}). This leads to $U^{238}_{phys} = 238.16$ and gives for the packing fraction

$$P_U = \frac{0.16}{238} = 0.00067$$

Dempster (1938) constructed a curve of packing fractions, in the region of high atomic weights, based on the known energies of radioactive disintegrations and on the assumption that the packing fraction of Pb^{206} is 0.00023. This curve indicates a value of 0.00055 for U^{238} (see Fig. 1.3). Comparison of this value with the empirical value calculated above (0.00067) shows that the accepted atomic weight might still be somewhat high. Since the empirical P value for Ra^{226} fits well on Dempster's curve, the experimental atomic weight of U^{238} is somewhat too high also compared with that of Ra^{226}. The latest determination by Hönigschmid of the atomic weight of radium gave 226.05 for Ra^{226}. By adding the mass equivalents of the energies lost by radiation in the transformation of U^{238} into Ra^{226}, $U^{238}_{chem} = 238.07$ is obtained. The corresponding packing fraction is

$$P_U = \frac{238.07 \times 1.00027 - 238}{238} = \frac{0.135}{238} = 0.00057$$

REFERENCES

1915 O. Hönigschmid and coworkers, Monatsh., 36: 51.
1916 O. Hönigschmid and coworkers, Monatsh., 37: 185.
1928 O. Hönigschmid and coworkers, Z. anorg. u. allgem. Chem., 170: 145.
1935 A. J. Dempster, Nature, 136: 180.
1936 O. Hönigschmid and coworkers, Z. anorg. u. allgem. Chem., 226: 289.
1938 A. J. Dempster, Phys. Rev., 53: 869-874.
1939a A. O. Nier, Phys. Rev., 55: 150-153.

1939b   A. O. Nier, Phys. Rev., 55: 153-163.
1946   O. Chamberlain, D. Williams, and P. Yuster, Phys. Rev., 70: 580.
1948   G. T. Seaborg and I. Perlman, Revs. Modern Phys., 20: 585-667.

Project Literature

British 1:   F. L. Clark, H. J. Spencer-Palmer, and R. N. Woodward, Report BR-521, October, 1944.
SAM Columbia 1:   M. Fox and B. Rustad, Report A-3828, Jan. 16, 1946.

Chapter 2

PROPERTIES OF THE URANIUM ATOM

1. X-RAY SPECTRUM

Uranium, being among the heaviest elements, has a most complex x-ray spectrum. The neutral uranium atom contains in its state of lowest energy the fully occupied electron shells K (2 electrons), L (8 electrons), M (18 electrons), and N (32 electrons) and the partially occupied shells O (21 electrons), P (9 electrons), and Q (2 electrons). The distribution of the six outermost electrons between the groups 5f (O_{VI-VII}), 6d (P_{IV-V}), and 7s (Q_I) probably corresponds to the formula f^3ds^2 (see Sec. 2.5). These six electrons are the valence electrons of uranium, and their excitation produces the optical spectrum. The other 86 electrons are "inner electrons," and their excitation produces the x-ray spectrum, which should thus contain the series K, L, M, N, O, and P. However, lines belonging to the last two "ultrasoft" series have not yet been observed.

1.1 <u>Absorption.</u> Table 2.1a gives a summary of several recent measurements of the wavelengths of the main absorption edges of uranium, made partly with uranium metal and partly with uranium oxide as absorber. The effect of chemical combination on the position of the main L edges, as observed by Manescu (1947), is illustrated by Table 2.1b. In addition to the measurements included in Table 2.1a, measurements of the main absorption edges also were carried out by de Broglie (1916a,b; 1919), Siegbahn (1919), Duane and coworkers (1920a,b,c; 1921), Stoner and Martin (1925), and Dauvillier (1927).

The fine structure of the absorption edges depends on chemical combination, and its investigation is a means of analysis of the valence electron bands in crystals. Dauvillier (1921a) reported the presence of a "white line" in the L_{III} edge; Cauchois and Manescu (1940b) saw a similar line in both the L_{III} and the L_{II} edge of uranium metal (at 720.4 and 590.4 X.U., respectively); no white line was noticeable in the L_I edge. Polaczek (1939) could see no fine structure in the M_{IV} and M_V edges of uranium oxide.

Table 2.1a—Absorption Edges of Uranium

Designation	Position of the edge X.U.	ν/R	Width, X.U.	Reference
K_I	106.58	8550	...	Mack and Cook (1927)
L_I	568.0	1604.3	4.5	Sandström (1930)
	568.2	1603.7	...	Cauchois and Manescu (1940b)
L_{II}	591.3	1541.0	4.3	Sandström (1930)
	590.7	1539.5		Cauchois and Manescu (1940b)
L_{III}	720.8	1264.2	4.2	Sandström (1930)
	720.7	1264.3	...	Cauchois and Manescu (1940b)
M_I	2228	408.9	...	Coster (1921a;
M_{II}	2385	382.1	...	1922a,b)
M_{III}	2873	317.18	...	Stenström (1919)
	2877	316.7	...	Lindberg (1929)
M_{IV}	3326	273.99	...	Stenström (1919)
	3327	273.9	...	Lindberg (1929)
	3326	273.99	9*	Polaczek (1939)
M_V	3491	261.03	...	Stenström (1919)
	3491	261.03	...	Lindberg (1929)
	3491	261.03	13†	Polaczek (1939)

*3322 to 3331 X.U.
†3487 to 3500 X.U.

Table 2.1b— Fine Structure of the L Absorption Edges of Uranium*

Edge	Minimum	α	A	β	B	γ

L_I

U metal	568.28					
UO_2	568.1_0					
UO_3	568.0_4					

L_{II}

U metal	590.71	590.1_7	$589._7$	$589._0$	$588._5$	$587._3$	$586._2$
UO_2	590.62	590.1_2	$589._7$	$588._7$	$588._3$	$586._9$	$585._6$
UO_3	590.50	590.0_0	$589._4$	$588._8$	$588._1$	$586._7$	$586._1$

L_{III}

U metal	720.76	720.1_7	719.50	$718._7$	$717._8$	$716._3$	$714._7$
UO_2	720.64	720.1_6	719.50	$718._6$	$717._6$	$716._2$	$714._5$
UO_3	720.54	720.0_3	719.30	$718._5$	$717._7$	$716._1$	$714._9$

*After Manescu (1947).

The first detailed measurements of the fine structure of the L_{II} and L_{III} absorption edges of uranium were made by Manescu (1947). Table 2.1b indicates the existence, on the short-wave side of these edges, first of a minimum (white line) and then of several maxima (black lines), α, β, and γ, separated by the minima A and B. Within the first minimum must lie the several electron transitions from the 2p group

Table 2.2 — Mass Absorption Coefficient (μ/ρ)
of Uranium Metal For X Rays

λ, X.U.	μ/ρ, cm^2/g*	λ, X.U.	μ/ρ, cm^2/g†
64	1.80	108_0	150
72	2.25	124_0	190
98	3.90	145_0	285
107.5	4.65 K edge	154_0	360
107.5	1.62	180_0	442
130	2.10	193_0	470
175	3.95	197_0	485
200	5.40	206_0	495
		218_0	516 M$_I$ edge
λ, X.U.	μ/ρ, relative units‡	224_0	525
		235_0	540 M$_{II}$ edge
400	0.34	249_0	560
450	0.45	260_0	570 M$_{III}$ edge
500	0.59	297_0	600
550	0.77 L$_{I,II}$ edges	304_0	600
600	0.64	318_0	600
631	0.73	322_0	600 M$_{IV}$ edge
708	1.00 L$_{III}$ edge	337_0	580
750	0.48	344_0	564 M$_V$ edge
800	0.56	366_0	410
		374_0	424
		393_0	490

*Allen, in Compton and Allison (1935a).
†Allen (1926).
‡Stoner and Martin (1925).

into the partially filled peripheral s and d groups; a fine structure of the white line corresponding to these transitions actually was observed in the spectrum of uranium metal taken with greater dispersion.

The wavelengths given in Table 2.1b for UO_2 are typical of the several uranium(IV) compounds investigated by Manescu; these included, in addition to UO_2, crystalline uranium(IV) phosphate and sulfate. Similarly, the UO_3 spectrum given in the table is typical of the uranium(VI) compounds investigated, among which were uranyl nitrate, acetate, and sulfate. In addition to the (theoretically predictable) shift

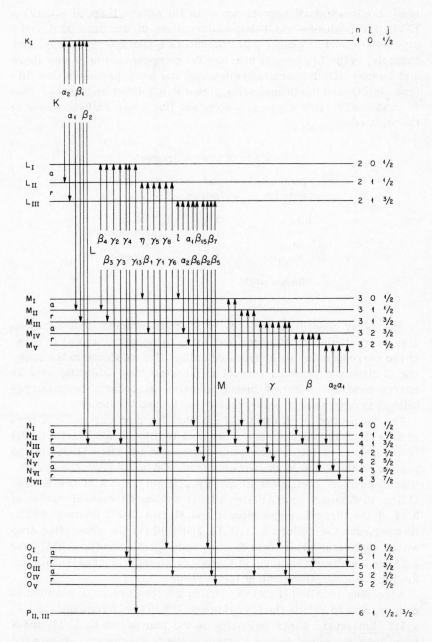

Fig. 2.1 — X-ray term diagram of uranium (after Siegbahn, 1931). (a) "Abschirmungs-dublett" (screening doublet); (r) "Relativitätsdublett" (relativistic, or spin doublet).

of all L edges toward shorter waves in the series U metal → U(IV) → U(VI), the positions and relative intensities of the lines α, β, and γ also were found to change with valence in a characteristic way. For example, in the L_{III} edge of uranium(IV) compounds, line α was sharp and intense, line β comparatively weak, and line γ prominent but diffuse. In UO_3 and the uranyl salts, line α was thinner and weaker than in uranium(IV), line β was stronger, and line γ was shifted closer to the main edge.

Table 2.3 — K Lines of Uranium*

Term combination	Line designation	Wavelength, X.U.
$K_I L_{II}$	2	130.95
L_{III}	1	126.40
$M_{II,III}$	1, 3	111.87
$N_{II,III}$	2	108.42

*Réchou (1925).

The x-ray absorption spectra of aqueous uranium salt solutions also were studied by Manescu and found to be quite similar to those of the corresponding solid hydrated salts. The existence in the spectra of dissolved uranium ions of a fine structure extending over an energy band about 200 ev broad indicates an orderly (pseudocrystalline) arrangement of water molecules around the ion.

The absorption coefficients of uranium for monochromatic x rays of different wavelengths were measured by Stoner and Martin (1925) in the region of the L absorption edges and by Allen (1926) in the regions of the K and M absorption edges. Table 2.2 gives the results. The absorption coefficient drops by a factor of 2.9 at the K edge (Allen, in Compton and Allison, 1935a) and by an over-all factor of 3.52 at the three L edges (Stoner and Martin, 1925; Küstner, 1932). Küstner gave the factors 1.11, 1.31, and 2.17 for the absorption drop at the three individual edges L_I, L_{II}, and L_{III}. Stephenson (1933) gave 2.27 for the absorption jump at the edge L_{III}, and Polaczek (1939) gave 2.4 for the absorption jump at the M_V edge.

According to Allen (1926) the absorption coefficient of uranium is proportional to $\lambda^{2.92}$ in the N region and $\lambda^{2.6}$ in the M region.

1.2 <u>Emission</u>. Lines belonging to the four series K, L, M, and N have been measured in the x-ray spectrum of uranium. Figure 2.1 shows the interpretation of the most important lines of the first three series, according to Siegbahn (1931). Tables 2.3 to 2.8 contain the wavelengths of all the measured lines.

Table 2.4 — L Lines of Uranium

Term combination	Line designation	Wavelength, X.U.				Relative intensity			Half-width Williams (1931)
		Schrör (1926)	Friman (1926)	Idei (1930)	Claësson (1936)	Allison* (1928, 1935b)	Allison and Andrews (1931)	Stephenson† (1933)	
L_IM_{II}	β_4	...	...	764.4	746.39	4.1 (3.2)	...	...	0.726
M_{III}	β_3	...	...	708.79	708.81	4.2 (3.3)	...	...	0.382
$M_{IV}\ddagger$	β_{10}	...	...	685.3	686.15	...	...	...	...
$M_V\ddagger$	β_9	...	...	679.5	679.6	...	...	...	...
N_{II}	γ_2	...	...	603.9	603.99	1.5 (1.1)	...	...	0.57
N_{III}	γ_3	...	...	597.1	597.30	1.4 (1.1)	...	...	0.47
$N_{IV}\ddagger$	...	...	...	...	589.7$_7$	...	...	...	...
$N_V\ddagger$	...	...	...	...	588.6$_5$	...	...	...	...
O_{II}	γ_4'	...	...	575.3	575.45	0 (0)	...	...	...
O_{III}	γ_4	...	...	573.6	573.90	0 (0)	...	...	...
$O_{IV,V}\ddagger$	...	...	...	...	571.3	...	...	...	...
$P_{II,III}$	γ_{13}	...	...	568.9	569.4$_1$	...	...	...	...
$L_{II}M_I$	η	...	...	803.5	803.59	1.0 (0.83)	...	...	...
$M_{III}\ddagger$	β_{17}	...	...	...	743.4$_7$	...	...	...	...
M_{IV}	β_1	718.55	718.51	...	718.54	49.4 (40.5)	36.7	...	0.299
N_I	γ_5	...	...	634.2	634.23	0 (0)	...	...	...
N_{III}	...	...	...	...	621.54	...	...	...	...
N_{IV}	γ_1	613.32	613.59	...	613.50	12 (9.7)	10.9	...	0.242
N_{VI}	...	...	...	...	601.9$_3$	...	...	...	...
O_I	γ_8	...	...	599.7	600.00	...	...	...	...
$O_{III}\ddagger$	...	...	...	...	596.0$_4$	...	...	...	...
O_{IV}	γ_6	...	...	593.4	593.61	2.2 (1.8)	...	...	0.233
					592.8§				
$P_{II,III}\ddagger$	...	...	...	...	591.9	...	...	...	...
$P_{IV,V}$	...	...	...	...	590.8$_1$	...	...	...	...
$L_{III}M_I$	l	...	...	1064.77	1064.89	2.4 (2.4)	...	...	...
$M_{II}\ddagger$	t	...	...	1032.9	1032.5$_4$	...	...	...	...
$M_{III}\ddagger$	s	...	...	961.4	961.8$_0$	...	...	...	...
M_{IV}	α_2	920.65	...	920.62	920.56	11.0 (11.0)	...⎫	3.73	0.494
M_V	α_1	908.91	908.7	...	908.69	100 (100)	100⎭		0.439
N_I	β_6	...	...	786.8	786.78	1.6 (1.6)	...	...	...
$N_{II}\ddagger$	...	...	...	...	778.4$_8$	...	...	...	...
$N_{III}\ddagger$	...	...	...	...	767.3$_8$	...	...	...	...
$N_{IV}\P$	β_{15}	...	...	755.1	755.10	...	...	...	...
N_V	β_2	753.23	753.07	...	753.14	28 (28)	...	1.00	0.369
$N_{VI,VII}\ddagger$	β_7'	...	...	737.3	737.07	...	...	...	...
O_I	β_7	...	...	734.6	734.64	0.4 (0.4)	...	...	...
$O_{II}\ddagger$	...	...	...	...	731.74	...	...	...	...
$O_{III}\ddagger$	...	...	...	...	729.3$_6$	...	...	...	...
$O_{IV,V}$	β_5	...	...	724.85	724.84	6.4 (6.4)	...	...	0.252
P_I**	...	...	...	...	723.7$_1$	...	...	...	...
$P_{II,III}$	...	...	...	...	722.7$_5$	...	...	...	...
$P_{IV,V}\dagger\dagger$	...	...	...	...	720.9$_0$	...	...	...	...

*These values were corrected by extrapolation to infinite exciting voltage; the actual values measured at 52.8 kv are in parentheses.

†Fluorescence spectrum.

‡These are "forbidden" (quadrupole) lines ($\Delta l \neq 1$); they are not shown in Fig. 2.1.

§Line found by Cauchois and Allais (1940a).

¶See also Allison (1929) who gave $\Delta\lambda = 1.86$ X.U. for the separation $\beta_2 - \beta_{15}$, and 11:1 for the intensity ratio of these two lines.

**Interpretation by Hulubei (1937).

††The $P_{IV,V}$ electrons are valence electrons; the interpretation of the line 720.9$_0$ is thus uncertain (since conductance niveaus, as well as the terms $O_{VI,VII}$ and Q_I, may also be involved).

(a) K Series. The wavelengths of the K lines of uranium are listed in Table 2.3.

(b) L Series. The wavelengths, intensities, and widths of the L lines of uranium are given in Table 2.4.

Earlier measurements of L emission lines were made by Coster (1921a,b; 1922a), Dauvillier and coworkers (1921b,c; 1923), and Siegbahn and Friman (1916a,b,c). Duane (1920c; 1922) summarized these

Table 2.5—Satellites in the L Series of Uranium

Line designation	Wavelength, X. U.			
	Idei (1930)	Richtmyer and Kaufman (1933)	Claësson (1936)	Cauchois (1946)
α_3	905.5	905.35*	905.45	...
...	...	...	...	782.5
...	...	...	...	778.9
β_2'	750.0	749.92	749.89	...
β_2''	748.9	748.93	748.93	...
...	...	...	723.71	...
μ	601.5	...	...	...

*Designated α^{IX} by Richtmyer and Kaufman.

measurements. In addition to the "diagram lines" listed in Table 2.4, several satellites have been observed, which are usually interpreted as "spark lines" (transitions following double ionization) (Table 2.5).

The width of L emission lines of uranium was measured by Williams (1931). The results are shown in the last column of Table 2.4.

(c) M Series. The M series is not so well-known as the L series. The best available data at the time of this writing are collected in Table 2.6.

Earlier wavelength determinations by Hjalmar (1923) and Stenström (1918) were summarized by Duane (1920c; 1922).

Satellites (spark lines) were measured in the M series by Lindberg (1928, 1931) and by Hirsch (1931) (Table 2.7).

Parratt (1932) used six uranium M lines in the measurement of half-width of lines reflected by a calcite crystal.

(d) N Series. The only available results at the time of this writing, those of Hjalmar (1923) and Dolejšek (1924), are rather uncertain, and the interpretations of the two investigators disagree in several cases (Table 2.8). Siegbahn and Magnusson (1934) measured several "ultrasoft" lines with wavelengths longer than 30 A.

Table 2.6 — M Lines of Uranium

Term combination	Line designation	Wavelength, X.U.			
		Hjalmar (1923)	Lindberg (1928, 1931)	Purdom and Cork* (1933)	Hirsch (1931)
$M_I N_{II}$	...	2909	...	...	...
N_{III}	...	2750	2745	...	...
O_{III}	...	2299	...	...	...
P_{III}	...	2248	...	...	...
$M_{II} N_I$	...	3321	3322	...	...
N_{IV}	...	2815	2813	...	...
O_{IV}	...	2439	2440	...	...
$M_{III} N_I$	...	...	4322	...	...
N_{IV}	...	...	3514	...	...
N_V	...	3472	3473	...	...
O_I	...	3107	3114	...	...
$O_{IV,V}$	...	2927	2941	...	...
$M_{IV} N_{VI}$	...	...	5040	...	...
N_{III}	...	...	4615	...	...
$N_{VI, VII}$	...	3709	3708	3708	3708.0
$O_{II,III}$	...	3570	3570	...	...
$M_V N_{III}$	...	...	4937	...	...
N_{VI}	α_2	3913	3916	3916	3914.6
N_{VII}	α_1	3901	3902	3902	3901.0
P_{III}	...	3514	...	...	...

*Glass grating.

Table 2.7 — Satellites in the M Series of Uranium

Designation	Wavelength, X.U.	Reference
α'	3886	
β'	3698	Lindberg (1928, 1931)
γ'	3463	
α_I	3892.2	
α_{II}	3884.3	
α_{IV}	3869.0	
β_I	3701.1	Hirsch (1931)
β_{II}	3693.3	
β_{III}	3683.1	
α''	3885	
β'	3684	
β''	3696	
β'''	3700	Hjalmar (1923)
γ'	3459	
γ''	3466	

(e) Fluorescence. The fluorescence emission of L and M lines in the uranium x-ray spectrum was studied by Stephenson (1933) and by Hevesy and Lay (1934a,b). Stephenson found, for absorption in the L_{III} niveau, a fluorescence yield of 0.67. The ratio of intensities of the lines $L\alpha_{1,2}$ and $L\beta_2$ in fluorescence was 3.73, which agrees with

Table 2.8 — N Lines of Uranium

Term combination,* Dolejšek (1924)		Wavelength, X.U.		
		Hjalmar (1923)	Dolejšek (1924)	Siegbahn and Magnusson (1934)
$N_I O_{II}$	$(N_{II} O_{IV})$	10,385	...	...
O_{III}	$(N_I O_{III})$	9,619	9873	...
P_{II}	$(N_I P_{III})$	8,691	8704	...
P_{III}		...	8605	...
$N_{II} O_I$		...	1285_0	...
O_{IV}		...	1008_0	...
O_V	$(N_{III} O_V)$	12,874	...	...
$N_{III} O_V$		...	1270_0	...
P_I	$(N_{III} P_{II})$	12,250	...	...
$N_{IV} N_{VI}$		...	...	3178_0
$N_V N_{VI,VII}$		...	...	3481_0
$N_{VI} O_{IV}$		...	...	4208_0
$N_{VII} O_V$		...	...	...
$N_{VI} O_V$		...	...	4328_0

*Hjalmar's interpretation is given in parentheses.

the value found for the characteristic emission spectrum. Hevesy and Lay (1934a,b) found a lower yield of fluorescence (0.45 in the L_{III} niveau and 0.06 in the M niveaus).

1.3 Photoelectric Effect. The electron emission from uranium under x-ray irradiation was observed by de Broglie (1922) and by Robinson (1925). Robinson analyzed magnetically the electrons emitted by uranium oxide irradiated by $CuK\alpha$ radiation. He found 10 electron groups attributable to the photoelectric effect in various electron groups of uranium (see Table 2.9).

1.4 Term System. Since the x-ray absorption edges are not sharp, the usual method of calculating x-ray terms consists in selecting one or a few most precisely known absorption edges and combining them with the much more sharply defined frequencies of the emission lines. Values obtained in this way by Bohr and Coster (1923), Dolejšek (1924), Lindberg (1931), Siegbahn (1931), Claësson (1936), and Cauchois (1940a,b) are listed in Table 2.10. The basic values from absorption measurements are underlined. Those taken from previous investigations and used as a basis for new calculations are bracketed. The two last columns contain term values calculated by Monk and Allison (MP Chicago 1) from general term equations, using the screening constants shown in Table 2.11, and the term values (in

Table 2.9—Photoelectric Emission of Uranium Oxide Irradiated with $CuK\alpha$ X Rays*

Group	Intensity†	Electron energy (ν/R)	Energy loss $(592.8 - \nu/R)$	Corresponding X term Designation	ν/R
1	3	276.9	315.9	M_{IV}	317
2	4	318.8	274.0	M_{IV}	274
3	5	332.4	260.4	M_V	261
4	2	486.8	106.0	N_I	106
5	3	500.3	92.5	N_{II}	93
6	5	517.3	75.5	N_{III}	76
7	6	540.4	52.4	$N_{IV,V}$	57,54
8	2−3	555.8	37.0	Oxygen K(?)	
9	3	566.8	26.0	$N_{VI,VII}O_I$	27
10	2	577.1	15.7	$O_{II,III}$	18,15
11	2−3	589.9	2.9	$O_{IV,V}P$	7,2

*Energy $\nu_0/R = 592.8$.
†Arbitrary scale 1 to 6.

electron volts) derived by Ruark and Maxfield (1935) from a combination of x-ray line frequencies with optical term values. (What optical values have been used in the case of uranium is not clear.) The K term in the last column is calculated from the absorption edge, since in this case the discrepancy between the result of direct absorption measurements ($\nu/R = 8550$, Table 2.1a) and the value derived from the L edge and the K emission lines ($\nu/R = 8477$) may be attributed to errors in the K-line measurements.

Russell (Los Alamos 1) attempted to use the term values given in Table 2.10 to find out which valence electron group, $P_{IV,V}$ or $O_{VI,VII}$, has the lower energy, in other words, whether the 6d or the 5f electron group is filled first in the uranium atom. He saw two arguments in favor of the assumption that in elements beyond thorium the 5f electrons are bound more strongly than the 6d electrons. In the first place, the term separation $O_I - O_{II}$ (5s-5p) is only slightly larger in uranium than in thorium $[\Delta(\nu/R) = 4.9$ for Th and 5.0 for U]. In the preceding period of the periodic system, the increase in the separation $N_I - N_{II}$ (i.e., 4s-4p) is somewhat larger and approximately con-

Table 2.10 — X Terms of Uranium*

Term	Missing electron	n	l	j	Term values (ν/R)						Theoretical values	Electron volts†
					Calculated from x-ray absorption limits and emission-line frequencies						Monk and Allison (MP Chicago 1)	Ruark and Maxfield (1935)
					Bohr and Coster (1923)	Dolejšek (1924)	Lindberg (1931)	Siegbahn‡ (1931)	Claësson (1936)	Cauchois (1940a,b)		
K	1s	1	0	1/2	8477.0	...	...	8474	...	...	8490	115,790
L_I	2s	2	0	1/2	1603.5	...	...	1602.6	...	...	1596	21,711
L_{II}	2p		1	1/2	1543.7	...	...	1542.7	...	1542.73	1535	20,900
L_{III}				3/2	1264.3	...	...	1264.2	1264.2	1264.32	1264	17,128
M_I	3s	3	0	1/2	408.9	...	...	408.5	408.7	...	408.5	5,537
M_{II}	3p		1	1/2	382.1	...	...	381.5	381.7	...	379.5	5,172
M_{III}				3/2	317.2	...	316.7	316.8	317.0	...	317.5	4,296
M_{IV}	3d		2	3/2	274.0	...	273.9	274.2	274.5	...	273.9	3,721
M_V				5/2	261.0	...	261.0	261.2	261.7	...	260.4	3,546
N_I	4s	4	0	1/2	106.6	[107.1]	105.8	106.0	105.9	...	...	1,439
N_{II}	4p		1	1/2	95.7	[96.0]	93.1	93.5	93.8	...	...	1,271
N_{III}				3/2	77.1	[77.4]	76.4	76.6	76.9	...	...	1,042
N_{IV}	4d		2	3/2	56.3	...	57.3	57.5	57.5	...	...	783
N_V				5/2	53.6	...	54.2	54.3	54.5	...	...	739
N_{VI}	4f		3	5/2	28.4	...	28.1	28.5	28.8	...	...	391
N_{VII}				7/2	27.6	...	27.4	27.6	...	...	...	382

Table 2.10 — (Continued)

Term	Missing electron	n	l	j	Bohr and Coster (1923)	Dolejšek (1924)	Lindberg (1931)	Siegbahn‡ (1931)	Claësson (1936)	Cauchois (1940a,b)	Monk and Allison (MP Chicago 1)	Ruark and Maxfield (1935) [Electron volts†]
					Calculated from x-ray absorption limits and emission-line frequencies [Term values (ν/R)]						Theoretical values	
O_I	5s	5	0	1/2	26.2	25.1		23.7	23.9			328
O_{II}	5p	5	1	1/2		19.4		18.3	19.0			258
O_{III}		5	1	3/2	15.4	12.3 or 14		13.9	14.7			193
O_{IV}	5d	5	2	3/2	5.8	5.6		7.0	7.4			102
O_V		5	2	5/2		6.1						
O_{VI}	5f	5	3	5/2		Optical terms						
O_{VII}		5	3	7/2								
P_I	6s	6	0	1/2		3.0				5.35		
P_{II}	6p	6	1	1/2		2.4		0.8	2.2§			17
P_{III}		6	1	3/2		1.2						
P_{IV}	6d	6	2	3/2					0.3¶			
P_V		6	2	5/2								
Q_I	7s	7	0	1/2		Optical terms						

*The table given by Russell (Los Alamos 1) is a combination of Siegbahn's values for the K and L terms, with Claësson's values for the M, N, O, and P terms, with the addition of Cauchois' value for the P_I term.
†Derived from optical terms and X lines.
‡Largely based on the values of Idei (1930).
§Although all $P_{II,III}$ lines are weak, the term value is definitely greater than 2.
¶Line interpretation as $L_{II}P_{IV,V}$ and $L_{III}P_{IV,V}$ uncertain.

stant (about $2\nu/R$ units per element) in the series Hf to Pt (where the 5d group is filled); it is near zero and changing irregularly in the rare earth series, where the 4f group is completed. The inference is that the elements from thorium to uranium resemble the rare earth series more than they do the hafnium-to-platinum sequence of elements. However, the data presented are not in themselves sufficient to make this conclusion convincing.

Table 2.11 — Screening Constants of Uranium*

Term	L_I	$L_{II,III}$	M_I	$M_{II,III}$	$M_{IV,V}$
σ_2	2.0	3.49	6.8	8.5	13.0
σ_1	21.4	22.4	38.9	40.7	44.2

*After Monk and Allison (MP Chicago 1).

The second argument is slightly better. Table 2.10 shows that the M_{IV} and M_V absorption edges (273.9 and 261.0) are smaller by about $0.5\nu/R$ units than the M_{IV} and M_V terms calculated from the L_{III} limit and the L emission lines (274.5 and 261.6). The interpretation of this difference, suggested by Russell, is that the absorption in both the L_{III} and the $M_{IV,V}$ levels leads to electron transfer into the incomplete valence groups, rather than to complete ionization. According to the l-selection rule, $\Delta l = \pm 1$, L_{III} (2p) electrons ($l = 1$) can be transferred into the $P_{IV,V}$ valence group (6d electrons, $l = 2$), but the $M_{IV,V}$ electrons (3d electrons, $l = 2$) can be transferred only into the valence group $O_{VI,VII}$ (group 5f, $l = 3$). If electrons in the latter group are bound more strongly (by about 5 ev) than those in the 6d group, the absorption in the $M_{IV,V}$ group must require less energy than that calculated from the difference between the absorption energy in the L_{III} group and the energy of the $L_{III}M_{IV,V}$ emission line. Thus the above-mentioned difference between the calculated and the observed $M_{IV,V}$ term value can be explained.

However, even if this difference appears to be beyond the limits of experimental error, its interpretation is by no means certain. An analysis of the electron levels of a crystal, particularly a metallic crystal, on the basis of the electronic states of free atoms cannot carry much weight because of the well-known transformation of these levels into "conductance bands," which belong to the crystal as a whole rather than to individual atoms.

It may be noted that the width of the M absorption edges (Table 2.1a) is such that the discrepancy between the calculated and the observed M_{IV} and M_V terms practically disappears if the term value is deter-

mined from the wavelength at which the absorption reaches its maximum, rather than from that at which the absorption begins to increase
or from the middle of the edge (see Polaczek, 1939).

The possibility of a measurable effect of nuclear spin on the x-ray
terms of the heaviest elements was suggested by Breit (1930). Breit

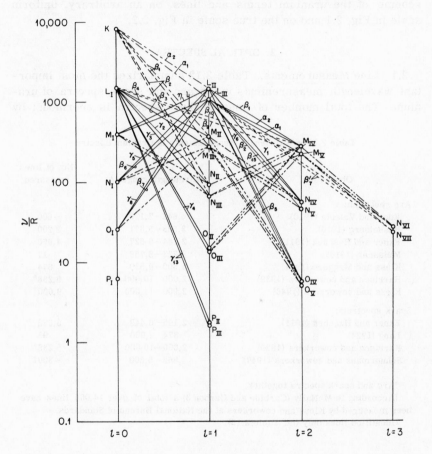

Fig. 2.2—X-ray spectrum of uranium (after Siegbahn, 1931). Ordinates are proportional to logarithms of energy levels.

calculated that, if uranium had a nuclear spin of $9/2$ units, its K level
should be split into two components with a separation of 22.5 ev and
its L_{II} level into two components with a separation of 4 ev. Williams
(1931) remarked that the natural breadth of the L_I line of uranium is
much too large to reveal a splitting of this order of magnitude. The
atom U^{238}, whose atomic number and atomic weight both are even,

probably has no nuclear magnetic moment; but the odd-atomic-weight nucleus U^{235} has a nuclear moment and possesses hyperfine structure, which has been resolved in some optical lines (see Sec. 2.1) and may become apparent also in the x-ray spectrum.

Figures 2.1 and 2.2, taken from Siegbahn's monograph, show a scheme of the uranium terms and lines, on an arbitrary, uniform scale in Fig. 2.1 and on the true scale in Fig. 2.2.

2. OPTICAL SPECTRA

2.1 <u>Line Measurements.</u> Table 2.12 summarizes the most important wavelength measurements in the arc and spark spectra of uranium. The total number of lines in these spectra is enormous; by

Table 2.12 — Wavelength Measurements in Uranium Spectra

Observers	Wavelength range, A	No. of lines measured
Arc spectrum:		
Eder and Valenta (1910)	4,642 – 7,130	~600
Hasselberg (1910)	3,583 – 5,871	2,200
Exner and Haschek (1911)	2,264 – 6,827	4,960
Meissner (1916)	8,223 – 8,758	17
Kiess and Meggers (1921)	5,500 – 9,530	674
Harrison and coworkers (1939)	2,000 – 10,000	5,238*
Kiess and coworkers (1946)	2,900 – 11,000	9,000†
Spark spectrum:		
Exner and Haschek (1911)	2,195 – 6,449	5,632
Lang (1924)	374 – 2,000	93
Harrison and coworkers (1939)	2,000 – 10,000	5,238*
Schuurmans and coworkers (1947)	2,865 – 5,600	~300‡

*Arc and spark spectra together.
†According to McNally (Carbide and Carbon 3) a total of over 14,000 lines have been measured by Kiess and coworkers at the National Bureau of Standards.
‡Identified lines only (cf. Table 2.18).

drawing upon the measurements made at the Spectroscopy Laboratory of the Massachusetts Institute of Technology (Harrison and coworkers, 1939) and at the National Bureau of Standards (Kiess and coworkers, 1946, and additional unpublished results), McNally (Carbide and Carbon 3) made a list (unpublished) of as many as 19,800 uranium lines between 11,908 and 2026 A.

Of the approximately twenty thousand known uranium lines, several thousand have been, at the time of this writing, identified as due to transitions between known terms in the spectra of U^+ ion and neutral U atom (spectra UII and UI, respectively).

Table 2.13 — Strongest Lines of Uranium*

Wavelength		Intensity				Classification		
		MIT†		Natl. Bur. Standards‡				
MIT†	Natl. Bur. Standards‡	Arc	Spark	Arc	Spark	Spectrum§	Low term¶	High term**
10555	(.02)	...	...	20	...	I	$^7M_6$	156_6
9276.24	(.43)	2	...	20	...	Unc.††	Unc.	Unc.
9093	(.68)	...	...	25	...	I	$^5K_5$	116_5
8951.97	(.94)	2	...	20	...	I	$^7K_5$	169_5
8757.75	(.77)	4	...	150	...	Unc.	Unc.	Unc.
8753.69	(.68)	2	...	25	...	I	D_7	218_6
8710.77	(.76)	$4h$	...	40	...	I	$^5H_4$	174_4
8691.28	(.29)	4	...	150	...	Unc.	Unc.	Unc.
8607.94	(.96)	8	...	600	6	I	$^5L_6$	116_5
8574.61	(.59)	2	...	30	...	I	$^7M_6$	179_5
8570.51	(.52)	4	...	120	...	I	C_7	197_6
8567.73	(.71)	2	...	40	...	I	$^5I_4$	161_4
8557.34	(.32)	2	...	30	...	I	B_8	223_7
8540.20	(.19)	4	...	100	2	I	$^7K_5$	174_4
8504.70	(.58)	4	...	80	4	II	$^6I_{4\frac{1}{2}}$	$103A_{4\frac{1}{2}}(?)$

*Compiled by McNally (Carbide and Carbon 3).

†Harrison and coworkers (1939). Intensities under "arc" are for standard d-c electric arc (220-volt line with stabilizing reactance and resistance); intensities under "spark" are for a standard type 20,000-volt condensed spark over a 5-mm gap. Abbreviations in intensity column: d = unresolved double (approximate coincidence of two lines); h = hazy, diffuse, nebulous; R = wide self-reversal. Intensities are determined by visual comparison of photographic densities, approximately corrected for varying plate sensitivity. The scale is expanded to approximate actual intensity ratios.

‡Kiess and coworkers (1946) and unpublished data communicated to McNally (Carbide and Carbon 3). Only last two digits are shown in wavelength column for comparison with the MIT data. Intensities are on arbitrary scale 1 to 500. Abbreviations: l = shaded or displaced to longer waves; R = wide self-reversal; s = shaded or displaced to shorter waves; h = hazy, diffuse, nebulous; d = unresolved double (approximate coincidence of two lines).

§I means line attributed to neutral U atom, and II means line attributed to monopositive U$^+$ ion.

¶All UI terms in this column are odd; low UII terms are partly odd (group A) and partly even (group B); for assignment of a term to one of these groups, see Table 2.18. Term values are given in Tables 2.16 (UI) and 2.18 (UII).

**Term values are given in Tables 2.17 (UI) and 2.19 (UII). See these for explanation of term symbols used.

††Unc. = unclassified.

Table 2.13 — (Continued)

Wavelength		Intensity				Classification		
		MIT†		Natl. Bur. Standards‡				
MIT†	Natl. Bur. Standards‡	Arc	Spark	Arc	Spark	Spectrum§	Low term¶	High term**
8496	(.09)	...	...	100	3	I	C_7	198_7
8450.03	(.02)	4	...	400	8	I	A_7	156_7
8445.35	(.37)	4	...	400	8	I	A_7	156_6
8441.25	(.20)	2	...	80	4	I	A_8	194_8
8381.86	(.86)	4	...	120	6	I	$^7K_6$	189_5
8346.74	(.75)	2	...	80	2	I	$^5H_4$	179_3
8318.34	(.34)	2	...	100	5	Unc.	Unc.	Unc.
8262.05	(.05)	4	...	150	8	I	C_7	202_6
8223	(.09)	...	...	200	3	I	$^7M_6$	184_5
8174.30	(.30)	4	...	250	5	I	$^5K_6$	165_6
7970.46	(.46)	8	...	100	10	I	C_7	206_6
7881.94	(.91)	15	...	400	20	I	$^7M_6$	189_5
7784.13	(.13)	20	...	500	25	I	$^5K_5$	134_5
7639.54	(.524)	4	...	200	6	I	$^5K_6$	173_6
7631.71	(.72)	8	...	250	8	I	A_7	169_7
7619.35	(.34)	8	...	300	10	I	A_8	207_7
7609.16	(.16)	4	...	100	10l(?)	I	B_7	204_7
7533.91	(.91)	20	...	250	25	I	A_7	170_6
7425.50	(.50)	18	...	150	15	I	$^5L_6$	134_5
7128.91	(.89)	20	...	200	20	I	$^5K_5$	146_6
7074.81	(.78)	25	...	80	8	I	$^5K_6$	184_5
6826.93	(.91)	25	...	400	40	I	$^5L_6$	146_6
6518.945	(.937)	15	...	80	8	I	$^7M_6$	215_6
6465.005	(4.97)	25	...	150	15	I	B_7	227_8
6449.167	(.158)	100	...	300	30	I	$^5K_5$	161_4
6395.446	(.422)	100	...	200	20	I	$^5L_6$	156_7
6392.781	(.745)	20	...	80	...	I	$^5L_6$	156_6
6389.804	(.774)	18	...	80	8	I	$^5H_4$	216_5
6372.469	(.434)	50	...	200	20	I	A_7	194_8
6359.305	(.284)	30	...	100	10	I	$^5L_6$	157_5

Table 2.13 — (Continued)

Wavelength		Intensity				Classification		
		MIT†		Natl. Bur. Standards‡			Low term¶	High term**
MIT†	Natl. Bur. Standards‡	Arc	Spark	Arc	Spark	Spectrum§		
6293.347	(.324)	15	...	60	6	I	$^5K_5$	165_6
6215.397	(.37)	12	...	100	4	I	A_7	198_7
6175	(.384)	...	...	200	4	I	$^5K_6$	204_7
6171.872	(.854)	30	...	250	5	I	A_8	238_9
6129.720	(.716)	10	...	100	5	I	$^5K_5$	169_5
6077.297	(.294)	40	...	200	20	I	$^5K_5$	170_6
5997.329	(.312)	25	...	150	8	I	$^7M_6$	229_7
5986.122	(.10)	25	...	150	15	I	A_3	205_4
5976.344	(.324)	50	...	200	20	I	A_7	205_8
5971.525	(.50)	50	...	250	25	I	$^5K_5$	173_6
5915.398	(.398)	125	...	600	30	I	$^5L_6$	169_7
5898.785	(.783)	8	...	80	1	I	$^7M_6$	231_7
5853.930	(.91)	10	...	20	20	II	$^6L_{6\frac{1}{2}}$	$104A_{5\frac{1}{2}}$
5845.272	(.25)	20	...	25	25	II	$^6L_{5\frac{1}{2}}$	$101A_{4\frac{1}{2}}$
5837.707	(.695)	30	1	30	30	II	$^6M_{6\frac{1}{2}}$	$114A_{6\frac{1}{2}}$
5836.047	(.034)	30	...	80	2	I	D_7	274_8
5798.552	(.540)	35	1	40	40	II	$^6I_{3\frac{1}{2}}$	$121A_{4\frac{1}{2}}$
5780.610	(.590)	40	...	100	10	I	$^7M_6$	235_7
5634.408	(.382)	10	...	80	...	I	A_8	253_8
5621.524	(.510)	12	1	100	...	I	A_7	215_6
5620.792	(.776)	30	1	200	...	I	$^5K_5$	184_5
5616.593	(.576)	4	...	80(?)	...	I	C_7	259_8
5610.905	(.886)	30	1	150	...	I	$^7M_6$	240_7
5602.913	(.897)	5	...	...	120	II	$^6I_{4\frac{1}{2}}$	$136A_{5\frac{1}{2}}$
5597.379	(.371)	6	1	...	150	II	$^6I_{4\frac{1}{2}}$	$138A_{4\frac{1}{2}}$
5581.610	(.604)	12	5	...	200	II	$^6L_{5\frac{1}{2}}$	$103A_{4\frac{1}{2}}$
5581.230	(.228)	10	4	...	150	II	$^6K_{4\frac{1}{2}}$	$104A_{5\frac{1}{2}}$
5580.819	(.810)	8	3	...	100	II	$^6I_{3\frac{1}{2}}$	$126A_{4\frac{1}{2}}$
5570.683	(.660)	15	15	...	200	Unc.	Unc.	Unc.
5564.187	(.168)	40	3	300	...	I	A_7	218_6

Table 2.13 — (Continued)

Wavelength		Intensity				Classification		
		MIT†		Natl. Bur. Standards‡				
MIT†	Natl. Bur. Standards‡	Arc	Spark	Arc	Spark	Spectrum§	Low term¶	High term**
5557.895	(.872)	10	1	150	...	Unc.	Unc.	Unc.
5551.441	(.424)	8	5	...	100	II	$^6I_{4\frac{1}{2}}$	$140A_{4\frac{1}{2}}$
5544.822	(.805)	8h	3h	...	100h,l	Unc.	Unc.	Unc.
5535.796	(.775)	6	2	...	100	II	$^6K_{5\frac{1}{2}}$	$107A_{5\frac{1}{2}}$
5531.282	(.262)	5	...	100	...	I	C_7	261_8
5527.848	(.827)	25	40	...	400	II	$B^6L_{5\frac{1}{2}}$	$102B_{4\frac{1}{2}}$
5511.501	(.492)	30	2	250	...	I	$^5K_5$	187_6
5510.435	(.414)	10	...	150	...	Unc.	Unc.	Unc.
5504.146	(.133)	12	10	...	150	II	$^6I_{4\frac{1}{2}}$	$142A_{4\frac{1}{2}}$
5500.702	(.688)	10	...	120	...	I	A_5	297_4
5496.444	(.430)	12	1	150	...	I	$^5K_6$	224_6
5492.970	(.954)	60	50	...	500	II	$^4L_{4\frac{1}{2}}$	$103A_{4\frac{1}{2}}$
5491.236	(.220)	8	6	...	150	Unc.	Unc.	Unc.
5488.906	(.892)	8	...	100	...	Unc.	Unc.	Unc.
5487.021	(.004)	12	8	...	200	II	$^4K_{5\frac{1}{2}}$	$132A_{5\frac{1}{2}}$
5482.548	(.536)	12	18	...	200	II	$B^6L_{6\frac{1}{2}}$	$110B_{5\frac{1}{2}}$
5481.223	(.213)	30	25	...	300	II	$^6I_{4\frac{1}{2}}$	$142A_{4\frac{1}{2}}$
5480.275	(.267)	15	25	...	300	II	$B^6L_{5\frac{1}{2}}$	$103B_{4\frac{1}{2}}$
5475.725	(.711)	20	18	...	250	Unc.	Unc.	Unc.
5465.690	(.683)	12	8	...	120	Unc.	Unc.	Unc.
5444.476	(.475)	10	6	...	200	II	$^4K_{5\frac{1}{2}}$	$134A_{5\frac{1}{2}}$
5409.096	(.080)	5	6	...	120	Unc.	Unc.	Unc.
5406.888	(.872)	5	...	100	...	I	C_7	266_7
5405.996	(.975)	5	6	...	150	II	$^6M_{7\frac{1}{2}}$	$169A_{7\frac{1}{2}}$
5403.204	(.193)	8	6	...	150	II	$^4K_{5\frac{1}{2}}$	$137A_{6\frac{1}{2}}$
5400.951	(.933)	10	6	150 (?)	...	II	$^6I_{3\frac{1}{2}}$	$131A_{4\frac{1}{2}}$
5386.209	(.193)	8	12	...	200	II	$B^6L_{6\frac{1}{2}}$	$112B_{5\frac{1}{2}}$
5385.559	(.538)	4	...	100h	...	I	A_8	261_7
5377.303	(.284)	5	3	...	100	Unc.	Unc.	Unc.
5368.432	(.414)	8	6	...	150	Unc.	Unc.	Unc.

Table 2.13 — (Continued)

Wavelength		Intensity				Classification		
		MIT†		Natl. Bur. Standards‡				
MIT†	Natl. Bur. Standards‡	Arc	Spark	Arc	Spark	Spectrum§	Low term¶	High term**
5363.820	(.808)	6	4	...	200	II	$^6H_{3\frac{1}{2}}$	$138A_{4\frac{1}{2}}$
5362.397	(.382)	3	8h	...	100	Unc.	Unc.	Unc.
5341	(.502)	...	...	120	...	I	A_3	225_4
5336.512	(.536)	6	...	120	...	I	$^5I_4$	231_4
5329.223	(.256)	10	1	150	...	I	$^5L_6$	187_6
5327.710	(.761)	8	6	...	200	Unc.	Unc.	Unc.
5321.605	(.603)	6	3	...	100	II	$^6H_{3\frac{1}{2}}$	$140A_{4\frac{1}{2}}$
5319.377	(.382)	6	8	...	100	Unc.	Unc.	Unc.
5315.279	(.268)	8	1	150	...	I	A_8	264_8
5311.881	(.874)	18	18	...	200	II	$^4I_{5\frac{1}{2}}$	$125A_{5\frac{1}{2}}$
5310.038	(.039)	10	8	...	150	II	$^4I_{4\frac{1}{2}}$	$104A_{5\frac{1}{2}}$
5308.544	(.542)	25	3	300R(?)	...	I	A_7	226_7
5300.587	(.566)	8	1	120	...	I	C_7	269_8
5299.470	(.439)	6	3	100(?)	...	I	B_8	295_9
5297.450	(.440)	8	...	150	...	I	$^5I_4$	233_5
5288.397	(.385)	6	6	...	100	II	$^6I_{3\frac{1}{2}}$	$138A_{4\frac{1}{2}}$
5280.389	(.378)	30	4	300R(?)	...	I	$^5L_6$	189_5
5278.180	(.164)	12	15	...	100	II	$^6I_{3\frac{1}{2}}$	$139A_{3\frac{1}{2}}$
5272.018	(.006)	8	2	100	...	I	A_8	266_7
5270	(.625)	...	...	100	...	I	C_7	271_8
5257.044	(.06)	15	18	40	80	II	$^6H_{3\frac{1}{2}}$	$142A_{4\frac{1}{2}}$
5247.749	(.75)	12	10	...	200	⎧ II	$^6M_{6\frac{1}{2}}$	$128A_{6\frac{1}{2}}$
						⎩ II	$^6I_{4\frac{1}{2}}$	$151A_{3\frac{1}{2}}$
5247.352	(.36)	6	6	...	100	II	$^6I_{3\frac{1}{2}}$	$140A_{4\frac{1}{2}}$
5238.613	(.621)	6	6	...	100	Unc.	Unc.	Unc.
5225.000(?)	(.116)	6	6	...	120	II	$^4I_{5\frac{1}{2}}$	$127A_{5\frac{1}{2}}$
5205.179	(.156)	10	6h	100	400	Unc.	Unc.	Unc.
5204.316	(.313)	10	10	...	150	II	$^4L_{6\frac{1}{2}}$	$150A_{5\frac{1}{2}}$
5184.587	(.580)	12	15	...	200	II	$^6I_{3\frac{1}{2}}$	$142A_{4\frac{1}{2}}$
5164.157	(.140)	15	1	200	20	I	C_7	274_8
5160.326	(.313)	18	20	...	200	II	$^4K_{5\frac{1}{2}}$	$145A_{6\frac{1}{2}}$

Table 2.13 — (Continued)

Wavelength		Intensity				Classification		
		MIT†		Natl. Bur. Standards‡				
MIT†	Natl. Bur. Standards‡	Arc	Spark	Arc	Spark	Spectrum§	Low term¶	High term**
5117.249	(.236)	12	10	...	200	II	$^6K_{5\frac{1}{2}}$	$115A_{4\frac{1}{2}}$
5107.342	(.329)	6	6	...	100	II	$^6M_{6\frac{1}{2}}$	135A
5088.301	(.286)	10	...	120	3	I	$^5L_6$	196_5
5085.863	(.856)	10	6	...	150	II	$^6K_{4\frac{1}{2}}$	$108A_{3\frac{1}{2}}$
5077.823	(.817)	5	4	...	150	Unc.	Unc.	Unc.
5063.773	(.766)	12	...	150	...	I	A_7	235_7
5047.416	(.406)	6	3	...	100	II	$^6K_{5\frac{1}{2}}$	$118A_{4\frac{1}{2}}$
5027.398	(.382)	40	4	400	40	I	$^5L_6$	198_7
5011.423	(.415)	8	...	120	...	I	$^5K_5$	205_4
5008.222	(.205)	30	25	...	300	II	$^6L_{6\frac{1}{2}}$	$114A_{6\frac{1}{2}}$
4992.940	(.934)	4	4	...	100	II	$^6M_{6\frac{1}{2}}$	$141A_{5\frac{1}{2}}$
4986.903	(.899)	8	6	...	150	II	$^6K_{4\frac{1}{2}}$	$109A_{3\frac{1}{2}}$
4972.100	(.109)	8	6	...	150	II	$^6K_{4\frac{1}{2}}$	$110A_{3\frac{1}{2}}$
4967.326	(.333)	5	...	120	...	I	A_7	239_8
4955.775	(.784)	8	...	120	...	I	A_8	278_8
4950.171	(.180)	8	8	...	100	II	$^4K_{5\frac{1}{2}}$	$159A_{6\frac{1}{2}}$
4942.640	(.643)	6	6	...	100	II	$^6L_{6\frac{1}{2}}$	$117A_{6\frac{1}{2}}$
4933.657	(.668)	8	8	...	150	II	$^4I_{5\frac{1}{2}}$	$142A_{4\frac{1}{2}}$
4928.447	(.442)	20	1	100	...	I	$^5K_6$	245_7
4924.644	(.644)	6	8	...	80	II	$^6H_{3\frac{1}{2}}$	$158A_{3\frac{1}{2}}$
4913.165	(.170)	8	5	...	150	II	$^6K_{5\frac{1}{2}}$	$121A_{4\frac{1}{2}}$
4911.668	(.674)	6	5	...	100	II	$^4I_{4\frac{1}{2}}$	$107A_{5\frac{1}{2}}$
4910.339	(.347)	15	1	120	...	I	$^7M_6$	266_7
4899.294	(.286)	25	25	40	120	II	$^6K_{4\frac{1}{2}}$	$111A_{3\frac{1}{2}}$
4885.126	(.148)	18	...	120	10	I	$^5L_6$	204_7
4861.015	(.992)	10	10	25	120	II	$^6I_{3\frac{1}{2}}$	$158A_{3\frac{1}{2}}$
4859.750	(.678)	8	8	15	80	II	$^4I_{4\frac{1}{2}}$	$108A_{3\frac{1}{2}}$
						II	$^6M_{6\frac{1}{2}}$	$144A_{5\frac{1}{2}}$
4858.085	(.087)	15	15	20	200	II	$^6M_{6\frac{1}{2}}$	$145A_{6\frac{1}{2}}$
4847.660	(.647)	10	10	25	150	II	$^6K_{5\frac{1}{2}}$	$124A_{5\frac{1}{2}}$
4819.544	(.548)	12	12	30	120	II	$^4I_{5\frac{1}{2}}$	$145A_{6\frac{1}{2}}$

Table 2.13 — (Continued)

Wavelength		Intensity				Classification		
		MIT†		Natl. Bur. Standards‡			Low term¶	High term**
MIT†	Natl. Bur. Standards‡	Arc	Spark	Arc	Spark	Spectrum§		
4819.251	(.253)	30	3	80	60	Unc.	Unc.	Unc.
4810.889	(.895)	6	...	100	40	I	A_7	245_8
4778.102	(.098)	6	...	100	25	Unc.	Unc.	Unc.
4772.701	(.693)	6	18	30	150	II	$^6K_{5\frac{1}{2}}$	$125A_{5\frac{1}{2}}$
4769.260	(.267)	6	15	20	100	II	$^4I_{4\frac{1}{2}}$	$109A_{3\frac{1}{2}}$
4756.803	(.808)	12	6	120	80	I	$^5K_5$	216_5
4755.729	(.743)	8	15	25	150	II	$^4I_{4\frac{1}{2}}$	$110A_{3\frac{1}{2}}$
4731.599	(.597)	40	50	50	150	II	$^6M_{6\frac{1}{2}}$	$153A_{6\frac{1}{2}}$
4722.726	(.712)	40	50	25	150	II	$^6L_{6\frac{1}{2}}$	$124A_{5\frac{1}{2}}$
4715.659	(.682)	10	5	120	60	I	$^5I_4$	256_4
4702.517	(.515)	10	20	100(?)	100(?)	II	$^6K_{5\frac{1}{2}}$	$127A_{5\frac{1}{2}}$
4689.074	(.077)	30	40	300(?)	300(?)	II	$^4I_{4\frac{1}{2}}$	$111A_{3\frac{1}{2}}$
4671.408	(.404)	20	30	150(?)	150(?)	II	$^6M_{6\frac{1}{2}}$	$159A_{6\frac{1}{2}}$
4666.856	(.855)	25	40	...	200	II	$^6L_{5\frac{1}{2}}$	$114A_{6\frac{1}{2}}$
4663.755	(.753)	18	3	120	40	I	$^5K_5$	220_6
4646.603	(.603)	25	40	50	250	II	$^6K_{4\frac{1}{2}}$	$120A_{4\frac{1}{2}}$
4641.658	(.657)	10	15	8	80	II	$^6M_{7\frac{1}{2}}$	$211A_{7\frac{1}{2}}$
4631.620	(.624)	30	3	100	50	I	$^5L_6$	215_6
4627.079	(.083)	30	60	40	200	II	$^6M_{6\frac{1}{2}}$	$161A_{6\frac{1}{2}}$
4622.427	(.428)	12	18	8	80	II	$^6L_{7\frac{1}{2}}$	$169A_{7\frac{1}{2}}$
4620.219	(.234)	25	12	300	200	I	$^7M_6$	278_7
4605.148	(.154)	12	25	12	120	II	$^4K_{5\frac{1}{2}}$	$178A_{5\frac{1}{2}}$
4603.665	(.660)	25	40	40	200	II	$^6K_{5\frac{1}{2}}$	$132A_{5\frac{1}{2}}$
4601.130	(.130)	18	25	25	100	II	$^6K_{4\frac{1}{2}}$	$121A_{4\frac{1}{2}}$
4584.847	(.850)	10	15	20	80	II	$^6L_{6\frac{1}{2}}$	$127A_{5\frac{1}{2}}$
4581.724	(.722)	8	18	8	80	II	$1A_{7\frac{1}{2}}(?)$	$230A_{7\frac{1}{2}}$
4576.635	(.643)	12	2	80	20	I	$^5K_5$	224_6
4573.687	(.682)	30	40	50	200	II	$^6K_{5\frac{1}{2}}$	$134A_{5\frac{1}{2}}$
4569.913	(.915)	25	40	50	200	II	$^6L_{5\frac{1}{2}}$	$119A_{4\frac{1}{2}}$
4567.687	(.688)	20	40	40	160	II	$^6L_{6\frac{1}{2}}$	$128A_{6\frac{1}{2}}$

Table 2.13 — (Continued)

Wavelength		Intensity				Classification		
		MIT[†]		Natl. Bur. Standards[‡]			Low term[¶]	High term[**]
MIT[†]	Natl. Bur. Standards[‡]	Arc	Spark	Arc	Spark	Spectrum[§]		
4555.095	(.096)	20	40	25	125	II	$^6M_{7\frac{1}{2}}$	$230A_{7\frac{1}{2}}$
4551.979	(.983)	15	1	80	20	I	$^5K_5$	225_6
4545.580	(.581)	20	25	80d	150	II	$^6K_{5\frac{1}{2}}$	$136A_{5\frac{1}{2}}$
4543.632	(.628)	50	80	100	500	II	$^6K_{4\frac{1}{2}}$	$124A_{5\frac{1}{2}}$
4538.190	(.185)	25	40	30	120	II	$^6L_{6\frac{1}{2}}$	$130A_{5\frac{1}{2}}$
4516.725	(.730)	15	1	80	20	I	$^5K_5$	227_6
4515.280	(.283)	25	40	50	250	II	$^6L_{5\frac{1}{2}}$	$120A_{4\frac{1}{2}}$
4510.320	(.310)	20	30	30	120	II	$^4I_{4\frac{1}{2}}$	$119A_{4\frac{1}{2}}$
4506.223	(.210)	6	12	10	150	II	$110B_{5\frac{1}{2}}$	$511_{4\frac{1}{2}}$
4490.835	(.830)	18	25	20	120	II	$^6L_{6\frac{1}{2}}$	$132A_{5\frac{1}{2}}$
4477.710	(.700)	20	25	15	75	II	$^6K_{4\frac{1}{2}}$	$125A_{5\frac{1}{2}}$
4472.335	(.322)	50	80	100	500	II	$^6L_{5\frac{1}{2}}$	$121A_{4\frac{1}{2}}$
4469.328	(.320)	12	1	100	20	I	$^5L_6$	223_7
4465.133	(.124)	20	25	30	120	II	$^6K_{5\frac{1}{2}}$	$142A_{4\frac{1}{2}}$
4462.974	(.972)	18	30	40	200	II	$^6K_{4\frac{1}{2}}$	$126A_{4\frac{1}{2}}$
4434	(.526)	...	...	30	150	II	$113B_{4\frac{1}{2}}$	$525_{4\frac{1}{2}}$
4433.889	(.890)	15	12	15	150	II	$^6M_{7\frac{1}{2}}$	$240A_{7\frac{1}{2}}$
4427.653	(.644)	12	15	20	100	II	$^6L_{5\frac{1}{2}}$	$123A_{4\frac{1}{2}}$
4393.588	(.600)	40	6	200	40	I	$^5L_6$	227_6
4371.760	(.764)	18	1	100	15	I	C_7	309_8
4362.05	(.051)	30	3	300	30	I	$^5L_6$	229_7
4355.741	(.748)	10	20	80	20	I	$^5K_5$	235_6
4347.192	...	18	18	...	...	II	$^6K_{4\frac{1}{2}}$	$131A_{4\frac{1}{2}}$
4341.688	...	50	50	...	...	II	$^6L_{5\frac{1}{2}}$	$126A_{4\frac{1}{2}}$
4328.742	(.728)	20	1	60	8	I	$^5K_5$	237_6
4313.147	(.134)	20	1	100	20	I	A_7	269_8
4297.112	...	18	18	...	...	II	$^6L_{5\frac{1}{2}}$	$127A_{5\frac{1}{2}}$
4288.841	(.838)	20	2	200	20	I	$^7M_6$	295_6
4269.613	...	20	30	...	...	II	$^6L_{6\frac{1}{2}}$	$145A_{6\frac{1}{2}}$
4252.426	...	15	20	...	...	II	$^4I_{5\frac{1}{2}}$	$184A_{5\frac{1}{2}}$

Table 2.13 — (Continued)

| Wavelength | | Intensity | | | | Classification | | |
MIT†	Natl. Bur. Standards‡	MIT† Arc	MIT† Spark	Natl. Bur. Standards‡ Arc	Natl. Bur. Standards‡ Spark	Spectrum§	Low term¶	High term**
4246.261	(.262)	30	2	100	10	I	$^5L_6$	235_7
4244.372	...	25	25	...	...	II	$^4I_{4\frac12}$	$127A_{5\frac12}$
4241.669	...	40	50	...	...	II	$^6M_{6\frac12}$	$185A_{5\frac12}$
4231.676	(.669)	25	...	60	20	I	A_8	312_9
4222.375	(.365)	18	8	150	30	I	A_7	274_8
4171.591	...	30	30	...	...	II	$^6L_{6\frac12}$	$153A_{6\frac12}$
4169.050	(.055)	20	$2h$	50	10	I	C_7	321_7
4162.430	(.431)	18	$3h$	60	15	I	A_7	278_8
4156.652	(.656)	15	$8h$	100	10	I	$^5K_5$	246_6
4153.976	(.972)	15	5	150	15	I	$^5L_6$	240_7
4141.228	...	20	30	...	...	II	$^6M_{7\frac12}$	$252A_{6\frac12}$
4128.336	...	18	20	...	...	II	$^4I_{5\frac12}$	$192A_{6\frac12}$
4124.725	...	30	25	...	...	II	$^6L_{6\frac12}$	$159A_{6\frac12}$
4116.097	...	25	35	...	...	II	$^4I_{4\frac12}$	$136A_{5\frac12}$
4106.931	(.906)	25	10	8	...	II	$^4I_{4\frac12}$	$139A_{3\frac12}$
4095.746	...	18	25	...	...	Unc.	Unc.	Unc.
4091.636	(.635)	12	...	50	3	I	$^5L_6$	244_6
4090.135	...	25	40	...	...	II	$^6L_{6\frac12}$	$161A_{6\frac12}$
4088.254	...	25	18	...	...	II	$^4I_{4\frac12}$	$140A_{4\frac12}$
4080.609	...	12	20	...	...	Unc.	Unc.	Unc.
4071.108	...	15	25	...	...	Unc.	Unc.	Unc.
4067.756	...	12	20	...	...	II	$^4L_{6\frac12}$	$237A_{5\frac12}$
4062.549	...	12	18	...	...	II	$^4I_{4\frac12}$	$141A_{5\frac12}$
4051.914	...	20	25	...	...	II	$^6L_{7\frac12}$	$211A_{7\frac12}$
4050.039	...	25	35	...	...	II	$^4I_{4\frac12}$	$142A_{4\frac12}$
4047.610	(.621)	18	3	80	8	I	$^5K_5$	253_5
4044.416	...	18	25	...	...	II	$^6M_{6\frac12}$	$198A_{5\frac12}$
						II	$^6L_{7\frac12}$	$214A_{6\frac12}$
						II	$^6I_{3\frac12}$	$224A_{4\frac12}$
4042.752	(.756)	4	10	150	15	I	$^5K_5$	253_6
4026.019	...	25	25	...	...	II	$^6K_{5\frac12}$	$170A_{4\frac12}$
4018.990	(9.04)	25	15	8(?)	...	II	$^6L_{5\frac12}$	$145A_{6\frac12}$

Table 2.13 — (Continued)

Wavelength		Intensity				Classification		
		MIT†		Natl. Bur. Standards‡				
MIT†	Natl. Bur. Standards‡	Arc	Spark	Arc	Spark	Spectrum§	Low term¶	High term**
4017.723	...	25	25	...	...	II	$^6I_{4\frac{1}{2}}$	$243A_{4\frac{1}{2}}$(?)
4005.698	(.696)	25	3	50	3	Unc.	Unc.	Unc.
4005	(.212)	...	...	80	4	I	$^5I_4$	294_3
4004.063	...	15	20	...	...	II	$^6L_{6\frac{1}{2}}$	$167A_{6\frac{1}{2}}$
3990.423	...	18	20	...	...	II	$^6K_{4\frac{1}{2}}$	$158A_{3\frac{1}{2}}$
3985.795	...	25	30	...	...	II	$^6L_{7\frac{1}{2}}$	$230A_{7\frac{1}{2}}$
3966.567	(.52)	20	30	...	6(?)	Unc.	Unc.	Unc.
3954.663	...	20	30	...	...	II	$B^4I_{4\frac{1}{2}}$	$110B_{5\frac{1}{2}}$
3943.820	(.820)	35	5	200	10	I	$^5L_6$	253_6
3932.026	...	35	50	...	...	II	$^6L_{5\frac{1}{2}}$	$153A_{6\frac{1}{2}}$
3930.982	...	12	35	...	...	Unc.	Unc.	Unc.
3926.216	(.218)	30	6	50	5	I	$^5L_6$	254_6
3915.884	...	20	30	...	...	II	$B^4I_{4\frac{1}{2}}$	$112B_{5\frac{1}{2}}$
3911.673	...	18	18	...	...	II	$^4I_{5\frac{1}{2}}$	$214A_{6\frac{1}{2}}$
3902.561	...	18	18	...	...	II	$^6L_{5\frac{1}{2}}$	157A
3896.779	...	20	25	...	...	II	$^6M_{6\frac{1}{2}}$	$228A_{5\frac{1}{2}}$
3894.123	(.121)	30	4	120	12	I	$^5L_6$	256_7
3892.684	...	20	30	...	...	II	$^6L_{7\frac{1}{2}}$	$240A_{7\frac{1}{2}}$
3890.364	...	35	30	...	...	II	$^6L_{5\frac{1}{2}}$	$159A_{6\frac{1}{2}}$
3882.361	...	18	18	...	...	II	$^6L_{6\frac{1}{2}}$	$178A_{5\frac{1}{2}}$
3881.461	...	30	20	...	...	II	$^6M_{6\frac{1}{2}}$	$230A_{7\frac{1}{2}}$
3871.042	(.035)	30	1	120	12	I	$^5L_6$	258_6
3865.923	...	20	25	...	...	II	$^6K_{5\frac{1}{2}}$	$185A_{5\frac{1}{2}}$
3859.580	...	20	30	...	...	II	$^6L_{5\frac{1}{2}}$	$161A_{6\frac{1}{2}}$
3854.655	...	20	30	...	...	II	$B^6I_{3\frac{1}{2}}$	$102B_{4\frac{1}{2}}$
3839.632	(.624)	30	2	120	12	I	A_7	299_7
3833.023	...	20R	15	...	...	Unc.	Unc.	Unc.
3831.465	...	25	25	...	...	II	$B^6I_{3\frac{1}{2}}$	$103B_{4\frac{1}{2}}$
3814.070	...	25	15	...	...	II	$^6K_{4\frac{1}{2}}$	$170A_{4\frac{1}{2}}$
3813.791	...	20	15	...	...	II	$^6K_{5\frac{1}{2}}$	$190A_{5\frac{1}{2}}$

Table 2.13 — (Continued)

Wavelength		Intensity				Classification		
		MIT†		Natl. Bur. Standards‡				
MIT†	Natl. Bur. Standards‡	Arc	Spark	Arc	Spark	Spectrum§	Low term¶	High term**
3811.999	(.995)	18	6	150	5	I	$^5L_6$	262_6
3783.840	...	20	25	...	...	II	$^4K_{5\frac{1}{2}}$	$250A_{6\frac{1}{2}}$
3782.841	...	25	30	...	...	II	$^6L_{5\frac{1}{2}}$	$167A_{6\frac{1}{2}}$
3780.716	...	15	20	...	...	II	$^6K_{4\frac{1}{2}}$	$173A_{5\frac{1}{2}}$
3763.266	(.270)	12	25	40	$4s(?)$	I	$^7K_6$	335_7
3748.678	...	15	25	...	...	II	$B^6I_{4\frac{1}{2}}$	$107B_{3\frac{1}{2}}$
3731.453	(.450)	12	...	100	$10l(?)$	I	$^5L_6$	269_6
3701.522	...	10	25	...	...	II	$^6K_{6\frac{1}{2}}$	$252A_{6\frac{1}{2}}$
3700.575	...	12	18	...	...	II	$^6K_{4\frac{1}{2}}$	$184A_{5\frac{1}{2}}$
3670.072	...	15	18	...	...	II	$^6K_{4\frac{1}{2}}$	$185A_{5\frac{1}{2}}$
3659.159	(.159)	15	1	100	10	I	$^5K_5$	278_6
3644.245	(.241)	18	2	60	4	I	$^5K_5$	280_6
3640.948	...	8	20	...	...	II	$^6L_{6\frac{1}{2}}$	$197A_{5\frac{1}{2}}$
3638.200	(.195)	5	2	150	10	I	A_7	312_8
3630.733	...	8	20	...	...	Unc.	Unc.	Unc.
3623.055	...	12	15	...	...	II	$^6K_{4\frac{1}{2}}$	$190A_{5\frac{1}{2}}$
3609.682	(.682)	15	12	10	...	Unc.	Unc.	Unc.
3608.96	...	18	10	...	...	Unc.	Unc.	Unc.
3591.747	(.743)	10	2	50	12	I	$^5K_5$	284_4
3584.879	(.880)	30	12	250	50	I	$^5L_6$	278_7
3569.062	(.080)	12	20	$8(?)$	...	Unc.	Unc.	Unc.
3566.598	(.595)	30	10	200	20	I	$^5K_5$	287_5
3561.800	(.804)	12	30	80	...	Unc.	Unc.	Unc.
3550.822	...	12	20	...	...	II	$^4I_{4\frac{1}{2}}$	$185A_{5\frac{1}{2}}$
3533.568	...	10	20	...	...	II	$^6K_{4\frac{1}{2}}$	$197A_{5\frac{1}{2}}$
3531.113	...	8	20	...	...	II	$^6L_{6\frac{1}{2}}$	$219A_{5\frac{1}{2}}$
3514.615	(.612)	18	5	200	20	I	$^5L_6$	284_5
3509.668	...	10	15	...	...	II	$^4I_{4\frac{1}{2}}$	$189A_{5\frac{1}{2}}$
3507.344	(.342)	10	3	80	8	I	$^5L_6$	285_5
3500.077	(.070)	15	2	80	8	I	$^5L_6$	286_5

Table 2.13 — (Continued)

Wavelength		Intensity				Classification		
		MIT[†]		Natl. Bur. Standards[‡]				
MIT[†]	Natl. Bur. Standards[‡]	Arc	Spark	Arc	Spark	Spectrum[§]	Low term[¶]	High term[**]
3490.242	...	12	20	...	...	II	$^6K_{4\frac{1}{2}}$	$206A_{5\frac{1}{2}}$
3489.371	(.366)	20	1	80	8	I	$^5L_6$	287_5
3424.557	(.55)	20	15	20	...	II	$^6L_{6\frac{1}{2}}$	$240A_{7\frac{1}{2}}$
3423.05	...	$12d$	$15d$	...	...	II	$^6K_{4\frac{1}{2}}$	$224A_{4\frac{1}{2}}$
						II	$^6I_{3\frac{1}{2}}$	$274A_{3\frac{1}{2}}$
3422.352	...	$18R$	$15R$	...	...	II	$^6K_{4\frac{1}{2}}$	$225A_{4\frac{1}{2}}$
						II	$^6I_{3\frac{1}{2}}$	$275A_{4\frac{1}{2}}$
3390.389	(.389)	18	10	25	30	Unc.	Unc.	Unc.
3341.661	...	12	15	...	...	II	$^6K_{5\frac{1}{2}}$	$250A_{6\frac{1}{2}}$
3322.118	...	18	12	...	...	II	$^6M_{6\frac{1}{2}}$	$277A_{5\frac{1}{2}}$
3293.590	(.590)	30	5	4	2	Unc.	Unc.	Unc.
3288.209	(.248)	25	20	3	2	II	$^6H_{3\frac{1}{2}}$	$298A_{4\frac{1}{2}}$
3270.124	...	20	25	...	...	II	$^6L_{5\frac{1}{2}}$	$237A_{5\frac{1}{2}}$
3265.806	...	25	18	...	...	II	$^6L_{5\frac{1}{2}}$	$239A_{4\frac{1}{2}}(?)$
3263.112	(.116)	25	8	$20s(?)$	...	I	$^5L_6$	306_6
3229.502	...	18	25	...	...	Unc.	Unc.	Unc.
3200.135	...	15	15	...	...	Unc.	Unc.	Unc.
3177.331	...	15	18	...	...	Unc.	Unc.	Unc.
3176.208	...	20	15	...	...	Unc.	Unc.	Unc.
3153.120	...	12	15	...	...	Unc.	Unc.	Unc.
3149.21	...	$18d$	$18d$	...	...	Unc.	Unc.	Unc.
3147.09	...	12	15	...	...	Unc.	Unc.	Unc.
3145.559	...	12	10	...	...	II	$^6K_{6\frac{1}{2}}$	$311A_{6\frac{1}{2}}$
3142.601	...	12	10	...	...	Unc.	Unc.	Unc.
3139.560	...	25	25	...	...	II	$^6H_{2\frac{1}{2}}$	$306A_{3\frac{1}{2}}$
3126.174	...	12	20	...	...	Unc.	Unc.	Unc.
3111.621	...	15	15	...	...	Unc.	Unc.	Unc.
3104.162	...	20	20	...	...	Unc.	Unc.	Unc.
3095.041	...	12	12	...	...	II	$^6L_{6\frac{1}{2}}$	$267A_{6\frac{1}{2}}$
3093.012	...	20	20	...	...	Unc.	Unc.	Unc.
3091.25	...	$15R,d$	$12R,d$	...	...	Unc.	Unc.	Unc.
3088.987	...	20	15	...	...	Unc.	Unc.	Unc.

Table 2.13 — (Continued)

Wavelength		Intensity				Classification		
		MIT†		Natl. Bur. Standards‡				
MIT†	Natl. Bur. Standards‡	Arc	Spark	Arc	Spark	Spectrum§	Low term¶	High term**
3084.238	...	15	12	...	...	II	$^6K_{5\frac{1}{2}}$	$278A_{6\frac{1}{2}}$
3080.741	...	12	12	...	...	II	$^6L_{6\frac{1}{2}}$	$268A_{5\frac{1}{2}}$
3073.812	...	15R	10R	...	...	II	$^6I_{4\frac{1}{2}}$	$326A_{5\frac{1}{2}}$
3072.783	...	20	20	...	...	Unc.	Unc.	Unc.
3062.536	...	12	15	...	...	II	$^6L_{6\frac{1}{2}}$	$270A_{5\frac{1}{2}}$
3057.91	...	20d	20d	...	...	Unc.	Unc.	Unc.
3047.571	...	15	20	...	...	Unc.	Unc.	Unc.
3039.263	...	15	12	...	...	II	$^6L_{7\frac{1}{2}}$	$320A_{7\frac{1}{2}}$
3031.991	...	15	15	...	...	Unc.	Unc.	Unc.
3029.131	...	20	15	...	...	Unc.	Unc.	Unc.
3025.034	...	12h	20h	...	...	II	$107B_{3\frac{1}{2}}$	$607_{4\frac{1}{2}}$
3024.510	...	18	15	...	...	Unc.	Unc.	Unc.
3024.385	...	25R	20R	...	...	Unc.	Unc.	Unc.
3016.956	...	15	12	...	...	II	$^6L_{6\frac{1}{2}}$	$280A_{5\frac{1}{2}}$
2967.894	...	15	20	...	...	Unc.	Unc.	Unc.
2966.12	...	15	25	...	...	Unc.	Unc.	Unc.
2960.942	...	15	25	...	...	II	$^6K_{4\frac{1}{2}}$	$277A_{5\frac{1}{2}}$
2956.060	...	10	60	...	...	Unc.	Unc.	Unc.
2954.39	...	12	15	...	...	II	$^6K_{5\frac{1}{2}}$	301A
2943.895	...	10	25	...	...	II	$^6K_{5\frac{1}{2}}$	$303A_{4\frac{1}{2}}$
2941.919	...	15	30	...	...	II	$^6K_{4\frac{1}{2}}$	281A
2932.61	...	10	25	...	...	Unc.	Unc.	Unc.
2931.414	(.41)	12	12	20	10	II	$^6L_{5\frac{1}{2}}$	$270A_{5\frac{1}{2}}$
2930.805	...	4	30	...	...	Unc.	Unc.	Unc.
2928.599	...	15	35	...	...	Unc.	Unc.	Unc.
2925.568	...	15	25	...	...	Unc.	Unc.	Unc.
2914.629	...	12	15	...	...	II	$^6K_{4\frac{1}{2}}$	286A
2914.253	...	18	25	...	...	II	$^6L_{5\frac{1}{2}}$	$273A_{5\frac{1}{2}}$
2908.275	...	12	30	...	...	II	$^6L_{5\frac{1}{2}}$	$276A_{6\frac{1}{2}}$
2906.913	...	18R	15h	...	...	Unc.	Unc.	Unc.

Table 2.13 — (Continued)

Wavelength		Intensity				Classification		
		MIT†		Natl. Bur. Standards‡				
MIT†	Natl. Bur. Standards‡	Arc	Spark	Arc	Spark	Spectrum§	Low term¶	High term**
2906.798	...	15	50	...	...	II	$^4I_{4\frac{1}{2}}$	$270A_{5\frac{1}{2}}(?)$
2894.512	...	15	15	...	...	Unc.	Unc.	Unc.
2889.627	...	30	50	...	...	II	$^6L_{5\frac{1}{2}}$	$280A_{5\frac{1}{2}}$
2887.252	...	25	25	...	...	Unc.	Unc.	Unc.
2882.741	...	18	20	...	...	II	$^6L_{5\frac{1}{2}}$	284A
2875.198	...	18	12	...	...	II	$^6K_{4\frac{1}{2}}$	$292A_{5\frac{1}{2}}$
2870.974	...	18	20	...	...	II	$^6L_{5\frac{1}{2}}$	285A
2865.679	...	30	50	...	...	II	$^4I_{4\frac{1}{2}}$	$280A_{5\frac{1}{2}}$
2864.276	...	18h	12h	...	...	Unc.	Unc.	Unc.
2860.466	...	35	30	...	...	II	$^4I_{4\frac{1}{2}}$	$283A_{4\frac{1}{2}}$
2858.903	...	35	25	...	...	II	$^4I_{4\frac{1}{2}}$	284A
2852.750	...	15	15h	...	...	Unc.	Unc.	Unc.
2849.480	...	18	15	...	...	II	$^6L_{5\frac{1}{2}}$	290A
2839.890	...	18	20	...	...	Unc.	Unc.	Unc.
2833.821	...	15	25	...	...	Unc.	Unc.	Unc.
2832.063	...	35	50	...	...	Unc.	Unc.	Unc.
2828.90	...	18	20	...	...	Unc.	Unc.	Unc.
2826.193	...	18	12h	...	...	II	$^4I_{4\frac{1}{2}}$	290A
2824.28	...	25	30	...	...	Unc.	Unc.	Unc.
2821.122	...	20	35	...	...	II	$^6K_{4\frac{1}{2}}$	$305A_{4\frac{1}{2}}$
2817.959	...	18	30	...	...	Unc.	Unc.	Unc.
2811.345	...	35	30	...	...	Unc.	Unc.	Unc.
2809.952	...	20	20	...	...	Unc.	Unc.	Unc.
2807.05	...	18	30	...	...	II	$^6L_{5\frac{1}{2}}$	$295A_{5\frac{1}{2}}$
2802.559	...	15	30	...	...	II	$^6L_{5\frac{1}{2}}$	$296A_{4\frac{1}{2}}$
2795.232	...	18	12	...	...	II	$^6K_{4\frac{1}{2}}$	307A
2793.937	...	25	30	...	...	II	$^6L_{5\frac{1}{2}}$	$298A_{4\frac{1}{2}}$
2762.850	...	15	20	...	...	Unc.	Unc.	Unc.
2754.155	...	20	35	...	...	Unc.	Unc.	Unc.
2748.451	...	18	15	...	...	Unc.	Unc.	Unc.

Table 2.13 — (Continued)

Wavelength		Intensity				Classification		
		MIT[†]		Natl. Bur. Standards[‡]				
MIT[†]	Natl. Bur. Standards[‡]	Arc	Spark	Arc	Spark	Spectrum[§]	Low term[¶]	High term[**]
2746.158	...	25	6	...	...	Unc.	Unc.	Unc.
2706.95	...	15d	20d	...	...	Unc.	Unc.	Unc.
2698.057	...	20	50	...	...	II	$^6L_{5\frac{1}{2}}$	312A
2695.490	...	12	30	...	...	II	$^6L_{5\frac{1}{2}}$	313A
2691.038	...	15	30	...	...	II	$^6L_{6\frac{1}{2}}$	324A
2683.279	...	25	25	...	...	Unc.	Unc.	Unc.
2664.153	...	18	20	...	...	Unc.	Unc.	Unc.
2645.47	...	20	25	...	...	II	$^4I_{4\frac{1}{2}}$	316A$_{4\frac{1}{2}}$
2635.528	...	25	50	...	...	II	$^4I_{4\frac{1}{2}}$	318A$_{4\frac{1}{2}}$
2608.197	...	25	6	...	...	Unc.	Unc.	Unc.
2597.689	...	25	15	...	...	Unc.	Unc.	Unc.
2591.252	...	18	12	...	...	II	$^4I_{4\frac{1}{2}}$	321A$_{4\frac{1}{2}}$
2565.406	...	30	30	...	...	II	$^4I_{4\frac{1}{2}}$	326A$_{5\frac{1}{2}}$
2500.864	...	18	12	...	...	Unc.	Unc.	Unc.
2403.423	...	...	30	...	...	Unc.	Unc.	Unc.
2378.16	...	...	35	...	...	Unc.	Unc.	Unc.
2318.47	...	...	25	...	...	Unc.	Unc.	Unc.
2306.91	...	...	25	...	...	Unc.	Unc.	Unc.
2282.78	...	...	25	...	...	Unc.	Unc.	Unc.
2276.05	...	...	20	...	...	Unc.	Unc.	Unc.
2248.03	...	...	25	...	...	Unc.	Unc.	Unc.
2219.28	...	15	2	...	...	Unc.	Unc.	Unc.
2158.61	...	...	20	...	...	Unc.	Unc.	Unc.

A list of 430 of the strongest lines of uranium, ranging in wavelength from 10,555 to 2158 A, is given in Table 2.13, compiled by McNally (Carbide and Carbon 3) from Massachusetts Institute of Technology and National Bureau of Standards measurements. More extensive tabulations can be found in Massachusetts Institute of Technology wavelength tables (Harrison and coworkers, 1939) (5,238 lines of UI and UII); in Kiess, Humphreys, and Laun (1946) (1,240 classified lines of UI); and van den Bosch (1949a) (743 classified lines of UII).

Of the strong uranium lines listed in Table 2.13, more than three-quarters (78 per cent) are classified as transitions between known U and U$^+$ terms. Of the remaining 84 lines, McNally estimated that 3 per cent may belong to U^{++}, 15 per cent to U$^+$, and 4 per cent to the neutral U atom.

Table 2.14—Persistent Lines of Uranium According to Different Observers

| | Harrison (1939) | | Kiess (Natl. Bur. Standards 1) | | Fred (MP Chicago 2) |
| | Intensity* in | | | | |
Angstroms	Arc	Spark	Intensity†	Order of appearance†	Order of appearance†
3019	...	...	...	...	...
3090	...	...	...	...	...
3102	...	...	...	...	...
3552.172	8	12	...	...	...
3672.579	8	15	...	...	...
3859.58	...	...	...	...	(5)
3932.02	...	...	...	...	(1)
3966.57	...	...	...	...	(4)
4090.14	...	...	...	...	(2)
4241.669	40	50	...	...	...
4341.69	...	...	...	...	(3)
4620.22	...	...	600	(5)	...
5027.38	...	...	400	(4)	...
5915.40	...	...	500	(1)	...
5919.61	...	...	...	...	...
6395.42	...	...	200	(3)	...
6449.10	...	...	300	(2)	...

*Photographic.
†Visual.

The use of uranium spectrum lines for the determination of this element, as well as for the determination of impurities in uranium, cannot be discussed here (see the analytical chemistry volumes of the National Nuclear Energy Series).

The most persistent uranium lines (which are not necessarily the strongest ones) are listed in Table 2.14. Earlier observations (Hartley and Moss, 1912; Meyer, 1921) were made with impure material; some of the lines listed by them appear not to belong to uranium at all.

Reproductions of the uranium arc and spark spectra, given in Eder and Valenta's atlas (Eder and Valenta, 1924), also are likely to contain many lines due to impurities.

2.2 Nuclear Effects in Uranium Spectra. Two types of nuclear effects occur in atomic spectra: (1) shifting of lines and (2) their splitting (so-called "hyperfine structure") due to interaction between nuclear spin momentum and the rotational momenta of the valence electrons.

(a) Line Shifts. Anderson and White (1947) gave a value of 0.426 cm^{-1} for the shift of the UI line 5027.40 A in U^{235} (compared to its position in U^{238}). McNally (Carbide and Carbon 1) compared the behavior of a number of UII lines and found the shifts to be particularly wide in lines whose emission involves the drop of an electron into one

Table 2.15 — Average Line Shifts $(U^{235} - U^{238})$
in UII Spectrum*

Electronic configuration involved (low term)	Line shift $(U^{235} - U^{238})$, cm^{-1}
$f^3 s^2$	0.9
$f^3 ds$	0.2
$f^3 d^2$	−0.2
$f^4 s$	0.3
$f^4 d$	−0.4

*After McNally (Carbide and Carbon 2).

of the penetrating 7s orbits. Isotopic shift measurements can therefore contribute to the term analysis of the spectrum. Table 2.15 shows the average line shifts for lines with different lower terms in the UII spectrum. Burkhart, Stukenbroeker, and Adams (1949) observed the isotopic shifts for both U^{235} and U^{233}. In the arc spectrum, over 900 lines in the region 2500 to 4800 A showed distinct isotopic shifts, and only two lines with λ greater than 4800 A (λ 6101.74 and 6465.0) exhibited a noticeable shift. The same lines showed shifts for both U^{235} and U^{233}. All shifts were toward the shorter wavelengths in the series $U^{238} \rightarrow U^{235} \rightarrow U^{233}$ (according to Table 2.15, McNally observed shifts also in the opposite direction). Aside from uniform direction, no quantitative relation appeared between the mass difference and the extent of the shift. The 11 tabulated typical shifts ranged from 0.14 to 1.37 cm^{-1} in the pair U^{238} and U^{235} and from 0.07 to 1.06 cm^{-1} in the pair U^{235} and U^{233}; there was no proportionality or even consistent parallelism between the two sets of shifts.

Comparison with Table 2.13 shows that all the 11 tabulated lines catalogued by Burkhart and coworkers (1949) belonged to the UII spectrum; in accordance with McNally's Table 2.15 the strongest shifts (0.93 to 1.37 cm^{-1}) were shown by lines with lower terms having the configuration $f^3 s^2$.

Burkhart and coworkers (1949) suggested that the isotopic shift can be used to determine spectrophotometrically the ratio of the uranium isotopes in a given sample and gave examples showing the reliability of this method.

Figure 2.3 shows the line 4244.4 A in three pure isotopes and three binary mixtures. The shifts are $\Delta\nu = 1.37$, 0.70, and $1.37 + 0.70$ or

Fig. 2.3 — Isotope shifts in the 4241.7 (A) and 4244.4 (B) lines of the uranium spectra. From top to bottom, these spectra are of (1) U^{238}; (2) U^{238} plus U^{235}; (3) U^{235}; (4) U^{235} plus U^{233}; (5) U^{233}; (6) U^{238} plus U^{233}.

2.07 cm^{-1}, respectively. The adjacent line, 4241.7 A, shows only a barely perceptible shift.

(b) Hyperfine Structure. The odd uranium isotopes, U^{235} and U^{233}, must have a nuclear spin different from zero, and their lines should therefore exhibit a hyperfine structure. First British observations (British 1) indicated a nuclear spin of $\frac{9}{2}$ for U^{235}, but more recent observations of Anderson and White (1947) at Berkeley were consistent with a spin value of $\frac{5}{2}$, leaving $\frac{7}{2}$ as a possible alternative. In the range 4000 to 8500 A, observations with a Fabry-Perot étalon revealed only two lines (5915 and 6926 A) with a resolvable and measurable hyperfine structure. The first line, which gave a 0.21 cm^{-1} broad hyperfine structure pattern, was used for the estimation of the nuclear spin.

2.3   Effects of Electric and Magnetic Fields on Spectrum.   The Stark effect of some uranium arc lines was measured by Nagaoka and Sugiura (1924); the pressure shift, essentially the Stark effect of the electric fields of neighboring atoms, was measured by Humphreys (1897).

The Zeeman effect of uranium lines was studied very extensively, since it provides the best opening wedge for the identification of spectral terms in complex spectra. The earliest measurements were made by Ross (1910). More recently, accurate measurements were carried out with a Weiss magnet (42,000 oersteds) at the Zeeman Laboratory at Amsterdam, with a Bitter electromagnet (85,000 oersteds) at the Massachusetts Institute of Technology, and with a water-cooled Weiss magnet (28,000 to 35,000 oersteds) at the National Bureau of Standards.

Only the results of the Dutch investigation (Schuurmans, van den Bosch, and Dijkwel, 1947; van den Bosch and van den Berg, 1949b) have been published in detail. The most recent compilation (at the time of writing of this chapter) is found in the paper by van den Bosch and van den Berg (1949b). It lists completely or partially resolved Zeeman patterns of 115 UI lines between 5511 and 2971 A and of 158 UII lines between 5504 and 3232 A and gives the J numbers and g values of the participating terms. These papers also contain photograms and photometer curves of the Zeeman effects of the following UII lines:

4646.61*†	4050.05†
4543.64*†	3932.03*†
4515.29*†	3890.37*
4472.34*†	3859.57*
4393.60†	3831.47*
4341.69*†	3670.08*†
	3270.13†

The Zeeman patterns observed at the Massachusetts Institute of Technology have not been published except for those of 19 lines given in an early report (MIT 2). A later report by McNally and Harrison (Carbide and Carbon 2) contains tracings of the Zeeman effect for the UII lines 4244.372, 4341.688, and 4543.632 A.

Kiess, Humphreys, and Laun (1946) determined the Zeeman patterns of several thousand lines in the UI spectrum. Most lines gave either triplets or unresolved patterns, but the few that gave well-resolved patterns provided essential clues for the analysis of the UI spectrum by identifying the g values of the terms and the quantum numbers J. The patterns of only seven lines were given as examples in the paper by Kiess and coworkers.

Figure 2.4 shows 10 examples of the magnetic splitting of uranium lines, taken from the work of van den Bosch and van den Berg (1949b). Figure 2.5 shows tracings of the σ and π components of the UII line

*Schuurmans et al. (1947).
†Van den Bosch and van den Berg (1949b).

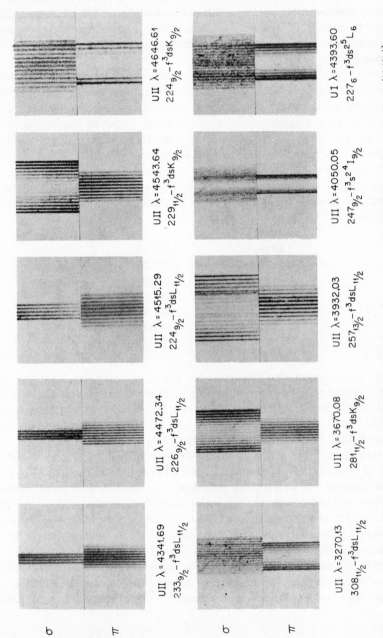

Fig. 2.4 — Zeeman effect of several lines in the U⁺ spectrum [after van den Bosch and van den Berg (1949b)].

4341.69 A, according to Massachusetts Institute of Technology measurements.

2.4 Term Analysis of the Uranium Spectra. The reason for the extreme complexity of the uranium spectra is the same as in the case of the rare earths. This complexity is due to the presence, in the most important states of the uranium atom and of the monopositive uranium ion, of several (usually 3 and sometimes 4) f electrons. The orbital

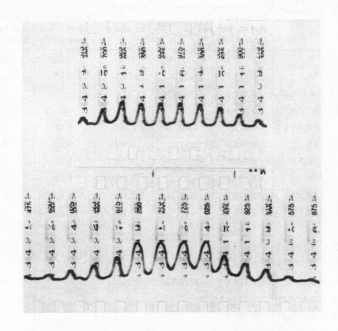

Fig. 2.5 — UII Zeeman pattern at wavelength 4341.688 A.

momentum of these electrons is high ($l = 3$), and permitted variations in their mutual orientation lead to a large variety of combined orbital momenta, L. For example, Table 2.16 shows UI terms of the f^3d^2s configuration with L values up to L = 9 (M terms). By aligning in parallel the orbital momenta of all six valence electrons, the L value could be raised to L = 13, but no terms of this type have as yet been observed in the uranium spectrum. Parallel alignment of the spins of all six valence electrons ($s = \frac{1}{2}$) leads to a total spin momentum S = 3 and thus permits septets ($7 = 2 \times 3 + 1$) as the highest multiplicity; in the monopositive uranium ion, the maximum value of S is $2\frac{1}{2}$, and the maximum multiplicity is therefore 6 ($2 \times 2\frac{1}{2} + 1$). So far, septet and quintet terms have been identified in the UI spectrum, and sextet and quartet terms have been identified in the UII spectrum. Terms of lower multiplicity have not yet been observed.

Table 2.16 — Low Odd Terms of UI

Electron configuration*	Term symbol*	Term value, cm^{-1}		g value		
				Theoretical for LS coupling	Observed	
		Kiess, Humphreys, and Laun (1946)	Schuurmans (1947)		Kiess, Humphreys, and Laun (1946)	Van den Bosch and van den Berg (1949b)
f³ds²	⁵L₆	0.0	0.00	0.714	0.75	0.75
f³ds²	⁵K₅	620.3	620.33	0.667	0.73	0.73
(f³ds²?)	A₇(=⁵L₇?)	3,800.8	3800.83	...	...	0.92
(f³ds²?)	A₃	3,868.4	...	...	...	...
f³ds²	⁵K₆	4,275.7	4275.75	0.905	0.93	0.91
f³ds²	⁵I₄	4,453.4	...	0.600	0.66	0.67
f³d²s	⁷K₅†	5,762.0	...	0.767	0.82	...
f³ds²	⁵H₄	5,991.3	...	0.900	0.86	...
f³d²s	⁷M₆	6,249.0	6249.04	0.571	0.62	0.62
f³d²s	⁷K₆†	7,005.5	...	0.976	0.97	...
?	B₃†	7,103.9	...	...	...	...
(f³ds²?)	B₇(=⁵K₇?)	7,326.1	...	...	...	...
(f³ds²?)	A₈(=⁵L₈?)	7,645.6	...	...	...	...
(f³d²s?)	C₇(=⁷M₇?)	8,118.6	8118.64	...	...	0.81
(f³d²s?)	D₇(=⁷K₇?)	10,347.3	...	...	...	...
(f³ds²?)	B₈(=⁵K₈?)	10,685.7	...	...	...	...
?	A₅	11,545.4	...	...	...	...
?	A₉	13,127.9	...	...	...	...

*Question marks indicate uncertainty of identification.
†McNally (Carbide and Carbon 3) suspects that these levels may be all from the electron configuration f³ds². In order of listing, they could then be ⁵I₅, ⁵I₆, and ⁵G₃, respectively.

In the case of spectra as complex as those of uranium, Zeeman patterns provide the first clues for the analysis by permitting the derivation of the g values of the participating terms. These can then be compared with theoretical values for terms of certain types (i.e., with certain combinations of L, S, and J values). Because of the high atomic weight of uranium, the theoretical g values calculated on the basis of ideal LS coupling (Russell-Saunders coupling) can, however, deviate considerably from the empirical g values, and the Jj coupling conditions can be approached more or less closely (this behavior is observed also in the spectra of the rare earth elements).

As mentioned in Sec. 2.2a, observations of the isotope shift also provide a useful clue in the analysis of the uranium spectrum, although its indications are much less specific than those derived from the Zeeman effect.

(a) Analysis of the UI Spectrum (Spectrum of Neutral Uranium Atom). A report on the analysis of the UI spectrum at the National Bureau of Standards was given by Kiess, Humphreys, and Laun (1946). This paper amplified the earlier unpublished reports (Natl. Bur. Standards 1,2). It gave a list of 18 low levels, which correspond to the odd-valence electron configurations $5f^3 6d7s^2$ (theoretically, this configuration alone could give a total of 158 terms with 386 components) and $5f^3 6d^2 7s$ (giving, theoretically, 1,122 terms with 3,256 components). These terms are shown in Table 2.16. Some of them have been given identifications, which are in part uncertain, on the basis of g values derived from Zeeman patterns. Table 2.16 also gives the term values found independently by Schuurmans (1947) in Amsterdam and the g values given by Kiess and coworkers (1946) and by van den Bosch and van den Berg (1949b). Changes in the term notation of certain of Kiess's identifications in Table 2.16, mentioned in a footnote, were suggested by McNally (Carbide and Carbon 3) because of the expected s-type splitting in the UII configuration $f^3 ds$ and because in Kiess's scheme another level from $f^3 d^2 s$ is supposed to be more stable than the lowest expected term component, $^7M_6$.

A second list (Table 2.17) contains the identified upper terms in the UI spectrum. They can be presumed to correspond to the lifting of a 6d or 7s electron into the 6p state and to belong to the even-valence electron configurations $f^3 dps$, $f^3 ps^2$, and $f^3 d^2 p$. Kiess, Humphreys, and Laun (1946) listed 275 such terms; 18 additional terms were found later (cf. Carbide and Carbon 3). Table 2.17 also contains the 30 term values determined by Schuurmans (1947); g values are given for 42 of these terms by Schuurmans (1947) and by van den Bosch and van den Berg (1949b).

The lowest identified UI term, ^5L, could belong to either of the two configurations $f^3 ds^2$ or $f^3 d^2 s$. Certain arguments were given by Kiess and coworkers in favor of the first assignment. In any case the occurrence of this term indicates the presence of three f electrons in the lowest known state of the free uranium atom. The lowest term of the configuration $d^4 s^2$ (which is present in tungsten and was first postulated also for uranium) would have been ^5D.

Van den Bosch and van den Berg (1949b) expressed doubts whether the lowest identified term of UI is the true ground term of the UI

Table 2.17—High Even Terms of UI

Term notation and J value*	Level,† cm⁻¹	g values‡	Term notation and J value*	Level,† cm⁻¹	g values‡
116_5	11,614.0		206_5	20,621.2	
122_5	12,227.8		206_6	20,661.5	
134_5	13,463.4(.42)	(0.66)	207_7	20,766.5	
146_6	14,643.9(.87)	(0.63)	208_5	20,851.6	
156_7	15,631.9(.84)	(0.90)	209_6	20,943.4	
156_6	15,638.4(.36)	(0.87)	210_4	21,062.3	
157_5	15,720.7		210_5	21,078.7	
161_4	16,121.9		212_6	21,265.1(.04)	
165_6	16,505.8(.80)	(0.90)	213_5	21,330.0	
169_7	16,900.4(.42)	(0.93)	215_4	21,545.1	
169_5	16,929.8		215_6	21,584.7(.71)	0.90(0.88)
170_6	17,070.5(.47)	(0.88)	216_5	21,637.0	
173_6	17,361.9(.84)	(0.86)	217_6	21,766.5	
174_4	17,468.2		218_6	21,768.0	
$178_{5,6}$	17,848.12§		219_3	21,940.6	
178_4	17,893.8		220_4	22,038.0	
179_5	17,908.2		220_6	22,056.3	
179_3	17,968.7		223_4	22,365.0	
181_4	18,186.0		223_7	22,368.4(.44)	1.04
182_6	18,253.9		223_5	22,377.7	
184_5	18,406.5(.51)	(0.80)	223_3	22,383.1	
185_4	18,526.9		224_4	22,383.4	
185_3	18,530.8		224_6	22,464.3	0.98
187_6	18,759.2(.20)	0.91(0.90)	225_6	22,582.7(.67)	0.97
189_5	18,932.8(.72)	(0.64)	225_4	22,584.5	
191_4	19,127.2		226_7	22,633.2	0.83
192_4	19,192.4		227_6	22,754.1(.08)	0.95(0.91)
194_5	19,471.9		227_4	22,774.1	
194_8	19,489.0		227_8	22,789.8	
195_4	19,552.5		228_6	22,862.4(.45)	
196_5	19,647.5	0.98	229_7	22,918.6(.51)	0.96(0.93)
196_3	19,668.4		230_7	23,057.7	0.97
197_4	19,740.7		231_8	23,110.8	
197_6	19,783.4	0.96	231_4	23,186.9	
197_7	19,826.7		231_7	23,197.0	
198_7	19,885.4(.46)	0.88(0.92)	233_5	23,325.2	
201_5	20,114.3		234_4	23,430.1	
201_4	20,148.0		234_5	23,432.8(.79)	
202_6	20,218.8(.83)	(0.80)	234_3	23,464.1	
203_5	20,311.5		235_5	23,486.7	
203_3	20,391.5		235_7	23,543.5(.50)	1.02(1.00)
204_6	20,420.5		235_4	23,560.6	
204_7	20,464.5		235_8	23,572.0	0.92
205_8	20,528.9		237_6	23,715.3	
205_4	20,569.2		237_7	23,779.2	

Table 2.17 — (Continued)

Term notation and J value*	Level,† cm⁻¹	g values‡	Term notation and J value*	Level,† cm⁻¹	g values‡
238_4	23,825.4		263_8	26,313.3	
238_9	23,843.7		263_7	26,391.3	
238_7	23,848.6		264_8	26,454.1	
239_8	23,926.7		265_8	26,492.1	
239_5	23,932.8		265_4	26,516.8	
240_6	24,026.2		265_6	26,550.4	
240_7	24,066.6(.54)	1.02(0.98)	265_5	26,566.9	
241_7	24,185.8		266_8	26,597.0	
243_7	24,333.8		266_7	26,608.5	
244_6	24,433.2	1.04	266_6	26,631.4	
244_5	24,448.0		267_6	26,715.5	
244_8	24,451.7		268_6	26,758.8	
245_9	24,517.3		269_6	26,791.6	
245_5	24,535.3		268_3	26,855.4	
245_7	24,560.4		269_5	26,920.7	
245_8	24,581.2		269_8	26,979.3	
246_6	24,671.4	0.95	270_8	26,989.1	
247_4	24,757.2		270_6	27,072.4	
249_5	24,906.8		271_8	27,086.4	
249_4	24,940.5		272_8	27,150.5	
250_5	25,017.1		271_4	27,184.1	
251_3	25,160.7		273_7	27,324.5	
251_5	25,178.1		273_5	27,381.7	
252_4	25,224.1		274_8	27,477.6	1.01
252_6	25,235.7	0.98	274_6	27,499.4	
252_5	25,255.4		275_6	27,605.7	
253_5	25,319.2		276_6	27,615.8	
253_6	25,349.0(8.77)	0.96(0.93)	276_5	27,682.2	
253_8	25,388.9		277_6	27,744.0	
254_6	25,462.6		277_5	27,778.0(.01)	
256_9	25,626.6		278_5	27,791.1	
256_4	25,653.3		278_8	27,818.5	1.05
256_7	25,672.5	0.96	278_7	27,887.0(6.98)	0.86(0.80)
257_8	25,789.0		278_6	27,941.2	
257_6	25,791.6		279_7	27,965.9	
258_5	25,805.8		279_9	27,969.7	
258_6	25,825.6	0.99	279_6	28,022.9	
259_8	25,918.1		280_7	28,048.2	
259_6	25,938.2		280_6	28,053.1	1.03
261_6	26,103.7		281_6	28,098.9	
261_8	26,192.4		281_7	28,118.8(.86)	
261_7	26,208.8		282_7	28,152.7	
262_6	26,225.5	1.03	281_5	28,188.3	
262_7	26,274.8		283_7	28,285.8	
263_5	26,305.0		283_4	28,430.5	

Table 2.17— (Continued)

Term notation and J value*	Level,† cm⁻¹	g values‡	Term notation and J value*	Level,† cm⁻¹	g values‡
284$_5$	28,444.5		303$_4$	30,303.7	
284$_7$	28,451.1		303$_5$	30,306.10§	
284$_4$	28,454.0		301$_6$	30,335.0	
284$_6$	28,470.2		b 303$_6$	30,353.30§	
285$_5$	28,503.5		b 304$_6$	30,435.92§	
285$_4$	28,543.4		302$_6$	30,451.4	
286$_5$	28,562.7		303$_6$	30,490.3	
285$_7$	28,566.4		304$_4$	30,499.1	
287$_5$	28,650.3(.29)	(0.82)	304$_6$	30,500.1	
287$_7$	28,798.9		b 305$_6$	30,538.68§	
288$_6$	28,860.9		305$_6$	30,586.7	
289$_6$	28,874.9		306$_5$	30,619.88§	
288$_7$	28,895.6		306$_8$	30,636.7	
289$_8$	28,993.7		306$_7$	30,642.8	
290$_6$	29,033.6		307$_5$	30,681.69§	
290$_8$	29,037.2		b 306$_6$	30,687.70§	
291$_7$	29,107.1		308$_5$	30,707.23§	
291$_6$	29,109.8		307$_7$	30,737.1	
292$_6$	29,126.1		307$_6$	30,875.6	
292$_5$	29,232.6		308$_6$	30,894.5	
292$_4$	29,250.5		309$_5$	30,936.6	
293$_9$	29,339.3		309$_4$	30,979.7	
294$_6$	29,400.9		309$_8$	30,986.3	
294$_8$	29,413.2		310$_5$	30,993.0	
294$_3$	29,413.7		310$_7$	31,024.8	
295$_3$	29,430.3		311$_5$	31,129.5	
295$_9$	29,550.3		311$_6$	31,135.0	
295$_6$	29,558.8	0.92	311$_7$	31,166.2(.26)	1.09
296$_8$	29,612.7		312$_4$	31,221.3	
296$_3$	29,644.6		313$_4$	31,243.5	
297$_4$	29,719.9		312$_9$	31,270.3	
297$_7$	29,790.7		312$_7$	31,276.0	
298$_7$	29,797.2		312$_8$	31,279.1	
299$_7$	29,837.6(.60)	1.04(1.03)	313$_7$	31,301.1	
298$_6$	29,865.5		313$_5$	31,339.8	
300$_7$	29,958.1		314$_7$	31,358.6	
b 299$_8$	29,978.06§		314$_6$	31,408.48§	
299$_6$	29,986.4		314$_8$	31,445.3	
299$_5$	29,999.58§		314$_5$	31,467.6	
300$_8$	30,027.2		315$_5$	31,488.2	
301$_5$	30,143.1		315$_6$	31,551.5	
302$_3$	30,222.4		315$_7$	31,568.40§	
b 302$_6$	30,245.62§		316$_5$	31,603.57§	
302$_7$	30,279.1		316$_4$	31,633.9	
300$_6$	30,294.3		317$_5$	31,744.2	

Table 2.17 — (Continued)

Term notation and J value*	Level,† cm⁻¹	g values‡	Term notation and J value*	Level,† cm⁻¹	g values‡
319_8	31,923.1		334_6	33,412.2	
319_5	31,946.0		335_7	33,570.6	
318_7	31,968.1		336_8	33,639.6	0.83
319_7	31,974.4		337_7	33,770.23§	
320_7	32,016.7		338_8	33,850.78§	
321_7	32,098.2		340_7	34,059.9	
321_9	32,108.4		340_3	34,065.4	
321_4	31,141.2		341_8	34,105.4	
323_5	32,317.8		342_6	34,285.19§	
324_8	32,477.8				
324_6	32,490.6				
325_6	32,495.7				
326_7	32,641.8				
327_7	32,774.2				
329_6	32,955.08§				

*Terms are in general designated by first three figures of their wave number, with J value as subscript.

†The first (6-digit) number is the level taken from Kiess, Humphreys, and Laun (1946); the number immediately following it in parentheses consists of the last two digits of the wave number from Schuurmans et al. (1947).

‡The g value not in parentheses is from van den Bosch and van den Berg (1949b); the g value in parentheses is from Schuurmans et al. (1947).

§Private communication to J. R. McNally, Jr., by C. C. Kiess (July, 1949). A "b" in the first column indicates a term (from Kiess) given a notation that duplicates an already existing notation.

atom. They pointed out that as yet no terms are known belonging to the even configuration f^4s^2 (which would combine with odd high terms and thus give rise to a second system of lines in the UI spectrum). Analogy with neodymium, whose lowest term has been found to belong to the configuration $4f^4 6s^2$ (Kiess and coworkers in 1946 quoted $4f^3 5d6s^2$ as the most stable configuration of neodymium), indicates the possibility that the neutral uranium atom may contain four 5f electrons in its ground state instead of three. Kiess and coworkers made the statement that practically all prominent UI lines have been fitted into the known system, but van den Bosch and van den Berg pointed out that about thirty of the strongest lines remain unclassified (cf. Table 2.13). They also said that the arc spectrum of uranium, as known now, shows a large number of spark lines and that further improvement of the excitation method may develop the arc spectrum in purer form and reveal additional lines.

The absolute value of the lowest known term of UI (i.e., the ionization potential of the uranium atom in the state f^3ds^2 $^5L_6$) is not known;

Table 2.18— Low Terms of U$^+$ Ion (UII)*

System	Configuration	Term symbol	Term value, cm^{-1}		g value		
					Observed		Theoretical for LS coupling
			MIT 2	Van den Bosch (1949a)	MIT 2	Van den Bosch and van den Berg (1949b)	
A	f^{3}s^2	$^4I_{4\frac{1}{2}}$	0.00	0.00	0.755	0.75	0.727
(odd)		$^4I_{5\frac{1}{2}}$	4,420.88	4,420.85	0.973	0.93	0.965
	f^{3}ds	$^6L_{5\frac{1}{2}}$	289.05	289.04	0.651	0.66	0.615
		$^6L_{6\frac{1}{2}}$	1,749.14	1,749.11	0.863	0.84	0.851
		$^6L_{7\frac{1}{2}}$	5,259.67	5,259.64	1.01	0.99	1.004
		$^4L_{6\frac{1}{2}}$	6,283.47	6,283.44	0.785	0.79	0.800
		$^6K_{4\frac{1}{2}}$	914.77	914.76	0.605	0.60	0.546
		$^6K_{5\frac{1}{2}}$	2,294.72	2,294.69	0.870	0.86	0.839
		$^6K_{6\frac{1}{2}}$	5,526.76	5,526.73	1.03	1.02	1.015
		$^4K_{5\frac{1}{2}}$	5,790.66	5,790.62	0.849	0.84	0.769
		$^6I_{3\frac{1}{2}}$	5,401.55	5,401.52	0.694	0.67	0.444
		$^6I_{4\frac{1}{2}}$	6,445.03	6,445.04	0.84	0.85	0.828
		$^6I_{5\frac{1}{2}}$	8,510.91	...	...	...	1.035
		$^6H_{2\frac{1}{2}}$	4,706.31	4,706.28	0.473	0.45	0.286
		$^6H_{3\frac{1}{2}}$	5,667.34	5,667.33	0.737	0.72	0.825
	f^{3}d^2	$^6M_{6\frac{1}{2}}$	4,585.48	4,585.43	0.780	0.79	0.667
		$^6M_{7\frac{1}{2}}$	8,394.36	8,394.40	0.975	0.95	0.879
	(?)	1A$_{7\frac{1}{2}}$	8,521.99(?)	...	1.05(?)	...	...
	(?)	2A$_{7\frac{1}{2}}$	9,291.77(?)	...	...	...	...
	(?)	3A$_{6\frac{1}{2}}$	12,645.67(?)	...	...	...	...
B	f^{4}s	$^6I_{3\frac{1}{2}}$	0.00	0.00	0.495	0.48	0.444
(even)		$^6I_{4\frac{1}{2}}$	1,052.65	1,052.64	0.826	0.83	0.755
		$^6I_{5\frac{1}{2}}$	3,683.82	3,683.88	...	1.03	1.035
		$^4I_{4\frac{1}{2}}$	3,759.55	3,759.59	...	0.81	0.727
	f^{4}d	$^6L_{5\frac{1}{2}}$	7,850.08	7,850.15	...	0.67	0.615
		$^6L_{6\frac{1}{2}}$	10,728.59	10,728.67	...	0.88	0.851
		$^6K_{4\frac{1}{2}}$	11,547.90	11,547.93	...	0.66†	0.546
		1B$_{6\frac{1}{2}}$	6,853.37(?)	...	...	...	...
		2B$_{5\frac{1}{2}}$	7,695.85(?)	...	...	...	...

*After McNally and Harrison (Carbide and Carbon 2), McNally (Carbide and Carbon 3), van den Bosch and van den Berg (1949b), and van den Bosch (1949a).

†Schuurmans et al. (1947).

no term series of the Balmer type which would permit extrapolation of the convergency limit has as yet been observed. However, from the fact that no lines of neutral uranium atom have been found below 2900 A, Kiess and coworkers concluded that the ionization potential of uranium must be as low as 4 ev. They pointed out that this conclusion is supported by the ease with which uranium is ionized in an electric arc.

(b) Analysis of the UII Spectrum (Spectrum of the U^+ Ion). The analysis of the U^+ spectrum has been initiated independently by Harrison and McNally at the Massachusetts Institute of Technology (MIT 1; Carbide and Carbon 1,2) and by Schuurmans and others at the Zeeman Laboratory in Amsterdam (Schuurmans et al., 1946a,b,c; 1947; van den Bosch and van den Berg, 1949b). The results obtained by Harrison and McNally have been first presented in a classified report (MIT 2), giving a list of 14 low terms and about 150 upper terms, with J values for all and g values for some of them. The lower terms were noted to form two separate systems of 12 and 2 terms, respectively; of the upper terms, 140 belonged to the first group and 10 to the second group. Low terms of the first system were at first attributed to the (even) configuration f^2d^2s. The lowest observed term in this system was an $I_{\frac{9}{2}}$ term, followed closely (at 289 cm^{-1}) by an $L_{\frac{11}{2}}$ term. Harrison first suggested that the latter may be the lowest component of the true ground term of the U^+ ion (a ^6L term is theoretically the lowest term of the configuration f^2d^2s) and that the lowest observed level, $I_{\frac{9}{2}}$, could be an accidentally depressed component of a ^6I term of the same configuration, whose center of gravity lies above that of the ^6L term.

This first tentative interpretation was changed in more recent reports (Carbide and Carbon 2,3), which attributed the above-mentioned group of 12 low terms (group A) to the (odd) configurations f^3s^2, f^3ds, and f^3d^2 (as indicated in Table 2.18). The two remaining low terms, identified at the Massachusetts Institute of Technology (group B), must then be even terms. Their position relative to the terms of system A could not be given, since no intercombination lines have been observed. In Table 2.18 the list of the low terms of group A is augmented by eight terms identified later at MIT and at Amsterdam, and the list of low terms of group B is augmented by seven new terms, five of which have been found first by the Amsterdam group.

Schuurmans, van den Bosch, and Dijkwel (1947) published a table of 16 low odd and 57 high even terms (system A) and a table of 7 low even and 5 high odd terms (system B); about 350 lines were identified as combinations of these terms. (The designations A and B are reversed in the Dutch papers.)

Van den Bosch (1949a) added to the analysis of the UII spectrum by identifying three new high odd and 17 "very high" even terms of system B. The correctness of the latter is, however, doubted by McNally (1950). Van den Bosch (1949a) also added 63 new high even terms to system A. Despite the addition of the "very high" group, still no intercombinations between the terms of the A and B systems could be detected.

Van den Bosch agreed with McNally in attributing the low terms of system A to the odd configurations f^3s^2, f^3ds, and f^3d^2; the high terms of this group then can be attributed to the even configurations f^2d^2s, f^2ds^2, f^3dp, and f^3sp. In system B the low terms can be interpreted as due to the even f^4s and f^4d configurations and the high terms to the odd configuration f^4p; the very high terms (if real) could perhaps be assigned to the even grouping f^47d.

As long as no intercombinations between terms of systems A and B are found (such intercombinations are known for the isoelectronic atom thorium), the question of the true ground term and thus of the most stable configuration of the isolated U^+ ion must remain open. The alternative is between $f^3s^2\ ^4I_{\frac{9}{2}}$ (system A) and $f^4s\ ^6I_{\frac{7}{2}}$ (system B). McNally (1950) conjectured that the first-named term may lie 4,000 cm^{-1} below the second one; the Amsterdam group was inclined to consider the second term as the ground term of U^+ and the $f^3s^2\ ^4I_{\frac{9}{2}}$ term as being situated higher by as much as 20,000 cm^{-1}.

Table 2.18 summarizes the data of both McNally and the Dutch investigators on the low levels of the U^+ ion. The terms of system A are ascribed to the configurations f^3s^2, f^3ds, and f^3d^2, with three 5f electrons; the terms of system B are ascribed to the configurations f^4s and f^4d, with four electrons in the 5f group. Table 2.19, compiled by McNally (Carbide and Carbon 3), summarizes in a similar way the data pertaining to the middle and very high terms of the two known systems in the UII spectrum. No attempt has been made to identify the L, S, or J values or the electronic configuration of the high UII terms, although, according to McNally (Carbide and Carbon 3) good indications as to the origin of several of them can be found in intensities, Zeeman effects, and isotope displacements. Examples are level 103A, which is probably $f^2ds^2\ ^4I_{4\frac{1}{2}}$, and level 161A, which is probably $f^3dp\ ^6M_{6\frac{1}{2}}$.

2.5 Electron Configuration. The chemical properties of the elements thorium to uranium seem more similar to the series Zr, Cb, Mo than to the series La, Ce, Pr. In other words, they appear at first to indicate a gradual completion of a d electron group rather than the filling of an f group, which is characteristic of the rare earth family. The first theoretical stability calculations by Sugiura and Urey (1926) seemed to confirm this conclusion by indicating stronger

Table 2.19—High Terms in UII Spectrum

Term designation*	J value	Level above $^4I_{4\frac{1}{2}}$,† cm^{-1}	g value		
			MIT 2	Van den Bosch and van den Berg (1949b)	Schuurmans (1947)
A. High Even Terms (System A)‡					
100A	$3\frac{1}{2},4\frac{1}{2}$	15,679.52(?)			
101A	$4\frac{1}{2}$	17,392.22	0.783 or 0.405		
102A	$5\frac{1}{2}$	17,434.37	0.810		
103A	$4\frac{1}{2}$	18,200.08	0.774	0.78	0.78
104A	$5\frac{1}{2}$	18,827.01	0.94		
105A	$4\frac{1}{2}$	19,097.56(?)			
106A	$3\frac{1}{2}$	19,517.73			
107A	$5\frac{1}{2}$	20,354.00	1.00		
108A	$3\frac{1}{2}$	20,571.73		0.97	0.92
109A	$3\frac{1}{2}$	20,961.72	0.855	0.92	0.86
110A	$3\frac{1}{2}$	21,021.40	0.885	0.86	0.88
111A	$3\frac{1}{2}$	21,320.22	0.827	0.85	0.80
112A	$4\frac{1}{2}$	21,555.28	1.023		
113A	$3\frac{1}{2},4\frac{1}{2},5\frac{1}{2}$	21,679.17(?)			
114A	$6\frac{1}{2}$	21,710.76	0.907	0.89	0.91
115A	$4\frac{1}{2}$	21,831.05	0.890		
116A	$3\frac{1}{2}$	21,860.06			
117A	$6\frac{1}{2}$	21,975.57	1.029 or 0.697		
118A	$4\frac{1}{2}$	22,101.37		0.86	0.87
119A	$4\frac{1}{2}$	22,165.18	0.885 or 0.621	0.92	0.90
120A	$4\frac{1}{2}$	22,429.84	0.929	0.94	0.93
121A	$4\frac{1}{2}$	22,642.49	0.872	0.86	0.87
122A	$5\frac{1}{2}$	22,764.93	0.980 or 0.760		
123A	$4\frac{1}{2}$	22,868.06	0.98	0.98	0.98
124A	$5\frac{1}{2}$	22,917.46	0.849	0.85	0.86
125A	$5\frac{1}{2}$	23,241.36	0.957	0.95	0.96
126A	$4\frac{1}{2}$	23,315.09	0.874	0.89	0.88
127A	$5\frac{1}{2}$	23,554.00	1.033	1.03	1.04
128A	$6\frac{1}{2}$	23,635.94	0.912	0.94	0.91
129A	$3\frac{1}{2},4\frac{1}{2}$	23,741.32			
130A	$5\frac{1}{2}$	23,778.14	0.865	0.89	0.86
131A	$4\frac{1}{2}$	23,911.66	1.056		
132A	$5\frac{1}{2}$	24,010.48	0.969	0.98	0.97
133A	$5\frac{1}{2}$	24,019.29			
134A	$5\frac{1}{2}$	24,152.82	0.906	0.92	0.91
135A	$5\frac{1}{2},6\frac{1}{2}$	24,159.73	0.959		
136A	$5\frac{1}{2}$	24,288.00	1.012	1.01	·1.01
137A	$6\frac{1}{2}$	24,293.13	1.019		
138A	$4\frac{1}{2}$	24,305.62	0.974		0.97
139A	$3\frac{1}{2}$	24,342.21	0.763 or 0.625		0.75

Table 2.19 — (Continued)

Term designation*	J value	Level above $^4I_{4\frac{1}{2}}$,† cm^{-1}	g value		
			MIT 2	Van den Bosch and van den Berg (1949b)	Schuurmans (1947)
140A	$4\frac{1}{2}$	24,453.45	1.100		1.10
141A	$5\frac{1}{2}$	24,608.16	0.904		0.89
142A	$4\frac{1}{2}$	24,684.16	0.918		0.93
143A	$6\frac{1}{2}$	24,923.59(?)	1.088		
144A	$5\frac{1}{2}$	25,156.89	0.962		
145A	$6\frac{1}{2}$	25,163.96	1.030		1.05
146A	$5\frac{1}{2}$	25,200.82			
147A	$5\frac{1}{2}$	25,357.02			
148A	$4\frac{1}{2}$	25,418.56			
149A	$4\frac{1}{2}$	25,437.59	0.94		
150A	$5\frac{1}{2}$	25,492.92	0.993		
151A	$3\frac{1}{2}$	25,495.47	0.958		
152A	$5\frac{1}{2}$	25,543.99			
153A	$6\frac{1}{2}$	25,714.06	1.010		1.01
154A	$2\frac{1}{2}$	25,737.60	0.923		
155A	$3\frac{1}{2}$	25,764.70(?)	1.05		
156A	$5\frac{1}{2}$	25,770.59			
157A	$4\frac{1}{2},5\frac{1}{2}$	25,906.06			
158A	$3\frac{1}{2}$	25,967.70	0.855		0.84
159A	$6\frac{1}{2}$	25,986.35	0.979		0.99
160A	$5\frac{1}{2}$	26,084.82			
161A	$6\frac{1}{2}$	26,191.34	0.900		0.90
162A	$5\frac{1}{2}$	26,285.18			
163A	$5\frac{1}{2}$	26,330.17			
164A	$6\frac{1}{2}$	26,397.89(?)	1.035		
165A	$6\frac{1}{2}$	26,415.18		1.03	1.01
166A	$5\frac{1}{2}$	26,595.51			
167A	$6\frac{1}{2}$	26,716.74	0.990	1.00	0.99
168A	$5\frac{1}{2}$	26,879.25			
169A	$7\frac{1}{2}$	26,887.29		1.05	1.08
170A	$4\frac{1}{2}$	27,126.11	1.012	1.06	1.01
171A	$5\frac{1}{2},6\frac{1}{2}$	27,143.73			
172A	$3\frac{1}{2}$	27,290.25	0.93 or 0.54	0.90	0.97
173A	$5\frac{1}{2}$	27,357.30			
174A	$4\frac{1}{2}$	27,361.05†			
175A	$5\frac{1}{2}$	27,390.51			
176A	$4\frac{1}{2}$	27,446.35			
177A	$5\frac{1}{2}$	27,453.53			
178A	$5\frac{1}{2}$	27,499.38	1.036 or 0.662	1.05	1.06
179A	$3\frac{1}{2}$	27,583.35(?)	0.730		
180A	$4\frac{1}{2}$	27,698.03	1.012		
181A	$4\frac{1}{2},5\frac{1}{2}$	27,708.38(?)			
182A	$5\frac{1}{2}$	27,725.02			
183A	$5\frac{1}{2}$	27,912.14			
184A	$5\frac{1}{2}$	27,930.28	0.997	1.01	1.01

Table 2.19 — (Continued)

Term designation*	J value	Level above $^4I_{4\frac{1}{2}}$,† cm^{-1}	g value		
			MIT 2	Van den Bosch and van den Berg (1949b)	Schuurmans (1947)
185A	$5\frac{1}{2}$	28,154.46	0.885	0.89	0.89
186A	$4\frac{1}{2}$	28,159.72			
187A	$3\frac{1}{2}$	28,347.28	1.02	0.98	1.02
188A	$5\frac{1}{2}$	28,444.51			
189A	$5\frac{1}{2}$	28,484.51			
190A	$5\frac{1}{2}$	28,507.96	1.020	1.03	1.04
191A	$5\frac{1}{2}$	28,587.30			1.07
192A	$6\frac{1}{2}$	28,636.89	1.064		
193A	$5\frac{1}{2}$	28,758.11			
194A	$4\frac{1}{2}$	28,763.02†			
195A	$5\frac{1}{2}$	29,162.14(?)	0.977		
196A	$4\frac{1}{2},5\frac{1}{2}$	29,205.39(?)			
197A	$5\frac{1}{2}$	29,206.71		1.03	1.05
198A	$5\frac{1}{2}$	29,303.92			
199A	$4\frac{1}{2},5\frac{1}{2}$	29,349.22			
200A	$6\frac{1}{2}$	29,352.89(?)			
201A	$4\frac{1}{2}$	29,388.68			
202A	$3\frac{1}{2},4\frac{1}{2}$	29,396.82			
203A	$5\frac{1}{2}$	29,402.00			
204A	$5\frac{1}{2}$	29,403.74			
205A	$6\frac{1}{2}$	29,476.76		1.09	1.10
206A	$5\frac{1}{2}$	29,557.96			
207A	$4\frac{1}{2}$	29,707.21†			
208A	$3\frac{1}{2}$	29,824.93†			
209A	$4\frac{1}{2}$	29,827.43		1.04	1.08
210A	$5\frac{1}{2}$	29,858.49			
211A	$7\frac{1}{2}$	29,932.41	1.06	1.04	1.06
212A	$3\frac{1}{2}$	29,933.94†			
213A	$5\frac{1}{2}$	29,936.47			
214A	$6\frac{1}{2}$	29,978.19			
215A	$6\frac{1}{2}$	30,000.16			
216A	$5\frac{1}{2},6\frac{1}{2}$	30,015.84			
217A	$5\frac{1}{2}$	30,037.11			
218A	$4\frac{1}{2}$	30,042.93			
219A	$5\frac{1}{2}$	30,060.72		1.07	1.15
220A	$5\frac{1}{2}$	30,061.77(?)	1.017		
221A	$5\frac{1}{2}$	30,066.59			
222A	$5\frac{1}{2},6\frac{1}{2}$	30,085.80			
223A	$4\frac{1}{2}$	30,086.78(?)			
224A	$4\frac{1}{2}$	30,120.14			
225A	$4\frac{1}{2}$	30,126.00			
226A	$6\frac{1}{2}$	30,129.88			
227A	$6\frac{1}{2}$	30,139.07			
228A	$5\frac{1}{2}$	30,240.45		1.06	1.06
229A	$4\frac{1}{2}$	30,263.98			

Table 2.19 — (Continued)

Term designation*	J value	Level above $^4I_{4\frac{1}{2}}$,† cm^{-1}	g value		
			MIT 2	Van den Bosch and van den Berg (1949b)	Schuurmans (1947)
230A	$7\frac{1}{2}$	30,341.70	1.019	1.01	1.02
231A	$5\frac{1}{2}$	30,374.11			
232A	$5\frac{1}{2}$	30,438.55			
233A	$5\frac{1}{2}$	30,468.75			
234A	$6\frac{1}{2}$	30,550.39			
235A	$5\frac{1}{2}$	30,561.51†			
236A	$6\frac{1}{2}$	30,716.23			
237A	$5\frac{1}{2}$	30,860.13	0.977	0.98	0.99
238A	$6\frac{1}{2}$	30,863.44†			
239A	$4\frac{1}{2}$	30,900.65(?)			
240A	$7\frac{1}{2}$	30,941.64	1.03	1.04	1.03
241A	$5\frac{1}{2}$	31,083.68			
242A	$5\frac{1}{2}$	31,219.22	1.040	1.00	1.04
243A	$4\frac{1}{2}$	31,327.76(?)	0.905		
244A	$6\frac{1}{2},7\frac{1}{2}$	31,346.11(?)			
245A	$4\frac{1}{2}$	31,670.33(?)			
246A	$3\frac{1}{2}$	31,784.81(?)	0.934		
247A	$6\frac{1}{2}$	31,936.69			
248A	$3\frac{1}{2}$	32,028.36(?)	1.023		
249A	$5\frac{1}{2}$	32,187.68			
250A	$6\frac{1}{2}$	32,211.36		1.03	1.02
251A	$5\frac{1}{2}$	32,416.69			
252A	$6\frac{1}{2}$	32,535.07	0.997	0.98	1.00
253A	$4\frac{1}{2}$	32,717.24†			
254A	$6\frac{1}{2}$	32,838.74			
255A	$5\frac{1}{2}$	33,018.86			
256A	$6\frac{1}{2}$	33,104.46			
257A	$3\frac{1}{2}$	33,184.00†			
258A	$7\frac{1}{2}$	33,224.14(?)	1.068(?)		
259A	$4\frac{1}{2}$	33,359.62†			
260A	$6\frac{1}{2}$	33,431.61			
261A	$6\frac{1}{2}$	33,475.06			
262A	$4\frac{1}{2}$	33,516.22†			
263A	$5\frac{1}{2}$	33,577.83			
264A	$3\frac{1}{2}$	33,648.16			
265A	$5\frac{1}{2}$	33,748.79			
266A	$3\frac{1}{2}$	33,795.02	0.977		
267A	$6\frac{1}{2}$	34,049.54			1.08
268A	$5\frac{1}{2}$	34,199.47			
269A	$6\frac{1}{2}$	34,268.16†			
270A	$5\frac{1}{2}$	34,392.34			
271A	$5\frac{1}{2},6\frac{1}{2}$	34,402.25			
272A	$5\frac{1}{2}$	34,439.33		1.05	1.03
273A	$5\frac{1}{2}$	34,593.06			
274A	$3\frac{1}{2}$	34,607.01			

Table 2.19 — (Continued)

Term designation*	J value	Level above $^4I_{4\frac{1}{2}}$,† cm^{-1}	g value		
			MIT 2	Van den Bosch and van den Berg (1949b)	Schuurmans (1947)
275A	$4\frac{1}{2}$	34,612.86	0.958 or 0.562		
276A	$6\frac{1}{2}$	34,663.67			
277A	$5\frac{1}{2}$	34,678.05			
278A	$6\frac{1}{2}$	34,708.25		1.12	1.04
279A	$7\frac{1}{2}$	34,791.40			
280A	$5\frac{1}{2}$	34,885.51	1.004	0.84	
281A	$4\frac{1}{2},5\frac{1}{2}$	34,896.18			
282A	$5\frac{1}{2},6\frac{1}{2}$	34,923.62			
283A	$4\frac{1}{2}$	34,949.12			
284A	$4\frac{1}{2},5\frac{1}{2}$	34,968.14			
285A	$4\frac{1}{2},5\frac{1}{2}$	35,110.22			
286A	$4\frac{1}{2},5\frac{1}{2}$	35,214.41			
287A	$4\frac{1}{2},5\frac{1}{2}$	35,234.08			
288A	$4\frac{1}{2},5\frac{1}{2}$	35,324.38			
289A	$3\frac{1}{2}$	35,326.16†			
290A	$4\frac{1}{2},5\frac{1}{2}$	35,372.90			
291A	$6\frac{1}{2}$	35,446.89†			
292A	$5\frac{1}{2}$	35,684.83			
293A	$5\frac{1}{2},6\frac{1}{2}$	35,790.18			
294A	$5\frac{1}{2},6\frac{1}{2}$	35,899.45			
295A	$5\frac{1}{2}$	35,903.18			
296A	$4\frac{1}{2}$	35,960.19			
297A	$4\frac{1}{2},5\frac{1}{2}$	36,029.20			
298A	$4\frac{1}{2}$	36,070.31			
299A	$4\frac{1}{2},5\frac{1}{2}$	36,089.94			
300A	$4\frac{1}{2},5\frac{1}{2}$	36,105.39			
301A	$4\frac{1}{2},5\frac{1}{2}$	36,132.82			
302A	$4\frac{1}{2}$	36,150.57			
303A	$4\frac{1}{2}$	36,253.41			
304A	$4\frac{1}{2},5\frac{1}{2}$	36,273.86			
305A	$4\frac{1}{2}$	36,351.21			
306A	$3\frac{1}{2}$	36,548.65†			
307A	$4\frac{1}{2},5\frac{1}{2}$	36,679.45			
308A	$6\frac{1}{2}$	36,707.40			
309A	$4\frac{1}{2}$	36,782.70			
310A	$4\frac{1}{2},5\frac{1}{2}$	37,149.71			
311A	$6\frac{1}{2}$	37,308.35†			
312A	$4\frac{1}{2},5\frac{1}{2}$	37,341.79			
313A	$4\frac{1}{2},5\frac{1}{2}$	37,377.10(?)			
314A	$4\frac{1}{2},5\frac{1}{2}$	37,659.53			
315A	$5\frac{1}{2}$	37,684.33			
316A	$4\frac{1}{2}$	37,789.18			
317A	$4\frac{1}{2},5\frac{1}{2}$	37,869.85			
318A	$4\frac{1}{2}$	37,931.77			
319A	$4\frac{1}{2},5\frac{1}{2}$	38,128.90			

Table 2.19 — (Continued)

Term designation*	J value	Level above $^4I_{4\frac{1}{2}}$,† cm^{-1}	g value		
			MIT 2	Van den Bosch and van den Berg (1949b)	Schuurmans (1947)
320A	$7\frac{1}{2}$	38,152.90			
321A	$4\frac{1}{2}$	38,579.87			
322A	$3\frac{1}{2}$	38,681.86			
323A	$5\frac{1}{2}$	38,788.39			
324A	$5\frac{1}{2},6\frac{1}{2}$	38,898.43			
325A	$3\frac{1}{2}$	38,903.29			
326A	$5\frac{1}{2}$	38,968.45			
327A	$5\frac{1}{2}$	39,108.95			
328A	$6\frac{1}{2},7\frac{1}{2}$	39,508.25†			
329A	$4\frac{1}{2}$	40,356.31			
330A	$4\frac{1}{2}$	40,391.36			
331A	$3\frac{1}{2}$	40,731.32			
332A	$5\frac{1}{2}$	41,317.04	1.059		
333A	$3\frac{1}{2},4\frac{1}{2}$	41,542.60			
334A	$6\frac{1}{2}$	44,174.01			
335A	$6\frac{1}{2}$	45,533.53			

B. High Odd Terms (System B)§

Term designation*	J value	Level above $^4I_{4\frac{1}{2}}$,† cm^{-1}	g value		
			MIT 2	Van den Bosch and van den Berg (1949b)	Schuurmans (1947)
100B	$4\frac{1}{2}$	25,805.44(?)			
101B	$4\frac{1}{2}$	25,869.43			
102B	$4\frac{1}{2}$	25,935.34	0.905	0.91	0.92
103B	$4\frac{1}{2}$	26,092.29	0.973	0.99	0.99
104B	$4\frac{1}{2}$	26,332.33			
105B	$3\frac{1}{2},4\frac{1}{2}$	27,065.42(?)			
106B	$3\frac{1}{2}$	27,297.27	0.576		
107B	$3\frac{1}{2}$	27,721.19	0.955	0.93	0.96
108B	$4\frac{1}{2},5\frac{1}{2}$	28,693.86			
109B	$5\frac{1}{2}$	28,777.79†			
110B	$5\frac{1}{2}$	28,963.30	1.01	1.01	1.00
111B	$4\frac{1}{2},5\frac{1}{2}$	29,058.92†			
112B	$5\frac{1}{2}$	29,289.40	1.08	1.08	1.07
113B	$4\frac{1}{2}$	29,970.82†			
114B	$4\frac{1}{2},5\frac{1}{2}$	32,586.02			
115B	$3\frac{1}{2}$	32,840.28			
116B	$5\frac{1}{2}$	33,056.48			
117B	$5\frac{1}{2}$	33,070.85			
118B	$3\frac{1}{2},4\frac{1}{2}$	33,593.76			
119B	$4\frac{1}{2}$	33,663.38			
120B	$4\frac{1}{2},5\frac{1}{2}$	33,960.47			
121B	$4\frac{1}{2},5\frac{1}{2}$	35,325.03			
122B	$3\frac{1}{2},4\frac{1}{2}$	36,353.41			
123B	$5\frac{1}{2}$	37,503.66			
124B	$5\frac{1}{2}$	37,785.87			
125B	$5\frac{1}{2}$	43,911.61			
126B	$5\frac{1}{2}$	47,210.23			

Table 2.19 — (Continued)

Term designation*	J value	Level above $^4I_{4\frac{1}{2}}$,† cm^{-1}	g value		
			MIT 2	Van den Bosch and van den Berg (1949b)	Schuurmans (1947)
C. Very High Even Terms ¶(System B)§					
511	$4\frac{1}{2}$	51,148.61†			
525	$4\frac{1}{2}$	52,514.61†			
538	$4\frac{1}{2}$	53,873.53†			
540	$4\frac{1}{2},5\frac{1}{2}$	54,082.25†			
541	$4\frac{1}{2}$	54,082.56†			
546	$4\frac{1}{2}$	54,691.55†			
549	$4\frac{1}{2}$	54,923.77†			
552	$4\frac{1}{2}$	55,237.96†			
556	$4\frac{1}{2}$	55,692.51†			
559	$4\frac{1}{2}$	55,935.17†			
560	$4\frac{1}{2}$	55,972.25†			
563	$4\frac{1}{2}$	56,306.80†			
575	$3\frac{1}{2},4\frac{1}{2}$	57,589.80†			
578	$4\frac{1}{2}$	57,840.15†			
589	$4\frac{1}{2}$	58,901.52†			
590	$4\frac{1}{2}$	59,055.97†			
607	$4\frac{1}{2}$	60,769.09†			

*Because of uncertainty of the true ground term, excited terms in parts A and B are designated by successive numbers from 100A and 100B upward, rather than by the first three figures in the term value as in Table 2.17. In part C, the latter designation is taken over from the paper by van den Bosch (1949a).

†These term values are from Massachusetts Institute of Technology measurements (MIT 2), with the exception of the terms followed by the symbol †, which are from van den Bosch (1949a).

‡System B of Dutch investigators.

§System A of Dutch investigators.

¶The reality of all these terms is doubted by McNally (1950).

binding of 6d as compared to the 5f electrons. However, subsequent calculations of Wu and Goudsmit (1933) using the Wentzel-Kramer-Brillouin approximation method led to the conclusion that at Z = 92 (uranium) the 5f electrons may be almost as strongly bound as the 6d electrons and that at Z = 93 at least one 5f electron is likely to be found in the ground state. Similarly, Goeppert-Mayer (1941) estimated by the Thomas-Fermi statistical approximation method that the 5f groups should begin to fill up at Z = 91 or 92.

Ephraim and Mezener (1933a,b) pointed to the analogy between the absorption spectra of rare earth compounds, with their sharp lines, and the spectra of uranium compounds, some of which also exhibit very narrow absorption bands. This can be taken as indicative of states in which the role of optical electrons is played by well-shielded

f electrons. The effect of anions on the cation spectrum also is similar in the rare earth salts and in uranium salts. However, the spectra
in question are those of the ions U(III) and U(IV), and highly charged
ions always have a stronger tendency than the neutral atoms for electron arrangements without vacant inner shells. Therefore, Ephraim's
suggestion that two f electrons are present in the neutral uranium
atom and that a "thoride" series in which the 5f shell is filled begins
with thorium was only a guess. Positive information about this point
has now been provided by the above-described analysis of the spectrum. This analysis shows that the free ion U^+ and the free neutral
uranium atom have the following ground terms:

$$U^+: \quad 5f^3 7s^2 \ {}^4I \quad (\text{or, perhaps, } 5f^4 7s \ {}^6I)$$
$$U : \quad 5f^3 6d7s^2 \ {}^5L$$

(As mentioned above, van den Bosch and van den Berg consider it a
possibility that the ground term of the U atom may turn out to be an
as yet unknown term of the configuration $5f^4 7s^2$.)

We note that both the U^+ ion and the U atom contain in the most
stable state (at least) three electrons in the 5f group.

Recent chemical experience tends to support the view that a series
similar to the rare earth series does exist in the last period of the
periodic table, even though the progressive increase in preferred
valence, which is limited in the "lanthanide series" to the first two
elements [La(III) and Ce(IV)], is more extended in the "actinide
series" [Ac(III), Th(IV), Pa(V), U(VI)]. However, the filling of an f
group does not necessarily produce a series of elements as similar
chemically as the rare earths. The close similarity of the lanthanide
elements is due to the narrow range of changes in their ionization
potentials and ionic radii. A qualitative analogy in electron distribution between the actinides and lanthanides does not necessarily imply
a similar quantitative uniformity in the properties of the first-named
group. The analogy between the actinide series and the rare earth
series becomes more pronounced in the "transuranic" region, where
the preferred valency declines [U(VI), Np(V), Pu(IV), Am(III)]. For a
comparative discussion of various physical and chemical properties
in the two series of elements, we refer here to discussions by Seaborg
(1949) and Zachariasen (1950).

REFERENCES

1897 W. G. Humphreys, Astrophys. J., 6: 169.

1910 J. M. Eder and E. Valenta, Sitzber. Akad. Wiss. Wien. Math. naturw. Klasse, Abt. IIA, 119: 39.

1910 B. Hasselberg, Kgl. Svenska Vetenskapsakad. Handl., 45 (1) (5): 39.

1910 A. D. Ross, Proc. Roy. Soc. Edinburgh, 30: 448.

1911 F. Exner and E. Haschek, "Die Spectren der Elemente bei normalen Druck," vol. 2, p. 282 and vol. 3, p. 256, F. Deuticke, Leipzig und Wien.

1912 W. N. Hartley and H. W. Moss, Proc. Roy. Soc. London, A87: 46.

1916a M. de Broglie, Compt. rend., 162: 597; 163: 354.

1916b M. de Broglie, J. phys., 6: 161.

1916 K. W. Meissner, Ann. Physik, 50: 727.

1916a M. Siegbahn and E. Friman, Phil. Mag., 31: 405.

1916b M. Siegbahn and E. Friman, Phil. Mag., 32: 46.

1916c M. Siegbahn and E. Friman, Physik. Z., 17: 17, 61.

1918 W. Stenström, Ann. Physik, 57: 347.

1919 M. de Broglie, Compt. rend., 168: 854; 169: 964.

1919 M. Siegbahn and E. Jönsson, Physik. Z., 20: 255.

1919 W. Stenström, dissertation Lund. See Coster (1921a,b; 1922a,b).

1920a W. Duane and R. A. Patterson, Proc. Natl. Acad. Sci. U. S., 6: 512.

1920b W. Duane, H. Frick, and W. Stenström, Proc. Natl. Acad. Sci. U. S., 6: 611.

1920c W. Duane, Bull. Natl. Research Council, 1(6): 389, 396.

1921a D. Coster, Compt. rend., 172: 1176.

1921b D. Coster, Z. Physik, 4: 184.

1921a A. Dauvillier, Compt. rend., 173: 37.

1921b A. Dauvillier, Compt. rend., 172: 915.

1921c A. Dauvillier, Compt. rend., 173: 647.

1921 W. Duane, Proc. Natl. Acad. Sci. U. S., 7: 270.

1921 C. C. Kiess and W. F. Meggers, Natl. Bur. Standards U. S., Sci. Papers, 16 (372).

1921 G. Meyer, Physik. Z., 22: 583.

1922 M. de Broglie, Compt. rend., 173: 1157.

1922a D. Coster, Z. Physik, 5: 143.

1922b D. Coster, Phys. Rev., 19: 20.

1922 W. Duane and R. A. Patterson, Proc. Natl. Acad. Sci. U. S., 8: 88.

1923 N. Bohr and D. Coster, Z. Physik, 12: 342.

1923 A. Dauvillier and P. Auger, Compt. rend., 176: 1298.

1923 E. Hjalmar, Z. Physik, 15: 79, 65.

1924 V. Dolejšek, Z. Physik, 21: 111.

1924 J. M. Eder and E. Valenta, "Atlas typischer Spektren," 2d ed., pp. 61, 118, plates 15, 21, 24, 29, 38, 50, and 52, A. Hölder, Vienna.

1924 R. J. Lang, Trans. Roy. Soc. London, A224: 419.

1924 H. Nagaoka and Y. Sugiura, Japan. J. Phys., 3: 71.

1925 G. Rechou, Compt. rend., 180: 1107.

1925 H. R. Robinson, Phil. Mag., 50: 241.

1925 E. C. Stoner and C. H. Martin, Proc. Roy. Soc. London, A107: 312.

1926 S. J. M. Allen, Phys. Rev., 27: 266; 28: 907.

1926 E. Friman, Z. Physik, 39: 825.

1926 J. Schrör, Ann. Physik, 80: 302.

1926 Y. Sugiura and H. C. Urey, Kgl. Danske Videnskab. Selskab. Math. fys. Medd., 7: 3.

1927 A. Dauvillier, Rev. sci., 65: 707.
1927 J. E. Mack and J. M. Cook, Phys. Rev., 30: 741.
1928 S. K. Allison, Phys. Rev., 32: 1.
1928 E. Lindberg, Z. Physik, 50: 83.
1929 S. K. Allison, Phys. Rev., 34: 176.
1929 E. Lindberg, Z. Physik, 54: 632.
1930 G. Breit, Phys. Rev., 35: 1447.
1930 S. Idei, Science Repts. Tôhoku Imp. Univ., First Ser., 19: 572.
1930 A. Sandström, Z. Physik, 65: 632.
1931 S. K. Allison and V. J. Andrews, Phys. Rev., 38: 452.
1931 F. R. Hirsch, Jr., Phys. Rev., 38: 919, 923.
1931 E. Lindberg, Nova Acta Regiae Soc. Sci. Upsaliensis, 7 (7): 21.
1931 M. Siegbahn, "Spektroskopie der Röntgenstrahlen," 2d ed., pp. 332, 335, 346,
       Verlag Julius Springer, Berlin.
1931 J. H. Williams, Phys. Rev., 37: 232, 1431.
1932 H. Küstner, Physik. Z., 33: 46.
1932 L. G. Parratt, Phys. Rev., 41: 561.
1933a F. Ephraim and M. Mezener, Helv. Chim. Acta, 16: 1257.
1933b F. Ephraim, J. Indian Chem. Soc., Ray-Memorial Vol., p. 243.
1933 E. G. Purdom and T. M. Cork, Phys. Rev., 44: 328, 975.
1933 F. K. Richtmyer and S. Kaufman, Phys. Rev., 44: 606.
1933 R. J. Stephenson, Phys. Rev., 43: 527-533.
1933 Ta-You Wu and S. Goudsmit, Phys. Rev., 43: 496.
1934a G. V. Hevesy and H. Lay, Nature, 134: 98.
1934b H. Lay, Z. Physik, 91: 533.
1934 M. Siegbahn and T. Magnusson, Z. Physik, 88: 559.
1935a A. H. Compton and S. K. Allison, "X-Rays in Theory and Experiment," pp. 801,
       803, D. Van Nostrand Company, Inc., New York.
1935b Ibid., p. 645.
1935 A. E. Ruark and F. A. Maxfield, Phys. Rev., 47: 107.
1936 H. Claësson, Z. Physik, 101: 499.
1937 H. Hulubei, J. phys. radium, 8: 260.
1939 G. R. Harrison and coworkers, Massachusetts Institute of Technology Wave-
       length Tables, John Wiley & Sons, Inc., New York.
1939 W. Polaczek, Sitzber. Akad. Wiss. Wien, Math. naturw. Klasse. Abt. IIa, 148: 81.
1940a Y. Cauchois and M. L. Allais, J. phys. radium, 1: 44.
1940b Y. Cauchois and I. Manescu, Compt. rend., 210: 172.
1941 M. Goeppert-Mayer, Phys. Rev., 60: 184.
1946 Y. Cauchois, Compt. rend., 222: 1484.
1946 C. C. Kiess, C. J. Humphreys, and D. D. Laun, J. Research Natl. Bur. Stand-
       ards, 37: 57.
1946a P. Schuurmans, Physica, 11: 419
1946b P. Schuurmans, Physica, 11: 475.
1946c P. Schuurmans, Physica, 12: 589.
1947 O. E. Anderson and H. E. White, Phys. Rev., 71: 911.
1947 I. Manescu, Compt. rend., 225: 537.
1947 P. Schuurmans, J. C. van den Bosch, and N. Dijkwel, Physica, 13: 117.
1949 L. E. Burkhart, G. Stukenbroeker, and S. Adams, Phys. Rev., 75: 83.
1949 G. T. Seaborg, Paper 21.1 of "The Transuranium Elements," National Nuclear
       Energy Series, Division IV, Volume 14B, McGraw-Hill Book Company, Inc.,
       New York, 1949; Nucleonics, 5 (5): 16.
1949a J. C. van den Bosch, Physica, 15: 503.
1949b J. C. van den Bosch and G. J. van den Berg, Physica, 15: 329-350.

1950   J. R. McNally, Jr., Phys. Rev., 77: 417.
1950   W. H. Zachariasen, Science, 111: 460.

Project Literature

British 1:   Personal communication through S. K. Allison, January, 1945.

Carbide and Carbon 1:   J. R. McNally, Jr., Report Y-232, Sept. 9, 1948.
Carbide and Carbon 2:   J. R. McNally, Jr., and G. R. Harrison, Report Y-340, Feb. 11, 1949.
Carbide and Carbon 3:   J. R. McNally, Jr., personal communication, January, 1950.

Los Alamos 1:   H. Russell, Jr., Report LA-145, Sept. 22, 1944.

MIT 1:   G. R. Harrison, personal communication, January, 1945.
MIT 2:   G. R. Harrison, unpublished data reported in a personal communication, January, 1945.
MP Chicago 1:   A. T. Monk and S. K. Allison, Report CP-2120, Sept. 9, 1944.
MP Chicago 2:   M. Fred, personal communication, February, 1945.

Natl. Bur. Standards 1:   C. C. Kiess, C. J. Humphreys, and D. D. Laun, Report A-1747, Feb. 7, 1944.
Natl. Bur. Standards 2:   Spectroscopy Section, Report [A]M-2187, undated. Received May 14, 1945.

Chapter 3

URANIUM IN NATURE

1. GENERAL SURVEY

1.1 <u>Igneous Rocks</u>. Uranium is now recognized as a ubiquitous element. The reason why more is known about its wide distribution in nature than about the distribution of many other elements is that the radioactive properties of uranium (and of its disintegration products, e.g., radium, which are always associated with it in nature) make it easy to detect and estimate even minute quantities of this element. The usual method of determination of small amounts of uranium in minerals is indirect. It consists of measuring by means of an electroscope the amount of radium emanation evolved by a given weight of material. From this measurement the amount of radium present can be derived; the amount of uranium can in turn be calculated by assuming the existence of radioactive equilibrium between the uranium and its transmutation product, radium (Kirsch, 1928). The constancy of the ratio of uranium to radium, 2.84×10^6 to 1, in unweathered rocks, which is the ratio expected on theoretical grounds, has been established experimentally by many independent workers.* Also, the uranium content of various minerals and rocks has been determined directly by a fluorescence method which provides an exceedingly sensitive analytical procedure (Urry, 1941; Umovskaja, 1940) (see Sec. 2.6; also Division VIII, Volume 4 of the National Nuclear Energy Series). Igneous rocks containing excess silicic acid ("persilicic" rocks, often designated as "acid" or "felsic") are found to contain significantly higher proportions of uranium than the "subsilicic" rocks (often called "basic" or "mafic"). This is shown in Table 3.1 in which rocks are listed roughly in order of decreasing

---

*In some cases the Ra/U ratio differs from that derived from the laws of radioactive equilibrium. This has been attributed to weathering and selective leaching (Starik, 1936, 1937; Segel, 1938).

SiO_2 content. In addition to uranium and thorium, the table lists potassium, which is of considerable geophysical importance as a third radioactive constituent of the earth (Hess and Roll, 1948).

The best available estimate of the mean uranium content in the surface of the earth's crust is 4×10^{-6} g per gram of rock. The

Table 3.1 — Radioactive Content of Igneous Rocks

Type of rock	SiO_2, approx. %	Amount per metric ton		
		Uranium, g	Thorium, g	Potassium, kg
Granite	70	9.0	20.0	34
Granodiorite	66	7.7	18.0	25
Diorite	60	4.0	6.0	17
Central basalts	50			
Continental		3.5	9.1	19
Oceanic		3.6	7.1	18
Plateau basalt		2.2	5.0	8
Gabbro	50	2.4	5.1	7
Eclogite		1.0	1.8	4
Peridotite	43	1.5	3.3	8
Dunite	40	1.4	3.4	0.3

concentration varies in normal rocks between about 0.2×10^{-6} and 25×10^{-6} g per gram. These values of Hevesy (1930, 1932) were accepted by Goldschmidt (1938a) (see also Evans and Goodman, 1941) as the best available estimates.*

The abundance of certain metallic elements in rocks is shown in Table 3.2. This table serves to illustrate the fact that metals such as cadmium, bismuth, mercury, and silver, which are not considered excessively rare, are present in the earth's crust in much smaller average amounts than uranium. There is, however, no simple relation between the mean concentration of an element in the earth's crust and the probability of finding economically important deposits of that element.

The question arises whether the mean concentration of uranium found near the surface can be considered typical of the earth's crust

*A word of caution is needed regarding some data in the frequently quoted standard work of Clarke and Washington (1924). Owing to a mathematical error and the use of an obsolete value for the ratio of uranium to radium, these authors gave for the average abundance of uranium a value that is too large by a factor of 20. A. E. Fersman ("Geochemistry," Leningrad, 1933) gives 9×10^{-6} g per gram, and J. S. Anderson [J. Proc. Roy. Soc. N. S. Wales, 76: 329-345 (1942)] gives 2×10^{-7} g per gram.

as a whole. The answer is that the upper part of the crust probably is enriched in uranium. As mentioned above, uranium is known to occur preferentially in persilicic igneous rocks. Since the main components of the deeper regions of the earth's crust are comparatively poor in silica, uranium would be expected to occur less abundantly

Table 3.2 — Abundance of Some Elements in Igneous Rock*

Element	Grams per metric ton	Atoms per 100 atoms Si	Element	Grams per metric ton	Atoms per 100 atoms Si
Li	65	0.091	In	0.1	0.000007
Be	6†	0.0067	Gd	6.36†	0.000394
Na	28,300	12.4	W	69	0.0038
Al	81,300	30.5	Pt	0.005	0.00000027
Si	277,200	100	Au	0.005	0.00000026
Cr	200	0.039	Hg	0.5†	0.000025
Cu	100	0.016	Pb‡	16	0.00080
Zr	40	0.0062	Bi	0.2	0.000009
As	5	0.00067	Th	11.5†§	0.00050
Ag	0.10	0.000009	U	4	0.00016
Cd	0.5	0.000045			

*Adapted from Goldschmidt (1938b).

†Value based on sedimentary rock analysis.

‡For an interesting estimate of the age of the sun based on the terrestrial Pb/U ratio, see Meyer (1937a).

§This value refers to analyses based mostly on sedimentary rocks. The ratio of thorium to uranium has been studied by a number of workers. Based on a hundred determinations, an average value of 3.2 may be assumed for this ratio (Keevil, 1938, 1944) in normal igneous rocks, exclusive of typical thorium or uranium minerals. Senftle and Keevil (1947) in later work found a Th/U ratio of from 2.40 to 3.98, average 3.39, in granitic rock; in intermediate rock an average ratio of 3.98 was observed. The geochemical distribution of thorium appears to be somewhat different from that of uranium (Meyer, 1937b; Kirsch and Hecht, 1938). Meyer (1937a) has derived an average Th/U ratio of from 6.5 to 7 for the whole of the earth; since a ratio of 3 to 4 is found near the surface, the relative concentration of thorium in the core of the earth must be much higher (see also Mulder, 1947).

there than near the surface. Another argument in favor of the same assumption can be derived from an analysis of the thermal balance of the earth. Radioactive disintegration produces heat. If it is assumed that radioactive elements uranium and thorium are present in concentrations found near the surface of the earth down to a depth of 16 km and that potassium is distributed fairly uniformly in the whole lithosphere, the heat produced by radioactivity would offset all the heat losses caused by thermal radiation from the earth into space. If significant amounts of uranium or thorium existed below this depth, the fact that the earth is in an approximately steady thermal state would become incomprehensible (Holmes, 1926a,b).

It should not be thought, however, that the deeper layers of the earth's crust are entirely devoid of radioactive elements; indeed, the presence of uranium in these layers was postulated by Joly (1909) to account for the occurrence of geological revolutions (periods of volcanism and mountain growth) in the history of the earth. It appears reasonable to assume that, while the bulk of uranium is contained within a depth of 16 or 20 km from the surface of the earth, some uranium may exist down to a depth of 40 to 48 km, which is the assumed thickness of the lithosphere.

The weight of the earth's crust to a depth of 20 km has been estimated at about 3.25×10^{19} tons. If an average uranium content of 4×10^{-6} g per gram is assumed, the weight of uranium contained in this crust must be about 1.3×10^{14} tons. The main quantity of uranium is contained in the silica-rich igneous rocks that comprise the continental shields. Only relatively small quantities are probably present in the igneous rocks (mostly of the subsilicic type) that form the floors of the oceans.

1.2 Sedimentary Rocks. Igneous and metamorphic rocks constitute 95 per cent of the weight of the outer 16 km of the earth's crust. The sedimentary rocks (shales, etc.) make up the balance. Except for the carnotites of Colorado and Utah, the sedimentary rocks appear to contain much less uranium than the igneous rocks (on the average perhaps only one-half as much).

1.3 Oceans, Rivers, and Thermal Springs. Uranium is present in measurable concentration in the water of the oceans. The older values for the uranium concentration in sea water (see Gmelin, 1936a, for literature citations) were based on determinations of radium. However, it is by no means certain that radioactive equilibrium between uranium and radium is maintained in the oceans. On the contrary, there is positive evidence that the radium concentration in sea water is perhaps only one-tenth of that which would be present in equilibrium. This is not unexpected since a large proportion of radium produced by disintegration of dissolved uranium is likely to be lost from sea water by precipitation as sulfate or carbonate. Therefore, the older data for the uranium concentration in sea water, which were derived from the radium content, are of little value. Applications of direct fluorescence analysis gave for the uranium content of sea water values from 0.36×10^{-6} to 2.3×10^{-6} g per liter (Hernegger and Karlik, 1935). The uranium content was found to vary in proportion to the total salinity. Ocean water of 3.5 per cent salinity contains about 2×10^{-6} g of uranium per liter (2×10^{-9} g per gram) (Föyn and coworkers, 1939a), or about one two-thousandth as much uranium as is present in an equal weight of rock. Assuming the volume of the oceans to be about 2×10^9 km^3, the total uranium content of the oceans must be of the order of 4×10^9 tons or 0.003 per cent of the earth's crust (1.3×10^{14} tons).

Numerous thermal springs are known to be radioactive. The activity, however, appears to be due mainly to radium; little is known of the uranium content of these waters. The uranium content of rivers is, on the basis of very fragmentary evidence, thought to be of the same order of magnitude as that of sea water.

A number of investigations have been carried out to determine the uranium content of oceanic sediments (Piggot and Urry, 1942; Hoffmann, 1942a). No radioactive equilibrium exists in these sediments; their uppermost layers in particular contain excessive amounts of radium. The reason for this was mentioned above, namely, the precipitation of insoluble radium sulfate and carbonate from sea water. Consequently the amount of uranium present in the sediments cannot be correctly evaluated from determinations of radium. Roughly, it has been estimated that a total of 10^6 tons of uranium is present in the oceanic sediments, a negligible quantity by comparison with that contained in continental rocks.

1.4 Living Matter. Uranium is a constituent of living matter (the "biosphere" in Vernadsky's terminology); the biological significance of the apparently universal distribution of uranium in plants and animals is discussed in the biological volumes of this Series. Uranium appears to be a normal component of protoplasm (Diobkov, 1937). It has been estimated that uranium occurs there in concentrations varying from 10^{-4} to 10^{-9} per cent by weight (Hoffmann, 1941a; 1942b,c; 1943a,b). The fixing of uranium by algae may have had some significance in the formation of certain uranium deposits (Hoffmann, 1941b) (see Sec. 3.3).

1.5 Extraterrestrial Occurrence. Numerous analyses of meteorites show that they contain uranium (Paneth, 1928, 1930, 1931; Quirke and Finkelstein, 1917; Noddack, 1930; Hoffmann, 1941c). However, in conformity with the rule that subsilicic terrestrial rocks in general have a low uranium content, it is found that stony meteorites (called "aerolites"), which have less SiO_2 than even the highly subsilicic terrestrial rocks, contain only about 3.6×10^{-7} g of uranium per gram. (This is an average value based on analyses of 20 aerolites.) The uranium content of iron meteorites (siderites) is even smaller. It has been established that meteorites contain on the average 9×10^{-8} g of uranium per gram, i.e., only about 0.5 per cent of the average uranium content of igneous rocks. This, too, is in agreement with our knowledge of the geochemical distribution of uranium, since the iron core of the earth is probably almost free of uranium. The relatively large helium content of small meteorites has been attributed in part to the nuclear disintegration of uranium and thorium by cosmic radiation (Bauer, 1948).

Spectroscopic observation has as yet been unable to establish with any certainty the presence of uranium in the sun or other stars.

The important application of the radioactivity of uranium to the determination of the age of rocks cannot be discussed here. We refer in this respect to the reviews by Lane (1924 and subsequent years), Kirsch (1928), Ellsworth (1932a), and Holmes (1931, 1937). The discovery of fission has brought new aspects into the problem of the role of radioactive elements in geochemistry and geophysics. For the first analysis of these aspects we refer to the review by Goodman (1942), "Geological Applications of Nuclear Physics," which contains an excellent bibliography (see also Siegl, 1947, and Khlopin and co-workers, 1947).

2. OCCURRENCE AND COMPOSITION OF URANIUM MINERALS*

At the time of this writing it is difficult to give a completely coherent picture of the origin of the bewildering array of uranium minerals found in nature. The difficulty arises from a variety of circumstances. In the first place, most of these minerals are so complex and exhibit such variability in composition that for only a few of them is even the chemical constitution known with any certainty. Furthermore, crystal structure analyses of most uranium minerals are as yet either fragmentary or nonexistent. Finally, certain aspects of the chemistry of uranium itself are only now becoming clear. This is particularly true of the uranium-oxygen system, a knowledge of which is important for understanding the nature of uraninite and pitchblende minerals.

Most uranium minerals are found in pegmatites, particularly granite pegmatites. The chief exceptions are pitchblende (which occurs in veins and seams), carnotite, and certain uranium deposits associated with carbonaceous matter, such as kolm (a kind of oil shale found in Sweden). The presence of uranium in granite pegmatites can be accounted for by applying the geochemical principles of Goldschmidt (1923a). According to his views, as the uniformly molten earth cooled,

*Tomkeieff (1946) has reviewed researches on the geochemistry of uranium.

its matter became separated into one vapor phase and three concentric condensed phases. The elements are supposed by Goldschmidt to have concentrated in the various phases in the way shown in Table 3.3.

The siderosphere constitutes the earth's core; the chalkosphere forms an intermediate shell; and the lithosphere is the outer crust of

Table 3.3 — Distribution of the Elements in the Various Zones of the Earth*

Elements of the siderosphere, iron phase	Elements of the chalkosphere, sulfide phase	Elements of the lithosphere, silicate phase	Elements of the atmosphere, gaseous phase
Fe, Co, Ni	[O], S, Se, Te	O, (S), (P), (H)	H, N,† (O), (Cl?)
P, C	Fe, (Ni), (Co), Mn?	Si, Ti, Zr, Hf, Th	He, Ne, A, Kr, Xe
Mo, (W?)	Cu, Zn, Cd, Pb	F, Cl, Br, I	
Pt, Ir, Os?, (Pd)	(Sn?), Ge, (Mo?)	B, Al, Sc, Y, La, Ce.	
Ru, Rh	As, Sb, Bi	Pr, Nd, Sm, Eu,	
		Gd, Tb, Dy, Ho,	
		Er, Tm, Yb, Lu	
		Li, Na, K, Rb, Cs	
		Be, Mg, Ca, Sr, Ba	
		(Fe), V, Cr, Mo,	
		[Ni], [Co]	
		Cb, Ta, W, U, Sn	
		(C)‡	

*Parentheses indicate minor amount; brackets indicate very small amount.
†Perhaps also as nitride in siderosphere under high temperature and pressure.
‡As carbonate.

the earth. The elements that concentrated in the lithosphere were those possessing the more thermochemically stable oxides. Uranium is one of these. As the liquid silicate magma cooled, the high-melting solid phases containing the main components of the mixture began to crystallize out. Where solid solubility between the crystallizing species and a minor constituent existed, the latter was coprecipitated at an early stage. If the crystallochemical constants of an element were such that its entry into the lattices of early formed minerals was not favored, the element was progressively concentrated in the residual liquid phase. In the third column of Table 3.3 we find some elements whose crystallochemical properties must have prevented their entry into the crystalline rocks formed from the bulk of the silicate magma; these are the rare earths, zirconium, thorium, hafnium, columbium, tantalum, tungsten, tin, lithium, beryllium, boron, and uranium. These rare elements must have become concentrated in the last portion of the liquid magma. Rocks formed from this residual silicate magma are well-defined geological entities rich in feldspar and are designated as "pegmatites." Thus a rational explanation is found for the presence in pegmatites of uranium together with thorium, tantalum, columbium, zirconium, hafnium, and the rare earths.

Many secondary reactions must have occurred after the primary deposition of uranium minerals from the magma. Oxidation was an ever-present probability. Selective leaching of the pegmatite could have occurred, or the pegmatite could have been suffused with hot solutions of various compounds. In this way many original minerals underwent considerable change. Consequently, unraveling the geological history of any particular uranium mineral often is a complicated problem, and only in isolated instances has a clear picture been obtained. However, with renewed interest in the subject the situation may improve.

In addition to the uranium minerals found in pegmatites, there are two other types that require special attention because of their economic importance. One is carnotite, which is examined in detail below. The other is pitchblende, veins of which have been found in Canada, Bohemia, and the Belgian Congo. The pitchblende veins are thought to be derived from the same residual magmas as the pegmatites. These magmas might have contained appreciable quantities of water; consequently, in addition to the fractions that crystallized out as pegmatites, aqueous fractions may also have been produced. At high temperature and high pressure most of the elements listed in the second column of Table 3.3, in so far as they were present in the residual magma, must have been concentrated in aqueous solutions: the same thing happened to uranium (many compounds of which are readily soluble in water). Streams of these hot solutions, impinging on igneous rocks, underwent chemical reactions which resulted in precipitation and the formation of so-called "hydrothermal" vein deposits. In this way uranium became associated in pitchblende veins with the chalkophilic elements copper, bismuth, silver, tin, and gold. Most of these elements were precipitated as sulfides, but, in accordance with its chemical properties, uranium was almost always deposited as oxide.

2.1 Classification of Uranium Minerals. The paucity of good x-ray crystallographic data makes any attempt at a classification of the uranium minerals rather arbitrary.

In the older mineralogy the terms "uraninite" and "pitchblende" were applied indiscriminately to almost any uranium-oxygen mineral. This should be avoided since distinct chemical differences exist between the two minerals and considerable confusion results from the use of the two terms as synonyms. Uraninite is found only in pegmatites and invariably contains significant quantities of thorium and rare earths. Pitchblende is found only in veins of hydrothermal origin and usually contains no thorium and only traces of rare earths. Uraninite probably was originally pure uranium dioxide, UO_2; all sexivalent uranium found in uraninite probably arose from subsequent oxidation. Pitchblende, however, probably had a composition close to U_3O_8 even when first formed. Uraninite is crystalline; pitchblende is usually amorphous. In the following presentation the term "uraninite" is used only for crystalline uranium oxide found in pegmatites* (see van Aubel, 1927).

The minerals discussed here are listed in Tables 3.15 and 3.16. The classification adopted for this discussion is as follows (cf. Tyler, 1930):

   A. Uranium minerals in pegmatites
      1. Uraninite
      2. Uranium-bearing columbates and tantalates
   B. Nonpegmatitic uranium minerals
      1. Pitchblende
      2. Uranium oxides formed by oxidation of pitchblende (often identical with those formed from uraninite)
   C. Secondary uranium minerals
      1. Uranates, silicates, carbonates, sulfates
      2. Uranium "micas," of the type

$$M(II)(UO_2)_2(XO_4)_2 \cdot nH_2O$$

where M is Ca, Cu, Fe, Pb, Mn, or UO_2, and X is P, V, or As
      3. Carbonaceous uranium-bearing substances

Secondary minerals resulting from oxidation or other chemical reactions may occur in both pegmatites and pitchblende veins. In the next sections some important members of these groups will receive more detailed discussion.

In order to facilitate an understanding of the composition of the uranium minerals, the ionic radii of a number of elements of frequent occurrence in uranium minerals are given in Table 3.4. A general rule is that ions whose radii differ by not more than 15 per cent are interchangeable in crystals [provided that certain restrictions, such as the requirement of conservation of charge, are not violated; for limitations and applications of this rule see Goldschmidt (1938a,b)]. As might be anticipated from the ionic radii, practically all minerals

---

*"Dana's System of Mineralogy" (1944) vol. I, pp. 611-621, includes pitchblende as a variety of uraninite; this usage is common in modern mineralogy. However, the chemical differences between the two minerals are such as to offer a certain justification for the treatment given here.

Table 3.4 — Ionic Radii* for Some Elements of Common Occurrence
in Uranium Minerals†

Ion	Ionic radii, Å	Ion	Ionic radii, Å	Ion	Ionic radii, Å
U(IV)	1.05	Cb(V)	0.69	P(V)	0.3 – 0.4
Th(IV)	1.10	Ta(V)	0.68	V(+V)	~0.4
Er(III)	1.02	Ti(IV)	0.64	As(V)	0.4
Yb(III)	1.00	Mo(IV)	0.68	Cr(VI)	0.3 – 0.4
Y(III)	1.06	Fe(III)	0.67	S(VI)	~0.34
Ca(II)	1.06				

*For coordination number 6.
†Selected from Evans (1939).

Table 3.5 — Uraninite Minerals

Mineral	Structure	Uranium, %	Optical and crystallographic properties	Typical occurrences
Uraninite	Essentially UO_2 with UO_3 present as a result of oxidation; fluorite structure, isomorphous with ThO_2, CeO_2; density 3.0 – 10.6 g/cc	65.2 – 74.5	Crystalline; opaque; octahedral or cubic isotropic crystals; color variable, may have semi-metallic luster; least altered specimens iron gray	Karelia, U.S.S.R. Masaki, Japan Gaya, India Grafton Center, N. H. Portland, Conn. Bedford, N. Y. Mitchell County, N. C. Pennington County, S. D. Quebec, Canada Cape Province, South Africa
Thorian uraninite (broggerite)	$(U, Th)O_2$; contains Th in substitution for U to at least 14% ThO_2; this variety is restricted to Th > Y, Ce, etc.; rare earth content 0 – 8%	48.7 – 74.9	Isotropic crystals	Anneröd Peninsula, Norway
Cerian and yttrian uraninite (cleveite, nivenite)	$[U(IV), R.E.(III)]O_2$; rare earth content 5.17%, with Y, Er, Ce, La present; UO_2/UO_3 varies from 0.5 – 0.7	53.3 – 66.4	Crystalline	Norway
	$[U(IV), R.E.(III)]O_2$; rare earth content, 11.2 – 11.8%; Y, Er, Ce, La present; UO_2/UO_3 ratio about 2.5; density 8.29 g/cc	57	Crystalline	Llano County, Texas Iasaka pegmatite on Honshu, Japan
Thorianite	$[Th(IV), U(IV)]O_2$; ThO_2/U_3O_8 varies from 6/1 to 2/1; density 8 – 9.7 g/cc	9.5 – 28.2	Cubic crystals; yellowish brown to black; refractive index 1.8	Ceylon and Japan(?) Madagascar Boshogoch River, Siberia Easton, Pa.

containing appreciable amounts of rare earth elements are likely to contain a little uranium.

2.2 Uraninite. (See Table 3.5.) Uraninite (Dana, 1944) may be considered the primary uranium mineral. It is found in granite and syenite pegmatites, where it is associated with zircon, tourmaline, monazite, mica, feldspar, etc. It is also often closely associated with minerals containing rare earths and columbium or tantalum. Uraninite was first shown to be essentially uranium dioxide, UO_2, by Goldschmidt and Thomassen (1923b), who examined natural uraninite crystals from southern Norway by x rays. According to these observers, uraninite is isomorphous with CeO_2 and ThO_2; it possesses a fluorite-type cubic structure. Its lattice constant is 5.460 A. Similar results were obtained by Schoep (1935), who found that natural uraninite gives the same x-ray pattern as synthetic uranium dioxide. Uraninites from different occurrences have been found to be crystallographically identical. The first column in Table 3.4 indicates which substituting ions are likely to occur in uraninite. It shows that similarity of ionic radii accounts to a certain extent for the invariable presence of rare earths in uraninite crystals.

As might be expected from the chemical properties of uranium dioxide, natural uraninite is always more or less strongly oxidized, so that its actual composition lies between UO_2 and $UO_{2.67}$ (U_3O_8), with tetravalent uranium usually predominant. This oxidation may be the result of weathering, but it may also result from the liberation of oxygen within the crystal in consequence of radioactive disintegration of uranium to lead.

$$UO_2 \rightarrow PbO + \tfrac{1}{2}O_2 \; (+ 6He)$$

The amount of oxygen presumably liberated by radioactive disintegration and bound in the crystal has been found experimentally to correspond very closely to the amount of PbO (RaG oxide) formed (Bakken and Gleditsch, 1939). Radioactivity thus plays an important role in the variations in composition of uranium oxide minerals. This is also proved by their generally metamict character.

Weathering effects in single crystals of uraninite have been studied by determining the U(IV)/U(VI) ratio in several layers of cleveite crystals as a function of the distance from the surface. The ratio increases toward the core (Bakken and Gleditsch, 1938; Alter and Kipp, 1936; Bakken, Gleditsch, and Pappas, 1948). The analytical work also indicates that uraninite may be subject to selective leaching, which particularly affects the radium and lead content of the crystals. The mineral becomes black by alteration; least altered specimens are iron gray. More pronounced alteration leads to coatings of scarlet, orange, yellow, green, gray, or brown decomposition products. Attempts have been made to reproduce the process of weathering of uraninite (Föyn, 1939b). When specimens of cleveite and bröggerite were heated in an autoclave with water to 200°C,

Table 3.6 — Per Cent Chemical Composition of Some Uraninite Minerals

Component	Bröggerite*	Nivenite†	Cleveite‡	Uraninite§	Uraninite¶
CaO	0.37	0.32	0.86	1.01	0.46
MnO	...	...	...	0.03	0.001
PbO	9.04	10.08	10.92	10.95	16.42
MgO	Trace	...	0.14	0.08	0.01
$(Y, Er)_2O_3$	1.11	9.46	9.99	2.14	1.01
$(Ce, La)_2O_3$	...	...	2.25	1.88	...
La_2O_3	0.27	2.36	...	...	0.80
CeO_2	0.18	0.34	...	...	0.265
UO_2	46.13	44.17	23.07	39.10	48.87
U_3O_8	...	...	...	...	...
UO_3	30.63	20.89	40.60	32.40	28.582
ThO_2	6.00	6.69	4.60	10.60	2.15
ZrO_2	0.06	0.34	...	...	0.22
SiO_2	0.22	0.46	...	0.19	0.055
Al_2O_3	...	...	...	0.09	...
Fe_2O_3	0.25	0.14	1.02	0.43	0.30
CO_2	...	...	...	...	...
H_2O	0.74	1.48	4.96	0.70	0.44
Insoluble	4.42	1.47	2.34	0.15	0.15
Remainder	0.19	0.08	...	0.31	0.39
Total	99.61	98.28	100.75	100.06	100.123
Density, g/cc	8.893	8.29	7.49	9.062	9.182

*From Gustav's Mine, Anneröd, Norway. F. W. Hillebrand, U. S. Geol. Survey Bull. 78: 43 (1891).

†From Baringer Hill, Llano County, Texas. F. W. Hillebrand, Am. J. Sci., 42: 390 (1891).

‡From Arendal, Norway. W. E. Hidden and J. B. Mackintosh, Am. J. Sci., 38: 474 (1889); recalculated by Hillebrand in the first reference above.

§From Wilberforce, Haliburton County, Ontario. H. V. Ellsworth, Natl. Research Council Can. Ann. Repts., App. H, Exhibit A, 1929-1930.

¶From Ingersoll Mine, Pennington County, S. D. C. W. Davis, Am. J. Sci., 11: 201 (1926).

cleveite (with a high UO_3/UO_2 ratio) underwent little change, whereas bröggerite (with a low UO_3/UO_2 ratio) showed extensive decomposition.

A number of typical analyses of uraninite minerals are given in Table 3.6. Numerous other analyses may be found in Doelter and Leitmeier's "Handbuch der Mineralchemie" (1929). An examination of the data reinforces Kirsch's view that all these minerals are prod-

Table 3.7 — Pegmatitic Cb, Ta, Ti Minerals Containing Uranium

Mineral	Composition and structure	Uranium, %	Optical and crystallographic properties	Occurrence	Described in* Doelter (1929)	Dana (1944)
			Pyrochlore, Microlite Series [$A_2B_2O_6(O,OH,F)$]			
			A = Na, Ca, K, Mg, Fe(II), Mn(II), Pb(?) Ce, La, Dy, Er, Y, Th, Zr, U			
			B = Cb, Ta, Ti, Sn(?), Fe(III)(?), W(?)			
Pyrochlore	Essentially $NaCaCb_2O_6F$	2.5 – 8.1	Lustrous, glassy to resinous; brown to nearly black	In nepheline syenite pegmatite, and granite pegmatites Urals and Caucasus, U.S.S.R.	III(1), 95	I, 748
Microlite	Essentially $(Na, Ca)_2 Ta_2$-$O_6(O, OH, F)$	0 – 1.3	Small octahedra; golden yellow to brown	Sweden Virginia Greenland W. Australia Massachusetts	III(1), 250	I, 748
Hatchettolite	Uranium pyrochlore	10 – 15	Transparent, isotropic octahedra; lustrous, resinous; $n = 1.98 \pm$	Hybla, Ontario Mitchell County, N. C. Madagascar	III(1), 250	I, 754
Ellsworthite	High-uranium calcium-iron pyrochlore; strongly altered; related to hatchettolite	15 – 17	Massive; amber yellow to chocolate brown; isotropic; $n = 1.89 \pm$	Hybla, Ontario Haliburton, Ontario	IV(2), 955	I, 755
			Fergusonite, Formanite Series (ABO_4)			
			A = Y, Er, Ce, La, Dy, U(IV), Zr, Th, Ca, Fe(II)			
			B = Cb, Ta, Ti, Sn, W			
Fergusonite	Essentially a columbate-tantalate of Y, Er	0.2 – 8.16	Tetragonal, C = 1.4643; pyramidal crystals; very dark brown	Greenland Sweden Norway Massachusetts Llano County, Texas Ceylon Terek River, Caucasus Mitchell County, N. C.	III(1), 252	I, 757
Rutherfordite	Altered fergusonite			Rutherford County, N. C.		I, 762
Yttrotantalite	Essentially an iron-yttrium-uranium-columbium tantalate	1.4 – 3.96	Orthorhombic, a:b:c = 0.5566:1.0: 0.5173; prismatic; yellowish brown to black; isotropic; $n = 2.15 \pm 0.02$	Norway Sweden Miask, Urals Alabama	III(1), 256	I, 763

Table 3.7 — (Continued)

Mineral	Composition and structure	Uranium, %	Optical and crystallographic properties	Occurrence	Described in* Doelter (1929)	Described in* Dana (1944)
Ishikawaite	Essentially a uranium(IV)-iron-rare earth columbate-tantalate; analogous to samarskite	20	Tabular orthorhombic crystals with a:b:c = 0.9451:1:1.1472; black; waxy luster; density 6.2–6.4	Iwaki Province, Japan		I, 766
Brannerite	$(U, Ca, Fe, Y, Th)_3Ti_5O_{16}$	40	Isotropic; n_{Na} = 2.30 ± 0.02; black, transparent in thin splinters	Custer County, Idaho		I, 774

AB_2O_6 Series

A = Y, Ce, Ca, U, Th, Er, La, Pb
B = Ti, Cb, Ta, Fe(III), Sn, W, Zr(?)

Mineral	Composition and structure	Uranium, %	Optical and crystallographic properties	Occurrence	Described in* Doelter (1929)	Described in* Dana (1944)
Euxenite-polycrase series, including variety lyndochite; yttrocrasite, khlopinite, eschwegite are similar, ill-defined minerals	$(Y, Ca, Ce, U, Th)(Cb, Ta, Ti)_2O_6$; the high-Ti end of this series of minerals is polycrase	2.3–14.5	Orthorhombic; very dark brown	Woodstock, W. Australia Nipissing, Ont. Norway Caucasus Swaziland Greenland North Carolina South Carolina Brazil Madagascar	III(1), 102	I, 787
Eschynite-priorite-blomstrandine series Samarskite† (wiikite and hjelmite are ill-defined minerals perhaps related to samarskite)	$[Y, Er, Ca, Fe(II), Th, U(IV)] (Ti, Cb)_2O_6$	1–5	Similar to euxenite	Norway Swaziland Wolhynia, U.S.S.R. Miask, Urals	III(1), 106	I, 793
	$(Y, Er, Ce, U, Ca, Fe, Pb, Th) (Cb, Ta, Ti, Sn)_2O_6$	3.5–14.0	Orthorhombic; a:b:c = 0.5456:1:0.5178; yellowish brown to black; isotropic; n = 2.20 ± 0.05	Berthier County, Que. Baltimore, Md. Mitchell County, N. C. Colorado Caucasus Madagascar	III(1), 256	I, 787

$A_mB_nX_x$ Series

m:n = 1.3
A = U, Ca, Th, Pb, Ce, Y, Er
B = Ti, Cb, Ta, Fe(III), Al(?)

Mineral	Composition and structure	Uranium, %	Optical and crystallographic properties	Occurrence	Described in* Doelter (1929)	Described in* Dana (1944)
Betafite group, including the related minerals betafite, samiresite, blomstrandite, mendeleyevite, djalamaite	$(U, Ca)(Cb, Ta, Ti)_3O_9 \cdot nH_2O$	Commonly 20–26 but down to 10	Greenish brown; isotropic; n = 1.91–1.97	Madagascar	III(1), 97	I, 803

Table 3.7 — (Continued)

Mineral	Composition and structure	Uranium, %	Optical and crystallo-graphic properties	Occurrence	Described in[*] Doelter (1929)	Dana (1944)
Ampangabeite	Oxide of Cb with rare earths and uranium; for-mula unknown; probably like betafite	10–17	Orthorhombic; isotropic; $n =$ 2.13 ± 0.03; all shades of brown	Ampangabé, Madagascar	IV(2), 957	I, 806
Delorenzite	An oxide of Ti, Y, U, Fe, and Sn, possibly AB_3O_8	~10	Rhombic; black; a:b:c = 0.3375: 1:0.3412	Craveggia, Italy	III(1), 52	I, 808
Zirkelite	$(Ca, Fe, Th, U)_2(Ti, Zr)_2$ O_5 (?)	1	Black; $n = 2.19 ± 0.01$	São Paulo, Brazil		I, 740

[*]The roman numeral refers to the volume, the arabic numeral in parentheses to the part, and the other arabic numeral to the page.

[†]For a recent analysis and age determination of a specimen of Australian samarskite, see Kleeman (1946). For analyses of a number of samples of samarskite from the Nellore District in India, see Karunakaran and Neelakantam (1947).

ucts of alteration and ion substitution in originally pure UO_2 (desig-nated by Kirsch as "ulrichite," a term sometimes applied in the literature to slightly weathered specimens of uraninite).

2.3 Other Uranium Minerals of Pegmatitic Occurrence; Colum-bates, Tantalates, and Titanates. Perhaps the most frequently oc-curring uranium minerals in pegmatites are the uranium-bearing columbates, tantalates, and titanates. Columbium, tantalum, and tita-nium are among the elements that crystallize during the last stages of magma solidification (see above). The ionic radii in Table 3.4 indicate the substitutions that may be encountered in the compounds containing these elements. The structure and composition of these minerals are not discussed here in detail but can be found in "Dana's System of Mineralogy" (1944). It is sufficient to indicate that these minerals are oxides of the isodesmic type. Since crystallographic data are for the most part lacking, these minerals can at present be classified only by chemical criteria. The chemical constitution of most of them can be expressed by the formula $A_m B_n O_{2(m+n)}$, where m/n is between 1 and 0.5. [A is a rare earth, U, Ca, Th, Fe(II), Na, Mn, or Zr; B is Cb, Ta, Ti, Sn, W(?), Zr(?), or Fe(III).] .These min-erals are extraordinarily complex, widely variable in composition, and difficult to analyze; the formulas suggested must therefore be regarded in many cases as highly provisional. In Table 3.7 are listed the more important uranium-bearing minerals of this type.

In Secs. 2.2 and 2.3, a number of the more important uranium minerals occurring in pegmatites have been discussed. Despite their wide distribution these minerals have practically no economic importance. Or, perhaps more correctly, they have had no such significance up to the time of the writing of this chapter. Only two of these minerals have been mined to a very limited extent for their uranium content, namely, betafite in Madagascar and euxenite-polycrase in Western Australia and Brazil. This situation may change in view of the new importance of uranium. Thus, by way of example, the Iasaka pegmatite in Honshu, Japan, contains about 100 g of uraninite per ton of rock; although unattractive for the extraction of radium, it may be worthy of exploitation for its uranium content (Iimori, 1941). On the whole, however, the total amount of uranium available in pegmatites is probably small.

2.4 Pitchblende and Its Alteration Products. Pitchblende is the most important uranium mineral from the point of view of richness of deposits. Its composition is variable, but in the absence of severe weathering conditions its formula approximates U_3O_8. Pitchblende occurs in metalliferous veins, together with sulfides and arsenides of Fe, Cu, Pb, Co, Ni, As, and Bi. It is found in hydrothermal (so-called "hyperthermal") tin veins as colloform crusts, associated with cassiterite, pyrite, galena, and Co-Ni-Bi-As minerals, especially at Cornwall, England. Another and more important mode of occurrence is in "mesothermal" Co-Ni-Bi-Ag-As veins (hydrothermal veins formed at moderate temperature). There the pitchblende is associated with pyrite, chalcopyrite, barite, fluorite, native bismuth, native silver, and Co-Ni-As compounds. Deposits at Jáchymov (Joachimsthal) in Bohemia, Johanngeorgenstadt in Saxony, and Great Bear Lake in Canada all are of this type. Pitchblende is also found as colloform crusts with pyrites, sphalerites, etc., in hydrothermal sulfide veins formed at moderate temperatures, in which Co-Ni minerals are absent. The deposit in Gilpin County, Colorado, exemplifies this last type of occurrence. The pitchblende deposits of the Great Bear Lake region and the Belgian Congo will receive more detailed consideration in Sec. 3.

As stated before, the formation of pitchblende deposits was associated with the formation of mineralized veins of the chalkophilic and siderophilic elements given in Table 3.3. The chemical reactions involved in the transport of uranium by hot aqueous streams and its precipitation as U_3O_8, which must have occurred in the last stages of magma crystallization, are as yet unknown. As already noted, pitchblende is found in association with sulfide deposits, but the uranium itself is present in the form of oxide.

Table 3.8—Oxidation Products of Pitchblende

Mineral	Composition	Uranium, %	Optical and crystallographic properties	Occurrence	Described in* Doelter (1929)	Dana (1944)
Becquerelite	$2UO_3 \cdot 3H_2O$?; formula uncertain; water lost at 500°C	74	Orthorhombic plates; canary yellow to orange	Kasolo, Katanga, Belgian Congo Wölsendorf, Bavaria	IV(2), 937	I, 625
Schoepite	$4UO_3 \cdot 9H_2O$?; complete analyses are lacking; water lost at 325°C	65–70	Small yellow crystals; orthorhombic; $a_0 = 14.40$, $b_0 = 16.89$, $c_0 = 14.75$; $a_0:b_0:c_0 = 0.852$: 1:0.873; each unit cell contains $U_{32}O_{96} \cdot 72H_2O$?	Kasolo, Katanga, Belgian Congo	IV(2), 939	I, 627
Ianthinite	$2UO_2 \cdot 7H_2O$; this must be incorrect; alters readily to becquerelite and schoepite	70	Orthorhombic; violet-black; semimetallic luster	Shinkolobwe, Kasolo, Katanga, Belgian Congo	IV(2), 941	I, 633
Gummite	$UO_3 \cdot nH_2O$ (see Sec. 2.4)	40–70	Yellow, orange, red, reddish brown to black	Found at many occurrences of uraninite and pitchblende Czechoslovakia Saxony Belgian Congo North Carolina Connecticut Quebec	IV(2), 950	I, 622
Clarkeite	A gummite; essentially a hydrous uranium oxide with Pb, alkalis, alkaline earths	~80	Reddish brown	Mitchell County, N. C.		I, 625
Fourmarierite	Perhaps $PbO \cdot 4UO_3 \cdot 5H_2O$	60–70	Orthorhombic; red to brown	Kasolo, Katanga, Belgian Congo	IV(2), 944	I, 628
Curite	A hydrated oxide of lead and uranium, perhaps $2PbO \cdot 5UO_3 \cdot 4H_2O$; water is completely lost at 450°C	60–70	Orange-red; orthorhombic; $a_0 = 12.52$, $b_0 = 12.98$, $c_0 = 8.35$; unit cell contains $Pb_6U_{15}O_{51} \cdot 12H_2O$	Kasolo, Katanga, Belgian Congo	IV(2), 942	I, 629

*The roman numeral refers to the volume, the arabic numeral in parentheses to the part, and the other arabic numeral to the page.

Pitchblende always occurs as a massive fine-grained characteristically black material which shows no signs of macroscopic crystallinity. The oxygen content is variable, although in many cases it is well approximated by the formula U_3O_8. These variations can be understood on the basis of crystallographic studies of the uranium-oxygen system. These studies (see Chap. 11) have shown that the U/O ratio can range between about $UO_{2.5}$ and UO_3 without phase transformation. Small amounts of iron, manganese, aluminum, calcium, magnesium, silicon, etc., together with lead and helium from radioactive disintegration, are usually found in pitchblende. A very important difference between pitchblende and uraninite is that pitchblende contains practically no thorium and less than 1 per cent rare earths. The uranium content of pitchblende varies between 40 and 76 per cent of the mineral. For numerous analyses, the compilations of Doelter (1929) and Hintze (1930) should be consulted.

The chemical properties of pitchblende are substantially those of the uranium oxide, U_3O_8 (see Chap. 11). The mineral liberates helium on heating (Hillebrand, 1889). Pitchblende from the Belgian Congo has been reported to yield on heating a sublimate of selenium also (Steinkuhler, 1923). All varieties of pitchblende are soluble in sulfuric or hydrochloric acid.

A considerable number of oxidation products of pitchblende are known. These occur either in complex mixtures of minerals or as alteration zones in pitchblende or uraninite. In most cases the analyses are not very satisfactory, and in the case of ianthinite they are almost certainly wrong. This circumstance makes it impossible to assign a definite constitution to these minerals. Table 3.8 lists a number of hydrated oxides that may have been formed by oxidation of pitchblende in situ. Since these minerals have been exposed to extensive leaching, small proportions of other elements are usually present. All these elements except lead are considered to be admixtures rather than essential components of the mineral.

The pitchblende deposits of the Belgian Congo are particularly rich in alteration products. In these deposits the minerals listed in Table 3.8 are found in complex mixtures. In addition to those listed in Table 3.8, Schoep and Stradiot (1947) have described two new species, paraschoepite and epiianthinite, of unknown composition, which are related to or derived from schoepite and ianthinite. Although the exact relations are unknown, these minerals represent various stages of oxidation. The term "gummite" is applied to minerals that are essentially hydrated oxides of uranium of unknown composition and probably represent the final stages of oxidation and hydration of pitchblende. Small percentages of alkali metals, alkaline earths, rare earths,

Table 3.9 — Uranates and Uranium Silicates*

Mineral	Composition	Uranium, %	Optical and crystallographic properties	Occurrence	Doelter† (1929)
Vandenbrandite (uranolepidite)	$2CuO \cdot 2UO_3 \cdot 5H_2O$	~60	Triclinic crystals; dark green to black $n_\alpha = 1.77$ $n_\gamma < 1.80$	Kalongwe, Katanga, Belgian Congo; along with kasolite, sklodowskite, etc.	
Uranosphaerite	$(BiO)_2U_2O_7 \cdot 3H_2O$	~50	Yellow orthorhombic crystals; optically positive $n_\alpha = 1.955$ $n_\beta = 1.985$ $n_\gamma = 2.05$	Schneeberg, Saxony; with walpurgite, trögerite, zeunerite, uranospinite	
Soddyite	$12UO_3 \cdot 5SiO_2 \cdot 14H_2O$ or $5UO_3 \cdot 2SiO_2 \cdot 6H_2O$	~70	Yellow orthorhombic crystals; a:b:c = 0.7959:1:1.6685; some prismatic crystals show strong pleochroism, colorless along α axis, dark violet along β axis, violet along γ axis	Kasolo, Katanga, Belgian Congo; along with curite	IV(2), 946
Uranotile‡ (uranophane; lambertite) (a dimorph, β-uranotile, also exists)	$CaO \cdot 2UO_3 \cdot 2SiO_2 - \cdot 6H_2O$	~50	Green or yellow; orthorhombic; a:b:c = 0.3705:1:1; optically negative; pleochroic	Schneeberg, Saxony Jáchymov, Czechoslovakia Arendal, Norway Villeneuve, Canada Mitchell County, N. C. Katanga, Belgian Congo Madrid, Spain Pennsylvania	II(2), 164
Sklodowskite (shinkolobwite)	$MgO \cdot 2UO_3 \cdot 2SiO_2 - \cdot 6H_2O$	~50−60	Orthorhombic yellow crystals $n_\alpha = 1.613$ $n_\beta = 1.635$ $n_\gamma = 1.657$ pleochroic; a:b:c = 0.3114:1:1.0554	Kasolo, Katanga, Belgian Congo	IV(2), 947
Kasolite§	$PbO \cdot UO_3 \cdot SiO_2 \cdot H_2O$	~40	Monoclinic; yellow; a:b:c = 1.8566:1:1.0811 $\beta = 103°40'$ $n_\alpha = 1.89$ $n_\beta = 1.90$ $n_\gamma = 1.967$	Kasolo, Katanga, Belgian Congo	IV(2), 949
Pilbarite	$PbO \cdot UO_3 \cdot ThO_2 - \cdot 2SiO_2 \cdot 4H_2O$	~27U ~30Th		Pilbara Goldfield, W. Australia	II(1), 804

*The dubious mineral species mackintoshite, maitlandite, and nicolayite are similar to the minerals treated in this table.

†The roman numeral refers to the volume, the arabic numeral in parentheses to the part, and the other arabic numeral to the page.

‡V. Billiet, Mineralog. Abstracts, 7: 108 (1938).

§Mineralog. Abstracts, 6: 429 (1937); 8: 15 (1941); Chem. Abstracts, 35: 7888 (1941).

alumina, silica, etc., are commonly present, but these appear to be extraneous gangue materials. The physical and chemical properties of gummite vary widely.

Table 3.10 — Uranium Carbonate Minerals

Mineral	Composition	Uranium, %	Optical and crystallographic properties	Occurrence	Doelter* (1929)
Rutherfordine	UO_2CO_3 (established for only one sample)	~75	Orthorhombic (?) yellow crystals	Morogoro, East Africa	I, 547
Uranothallite	$2CaO \cdot UO_3 \cdot 3CO_2 \cdot 10H_2O$	~30	Apple-green orthorhombic crystals; a:b:c = 0.954:1: 0.783; strongly birefringent $n_\alpha = 1.50$ $n_\beta = 1.503$ $n_\gamma = 1.537$	Jáchymov, Czechoslovakia Johanngeorgenstadt, Saxony Adrianople, Turkey	I, 545
Liebigite	Hydrated carbonate of Ca and U, exact composition uncertain	~30			I, 546
Schroeckingerite (dakeite)	A sodium calcium uranium carbonate-sulfate, $3CaO \cdot Na_2O \cdot UO_3 \cdot 3CO_2 \cdot SO_3 \cdot 10H_2O$	25	Pseudohexagonal plates; greenish yellow $n_\alpha = 1.49$ $n_\beta = 1.54$ neg., 2V 0−25 deg	Jáchymov, Czechoslovakia Wyoming	I, 546
Voglite	May be $Cu(UO_2) \cdot (CO_3)_2 \cdot 10H_2O$	~30	Apple green to dark green; probably orthorhombic; optically positive; pleochroic	Jáchymov, Czechoslovakia Ferghana, Turkestan	I, 546
Sharpite†	$6UO_3 \cdot 5CO_2 \cdot 8H_2O$	67	Orthorhombic (?); yellowish green; extinction parallel; elongation positive; $\alpha = 1.633, \gamma \sim 1.72$; birefringence very high; pleochroism feeble, x = brownish, z = clear yellow to slightly greenish	Belgian Congo	

*The roman numeral refers to the volume, and the arabic numeral to the page.
†Mineralog. Abstracts, 7: 225 (1939); Am. Mineral., 24: 658 (1939); J. Melon, Bull. Inst. roy. colonial belge, 9: 333 (1938).

2.5 Secondary Uranium Minerals. (For references and details see Gmelin, 1936b.) The formation of uranates (as well as silicates, carbonates, phosphates, vanadates, arsenates, and sulfates of uranium) presumably involved complete dissolution of uraninite or pitchblende and the subsequent deposition of the secondary compounds. Some, but not all, of these secondary minerals may have formed in

THE CHEMISTRY OF URANIUM

Table 3.11—Uranium Sulfate Minerals

Mineral	Composition	Uranium, %	Optical and crystallographic properties	Occurrence	Doelter* (1929)
Johannite (gilpinite)	$CuO \cdot 2UO_3 \cdot 2SO_3 \cdot 7H_2O$	~40−50	Triclinic; a:b:c = 1.218:1:0.6736 $\alpha = 69°24'$ $\beta = 124°56'$ $\gamma = 132°56'$	Gilpin County, Colo. Johanngeorgenstadt, Saxony Jáchymov, Czechoslovakia Cornwall, England Middletown, Conn.	IV(2), 649
Uraconite	Hydrous sulfate of uranium and copper	~60−70	Amorphous; orange or yellow	Jáchymov, Czechoslovakia Cornwall, England	IV(2), 651
Uranopilite†	$6UO_3 \cdot SO_3 \cdot 16H_2O$; dehydrates to β-uranopilite, $6UO_3 \cdot SO_3 \cdot 10H_2O$	64	May be monoclinic; lemon yellow $n_\alpha = 1.621$ $n_\beta = 1.623$ $n_\gamma = 1.631$ β-uranopilite $n_\alpha = 1.72$ $n_\beta = n_\gamma = 1.76$	Jáchymov, Czechoslovakia Johanngeorgenstadt, Saxony St. Just, Cornwall, England	IV(2), 651
Voglianite	Probably a variety of uranopilite; both a calcium variety and a copper variety of voglianite are known	67	Apple green	Jáchymov, Czechoslovakia	IV(2), 652
Uranochalcite	Hydrous sulfate of uranium, copper, and calcium	~30	Grass green; optically positive $n_\alpha = 1.655$ $n_\gamma = 1.662$ pleochroic	Jáchymov, Czechoslovakia	IV(2), 654
Zippeite	$2UO_3 \cdot SO_3 \cdot nH_2O$, n varying from 3 to 8; some samples contain Cu (see voglianite)	~50−60	May be monoclinic for trihydrate $n_\alpha = 1.630$ $n_\beta = 1.689$ $n_\gamma = 1.739$ pleochroic; optical properties vary with degree of hydration	Jáchymov, Czechoslovakia Fruita, Utah Great Bear Lake, Canada	IV(2), 655

*The roman numeral refers to the volume, the arabic numeral in parentheses to the part, and the other arabic numeral to the page.

†See Nováček (1946) for a discussion of the structure and dehydration of uranopilite.

aqueous solutions near the critical temperature of water, but little is known at present about chemical reactions that take place under such conditions.

(a) Silicates and Uranates. For convenience, and following Ells-worth's (1932a) classification, these two classes of minerals are treated together. Their composition is often ill-defined. A number of characteristic minerals of these types are described in Table 3.9. Vaes (1946) has described billietite, a hydrated barium uranate, and vandendriesscheite, masuyite, and richetite, hydrated lead uranates found in the Belgian Congo deposits.

The structure of uranotile, $CaO \cdot 2UO_3 \cdot 2SiO_2 \cdot 6H_2O$, has been determined by x-ray investigation, and it has been shown that it forms orthorhombic crystals that are isomorphous with sklodowskite (Billiet, 1936).

(b) Carbonates. For the most part these minerals are poorly characterized. Uranothallite appears to be the most common mineral of this class. Little is known of their genesis, but it is likely that these minerals are not formed by a simple weathering process in situ but rather by solution and reprecipitation from aqueous solution. In Table 3.10 are listed a number of uranium carbonate minerals. The chemical formulas of these minerals are uncertain. The same is true of the hydrated uranium carbonates studtite and diderichite described by Vaes (1946).

(c) Sulfates. Minerals of this type (sometimes called "uranium ochers") are of rather frequent occurrence but are usually found only in very small amounts. Many of these minerals are water-soluble and undergo alteration readily, which has made the assignment of chemical formulas difficult. The chemical identity of the minerals listed in Table 3.11 is therefore doubtful. Uranopilite appears to be the principal member.

(d) Phosphates, Arsenates, and Vanadates. A large number of uranium minerals of secondary origin have a composition that can be expressed by the general formula $M(II)(UO_2)_2(XO_4)_2$, where M = Cu, Fe, Mn, UO_2, Ca, or Pb, and X = P, As, or V (see Table 3.4). It would appear reasonable for sulfates to be members of the same class, but the analytical data available at the time of this writing do not allow this classification. For convenience (and because of their importance) the phosphates, arsenates, and vanadates are listed separately in Tables 3.12, 3.13, and 3.14, respectively.

The optical and morphologic crystallographic properties of a number of phosphates and arsenates present a peculiar problem. Torbernite, zeunerite, and artificial uranospinite are optically uniaxial. Autunite, uranocircite, and natural uranospinite are biaxial with orthorhombic symmetry. Trögerite appears to be monoclinic according to its optical properties but tetragonal according to its morphology and structural crystallography. It has been conjectured that all these minerals are tetragonal, despite the anomalies that result from such an assumption (Goldschmidt, 1899). This view is reinforced by the

Table 3.12—Uranium Phosphate Minerals

Mineral	Composition	Uranium, %	Optical and crystallographic properties	Occurrence	Doelter* (1929)
Phosphor- uranylite	$(UO_2)_3(PO_4)_2 \cdot nH_2O$ (n usually 6–12)	60–64	Deep lemon yellow; probably ortho- rhombic; optically negative $n_\alpha = 1.691$ $n_\beta = 1.720$ $n_\gamma = 1.720$ pleochroic	Mitchell County, N. C.	III(1), 573
Autunite	$Ca(UO_2)_2(PO_4)_2 \cdot nH_2O$ (n usually 8–12)	~50	Small green to yel- low crystals; ortho- rhombic, a:b:c = 0.9876:1:2.8530, may actually be tetragonal; opti- cally negative; weak double refrac- tion $n_\alpha = 1.557$ $n_\beta = 1.575$ $n_\gamma = 1.577$	Autun, France Madagascar Falkenstein, Saxony Connecticut Massachusetts Pennsylvania North Carolina South Carolina Utah China Cornwall, England	III(1), 573
Bassetite	$Fe(UO_2)_2(PO_4)_2 \cdot nH_2O$	~50	Yellow monoclinic crystals; a:b:c = 0.3473:1.0:0.3456 $\beta = 89°17'$; opti- cally biaxial, 2E = 110° $n_\beta = 1.574$ $n_\gamma = 1.580$ pleochroic	Cornwall, England	IV(2), 959
Uranocircite	$Ba(UO_2)_2(PO_4)_2 \cdot 8H_2O$	~50	Orthorhombic; yel- low-green, resem- bling autunite; doubly refracting; optically negative $n_\alpha = 1.61$ $n_\beta = 1.623$ $n_\gamma = 1.623$ pleochroic	Schneeberg, Saxony	III(1), 575
Saléeite	$Mg(UO_2)_2(PO_4)_2 \cdot 8H_2O$	~60	Yellow; monoclinic with pseudotetra- gonal symmetry; optically negative $n_\alpha = 1.559$ $n_\beta = 1.570$ $n_\gamma = 1.574$	Shinkolobwe, Katanga, Belgian Congo	
Fritzcheite	$Mn(UO_2)_2(PO_4)_2 \cdot 8H_2O$; also contains significant amounts of vanadium		Reddish-brown quad- ratic tables; resembles autunite; apparently tetra- gonal-prismatic	Erz Gebirge, Saxony Autun, France Neudeck, Czechoslovakia	III(1), 575
Torbernite	$Cu(UO_2)_2(PO_4)_2 \cdot nH_2O$ (n usually 8–12)	~50	Tetragonal, a:c = 1: 2.97; green; weakly doubly refracting; optically uniaxial;	Found frequently with autunite Johanngeorgen- stadt, Saxony	III(1), 576

Table 3.12 — (Continued)

Mineral	Composition	Uranium, %	Optical and crystallographic properties	Occurrence	Doelter* (1929)
			negative $n_\omega = 1.63$ $n_\epsilon = 1.57$ pleochroic	Schneeberg, Saxony Cornwall, England Vielsalm, Belgium Jáchymov, Zinnwald, Czechoslovakia Portugal	
Metatorbernite I	$Cu(UO_2)_2(PO_4)_2\cdot 8H_2O$ (see Sec. 2.5d)	30−50	Light green to apple green; tetragonal, a:c = 1:2.28; optically positively uniaxial; pleochroic; for Na light $n_\omega = 1.623$ $n_\epsilon = 1.625$	Gunnislake and Tincroft, Cornwall, England Katanga, Belgian Congo Temple Mountain, Utah Spain	IV(2), 960
Uranospathite	Composition unknown; no analyses; perhaps a hydrated uranium phosphate		Rhombic, pseudotetragonal; yellow to green; biaxial $n_\beta = 1.510$ $n_\gamma = 1.521$ pleochroic	Redruth, Cornwall, England	IV(2), 959
Parsonsite	$2PbO\cdot UO_3\cdot P_2O_5\cdot H_2O$	~25	Monoclinic or triclinic; light brown; weak double refraction $n_\alpha = 1.85$ $n_\gamma = 1.862$	Kasolo, Katanga, Belgian Congo	IV(2), 962
Dewindtite (stasite)	$3PbO\cdot 5UO_3\cdot 2P_2O_5\cdot 12H_2O$	~50	Canary yellow; rhombic; doubly refracting; optically positive $n_\alpha = 1.762$ $n_\beta = 1.763$ $n_\gamma = 1.765$	Kasolo, Shinkolobwe, Katanga, Belgian Congo	IV(2), 964
Dumontite	$2PbO\cdot 3UO_3\cdot P_2O_5\cdot 5H_2O$	45	Ocher yellow in large crystals; prismatic; b:c = 1:1.327; strong pleochroism; doubly refracting; optically positive $n_\alpha = 1.88$ $n_\beta = 1.89$	Closely associated with torbernite at Shinkolobwe, Katanga, Belgian Congo	IV(2), 963
Renardite†	$PbO\cdot 4UO_3\cdot P_2O_5\cdot 9H_2O$	13	Rhombic; optically negative; indices of refraction $\alpha = 1.715 \pm 0.003$ $\beta = 1.736 \pm 0.003$ $\gamma = 1.739 \pm 0.003$	Kasolo, Katanga, Belgian Congo	

*The roman numeral refers to the volume, the arabic numeral in parentheses to the part, and the other arabic numeral to the page.

†Bull. soc. franc. minéral., 51: 247 (1928).

Table 3.13 — Uranium Arsenate Minerals

Mineral	Composition	Uranium, %	Optical and crystallographic properties	Occurrence	Doelter* (1929)
Trögerite	$(UO_2)_3(AsO_4)_2\cdot 12H_2O$	56.2	Golden yellow; tetragonal; a:c = 1:2.16, or monoclinic, a:b:c = 0.463:1:0.463 (see Sec. 2.5d); optically negative; biaxial $n_\alpha = 1.585$ $n_\beta = 1.630$ $n_\gamma = 1.630$ other samples are uniaxial $n_\omega = 1.624$ $n_\epsilon = 1.580$	Schneeberg, Saxony Bald Mountain, Black Hills, S. D.,	III(1), 730
Uranospinite	$Ca(UO_2)_2(AsO_4)_2\cdot nH_2O$ (n is usually $8-12$); isomorphous with autunite	50	Rhombic or tetragonal; yellow to apple green; biaxial; negative; optical anomalies (see Sec. 2.5d)	Schneeberg, Saxony Utah	III(1), 730
Zeunerite	$Cu(UO_2)_2(AsO_4)_2\cdot nH_2O$ (n is usually $8-12$); isomorphous with torbernite	$50-53$	Grass-green tabular tetragonal crystals; uniaxial; negative $n_\omega = 1.643-1.635$ $n_\epsilon = 1.623-1.615$ pleochroic	Schneeberg and Erz Gebirge, Saxony Cap Garonne, France Cornwall, England	III(1), 731
Walpurgite	$Bi_{10}(UO_2)_3(OH)_{24}(AsO_4)_4$	16.5	Orange, honey, and straw yellow; triclinic, pinacoidal, or monoclinic; biaxial, negative; very high refractivity $n_\alpha = 1.90$ $n_\beta = 2.00$ $n_\gamma = 2.05$	Schneeberg, Saxony Portugal	III(1), 729

*The roman numeral refers to the volume, the arabic numeral in parentheses to the part, and the other arabic numeral to the page.

fact that uniaxial specimens of trögerite and uranospinite have occasionally been found in nature. The anomalous optical properties of trögerite have been explained by Goldschmidt as having resulted from distortion of the crystal lattice of this mineral by zeunerite and from distortion of the crystal edges and faces by small flakes produced by displacement of the cleavage lamellae. The problem would benefit by a renewed investigation.

Few chemical studies have been made on the minerals of this class. Solubilities have been determined for most of them, and some studies on the dehydration of the hydrated minerals have been made. A study of the alteration of autunite and zeunerite has also been made (Starik,

Table 3.14 — Uranium Vanadate Minerals

Mineral	Composition	Uranium, %	Optical and crystallographic properties	Occurrence	Doelter* (1929)
Ferghanite	$UO_3 \cdot V_2O_5 \cdot 6H_2O$	65	Apparently rhombic; a:b = 0.75:1; sulfur-yellow; uniaxial; weak double refraction and low refractive index; practically no pleochroism	Ferghana, Turkestan	III(1), 849
Tyuyamunite (randite)	$CaO \cdot 2UO_3 \cdot V_2O_5 \cdot nH_2O$ (n usually 9–10 but down to 4)	50–60	Rhombic; a:b = 0.77:1; canary-yellow to green; biaxial; optically negative; very high index of refraction and double refraction $n_\alpha = 1.67$ $n_\beta = 1.87$ $n_\gamma = 1.895$ pleochroic	Ferghana, Turkestan Paradox Valley, Colorado Henry Mts., Utah	III(1), 848
Rauvite	$CaO \cdot 2UO_3 \cdot 6V_2O_5 \cdot 2OH_2O$	17	Reddish purple; anisotropic; index of refraction ~1.88	Green River, Utah	IV(2), 968
Carnotite	$K_2O \cdot 2UO_3 \cdot V_2O_5 \cdot nH_2O$ (n usually 1–3)	~50	Rhombic; a:b = 0.81:1; yellow; birefringent; optically negative $n_\alpha = 2.06$ $n_\beta = 2.06$ $n_\gamma = 2.06–2.08$ pleochroic	Colorado Utah S. Australia (as a powdery impregnation in sandstone) Pennsylvania Katanga, Belgian Congo	III(1), 844
Uvanite	$2UO_3 \cdot 3V_2O_5 \cdot 15H_2O$	33	Brownish yellow; rhombic; biaxial $n_\alpha = 1.817$ $n_\beta = 1.879$ $n_\gamma = 2.057$ pleochroic	Temple Mt., Utah	IV(2), 967

*The roman numeral refers to the volume, the arabic numeral in parentheses to the part, and the other arabic numeral to the page.

1941). The relations encountered in torbernite deserve special mention (Rinne, 1901).

$$\text{Torbernite} \underset{}{\overset{60\text{-}65^\circ C}{\rightleftharpoons}} \text{Metatorbernite I} \underset{}{\overset{100^\circ C}{\rightleftharpoons}} \text{Metatorbernite II}$$

$Cu(UO_2)_2(PO_4)_2 \cdot 12H_2O$	$Cu(UO_2)_2(PO_4)_2 \cdot 8H_2O$	$Cu(UO_2)_2(PO_4)_2 \cdot 4H_2O$
Tetragonal, uniaxial, optically negative	Tetragonal, optically positive	Rhombic, biaxial, optically negative

This transformation is, as might be expected, also a function of pressure (Hallimond, 1916). The minerals that contain calcium in place of copper (autunites) behave in a similar fashion.

(e) Carbonaceous Uranium Minerals. A number of occurrences of uranium-rich carbonaceous materials are known. Kolm is a coallike substance of high ash content which occurs as disk-shaped lenses (in an alum shale) (Wickman, 1942; Wells and Stevens, 1931) in Cambrian

Table 3.15 — An Alphabetical Reference List of Tabulated Uranium Minerals

Mineral	Table	Mineral	Table	Mineral	Table
Ampangabeite	3.7	Ishikawaite	3.7	Shinkolobwite	3.9
Autunite	3.12	Johannite	3.11	Sklodowskite	3.9
Bassetite	3.12	Kasolite	3.9	Soddyite	3.9
Becquerelite	3.8	Khlopinite	3.7	Stasite	3.12
Betafite	3.7	Kolm	Sec. 2.5e	Thorianite	3.5
Blomstrandine	3.7	Lambertite	3.9	Torbernite	3.12
Blomstrandite	3.7	Liebigite	3.10	Trögerite	3.13
Brannerite	3.7	Lyndochite	3.7	Tyuyamunite	3.14
Bröggerite	3.5,	Mackintoshite	3.9	Uraconite	3.11
	3.6	Maitlandite	3.9	Uraninite	3.5,
Carnotite	3.14	Mendeleyevite	3.7		3.6
Clarkeite	3.8	Metatorbernite	3.12	Uranochalcite	3.11
Cleveite	3.5,	Microlite	3.7	Uranocircite	3.12
	3.6	Nicolayite	3.10	Uranolepidite	3.9
Curite	3.8	Nivenite	3.5,	Uranophane	3.9
Dakeite	3.10		3.6	Uranopilite	3.11
Delorenzite	3.7	Parsonsite	3.12	Uranospathite	3.12
Dewindtite	3.12	Phosphoruranylite	3.12	Uranosphaerite	3.9
Djalamaite	3.7	Pilbarite	3.9	Uranospinite	3.13
Dumontite	3.12	Pitchblende	Sec. 2.4	Uranothallite	3.10
Ellsworthite	3.7	Priorite	3.7	Uranotile	3.9
Eschwegite	3.7	Pyrochlore	3.7	Uvanite	3.14
Eschynite	3.7	Randite	3.14	Vandenbrandite	3.9
Euxenite-polycrase	3.7	Rauvite	3.14	Voglianite	3.11
Ferghanite	3.14	Renardite	3.12	Voglite	3.10
Fergusonite	3.7	Rutherfordine	3.10	Walpurgite	3.13
Fourmarierite	3.8	Rutherfordite	3.7	Wiikite	3.7
Fritzcheite	3.12	Saléeite	3.12	Yttrocrasite	3.7
Gilpinite	3.11	Samarskite	3.7	Yttrotantalite	3.7
Gummite	3.8	Samiresite	3.7	Zeunerite	3.13
Hatchettolite	3.7	Schoepite	3.8	Zippeite	3.11
Hjelmite	3.7	Schroeckingerite	3.10	Zirkelite	3.7
Ianthinite	3.8	Sharpite	3.10		

slates of the Västergotland plains in Sweden. On combustion it leaves a residue of about 27 per cent ash, which has a uranium content up to 2.87 per cent (average 1.87 per cent U_3O_8) and has been used as a commercial source of radium. Numerous asphalts are also known which contain small amounts of uranium (Longobardi and coworkers, 1947; Lexow and Maneschi, 1948; Fohs, 1948). Thucholite is a substance of unknown chemical structure which contains C, H, O, H_2O, Th, U, Ce, Y, Er, V, P, Ca, Si, and many other common elements. It is found in Canada and Sweden (Grip and Ödman, 1944). Despite the very wide distribution of the large number of minerals described in this section, only carnotite has been of considerable

Table 3.16 — Minerals Reported to Contain
Some Uranium but Not Discussed
in the Text*

Allanite	Rowlandite
Corvusite	Tscheffkinite
Cyrtolite	Xenotime
Gadolinite	Yttrialite
Monazite†	Zircon
Naegite	

*Many other rare earth minerals un-doubtedly contain small amounts of uranium.

†Since monazite is being used for thorium extraction, by-product uranium from this source (and from xenotime) may become important.

economic importance in the past. This mineral is discussed further in Sec. 3.3.

2.6 Fluorescence of Uranium Minerals. (Meixner, 1939, 1940a,b; Haberlandt, 1935.) It has been known for a long time that many uranium minerals fluoresce. With rare exceptions, strong luminescence is observed only with secondary uranium minerals. Few uranium minerals of pegmatitic origin show luminescence. The strongly luminescent uranium minerals are phosphates, sulfates, and arsenates, which show an intense and characteristic yellow-green color under ultraviolet light. These minerals include autunite, uranospinite, uranocircite, and uranopilite. The uranium carbonates and uranothallite exhibit an intense green fluorescence; schroeckingerite fluoresces bright yellow-green. Weakly luminous with a yellowish color are carnotite, dewindtite, soddyite, becquerelite, and some specimens of zippeite, uranotile, and gummite. The uranium silicates and the uranium micas containing Cu, Pb, Bi, Mn, and Fe are nonluminescent; this is true of zeunerite, fritzcheite, bassetite, trögerite, torbernite, phosphoruranylite, tyuyamunite, walpurgite, cuprosklodowskite, ianthinite, kasolite, curite, fourmarierite, johannite (gilpinite), uranosphaerite, betafite, hatchettolite, and ellsworthite. No definite relation between fluorescence and chemical structure can be established at the time of this writing.* It has been observed that spraying non-luminous uranium compounds with acids, such as sulfuric, nitric, hydrochloric, acetic, or phosphoric acids, causes them to luminesce under ultraviolet light. Of these acids, phosphoric acid has only a weak effect, and acetic acid produces the most intense fluorescence (Melkov and Sverdlov, 1941).

These observations are of practical importance in prospecting. In addition to fluorescence techniques, methods depending on the detection of gamma rays also are of increasing importance in prospecting for uranium deposits. For examples see Dreblow (1942), Sill and Peterson (1945, 1947), Northup (1945), De Ment (1946), Haberlandt and Hernegger (1946), Przibram (1946), and Gibb and Evans (1947).

3. ECONOMIC MINERAL DEPOSITS OF URANIUM

Before 1942 no mining operations were conducted solely for the sake of obtaining uranium. Uranium ores were mined primarily for their radium content. Appreciable quantities of uranium were obtained incidentally to vanadium production from carnotites. No large market existed for uranium compounds; the economy of mining any particular ore was determined exclusively by the market value of its radium or vanadium content. With the discovery of nuclear fission and its technical applications, uranium became a material of tremendous importance. The economic criteria formerly applied became irrelevant, and occurrences of uranium that were not exploited in the past acquired a new significance. It is not possible to describe here the mining operations in effect at the time of this writing or to give the quantity of ore produced, the estimated reserves, or the results of the very intensive prospecting that has been carried out since 1940.

Although uranium is not an excessively rare component of the earth's crust, relatively few uranium deposits were found worth working in the past. Two rich deposits are those at Great Bear Lake in Canada and at Katanga in the Belgian Congo. There also are deposits of importance in Czechoslovakia and in the United States.

3.1 <u>Great Bear Lake Pitchblende Deposits, Northwest Territories, Canada.</u> (Kidd, 1932, 1936; Krusch, 1937a; Lemaire, 1946a; Déribéré, 1947.) Great Bear Lake is exactly on the Arctic Circle, east of the Mackenzie River, 1,380 miles by boat or 800 miles by air from the railhead at Waterways, Alberta. The deposits were discovered in 1930 at Echo Bay on the east side of the lake.

---

*A survey of fluorescent uranium minerals given by De Ment (1945) differs in some details from the data of Meixner quoted above.

The uranium occurs as pitchblende in a highly mineralized area. The ore is found in replacement lodes and stockworks along fracture and shear zones that traverse pre-Cambrian sedimentary rocks and altered volcanic rocks cut by granodiorite. Several zones are known, ranging from 30 to 800 ft wide. The process of mineralization was apparently very complex; some 40 minerals have been recognized. The principal minerals of the area are pitchblende, native silver, pyrite, and chalcopyrite. Compounds of iron, cobalt, nickel, copper, lead, zinc, silver, molybdenum, bismuth, and manganese are also found. Kidd and Haycock (1935) consider this deposit to have been formed from hydrothermal solutions released by a congealing granite magma. The mineralization was in several stages. The pitchblende appears to have been deposited in the earliest phase of the hydrothermal process, followed by Co-Ni, Pb-Zn-Cu, and finally by Cu-Ag minerals. The pitchblende may represent colloidal deposition in cavities. A similar sequence is postulated by Furnival (1939) for the Contact Lake deposits in the Great Bear Lake area. The pitchblende content of the ore shipped from the Eldorado mines at Great Bear Lake varies from 30 to 62 per cent U_3O_8 (Parmelee, 1938).

The only deposit other than the Great Bear Lake that has been worked extensively in Canada is at Wilberforce Station, Cardiff Township, in Haliburton County, Ontario. The mineral is a uraninite which occurs in pegmatites, and the ore has an average uranium content of only about 0.1 per cent (Ellsworth, 1932a). Numerous small pegmatite deposits are described by Ellsworth.

3.2 Belgian Congo Deposits. (Hess, 1934; Legraye, 1944; Charrin, 1947.) Until the development of the Canadian Great Bear Lake deposits, the pitchblende deposits at Shinkolobwe, Katanga, in the Belgian Congo were the world's leading source of uranium and radium. The deposits were discovered in 1915. Systematic mining began in 1921. The deposit is in isolated silicified breccia 15 miles south of Kambove on the divide between the drainage basins of the Mura and Panda rivers. The rocks form a ridge 40 to 45 ft high and 250 ft long. The uranium is in some cases very close to the surface; solid masses of pitchblende have been found under only a few centimeters of topsoil. Presumably the fresh-vein material is covered by thin transported soil.

Since the entire region is heavily mineralized, a magmatic source for the solutions from which the deposits formed cannot be doubted. The host rock containing the ore deposits resembles that observed in numerous copper deposits in Katanga. It consists of altered carbonate rocks of the "Série des Mines," which is surrounded by comparatively undisturbed argillaceous and talcose schists (Kundelungu formation). In addition to uranium there are found at Shinkolobwe copper, cobalt, nickel, vanadium, iron, and precious metals. Molybdenum (as MoS$_2$ and wulfenite) has also been observed. None of these elements is found in the Kundelungu formation itself. The Série des Mines in which the uranium is found is about 200 meters (656 ft) thick and consists mainly of dolomitic limestone. The pitchblende-bearing veins tend to parallel the main faults that cut the dolomite although they are not in the faults. The veins are capricious and swell in short distances from a few centimeters in width to a meter and yield masses of compact pitchblende weighing several tons. The uranium minerals are free of visible gangue materials. To minimize losses, the entire mass of rock is mined, and the ore is hand-sorted. The veins are so irregular that it is difficult to estimate the reserves.

Besides the veins, uranium also occurs disseminated in the wall rocks adjacent to them. In the wall rock it is present in various minerals derived from the complete alteration of pitchblende. Other disseminations made up principally of torbernite have no apparent relation to the veins. More than half the oxidized uranium is in the form of torbernite, but numerous other oxidation products of pitchblende also have been observed here (Buttgenbach, 1935). Kasolite and sklodowskite occur as disseminated minerals, with the sklodowskite lining cavities in the rock similarly to the torbernite. Uraninite is also present (Hitchen, 1934). Small quantities of copper, cobalt, and nickel sulfides, which are present in considerable quantities in other parts of the Série des Mines, also are found in the uranium veins.

The mineralization sequence is complex. According to Thoreau and de Terdonck (in Hess, 1934), the ores were formed by siliceous solutions entering carbonate rocks and producing a series of silicate minerals; pitchblende was deposited next, and this was followed by a succession of sulfide minerals. Later processes altered the ores and yielded many secondary minerals. The deposition of pitchblende is thus considered an early high-temperature phase of the copper mineralization. This mineralization sequence appears to be common in this region. Pitchblende or its alteration products have been found in seven deposits from Ruashi on the southeast to the extreme northwest of the copper zone, passing through Lurshia and Kambove (Vandendriessche, 1935).

3.3 Colorado and Utah Deposits. (Hess, 1933; Krusch, 1937c; Stokes, 1944.) The carnotite deposits in Colorado and Utah are the only deposits in the United States from which uranium has hitherto been obtained commercially. There are uraninite occurrences in North Carolina, Connecticut, New Hampshire (Shaub, 1937), the Black Hills of South Dakota, at Baringer Hill in Texas, and various other places; but the limited size of these deposits seems to have precluded commercial development in the past.

Carnotite deposits are found from Coal Creek near Meeker, Colo., to Carizo Mountain on the line between New Mexico and Arizona, and from Huerfano County, Colorado, westward to Silver Reef in the southwestern corner of Utah, but the principal deposits are distributed over an area 130 miles long and 50 miles wide in southwestern Colorado and southeastern Utah. The deposits are spotty in distribution but contain in the aggregate large quantities of uranium and vanadium minerals. The vanadium is much in excess of the uranium. The ore occurs in seams and pockets, frequently as an incrustation or impregnation in sandstone. The greatest deposits are reputed to be in the area between Paradox Valley and the San Miguel River. Here the main ore-bearing sandstone forms lenses about 100 ft thick and is only part of a very much thicker formation (Morrison) of shale and sandstone.

The uranium minerals found in this general region are principally carnotite and tyuyamunite, but other complex vanadium-uranium minerals (uvanite, rauvite, zeunerite) also are found. The deposits appear to be of sedimentary origin (Fischer, 1937). Much of the ore is associated with fossil logs, plant remains, and saurian bones, or with asphalt. According to Hess, the process of formation of the Colorado Plateau deposits may be reconstructed as follows: It can be assumed that the uranium and vanadium in this region originally existed in veins that also contained pyrite. Oxidation resulted in the formation of sulfuric acid and ferric sulfate, which then dissolved the uranium and vanadium as sulfates. The nature of the Morrison sandstones, which contain most of the deposits, indicates an origin in shallow waters with shifting islands, spits, and shores. The sulfate solutions of uranium and vanadium entered these shallow waters, where abundant algal vegetation was macerated by wind and waves. Logs were swept into the shallows by rivers and stranded there. Around them were packed sand and macerated vegetation. The organic matter is thought to have concentrated the uranium and vanadium from the very dilute solutions either by reduction or by ion displacement. In time, after organic remains became petrified, replacement of organic material by calcite and then by uranium-vanadium minerals may also have occurred. The importance of the vegetable matter in the formation of carnotites is indicated by the exceedingly high concentration of uranium and vanadium in some petrified logs found near the San Miguel River. For example, one petrified log 100 ft long and 4 ft in diameter contained roughly 105 tons of carnotite. Fischer (1937) suggests the possibility that microorganisms may also have played a role in the concentration of the uranium and vanadium from dilute solutions (see also Frederickson, 1948).

Richer ore selectively mined in this region for its U_3O_8 content has averaged from 1.25 to 1.5 per cent U_3O_8 and about 3.5 per cent V_2O_5. The composition of typical carnotite ore is given in Table 3.17. Undoubtedly, considerable quantities of ore that was formerly thought to be too low grade to justify mining exist in this region.

Table 3.17 — Composition of Carnotite Ore

Component	Range,%
V_2O_5	2.25 − 10.5
U_3O_8	0.25 − 3.0
SiO_2	70 − 80
$Fe_2O_3 + Al_2O_3$	5 − 12
$CaCO_3 + CaSO_4$	Trace to 3
$BaCO_3 + BaSO_4$	Trace to 2
$CuCO_3 + CuS$	Trace to 2
Pb	~1
Na, K, Mg, Ra, etc.	Small amounts
H_2O	5 − 25

The carnotite deposits are a geological rarity. On the other hand, the largest known vanadium deposits in the world (at Minasragra, Peru) contain no uranium despite the presence of organic matter such as asphalt. The Ferghana deposit of carnotite and tyuyamunite in Turkestan appears to be similar to the American carnotite deposits (Chirvinsky, 1925; Fersman, 1930).

3.4 Uranium Deposits of Minor Significance. In addition to the three major deposits discussed above, there are many localities scattered all over the earth where uranium has been obtained at one time or another. These are briefly enumerated below.

(a) United States of America. Prior to the development of the carnotite deposits, small amounts of uranium were mined in Gilpin County, Colorado, where pitchblende is found associated with pyrite and zinc and lead sulfides (Bastin, 1915; Alsdorf, 1916).

(b) Czechoslovakia and Germany. (Moore and Kittril, 1913a,b; Krusch, 1937d.) The Erz Gebirge is a mountainous area along the boundary between Saxony in Germany and Bohemia in Czechoslovakia. In this area pitchblende and many other uranium minerals have been found at Schneeberg, Annaberg, and Johanngeorgenstadt in Germany and particularly at Jáchymov in Czechoslovakia. The deposits appear to be similar in nature and origin in all these places. The most important area is that at Jáchymov; it furnished the uranium from which radium was first extracted by M. and Mme. Curie. The geology of this region has been much studied.

Silver and cobalt have been mined at Schneeberg in Saxony since the end of the fifteenth century. Here pitchblende is associated chiefly with native bismuth and cobalt and nickel minerals. The minerals are found in the mountains near the towns of Schneeberg, Annaberg, and Johanngeorgenstadt in a network of veins in altered slate cut by granite masses. The gangue minerals are calcite, ankerite, barite, fluorite, and quartz. The mineralization of this region is very complex (Kohl, 1941).

At Jáchymov the ore deposits are in a mica schist interbedded with lime schist and crystalline limestone, with the whole intruded by large masses of late Paleozoic granite and cut by numerous dikes of quartz porphyry and basalt (Schneiderhöhn, 1938, 1939). The mineral veins of hydrothermal origin cut the quartz porphyry dikes and are themselves cut by the basalt. These mineral veins are usually 6 in. to 2 ft thick. The character of the mineralization varies widely. Silver, nickel, bismuth, and arsenic minerals are present together with copper, lead, zinc, cobalt, and iron. Deposition occurred in three stages: first cobalt and nickel, then uranium, and lastly silver.

Other small uranium deposits in Germany are described by Kohl (1933).

(c) Portugal. (Moore and Kittril, 1913b; Krusch, 1937e; Bernardo Ferreira and Cotelo Neiva, 1945.) The uranium-bearing zone lies in the area of massive granite that occupies nearly the entire northern part of Portugal between the desert of Gallice and Castello Branco and reaches into the provinces of Minho, Trás-os-Montes, and Beira. The richest parts of the district are between the towns of Guarda and Sabugal and in the region of Villar-Formosa. The veins near Guarda are especially rich in wolframite.

The uranium minerals occur in narrow pegmatite dikes 0.5 to 1 meter in width. In the Rosmaneira region, tungsten and tin in addition to uranium are present in commercially valuable amounts. Autunite, $Ca(UO_2)_2(PO_4)_2 \cdot 12H_2O$, seems to be the oldest uranium mineral there, accompanied by numerous alteration products (Lepierre, 1933). The autunite is present in small groups of square tablets of an intense yellow color, in small plates, as a yellow coating on the rock, or still oftener as bright yellow specks disseminated throughout the dull yellowish rock. In the clay parts of the veins the uranium mineral often is invisible and can be detected only by its radioactivity. There are also blotches on the surface of the granite that give it an intense yellow color, although the actual uranium content of the granite is very low. The uranium content of the veins varies within very wide limits. Ore containing 2 per cent U_3O_8 is considered excellent, and that carrying 1 per cent is good average ore (Dörpinhaus, 1914; see also Ebler and Bender, 1915).

(d) England. (Dines, 1930; Ellsworth, 1932b.) The mineral deposits in Cornwall and South Devon have been mined from antiquity for tin and copper. The main occurrences of uranium are at Wheal Trenwith, St. Ives, and the South Terras Mine, St. Stephen. This last deposit is the only known vein in Cornwall in which the chief minerals are the uranium compounds. This vein contained enough uranium to warrant mining. The South Terras vein lies in slate intruded by granite. The adjacent veins are mainly tin- and copper-bearing, and others carry such minerals as ores of cobalt, nickel, lead, uranium, and iron and occasionally some arsenic and chalcopyrite. It is likely that the veins high in uranium, cobalt, and nickel are not contemporary with those carrying iron.

(e) Union of Soviet Socialist Republics. (Krusch, 1937f.) Numerous small deposits of uranium have been reported in Russia. Uraninite has been found in northern Karelia, near Zhitomir, at Ekaterinoslav (Dnepropetrovsk) in the Ukraine, and in a number of other localities (for references see Gmelin, 1936c). These are in all probability small deposits which have never been exploited. The most important known deposit of uranium in Russia is in the province of Ferghana in Russian Turkestan (the region now included in the Kirghiz Soviet Republic). The principal mineral in this deposit is tyuyamunite, which is closely related to carnotite. These deposits are reputed to resemble those of Utah and Colorado but are in limestone instead of sandstone (Melkov, 1945).

Although intensive prospecting throughout the U.S.S.R. has been carried out, little has been reported in the literature concerning the results of this search. A deposit of uranium-vanadium ore has been discovered at Taboskov 69 km north of Andizhan in Kirghizia (about 100 miles northeast of Ferghana). The vein is about 1 meter thick and 250 meters long, and it averages about 3 per cent U_3O_8.[*] Large deposits of uranium are also reputed to have been uncovered in Trans-caucasia. The ore resembles carnotite and averages 3 per cent U_3O_8. The Caucasian deposit is said to be larger than those at Tjujamujan and Taboskov.[†] The opinion has been expressed (Tyurin, 1944) that no uranium will be found in the Karatau vanadium deposits.

---

[*] "Economic Review of the Soviet Union," vol. IX, p. 276, Amtorg Trading Corporation, New York, 1934.

[†] U. S. Bur. Foreign and Domestic Com., "Russian Economy," p. 13, Sept. 15, 1934, New York (1935).

The Radium Institute of the Academy of Sciences of the U.S.S.R. has been very active, and Russia is reputed to possess a considerable radium industry about which very few details are available in this country (Monsavoff, 1943).

(f) Bulgaria. There are two known deposits of uranium in Goten, Bulgaria. The ores consist of torbernite and metatorbernite I and II, which must have originated through hydrothermal action (Konjarov, 1938). Extensive deposits of autunite have been reported near the village of Streltsch.*

(g) Norway and Sweden. The pegmatite deposits in Norway north of Christiansund and Evji have been known and studied for a long time.

In Sweden† the state-owned oil-shale plant at Kvantorp in the province of Narke utilizes kolm-carrying slates that yield a residue containing uranium and other metals (Krusch, 1937g).

(h) Madagascar. (Krusch, 1937h.) Madagascar is one of the few places where pegmatitic uranium minerals have been mined in the past. The majority of these occur in the older rock complex in the middle of the island. The uranium minerals found there are euxenite-polycrase, betafite, ampangabeite, fergusonite, samarskite, and blom-strandite. Those of economic interest occur mainly in potassium-rich and only occasionally in sodium-rich pegmatites. Autunite and uranocircite are found in an ancient lake bed (Lake Antsirabe) that presumably received the drainage from weathering pegmatites that contained betafite and other complex uranium-columbium-tantalum minerals.

(i) Africa (Other than the Belgian Congo). (Krusch, 1937b.) Pitch-blende has been found at Ulugura, Morogoro District, East Africa, and in the Loldaiga region, Nangaki District, British East Africa. Deposits have also been reported near Messina, North Transvaal, and in the Gordonia District, Cape Province.

(j) Australia.‡ (Krusch, 1937i; Mawson, 1944; Carroll and Rowledge, 1945.) Uranium deposits have been reported from Australia. One is at Radium Hill (Mt. Painter) near Olary in South Australia, and the other is in the Pilbara gold field in Western Australia. At Radium Hill (20 miles southeast of Olary and 275 miles northeast of Adelaide) are found torbernite, autunite, uranophane, gummite, and

---

*U. S. Bur. Mines, Minerals Yearbook, 1935.

†Ibid., 1943.

‡U. S. Dept. Commerce, Bur. Mines, "Mineral Resources of the U. S.," U. S. Government Printing Office, Washington, D. C., 1925, p. 617; 1926, pp. 267-268; 1927, pp. 441-442; 1928, p. 136; 1929, pp. 108-109; 1931, p. 188.

U. S. Bur. Mines, Minerals Yearbook, 1932-1933, pp. 188, 333; 1934, p. 506; 1935, p. 558.

fergusonite. These appear to have originated from uranium-titanium-tantalum-columbium minerals in nearby pegmatites.

Pilbarite, a complex hydrated lead-uranium-thorium silicate containing small amounts of cerium and yttrium (Simpson, 1910), has been mined at Pilbara in West Australia.

(k) <u>Mexico</u>. (Krieger, 1932; Krusch, 1937j.) The deposits are said to be mainly 120 miles northeast of Guadalupe in the state of Chihuahua. Gold is found with the uranium. The vein varies between a few centimeters and 5 meters in width (González Reyna, 1946).

Table 3.18 — Uranium Ore and Compounds Imported for Consumption in the United States, 1939 to 1943

Year	Uranium ores		Uranium oxide and salts	
	Weight, lb	Value, dollars	Weight, lb	Value, dollars
1939	5	10	1,439,324	1,197,786
1940	2,400,198	2,110,927	240,199	388,355
1941	...	...	387,505	501,370
1942	541,307	806,919	377,398	851,098
1943	...	...	211,348	431,410

(l) <u>Brazil</u>. Recently, uranium has been found at Proina, Minas Geraes. Ore containing 13 to 18 per cent U_3O_8 after concentration has been obtained from pegmatites exploited for mica, tourmaline, and beryl. The ore at Uba, Minas Geraes, can be concentrated to 75 per cent samarskite, 15 per cent monazite, and 10 per cent columbite (de Araujo, 1945). Euxenite was found at Pomba, Minas Geraes, as long ago as 1911 but has never been mined (Krusch, 1937k). A survey of Brazilian uranium resources has been made fairly recently (Leonardos, 1936). The bulk of the Brazilian uranium occurrences are in the state of Minas Geraes (Hess and Henderson, 1925; Silva Pinto, 1947).

(m) <u>France</u>. A discussion of French deposits is given by Lemaire and Charrin (1946b).

3.5 <u>Statistics</u>. At this time there is little point in attempting to discuss the statistics of uranium production and price. A very rough idea of the economic history of uranium can be gained from the successive volumes of the Minerals Yearbook. By way of illustration, data from recent editions of the Minerals Yearbook are given in Table 3.18.

REFERENCES

1889   W. F. Hillebrand, Am. J. Sci., (3) 38: 329.

1899   V. M. Goldschmidt, Z. Kryst. Mineral., 31: 468.

1901   F. Rinne, Centr. Mineral. Geol., 1901: 618.

1909   J. Joly, "Radioactivity and Geology," Chap. VII, Constable & Co., Ltd., London.

1910   E. S. Simpson, Chem. News, 102: 283.

1913a R. B. Moore and K. L. Kittril, U. S. Bur. Mines Bull. 70, p. 47.

1913b R. B. Moore and K. L. Kittril, U. S. Bur. Mines Bull. 70, p. 49.

1914   W. T. Dörpinhaus, Met. Engr., 11: 302.

1915   E. S. Bastin, Econ. Geol., 10: 262.

1915   E. Ebler and W. Bender, Z. angew. Chem., 28: 30.

1916   P. C. Alsdorf, Econ. Geol., 11: 266.

1916   A. F. Hallimond, Mineralog. Mag., 17: 326.

1917   T. T. Quirke and L. Finkelstein, Am. J. Sci., (4) 44: 237.

1923a V. M. Goldschmidt, "Geochemische Verteilungsgesetze der Elemente," No. 3, J. Dybwad, Oslo.

1923b V. M. Goldschmidt and L. Thomassen, Videnskapsselskapets-Skrifter I. Mat.-naturv. Klasse Kristiana, 2: 12.

1923   W. Steinkuhler, Bull. soc. chim. Belg., 32: 233.

1924   F. W. Clarke and H. S. Washington, U. S. Geol. Survey, Profess. Paper 127.

1924   A. C. Lane, "Report of the Committee on the Measurement of Geologic Time," Natl. Research Council, Washington, D. C.

1925   P. N. Chirvinsky, Mineralog. Mag., 20: 287-295.

1925   F. L. Hess and E. P. Henderson, J. Franklin Inst., 200: Z35.

1926a A. Holmes, Phil. Mag., December, p. 1225.

1926b A. Holmes, Geol. Mag., July.

1927   R. van Aubel, Compt. rend., 185: 586.

1928   G. Kirsch, "Geologie und Radioaktivität," pp. 35-42, Verlag Julius Springer, Leipzig.

1928   F. Paneth, Z. Elektrochem., 34: 645.

1929   C. Doelter and H. Leitmeier, "Handbuch der Mineralchemie," Bd. 4, 2 Tl., 909, Th. Steinkopff, Dresden-Leipzig.

1930   H. G. Dines, Mining Mag., 42: 213-217.

1930   A. E. Fersman, Abhandl. prakt. Geol., 19 (II): 1-52.

1930   G. Hevesy, E. Alexander, and K. Würstlin, Z. anorg. u. allgem. Chem., 194: 316.

1930   C. Hintze, "Handbuch der Mineralogie," Bd. 1, 3 Abt., 2 Hälfte, p. 4152; 4 Abt., 1 Hälfte, p. 971, Berlin and Leipzig.

1930   I. Noddack and W. Noddack, Naturwissenschaften, 18: 757.

1930   F. Paneth, Z. Elektrochem., 36: 727.

1930   P. Tyler, U. S. Bur. Mines Inform. Circ. 6312, p. 16.

1931   A. Holmes and A. Kovarik, The Age of the Earth, IV, Bull. Natl. Research Council, No. 80.

1931   F. Paneth, Naturwissenschaften, 19: 164.

1931   R. C. Wells and R. E. Stevens, J. Wash. Acad. Sci., 21: 409.

1932a H. V. Ellsworth, "Rare Element Minerals of Canada," Can. Dept. Mines Resources Geol. Survey, Econ. Geol. Ser. No. 11, Chap. X, pp. 136-257.

1932b H. V. Ellsworth, "Rare Element Minerals of Canada," Can. Dept. Mines Resources Geol. Survey, Econ. Geol. Ser. No. 11, Chap. X, p. 132.

1932   G. Hevesy, "Chemical Analysis by X-rays and Its Applications," Chap. XVI, McGraw-Hill Book Company, Inc., New York.

1932 D. F. Kidd, Can. Dept. Mines Resources Geol. Survey, Econ. Geol. Ser. No. 11.

1932 P. Krieger, Econ. Geol., 27: 651.

1933 F. L. Hess, Ore Deposits of the Western States, Am. Inst. Mining Engrs., pp. 455-480.

1933 E. Kohl, Chem. Tech. Z., 7: 1.

1933 C. Lepierre, Bull. soc. chim. France, (4) 53: 72.

1934 F. L. Hess, U. S. Bur. Mines, Minerals Yearbook, p. 449; from J. Thoreau and R. du Trieu de Terdonck, L'Institut colonial Belge, section sci. nat. et med., Tome I, fasc. 8, coll. 4.

1934 C. Stanfield Hitchen and R. van Aubel, Compt. rend., 199: 1133-1135.

1935 H. Buttgenbach, Chimie & industrie, 35: 79.

1935 H. Haberlandt, B. Karlik, and K. Przibram, Sitzber. Akad. Wiss. Wien, Math. naturw. Klasse, Abt. IIa, 144: 135-140.

1935 F. Hernegger and B. Karlik, Sitzber. Akad. Wiss. Wien, Math. naturw. Klasse, Abt. IIa, 144: 217-226.

1935 D. F. Kidd and M. H. Haycock, Bull. Geol. Soc. Am., 46: 879-960.

1935 A. Schoep and V. Billiet, Ann. soc. géol. Belg. Bull., 58: B198-206.

1935 A. Vandendriessche, Natuurw. Tijdschr., 17: 197-203.

1936 C. M. Alter and E. M. Kipp, Am. J. Sci., (5) 32: 120-128.

1936 V. Billiet, Natuurw. Tijdschr., 18: 79; from Chem. Abstracts, 31: 4236.

1936a L. Gmelin, "Handbuch der anorganischen Chemie," System No. 55, p. 11, Verlag Chemie, Berlin.

1936b Ibid., pp. 17-37.

1936c Ibid., p. 7.

1936 D. F. Kidd, Can. Dept. Mines Resources Geol. Survey Mem. 187.

1936 O. Henry Leonardos, Ministry Agr. (Rio de Janeiro) Servico fomento produccao mineral, Bull. 11; from Chem. Abstracts, 30: 8093.

1936 I. E. Starik, Akad. V. I. Vernadskomu Pyatidessyatiletiyu Nauch. Deyatelnosti 1: 455-462; from Chem. Abstracts, 33: 9211.

1937 A. A. Diobkov, Compt. rend. acad. sci. U.S.S.R., 17: 229-232.

1937 R. P. Fischer, Econ. Geol., 32: 197-198; 906-951.

1937 A. Holmes, "The Age of the Earth," Thomas Nelson & Sons, New York.

1937a P. Krusch, "Die metallischen Rohstoffe," vol. 1, pp. 97-103, F. Enke, Stuttgart.

1937b Ibid., p. 93.

1937c Ibid., pp. 95-97.

1937d Ibid., pp. 104-110.

1937e Ibid., pp. 113-116.

1937f Ibid., pp. 117-119.

1937g Ibid., pp. 119-120.

1937h Ibid., pp. 94-95.

1937i Ibid., p. 121.

1937j Ibid., p. 103.

1937k Ibid., p. 104.

1937a S. Meyer, Naturwissenschaften, 25: 764-765.

1937b S. Meyer, Sitzber. Akad. Wiss. Wien, Math. naturw. Klasse, Abt. IIa, 146: 175-197.

1937 B. M. Shaub, Am. Mineral., 22: 207 (1937); 23: 334-341 (1938).

1937 I. E. Starik, Trav. inst. état radium U.S.S.R., 3: 211-217.

1938 R. Bakken and E. Gleditsch, Am. J. Sci., (5) 36: 95-106.

1938a V. M. Goldschmidt, "Geochemische Verteilungsgesetze der Elemente," 9: 60, J. Dybwad, Oslo.

1938b V. M. Goldschmidt, "Geochemische Verteilungsgesetze der Elemente," 9: 99, J. Dybwad, Oslo.

1938   N. B. Keevil, Econ. Geol., 33: 685-696.

1938   G. Kirsch and F. Hecht, Z. anorg. u. allgem. Chem., 236: 157-164.

1938   G. Konjarov, Trudy podz. bogat. i. min. ind. Bulgarija, 8: 236-244; from Chem. Abstracts, 34: 2742 (1939).

1938   H. C. Parmelee, Eng. Mining J., 139: 31-35.

1938   H. Schneiderhöhn, Umschau, 42: 951.

1938   N. M. Segel, Trav. inst. état radium U.S.S.R., 4: 350-383.

1939   R. Bakken and E. Gleditsch, Nord. Kemi Kermode Forh., 5: 200-201; from Chem. Zentr., 1942 (II): 2572.

1939   R. C. Evans, "An Introduction to Crystal Chemistry," p. 171, Cambridge University Press, London.

1939a E. Föyn, B. Karlik, H. Pettersson, and E. Rona, Nature, 143: 275-276.

1939b E. Föyn, Norsk Geol. Tids., 17: 197-292; from Chem. Abstracts, 32: 6583.

1939   G. M. Furnival, Econ. Geol., 34: 739-776.

1939   H. Meixner, Naturwissenschaften, 27: 454.

1939   H. Schneiderhöhn, Neues Jahrb. Mineral. Geol., (II) 1939: 243-244.

1940a H. Meixner, Chem. Erde, 12: 433-450.

1940b H. Meixner, Z. Krist. Mineral. Petrog., Abt. A, 52: 275-277.

1940   V. Umovskaja, Compt. rend. acad. sci. U.R.S.S., 29: 380-383.

1941   R. D. Evans and C. Goodman, Bull. Geol. Soc. Am., 52: 459.

1941a J. Hoffmann, Wien klin. Wochschr., 54: 1055-1059.

1941b J. Hoffmann, Naturwissenschaften, 29: 403-404.

1941c J. Hoffmann, Zentr. Mineral. Geol., 1941A: 31-37.

1941   T. Iimori, Am. J. Sci., 239: 819-821.

1941   E. Kohl, Z. prakt. Geol., 49: 99-107.

1941   V. G. Melkov and Z. M. Sverdlov, Compt. rend. acad. sci. U.R.S.S., 31: 361-362.

1941   I. E. Starik, A. G. Samartseva, and M. L. Yashchenko, Compt. rend. acad. sci. U.R.S.S., 31: 909-910.

1941   W. D. Urry, Am. J. Sci., 239: 191-203.

1942   W. D. Dreblow, Z. Instrumentenk., 62: 60-66; 85-93.

1942   C. Goodman, J. Applied Phys., 12: 276-289.

1942a J. Hoffmann, Chem. Erde, 14: 239-252.

1942b J. Hoffmann, Chem. Ztg., 66: 181-183.

1942c J. Hoffmann, Z. physiol. Chem., 276: 275-279.

1942   C. S. Piggot and W. D. Urry, Am. J. Sci., 240: 1-12; 93-103.

1942   F. E. Wickman, Geol. Fören Förh., 64: 466.

1943a J. Hoffmann, Biochem. Z., 313: 277-287.

1943b J. Hoffmann, Bodenkunde u. Pflanzenernähr., 32: 295-336.

1943   B. Monsavoff, Can. Chem. Process Inds., 27: 710.

1944   "Dana's System of Mineralogy," vol. I, by C. Palache, H. Berman, and C. Frondel, pp. 611-620, John Wiley & Sons, Inc., New York.

1944   E. Grip and O. H. Ödman, Sveriges Geol. Undersökn. Årsbok, 38: No. 6.

1944   N. B. Keevil, Am. J. Sci., 242: 309-320.

1944   M. Legraye, Ann. soc. géol. Belg. Bull., 68: B157-174.

1944   D. Mawson, Trans. Roy. Soc. S. Australia, 68: 334-357.

1944   W. L. Stokes, Bull. Geol. Soc. Am., 55: 951-992.

1944   B. A. Tyurin, Bull. acad. sci. U.R.S.S., Sér. géol., 1944: 99-105.

1945   A. Bernardo Ferreira and J. M. Cotelo Neiva, Portugal, Direc. geral minas e serv. geol., 1: 177-189.

1945   D. Carroll and H. P. Rowledge, Western Australia, Dept. Mines, Mineral Resources W. Australia Bull., 3: 71-150.

1945   J. De Ment, "Fluorochemistry," pp. 476-482, Chemical Publishing Company, Inc., Brooklyn, N. Y.

1945   V. G. Melkov, Mém. soc. russe minéral., 74: No. 1, 41-47.

1945   C. E. Nabuco de Araujo, Jr., Chem. Eng. News, 23: 1900.

1945   M. A. Northup, Ind. Eng. Chem. Anal. Ed., 17: 664-670.

1945   C. W. Sill and H. E. Peterson, U. S. Bur. Mines Inform. Circ. 7337.

1946   J. De Ment, J. Chem. Education, 23: 213-219.

1946   J. González Reyna, Comite direct. invest. recursos mineral. Mex., Bol. 5.

1946   H. Haberlandt and F. Hernegger, Anz. Akad. Wiss.Wien, Math. naturw. Klasse No. 13: 116.

1946   A. W. Kleeman, Trans. Roy. Soc. S. Australia, 70: 175-177.

1946a   E. Lemaire, Génie civil, 123: 201-204; 213-215.

1946b   E. Lemaire and V. Charrin, Génie civil, 123: 86-87.

1946   R. Nováček, Mineralog. Abstracts, 9: 211.

1946   K. Przibram, Bull. classe sci., Acad. roy. Belg., 32: 363-369.

1946   S. I. Tomkeieff, Science Progress, 34: 696-712.

1946   J. F. Vaes, Ann. soc. géol. Belg. Bull., 70: B212-225.

1947   V. Charrin, Génie civil, 124: 386-389.

1947   M. Déribéré, La Nature, 1947: 41-43.

1947   T. R. P. Gibb, Jr., and H. T. Evans, Jr., Science, 105: 72-73.

1947   C. Karunakaran and K. Neelakantam, Proc. Indian Acad. Sci., 25A: 404-407.

1947   V. G. Khlopin, E. K. Gerling, and N. V. Baranovskaya, Bull. acad. sci. U.R.S.S., Classe sci. chim., 1947: 599-604.

1947   E. Longobardi, N. Florentine, and A. Mercader, Anales asoc. quím. argentina, 35: 131-136.

1947   C. J. Mulder, Atoom, 1: 133-139; 161-163.

1947   A. Schoep and S. Stradiot, Am. Mineral., 32: 344-350.

1947   F. E. Senftle and N. B. Keevil, Trans. Am. Geophys. Union, 28: 732-738.

1947   W. Siegl, Berg- u. hüttenmänn. Monatsh. montan. Hochschule Leoben, 92: 112-114.

1947   C. W. Sill and H. E. Peterson, Anal. Chem., 19: No. 9, 646-651.

1947   M. da Silva Pinto, Mineralog. Abstracts, 10: 115.

1948   R. Bakken, E. Gleditsch, and A. C. Pappas, Bull. soc. chim. France, 1948: 515-517.

1948   C. A. Bauer, Astron. J, 53: 110.

1948   F. J. Fohs, Bull. Am. Assoc. Petroleum Geol., 32: 317-350.

1948   A. F. Frederickson, Science, 108: 184-185.

1948   V. F. Hess and J. D. Roll, Phys. Rev., 73: 916-918.

1948   S. G. Lexow and E. P. P. Maneschi, Univ. nacl. Cuyo, Facultad ing. y cienc. exact., fis. y nat., Inst. petrol. (Mendoza), Pub. 1, pp. 5-11.

Part 2

URANIUM METAL

Chapter 4

EXTRACTION OF URANIUM FROM ORES AND PREPARATION OF URANIUM METAL

1. EXTRACTION OF URANIUM FROM ORES

Uranium ores until very recently were processed mainly for their radium content; this situation has now been altered, and radium is likely to become a by-product of uranium production. The older literature was concerned almost exclusively with the efficient isolation of radium, and little attention was devoted to the best methods of uranium recovery.

Only two uranium ores have been processed extensively, pitchblende and carnotite. A variety of methods have been employed which depend on the nature of the ore and the character of the other elements present in it. The general procedure for all ores is similar to the following: (1) leaching the ore with sulfuric, nitric, or hydrochloric acid to solubilize the uranium (occasionally, alkaline extraction or fusion has been used to open up the ore); (2) converting the uranium to a soluble complex carbonate, a treatment that removes iron, aluminum, and manganese; (3) precipitation of PbS and CuS from the uranyl solutions; (4) recovery of the uranium as $Na_2U_2O_7$ or $(NH_4)_2U_2O_7$. In the case of carnotite, special steps must be taken to separate vanadium and phosphate from the uranium. The details can best be followed by considering a number of specific processes. Bachelet (1947a) has described the extraction of uranium from betafite (Chap. 3, Sec. 2.3) by leaching the uranium with concentrated sulfuric acid, followed by a conventional procedure for separating the uranium from columbium, tantalum, and titanium.

1.1 Recovery of Uranium from Canadian Pitchblende. (MP Chicago 1; Pochon, 1937; Safford and Kuebel, 1943; Kuebel, 1944.) The recovery process is complicated by the presence in the ore of considerable quantities of gold and silver. Carbonates and sulfides, which are also present, must be destroyed before the acid treatment; otherwise, considerable frothing occurs. In addition, considerable amounts of arsenic and copper must be removed.

The ore mined at Radium City on Great Bear Lake is concentrated in the ratio 50/1 by mechanical separation and flotation. The mined rock contains about 1 per cent U_3O_8. The concentrate contains about 50 per cent U_3O_8 and from 1 to 7 per cent silver, depending on the section of the mine. This concentrate is shipped to the Port Hope (Ontario) refinery. The process carried out there is outlined in Fig. 4.1 (cf. Mactaggert, 1943).

The concentrate is first pulverized in a ball mill and sent through a magnetic separator to remove magnetite. A series of flotation cells removes the lighter components of the ore. The product from the flotation cells is dried in a furnace at 600°C, which decomposes the sulfides and carbonates and volatilizes part of the arsenic and antimony. Sodium chloride is then added, and the temperature is increased to 800°C; this converts silver to AgCl. After cooling, the roasted material is leached with sulfuric acid to remove the uranium, manganese, copper, and iron. At the leaching stage it is customary to add barium chloride to provide a carrier for the radium present and ensure that it remains in the undissolved portion of the ore. The pH of the acid extract is adjusted to 2.8 by the addition of calcium hydroxide, and ferric chloride is added to remove arsenic as insoluble ferric arsenate. After filtration, a sufficient excess of sodium carbonate is added to solubilize the uranium and to precipitate ferric, aluminum, and manganese hydroxides, etc. After decantation, sodium hydroxide is added to the uranium tricarbonate liquor to precipitate $Na_2U_2O_7$. The sodium diuranate is purified by dissolving it in hydrochloric acid and saturating the solution with hydrogen sulfide, which precipitates the sulfides of copper and arsenic. After the excess hydrogen sulfide is removed by boiling, ammonium hydroxide is added, and the uranium is recovered as $(NH_4)_2U_2O_7$. This is converted to U_3O_8 by ignition at 1000°C. It was at one time the practice to leach the U_3O_8 three times with hot hydrochloric acid to remove acid-soluble impurities, but this has been abandoned because of the high uranium losses. The oxide so produced has a U_3O_8 content of 97 to 99 per cent.

The sodium chloride treatment described above is intended to separate the silver present in the ore. As an alternative the pulverized ore, after magnetic concentration, may be subjected to a cyanide treatment which removes not only silver but also any gold that may be present. This procedure is preferred when gold is present in appreciable amounts.

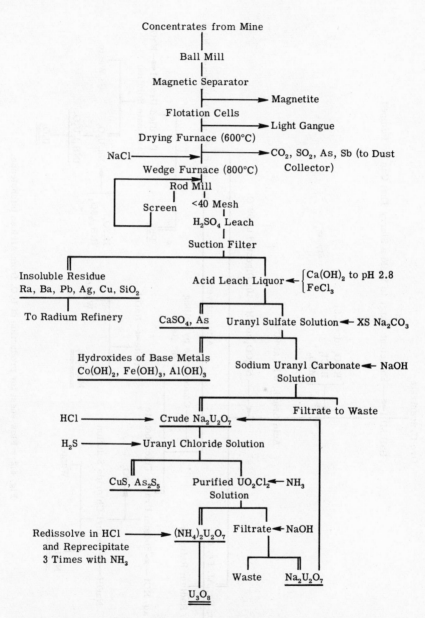

Fig. 4.1 — Flow sheet for production of U_3O_8 from Canadian pitchblende at Port Hope (Ontario) refinery.

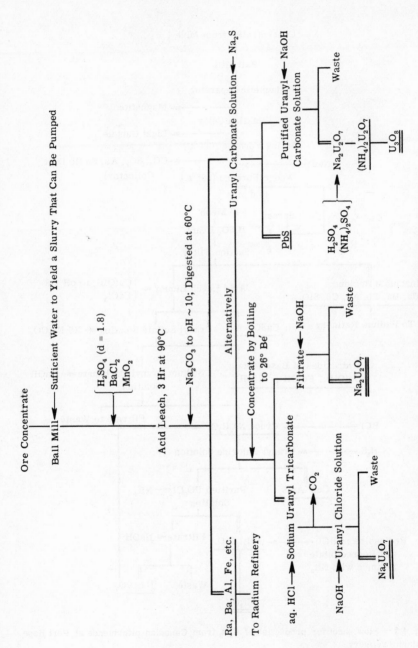

Fig. 4.2 —Flow sheet for production of U_3O_8 from African pitchblende.

The sulfuric acid leach is carried out in stoneware apparatus. As has been pointed out above, it is customary to add barium chloride to provide a carrier for the radium. For 375 lb of roasted ore there are used 350 lb of H_2SO_4 (sp. gr. = 1.8), 300 lb of water, and 20 lb of sodium nitrate. The sodium nitrate is added to ensure complete oxidation of the uranium to the sexivalent state.

1.2 Recovery of Uranium from African Pitchblende. The chemical problems associated with uranium recovery from African pitchblende are on the whole simpler than those encountered in its recovery from Canadian ores. Since gold and silver are absent, the roasting with sodium chloride (or the alternative cyanide treatment) can be omitted. Some other minor modifications of the Canadian process are usually introduced in processing African ore, particularly in the extraction step. For example, manganese dioxide is often used as the oxidizing agent instead of sodium nitrate. The processes used for the various types of pitchblende become identical after the leading step. A typical flow sheet for African ore is given in Fig. 4.2.

1.3 Extraction of Uranium from Jáchymov Pitchblende. Numerous processes have been devised by successive generations of chemists for the recovery of uranium and radium from Jáchymov pitchblende. The minor present-day significance of these deposits makes it unnecessary to give a detailed description of these processes, which are very complex because of the presence of about thirty metals in the ore. The ore is usually concentrated by hand-picking and flotation. The concentrate is ground and roasted with a mixture of soda ash and sodium sulfate at 800°C for 10 hr. Sulfur is eliminated as SO_2, and uranium, antimony, tungsten, molybdenum, and vanadium are converted to sodium salts. The roasted ore is then reground and leached, first with sulfuric acid and then with nitric acid. The bulk of the uranium is thus converted to water-soluble uranyl sulfate. The further purification procedure is dependent on the nature of the other elements present in any particular case (Henrich, 1918; Ulrich, 1923a). Extraction of the ore directly with sulfuric and nitric acids has also been employed (Marckwald, 1908, 1911).

1.4 Extraction of Uranium from Carnotite.* Extraction of uranium from carnotite ores presents a considerably more difficult problem than extraction from pitchblende because of the necessity of processing relatively large amounts of crude ore. The quantitative removal of vanadium and phosphate from the uranium has also proved troublesome.

---

*See Doerner (1930a) for a bibliography of patents relating to radium production.

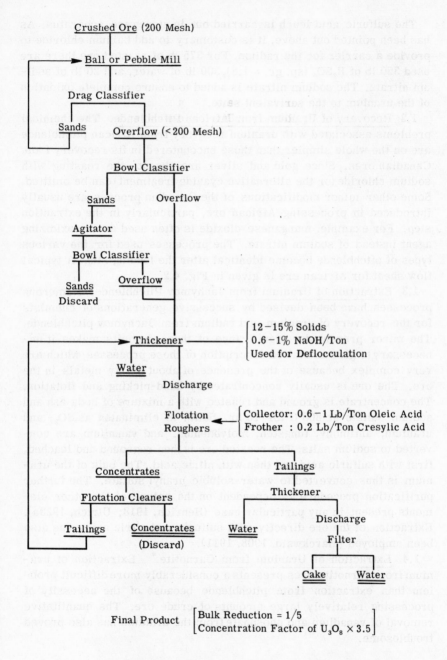

Crushed Ore (200 Mesh)

Ball or Pebble Mill

Drag Classifier

Sands

Overflow (< 200 Mesh)

Bowl Classifier

Sands   Overflow

Agitator

Bowl Classifier

Sands   Overflow
Discard

Thickener — {
12−15% Solids
0.6−1% NaOH/Ton
Used for Deflocculation
}

Water

Discharge

Flotation
Roughers — {
Collector: 0.6−1 Lb/Ton Oleic Acid
Frother : 0.2 Lb/Ton Cresylic Acid
}

Concentrates   Tailings

Flotation Cleaners   Thickener

Discharge

Tailings   Concentrates   Water   Filter
                  (Discard)

                                              Cake   Water

Final Product [
Bulk Reduction = 1/5
Concentration Factor of $U_3O_8 \times 3.5$
]

Fig. 4.3 —Suggested flow sheet for concentration of carnotite ores.

Although little attention was devoted in the past to producing uranium-rich concentrates from low-grade carnotite, considerable effort has been made in this direction recently. The carnotite occurs in small aggregates of clear friable particles, as a fine-grained component of the clayey matrix (limonite) which cements the individual grains of quartz together, and also as a thin tenacious coating on the quartz grains. By grinding and passing through a 200-mesh sieve (roughly the particle size of the individual quartz grains) much of the carnotite can be recovered (DeVaney, 1938). The larger the amount of fines produced, the better is the recovery; 65 to 78 per cent recoveries have been achieved, with a concentration factor of 2 to 4 in the sieved material compared to the original ore (Nye and Demorest, 1939; Shelton and Engel, 1942). Wet grinding appears superior to dry grinding. It has been suggested that the remnants of carnotite adhering to the quartz particles might be freed by agitating the ground ore with a jet of air so that the individual particles undergo attrition; the fines could then be collected in a cyclone separator (Dunn, 1939, 1940). Preliminary studies have been made on various ore-dressing procedures, and a tentative flow sheet for a concentration procedure based on deflocculation and flotation has been proposed (Fig. 4.3) (Davis, 1938; Engel and Shelton, 1942; McCoy, 1916).[*]

A preliminary roasting of carnotite ores has been advocated by Doerner. Many advantages are claimed. Frequently the carnotite contains appreciable amounts of carbonaceous materials (asphalts, etc.) that make grinding very difficult and interfere with subsequent acid treatment. Roasting destroys the organic matter and simultaneously converts iron to an acid-insoluble oxide. It is claimed that exceptionally high recoveries of the mineral values can be obtained by utilizing this roasting procedure (Doerner, 1928).

Various methods are available for the extraction of uranium, vanadium, and radium from carnotite. Perhaps the most widely employed procedure involves the treatment of the ore with dilute or concentrated sulfuric acid at either normal (Fleck, 1907) or elevated temperatures (100 to 300°C) (Bredt, 1914; McCoy, 1914; Danforth, 1915; Schlundt, 1916a,b). When the operation is carried out at normal temperature, the pulverized ore is moistened with 5 to 10 per cent by weight of sulfuric acid (60° Baumé) and allowed to age for 20 to 90 days. At elevated temperatures the extraction is much more rapid. After the acid treatment, extraction is completed by leaching with

---

[*]The dressing of ores of the lesser metallic minerals, including uranium, is discussed by Mitchell (1947).

dilute sulfuric acid. Uranium and vanadium are obtained as an aqueous solution of the sulfates.

The U. S. Bureau of Mines at one time advocated leaching with hydrochloric or nitric acid instead of the cheaper sulfuric acid (Parsons, 1915a). Most ores have been found to require a preliminary treatment with sodium carbonate or sodium hydroxide before hydrochloric acid extraction can be performed satisfactorily. Although numerous methods specifying hydrochloric acid have been patented, they do not appear to have been much used (Moore, 1916; Bell, 1922). It has also been proposed to use a mixture of hydrochloric and oxalic acids (Bell, 1921). In general, hydrochloric acid treatment is unsuitable if radium recovery is desired.

Nitric acid treatment is not effective for ores containing carbonaceous matter (Plum, 1915; Moore, 1916; Parsons, 1916; Viol, 1916). The recovery of vanadium from such ores is very poor, and considerable amounts of uranium and vanadium are lost by precipitation during the subsequent filtrations, which are very slow and are difficult to perform. Nitric acid solution is considered an unsatisfactory method, especially for dust concentrates (Doerner, 1928, 1930b). The nitric acid procedure can be much improved by roasting, as described by Doerner (Fig. 4.4).

Small amounts of hydrofluoric acid in other acids aid the extraction of all the mineral values from the ore.

By and large, sulfuric acid remains the most favored acid leaching agent.

Leaching can also be effected by treatment with solutions of alkali carbonates or alkalis (Haynes, 1905; Bleecker, 1913a; Thews, 1923). For example, the finely pulverized ore may be heated with sodium or potassium carbonate solution at 90°C, whereupon as much as 80 per cent of the uranium and 60 to 65 per cent of the vanadium may go into solution. The concentration of the carbonate solution is varied with the amounts of uranium and vanadium present in the ore and their ratio to each other. Alkali leaching has also been performed under pressure (Bleecker, 1919).

When a mixture of carbonates and alkali hydroxides is used, vanadium is extracted as a soluble vanadate, whereas the uranium remains in the residue as insoluble uranate. After removal of the bulk of the vanadium, the uranium can be leached out with acid (Bleecker, 1913b; Moore, 1913). Treatment of the ore with sodium hydroxide solution at 200 to 300°C has also been proposed (Fischer, 1912; Ebler, 1915; Schlesinger, 1918). It is doubtful that a clean-cut separation of the uranium and vanadium can be effected in this way; however, the method may be useful when the vanadium is present in much greater

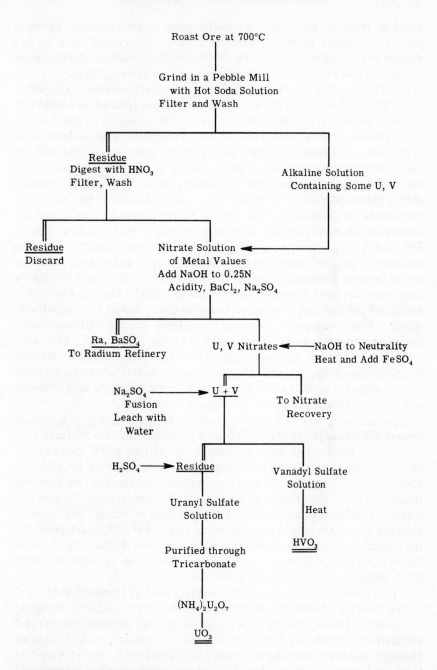

Fig. 4.4 — Flow sheet for Bureau of Mines nitric acid process for carnotite ore (modified by Doerner, 1928, 1930a).

quantity than the uranium. A modification of the alkaline leaching process has been proposed wherein a "per" compound (such as hydrogen peroxide) is added to the $NaOH-Na_2CO_3$ solution. Under these conditions the uranium and vanadium dissolve (Gibbs, 1935).

In addition to the extraction processes discussed above, all of which operate in aqueous solution at moderate temperatures, a number of high-temperature fusion methods have been proposed to open up the ore. Thus the finely powdered ore (or concentrate) may be fused with a mixture of sodium chloride and sodium hydroxide. The sodium vanadate is leached out with water, and the uranium is removed from the residue with sulfuric acid (Bleecker, 1909; Vogt, 1915). Fusion with a threefold excess of sodium acid sulfate ($NaHSO_4$) has also been recommended (Radcliffe, 1911, 1914), as has fusion with solid sodium carbonate (Parsons, 1915b). Recently the use of ammonium sulfate at 380 to 430°C has been proposed. This method is attractive because the ore is roasted and decomposed in a single operation. At these temperatures ammonium sulfate is equivalent to SO_3, and the latter may also be used directly (McCormack, 1939). Both uranium and vanadium go into solution when the reaction product is treated with water. Few comparative data are available on the efficiencies of these procedures, which appear to have been little used. They may be valuable for rich, concentrated material but are obviously difficult to apply when large amounts of low-grade material are to be treated. Large amounts of silica are likely to be solubilized by alkali fusion and may cause difficulties in subsequent operations.

Separation of uranium and vanadium by heating the mixture of uranyl and vanadyl sulfates obtained by evaporation of a sulfuric acid leach of carnotite has been proposed. At 500 to 650°C the vanadyl sulfate decomposes and forms insoluble iron vanadate by reaction with the iron present. The uranyl sulfate redissolves on leaching (Potter, 1939). Separation of uranium and vanadium can also be effected by adjusting the acidity of the solution of uranyl and vanadyl sulfates (from the sulfuric acid leach) to pH 1.0 to 2.0 and oxidizing the vanadium with sodium chlorate or manganese dioxide. The bulk of the vanadium precipitates under these conditions as vanadic acid, HVO_3 (Stamberg, 1939; Fleck, 1940).

A procedure for autunite has been described by Glaser (1912).

The methods discussed above are for the most part experimental or from the patent literature. A process that has enjoyed commercial utilization consists in roasting the raw carnotite ore with sodium chloride and then leaching with water to remove the sodium vanadate formed during roasting. The tailings are then leached with dilute sulfuric acid to yield a solution containing uranium and the water-insoluble vanadium. The uranium is then isolated by procedures already discussed.

1.5 Special Extraction Procedures for Low-concentration Ores. (Ulzer, 1908; Ebler, 1913; Loomis, 1916; Ulrich, 1923b.) A number of ingenious procedures have been suggested for low-grade uranium minerals. While these procedures have been of little significance in the past, they are discussed briefly here since they may be useful in the development of new methods in the future.

(a) Chlorination. This method has been used on betafite from Madagascar. The finely pulverized mineral is mixed with wood charcoal and treated with chlorine at red heat. The volatile chlorides UCl_5, $CbCl_5$, and $TaCl_5$ are formed and distilled from the reactor (Curie, 1925). An analogous method has been applied to American pitchblende containing pyrite. Treatment with chlorine at elevated temperatures results in distilling uranium and iron chlorides from the reaction mixture (Cable, 1918). The chlorinating agent is presumably SCl_2 (see Chap. 14). Along similar lines it has been shown that phosgene reacts with autunite (Barlot, 1913).

$$Ca(UO_2)_2(PO_4)_2 + 10COCl_2 \xrightarrow{800°C} 2UCl_4$$
$$+ 2POCl_3 + 10CO_2 + CaCl_2 + 2Cl_2$$

Other chlorinating reagents, such as thionyl chloride and carbon tetrachloride, should also be of interest in devising efficient methods for treating low-grade ores.

(b) Extraction with Sulfur Dioxide. On treatment of pulverized carnotite with a saturated aqueous solution of sulfur dioxide at normal temperature and pressure, uranium and vanadium are reported to go into solution (Burfeind, 1914; Loomis, 1916). The use of liquid sulfur dioxide for this purpose has also been tried (Hedström, 1922).

(c) Solvent Extraction. A number of patents have been issued to Hixson (1941) covering a process for extracting uranium (and other metals) from ores by a solvent-extraction method. In brief, the ore is leached to provide an aqueous solution of the mineral values, and these are then separated and recovered by countercurrent extraction with a solvent immiscible with water. Recycling may be employed for increased efficiency. This is a very promising approach and will no doubt be the subject of intensive investigation in the future. Bachelet, Cheylan, and LeBris (1947b) have examined the solvent extraction of uranium ores; they observed that uranium could readily be purified from silicon, titanium, vanadium, columbium, tantalum, molybdenum, and tungsten by solvent extraction of a nitrate solution with ether. Boron and phosphorus were incompletely separated.

A similar procedure has been in use for some time for purifying crude U_3O_8. For certain purposes it is necessary to remove all traces of rare earths from the uranium. This has been accomplished by a solvent-extraction process that depends on the solubility of uranyl nitrate in ethyl ether, in which the rare earth nitrates are insoluble (Péligot, 1842). A very pure uranium oxide can be thus obtained.

2. METAL PREPARATION

Although metallic uranium was first prepared more than a hundred years ago, it is only since 1940 that it has become possible to produce the pure metal on a large scale. Prior to the development of the procedures described in this chapter, uranium metal had been prepared principally for scientific study and only on a small scale. The processes were inefficient, and the metal produced was not of the highest purity. The modern metallurgy of uranium is treated in detail in other volumes of the National Nuclear Energy Series. In this chapter it is proposed to present only a brief survey of uranium-metal preparation.

2.1 Historical Survey. In 1789 Klaproth, the discoverer of the element uranium, reduced uranium trioxide with carbon at high temperatures. The product had a metallic appearance and was assumed by him to be the free metal. This claim was generally accepted until 1840, when Péligot demonstrated that the product obtained by Klaproth's procedure was a lower oxide (UO_2). Péligot succeeded in preparing the metal for the first time by the reduction of uranium tetrachloride with potassium metal.

Following Péligot's discovery, other workers prepared uranium metal on a laboratory scale. The methods used fall into four classes: (a) reduction of uranium oxides with carbon; (b) reduction of uranium oxides with aluminum, calcium, or magnesium; (c) reduction of uranium halides with alkali metals or alkaline earth metals; (d) electrolytic reduction of uranium halides. With few exceptions, these methods yielded metal in the form of powder.

(a) Reduction of Uranium Oxides with Carbon. In order to reduce uranium oxide to the metal with carbon it is necessary to use the very high temperatures of the electric arc. Moissan was able to achieve this reduction in the electric-arc furnace (Moissan, 1883; 1894; 1896a,b,c; 1901; 1904). A mixture of U_3O_8 and carbon (obtained from sugar) was strongly compressed in a carbon crucible. Reduction took place in a few minutes (in an arc of 450 amp, 60 volts). Moissan thought that essentially carbon-free metal could be obtained in this way, but, since uranium reacts easily with carbon at elevated temperatures to form uranium carbide, this statement cannot be correct. Moissan emphasized the importance of eliminating nitrogen as well as oxygen from the reaction system. If smaller amounts of carbon (40 g of carbon per 500 g of U_3O_8) are used to avoid the formation of carbide, the metal formed contains oxide (MP Ames 1). Using a carbon crucible makes formation of uranium carbide after long-continued heating at elevated temperatures inevitable. Aloy (1901) repeated this work and reported that milder conditions than those used by Moissan could be used to effect reduction. Recently Moissan's work was repeated using UO_2 and U_3O_8 with sugar charcoal, acetylene black, graphite, and other forms of carbon. In all cases uranium carbide was formed in appreciable amounts. Variation of the oxide/carbon ratio over a wide range did not eliminate carbide formation (MP Ames 1).

(b) Reduction of Uranium Oxides with Aluminum, Calcium, or Magnesium. Thermodynamic calculations show that uranium dioxide is reducible by calcium, magnesium, or aluminum at sufficiently elevated temperatures. Reactions of this kind were usually carried out in an evacuated steel apparatus. Jander (1924) reduced UO_2 with calcium at 950 to 1250°C in a tightly sealed iron crucible. Finely powdered metal of a fair degree of purity was reportedly obtained. Essentially the same procedure was employed by Botolfsen (1929); U_3O_8 and calcium were heated in a steel tube which was inserted into an evacuated quartz tube. In this, as well as in the earlier work of Wedekind (1911, 1914), who used a similar procedure, the metal obtained was in the form of very fine powder. While these workers claimed that metal produced by reduction of uranium oxide with calcium was of a high degree of purity, others were unable to confirm this. Thus when James and coworkers (1926) heated U_3O_8 with the theoretical quantity of calcium at 1000°C for 45 min in an evacuated steel bomb, the main product was a brown powder. Some globules of metallic uranium were present, but even these contained considerable amounts of iron and oxygen. These observations have been confirmed by more recent work (MP Ames 2).

Reduction with calcium metal has been modified by carrying it out in the presence of alkali halides or alkaline earth halides. These procedures are described in a number of patents assigned principally to the Westinghouse Lamp Division (Marden, 1923; Rich, 1927; Cachemaille, 1924). The addition of calcium chloride or magnesium chloride is reported to facilitate the reaction by acting as a flux, and to produce the metal in a coarser form. Magnesium metal may also be used to reduce uranium oxide, but owing to its high vapor pressure it tends to distill from the reaction mass and must therefore be used under pressure (Marden, 1926). Once initiated, the reaction is violent and difficult to control. The use of a calcium-magnesium alloy of the approximate composition Ca_3Mg_4 was reported to be advantageous (Rich, 1927).

Uranium oxide can also be reduced with calcium hydride.

$$UO_2 + 2CaH_2 \rightarrow U + 2CaO + 2H_2$$

(Alexander, 1937; MP Chicago 2,3.) A mixture of uranium oxide and calcium hydride is placed in a nichrome cylinder, which is then placed in a gas-heated retort. At 960°C reaction occurs. The charge is allowed to cool under vacuum to eliminate hydrogen. It is then removed, crushed, and ground, and the lime is leached out with dilute acetic acid. The wet uranium powder is washed with water, dried, pressed into compacts, and sintered.

Numerous workers have investigated the reduction of uranium oxide with aluminum metal (Moissan, 1896b; Stavenhagen, 1899, 1902; Aloy, 1901; Giolitti, 1908; Marden, 1927). At 600°C, U_3O_8 is reduced by aluminum powder to a lower oxide, but apparently a much higher temperature is required to obtain reduction to uranium metal. Stavenhagen attempted to achieve this by employing liquid air as oxidant. He claimed to have obtained a molten regulus of uranium metal, but, in view of the tendency of uranium metal to react with oxygen and nitrogen, it is not surprising that his results have not been confirmed. The aluminothermic reduction of metal oxides, including uranium oxides, in a centrifugal apparatus is described by Merle (1946), who claims various advantages for such a procedure. Although aluminum can be separated from uranium-aluminum alloys by vacuum fusion, the tendency of uranium to form such alloys during aluminothermic reduction of oxide is a complicating feature.

(c) Reduction of Uranium Halides with Alkali Metals or Alkaline Earth Metals. As mentioned above, the first successful preparation of uranium metal was achieved by Péligot (1842) by the reduction of uranium tetrachloride with metallic potassium. Since then the reduction of uranium halides with alkali metals or alkaline earth metals has been frequently employed. The metal obtained by Péligot was contaminated by platinum, since the reduction was carried out in a platinum crucible. Replacement of potassium by sodium did not improve the procedure greatly (Péligot, 1869). In Péligot's method, potassium chloride is added as a flux, and the platinum crucible is protected from the air by a layer of charcoal. Zimmermann (1883) modified Péligot's procedure by the use of an iron bomb (Wilson, 1878). A layer of fused sodium chloride was employed in an effort to preserve the bomb from attack; uranium tetrachloride, sodium chloride, and metallic sodium were charged into the bomb, and the reactants were heated to white heat after screwing the bomb cover into place. Zimmermann claimed that the metal so produced was pure and massive. Moissan (1896c) repeated Zimmermann's work and reported that the uranium metal obtained in this way contained up to 2 per cent iron. Moissan preferred to use the double salt Na_2UCl_6 (rather than UCl_4) because of its less hygroscopic nature. He also carried out the reduction with metallic sodium in an iron cylinder closed with a screw stopper, but the product was a finely divided powder of rather doubtful purity. A number of other workers employed the same bomb technique without obtaining substantially better results. Mixter (1912) followed Moissan's procedure very closely, using Na_2UCl_6, and obtained 97 to 99 per cent pure uranium powder.

Roderburg (1913) conducted an extensive investigation into the methods for preparing pure uranium with particular reference to the reduction of uranium halides by alkali metals. In an effort to improve the purity of the product, bombs of various alloy steels were investigated, but in all cases the uranium metal produced contained substantial amounts of iron. Roderburg also attempted the reduction of uranium tetrafluoride with sodium and potassium but obtained only partial reaction. Fischer (1913) and Rideal (1914) also investigated various methods for the preparation of uranium metal. Of a considerable number of possibilities explored, they preferred the reduction of uranium tetrachloride with magnesium or sodium in the presence of calcium chloride. The metal obtained in this way, however, was of an uncertain degree of purity, as evidenced by variations in the electrical conductivity. In an effort to minimize contamination by oxygen and nitrogen, Lely and Hamburger (1914) performed the reduction of uranium tetrachloride with sodium in an evacuated metal bomb. The product was in the form of a fine powder and still contained significant amounts of oxygen. A description of such a vacuum reduction is given in considerable detail by Moore (1923). The reduction of uranium halides by calcium, magnesium, calcium-magnesium alloys, and aluminum has been investigated by Marden and made the subject of a series of patents assigned to the Westinghouse Lamp Division (1922). In these procedures UF_4, KUF_5, or some other nonvolatile halide or double salt of uranium is reduced by one of the previously mentioned metallic reducing agents, usually in the presence of a flux such as calcium chloride. After the reaction mass has been leached and washed, the product is a fine powder.

James and coworkers (1926) appear to have been successful in preparing massive uranium metal with a fair degree of purity. These workers employed the reaction

$$UCl_4 + 2Ca \rightarrow 2CaCl_2 + U$$

which was carried out in an evacuated steel bomb. By using redistilled calcium, a product quite free of iron could be obtained in massive form. The success of these workers can be attributed, at least in part, to the fact that experiments were carried out on a fairly large scale, which served to minimize contamination and the heat loss from the system. This work is probably the best that is recorded in the older literature. A recent study of the reduction of uranium halides by metallic potassium (generated by reaction of calcium carbide on potassium chloride) has been made by Lautié (1947). In the course of this work it was observed that uranium oxide can be reduced by ferrosilicon at high temperatures.

(d) Electrolytic Reduction of Uranium Halides. The last of the methods employed by earlier workers in this field was electrolytic reduction of uranium halides in fused salt baths. This method as invented by the authors is disclosed in the literature by Driggs and Lilliendahl (1930b). Apart from the pioneering studies of Moissan (1904), the most significant work in the electrochemical preparation of uranium metal has been carried out at the Westinghouse Electric Corporation[*] (Westinghouse 1,2) at Bloomfield, N. J., by Driggs (1928, 1930a) (Lilliendahl and Highriter, 1935; Marden, 1934). In the Westinghouse process the electrolytic bath consisted of the double salt, potassium uranium fluoride, KUF_5, or of uranium tetrafluoride, UF_4, dissolved in a molten mixture of 80 per cent calcium chloride and 20 per cent sodium chloride. The bath was contained in a graphite crucible which served as the anode; the cathode was made of molybdenum. The bath was operated at 900°C at a current density of about 150 amp/dm². The uranium deposit was granular and pyrophoric and required an elaborate sequence of washing and drying operations in order to eliminate occluded salts.

Numerous workers have attempted to deposit metallic uranium from an aqueous solution, but there is no evidence to indicate the feasibility of such a procedure. In all cases, complex hydrated oxides were obtained on the cathode. This failure of metallic uranium to deposit from an aqueous solution is not surprising, since manganese

---

[*]Methods of producing metallic thorium and uranium developed at Westinghouse are outlined in Steel, 120: 93-94, 124, 126, 130, 133 (1947).

is the most electropositive metal that has yet been plated from aqueous solution, and uranium is almost certainly above manganese in the electrochemical series. There is some indication that dilute uranium amalgams may form when aqueous solutions are electrolyzed at high current densities with a mercury cathode (Férée, 1901). Repeated

Table 4.1 — Possible Reactions for the Preparation of Uranium Metal

Reaction	ΔH/mole of U at 298°C	M.p.[*] of slag, °C	B.p.[*] of slag at 760 mm, °C
$UO_2 + 2Ca \rightarrow U + 2CaO$	−47	2572	2850
$UO_2 + 2Mg \rightarrow U + 2MgO$	−35	2500−2800	...
$UO_2 + \frac{4}{3}Al \rightarrow U + \frac{2}{3}Al_2O_3$	−11.4	2050	2250
$U_3O_8 + 8Ca \rightarrow 3U + 8CaO$	−123.1	2572	2850
$U_3O_8 + 8Mg \rightarrow 3U + 8MgO$	−108.1	2500−2800	...
$U_3O_8 + \frac{16}{3}Al \rightarrow 3U + \frac{8}{3}Al_2O_3$	−76.4	2050	2250
$UF_4 + 4Na \rightarrow U + 4NaF$	−98	980−997 (1040)	1700
$UF_4 + 2Ca \rightarrow U + 2CaF_2$	−134	1330	2500
$UF_4 + 2Mg \rightarrow U + 2MgF_2$	−82	1225	2260
$UF_4 + \frac{4}{3}Al \rightarrow U + \frac{4}{3}AlF_3$	6	1040	...
$UCl_4 + 4Na \rightarrow U + 4NaCl$	−141	801	1413 (1490)
$UCl_4 + 2Ca \rightarrow U + 2CaCl_2$	−131	772	1925
$UCl_4 + 2Mg \rightarrow U + 2MgCl_2$	−55	708	1420
$UCl_4 + \frac{4}{3}Al \rightarrow U + \frac{4}{3}AlCl_3$	28	190 at 2.5 atm	182.7; subl. 177.8
$UBr_4 + 4Na \rightarrow U + 4NaBr$	−144	755	1390
$UBr_4 + 2Ca \rightarrow U + 2CaBr_2$	−124	765	1200 (?) (806−812)
$UBr_4 + 2Mg \rightarrow U + 2MgBr_2$	−44	695	1125
$UBr_4 + \frac{4}{3}Al \rightarrow U + \frac{4}{3}AlBr_3$	31	97.5	263.3
$UCl_3 + 3Li \rightarrow U + 3LiCl$	...	613	1353

[*]Values in parentheses are alternate values also to be found in the literature and are quoted where there is no good basis for a definitive value.

attempts have also been made to electrodeposit uranium from solutions in nonaqueous solvents (Pierlé, 1919; Audrieth, 1931). Saturated solutions of uranium tetrabromide in benzene, ethyl bromide, diethyl ether, dioxane, acetone, acetone saturated with sulfur dioxide, nitrobenzene, formamide, and formamide containing hydrogen bromide have been subjected to electrolysis (MP Berkeley 1). Only in the formamide solution was there any evidence of cathodic deposition, and it is not at all certain that the deposit was uranium metal. The Berkeley results have therefore not clarified the situation. It may be concluded that at present electrolysis of either aqueous or nonaqueous solutions of uranium salts does not appear to be a practical method of making uranium. However, relatively little work has been done in this field, and future research may change these conclusions.

(e) Summary. Table 4.1 summarizes a number of possible reactions for the preparation of uranium metal. In comparison with the reduction of uranium oxides, the reduction of uranium halides with alkali metals or alkaline earth metals presents a much more favorable picture. This yields slags, which for the most part melt at relatively low temperatures. Some of these reactions are sufficiently exothermic to produce completely molten reaction mixtures. The heats of reaction indicate that UF_4, UCl_4, and UBr_4 can be reduced by either Na, Ca, or Mg but that reduction by Al would not occur spontaneously. Sodium has the largest heat of reaction when used with chlorides or bromides, and calcium is superior in this respect in the reduction of fluorides. Reduction by magnesium is much less exothermal than that by calcium in all cases but may still, according to Table 4.1, be adequate.

2.2 Preparation of Uranium Metal by Thermal Decomposition of Uranium Halides. (UCRL 1,2,3; MP Chicago 4,5.) A number of metals have been prepared by thermal decomposition of the metal halide vapors on a hot filament. The method has been applied particularly to the preparation of zirconium, titanium, tungsten, and thorium (van Arkel, 1939). Very pure metal can be obtained in this way.

In applying this method to the preparation of pure uranium metal, a number of difficulties arose. Compared to metals that had been successfully prepared by the hot-wire technique, uranium has a low melting point (1133°C). The process was first carried out in such a way that molten uranium was allowed to drip off the wire (MP Chicago 6). However, this was not considered a satisfactory procedure; a better solution is to use such low decomposition temperatures that the uranium remains solid (MP Chicago 7).

The most satisfactory halide for this process is uranium tetraiodide. The other halides of uranium either require too high a filament temperature to have much practical significance or are not volatile enough. The equilibrium (see Chap. 15)

$$UI_4 \rightleftharpoons UI_3 + \tfrac{1}{2}I_2 \tag{1}$$

has an important bearing on the problem. A certain partial pressure of iodine is required to prevent uranium tetraiodide from dissociating in the vapor phase. (Uranium triiodide is substantially nonvolatile.) The reaction that occurs on the filament is

$$UI_4 \rightleftharpoons U + 2I_2 \tag{2}$$

A high partial pressure of iodine favors the forward reaction, and thus the metal precipitation, by increasing the amount of uranium tetraiodide in the vapor phase (according to Eq. 1), but at the same time it also favors the back reaction in Eq. 2 and thus slows down the metal deposition. As a result of this, the rate of deposition reaches a maximum at a certain optimum iodine pressure. This pressure is a function of the temperature of the filament and can be calculated from the equilibrium constants of the two reactions involved.

The preparation of metallic uranium by hot-wire decomposition of uranium tetraiodide has been successfully carried out at a filament temperature below the melting point of uranium. The reduction is carried out in a glass bulb immersed in an air reflector furnace. Solid UI_3 is introduced, and the system is evacuated. The region where the UI_3 rests is surrounded by a copper cooling coil which is used to maintain the UI_3 at any desired temperature. The vessel containing iodine is connected with the bulb through a long vertical tube which allows the iodine to be maintained at any desired temperature, thus regulating its partial pressure in the system. A suitable set of operating conditions is as follows: filament 1030 to 1100°C, bulb temperature 520 to 560°C, UI_3 temperature 500 to 540°C, $P_{I_2} = 7 \times 10^{-3}$ mm Hg. Under these conditions solid uranium is deposited on the filament.

The thermal dissociation of uranium tetraiodide has been discussed in some detail. Prescott and Holmes (UCRL 2) have shown that, for a bulb temperature of 800°K, metal is deposited when the iodine pressure in the communicating reservoir is below the critical value given by the empirical equation

$$\log P_{I_2} = 7.860 - 1.234 \, \frac{10^4}{T_2}$$

A theoretical analysis of the thermal dissociation of UI_4 has been made (UCRL 4), and further analysis has been made of the detailed mechanisms (UCRL 5). In earlier treatments it was assumed that all iodine leaving the filament was in the form of iodine molecules, I_2; in actuality, complete dissociation to iodine atoms occurs at the filament temperature. In addition, the tetraiodide is converted to triiodide on contact with the filament, and, if not decomposed to metal, is reconverted to tetraiodide at the bulb temperature. The relation given above for the critical conditions for the metal deposition appears to be in accord with currently accepted measured and estimated values for the equilibria involved.

REFERENCES

1842   E. Péligot, Ann. chim. et phys., (3) 5: 7, 42; Ann., 43: 257, 283.
1869   E. Péligot, Ann. chim. et phys., (4) 17: 368.
1878   L. F. Nilson and O. Pettersson, Ber., 11: 383.
1883   H. Moissan, Compt. rend., 116: 347.
1883   C. Zimmermann, Ann., 216: 14.
1894   H. Moissan, Bull. soc. chim. Paris, (3) 11: 11.
1896a   H. Moissan, Compt. rend., 122: 1088.
1896b   H. Moissan, Compt. rend., 122: 1302.
1896c   H. Moissan, Ann. chim. et phys., (7) 9: 264.
1899   A. Stavenhagen, Ber., 32: 3065.
1901   J. Aloy, Bull. soc. chim. Paris, (3) 25: 344.
1901   M. J. Férée, Bull. soc. chim. Paris, (3) 25: 622.
1901   H. Moissan, Bull. soc. chim. Paris, (3) 25: 344.
1902   A. Stavenhagen and E. Schuchard, Ber., 35: 909.
1904   H. Moissan, "The Electric Furnace," pp. 162-167, E. Arnold & Co., London.
1905   J. H. Haynes and W. D. Engle, U. S. Patent No. 808839.
1907   H. Fleck, W. G. Haldine, and E. L. White, U. S. Patent No. 890584.
1908   F. Giolitti and G. Tavanti, Gazz. chim. ital., 38 (II): 239.
1908   W. Marckwald, Ber., 41: 1529.
1908   F. Ulzer and R. Sommer, German Patent D. R. P. No. 254,241.
1909   W. F. Bleecker, U. S. Patent No. 1015469.
1911   W. Marckwald and A. S. Russell, Ber., 44: 772.
1911   S. Radcliffe, U. S. Patent No. 1049145.
1911   E. Wedekind, Z. angew. Chem., 24: 1179.
1912   S. Fischer, U. S. Patent No. 1054102.
1912   F. Glaser, Chem. Ztg., 36: 1167.
1912   W. G. Mixter, Z. anorg. Chem., 78: 231.
1913   J. Barlot and E. Chauvenet, Compt. rend., 157: 1153.
1913a   W. F. Bleecker, U. S. Patent No. 1065581.
1913b   W. F. Bleecker, U. S. Patent No. 1068730.
1913   E. Ebler, German Patent D. R. P. No. 296,132.
1913   A. Fischer, Z. anorg. Chem., 81: 189.
1913   R. B. Moore, U. S. Patent No. 1165692.
1913   A. Roderburg, Z. anorg. Chem., 81: 122.
1914   O. P. C. Bredt, U. S. Patent No. 1154231.
1914   J. H. Burfeind, U. S. Patent No. 1095377.
1914   D. Lely, Jr., and L. Hamburger, Z. anorg. Chem., 87: 220.
1914   H. N. McCoy, U. S. Patent No. 1098282.
1914   S. Radcliffe, J. Soc. Chem. Ind., 33: 229.
1914   E. Rideal, J. Soc. Chem. Ind., 33: 673.
1914   E. Wedekind, U. S. Patent No. 1088909.
1915   C. W. Danforth, W. Samuels, and W. Martersteck, U. S. Patent No. 1126182.
1915   E. Ebler and W. Bender, Z. angew. Chem., 28 T: 34.
1915a   C. L. Parsons, R. B. Moore, S. C. Lind, and O. C. Schaefer, U. S. Bur. Mines Bull. No. 104, pp. 17, 27, 30.
1915b   C. L. Parsons, R. B. Moore, S. C. Lind, and O. C. Schaefer, U. S. Bur. Mines Bull. No. 104, p. 24.
1915   H. M. Plum, J. Am. Chem. Soc., 37: 1797.
1915   L. Vogt, U. S. Patent No. 1129029.
1916   A. G. Loomis and H. Schlundt, Ind. Eng. Chem., 8: 990.
1916   H. N. McCoy, U. S. Patent No. 1195698.

1916   R. B. Moore, U. S. Patent No. 1165693.
1916   C. L. Parsons, Ind. Eng. Chem., 8: 469.
1916a H. Schlundt, J. Phys. Chem., 20: 485.
1916b H. Schlundt, U. S. Patent Nos. 1181411 and 1194669.
1916   C. H. Viol, Ind. Eng. Chem., 8: 284, 660.
1918   R. Cable and H. Schlundt, Chem. & Met. Eng., 18: 1.
1918   F. Henrich, "Chemie radioaktiver Stoffe," pp. 298ff., Verlag Julius Springer, Berlin.
1918   W. A. Schlesinger, U. S. Patent No. 1435180.
1919   W. F. Bleecker, U. S. Patent No. 1399246.
1919   C. Pierlé and L. Kahlenberg, J. Phys. Chem., 23: 517.
1921   W. A. J. Bell, U. S. Patent No. 1526943.
1922   W. A. J. Bell, U. S. Patent No. 1522040.
1922   H. O. Hedström, Chem. Zentr., 1922 (II): 936.
1922   J. W. Marden (assigned to Westinghouse Lamp Division) U. S. Patent Nos. 1437984, 1646734 (1927), and 1814721 (1931).
1923   J. W. Marden, U. S. Patent No. 1659209.
1923   R. W. Moore, Trans. Am. Electrochem. Soc., 43: 319.
1923   K. B. Thews and F. J. Heinle, Ind. Eng. Chem., 15: 1159.
1923a C. Ulrich, Z. angew. Chem., 36: 41, 50.
1923b C. Ulrich, Z. angew. Chem., 36: 51.
1924   A. Cachemaille, English Patent No. 238663.
1924   W. Jander, Z. anorg. u. allgem. Chem., 138: 321.
1925   Maurice Curie, "Le radium et les radioéléments," p. 205, J. B. Baillière & fils, Paris.
1926   C. James, J. F. Goggins, J. J. Cronin, and H. C. Fogg, Ind. Eng. Chem., 18: 114.
1926   J. W. Marden (assigned to Westinghouse Lamp Division) U. S. Patent No. 1602542.
1927   J. W. Marden, U. S. Patent No. 1648954.
1927   M. N. Rich, U. S. Patent No. 1738669.
1928   H. A. Doerner, U. S. Bur. Mines Repts. Invest., No. 2873.
1928   F. H. Driggs and coworkers, U. S. Patent Nos. 1821176, 1842254, and 1861625 (1932).
1929   E. Botolfsen, Bull. soc. chim. France, 45: 626.
1930a H. A. Doerner, U. S. Bur. Mines Repts. Invest., No. 3057.
1930b H. A. Doerner, Ind. Eng. Chem., 22: 185-188.
1930a F. H. Driggs, Eng. Mining J., 130: 119.
1930b F. H. Driggs and W. C. Lilliendahl, Ind. Eng. Chem., 22: 516.
1931   L. F. Audrieth and H. W. Nelson, Chem. Revs., 8: 338.
1934   J. W. Marden, Trans. Electrochem. Soc., 66: 8.
1935   H. L. Gibbs, U. S. Patent No. 1999807.
1935   W. C. Lilliendahl and H. W. Highriter, Am. Inst. Mining Met. Engrs. Tech. Pubs., 630.
1937   P. P. Alexander, Metals & Alloys, 8: 263-264; 9: 45-48 (1938).
1937   M. Pochon, Chem. & Met. Eng., 44 (No. 7): 362-365.
1938   C. W. Davis and coworkers, U. S. Bur. Mines Repts. Invest., No. 3370, p. 92.
1938   F. E. DeVaney, Eng. Mining J., 139: 43-45 (November).
1939   H. E. Dunn, C. P. Rees, and A. A. Sproul, U. S. Patent No. 2175484.
1939   H. McCormack, U. S. Patent No. 2176609.
1939   R. D. Nye and D. J. Demorest, U. S. Patent No. 2173523.
1939   J. S. Potter, U. S. Patent No. 2180692.
1939   C. J. Stamberg, U. S. Patent No. 2176610.
1939   A. E. van Arkel, "Reine Metalle," pp. 183, 193, 269; Verlag Julius Springer, Berlin; Edwards Brothers, Inc., Ann Arbor, Mich., 1943.

1940   H. E. Dunn, U. S. Patent No. 2175457 to Vanadium Corp. of America.
1940   H. Fleck, U. S. Patent No. 2199696.
1941   A. W. Hixson and R. Miller, U. S. Patent Nos. 2227833, 2202525 (1940), and
       2211119 (1940).
1942   A. L. Engel and S. M. Shelton, U. S. Bur. Mines Repts. Invest., No. 3628.
1942   S. M. Shelton and A. L. Engel, U. S. Bur. Mines Repts. Invest., No. 3636.
1943   E. F. Mactaggert, Chem. & Met. Eng., 50 (No. 7): 178-181.
1943   W. H. Safford and A. Kuebel, J. Chem. Education, 20: 88-91.
1944   A. Kuebel, J. Chem. Education, 21: 148-149.
1946   J. M. Merle, U. S. Patent No. 2395286.
1947a M. Bachelet, Bull. soc. chim. France, 1947: 628-632.
1947b M. Bachelet, E. Cheylan, and J. LeBris, J. chim. phys., 44: 302-305.
1947   R. Lautié, Bull. soc. chim. France, 1947: 974-977.
1947   F. B. Mitchell, Mine & Quarry Eng., 13: 330-305, 334-342.

Project Literature

MP Ames 1:   F. H. Spedding, Reports CN-127, June 13, 1942, and CC-177, July 8, 1942.
MP Ames 2:   H. A. Wilhelm and W. H. Keller, Report CC-238, Aug. 15, 1942.

MP Berkeley 1:   E. D. Eastman and B. J. Fontana, Electrolysis of Non-aqueous Solu-
      tions of Uranium Bromides, and Reaction of Alkali Metals with Uranium Bromides in
      Liquid Ammonia, in National Nuclear Energy Series, Division VIII, Volume 6.

MP Chicago 1:   H. N. McCoy and H. C. Anderson, Report CS-70, undated.
MP Chicago 2:   T. W. Davis and R. Penneman, Report CC-276, Sept. 15, 1942.
MP Chicago 3:   F. Foote, Reports CA-278, Sept. 26, 1942, and CT-422, Jan. 15, 1943.
MP Chicago 4:   T. T. Magel, Reports CK-1240, Jan. 19, 1944; CK-1130, Dec. 10, 1943;
      and CK-1040, Nov. 6, 1943.
MP Chicago 5:   T. T. Magel and L. S. Foster, Reports CK-963, Nov. 11, 1943, and CK-
      897, July 28, 1943.
MP Chicago 6:   T. T. Magel, Report CK-1130, Dec. 10, 1943.
MP Chicago 7:   F. Foote, Reports CT-2616, December, 1944, and CT-2668, January,
      1945.

UCRL 1:   C. H. Prescott, Jr., and F. L. Reynolds, Report RL-4.6.206, Sept. 29, 1943.
UCRL 2:   C. H. Prescott, Jr., and J. A. Holmes, Report RL-4.6.260, Apr. 27, 1944.
UCRL 3:   F. L. Reynolds and J. A. Holmes, Report RL-4.6.265, June 5, 1944.
UCRL 4:   C. H. Prescott, Jr., Report RL-4.6.273, July 27, 1944.
UCRL 5:   C. H. Prescott, Jr., F. L. Reynolds, and J. A. Holmes, The Preparation of
      Uranium Metal by Thermal Dissociation of the Iodide, in National Nuclear Energy
      Series, Division VIII, Volume 6.

Westinghouse 1:   J. W. Marden, Report A-605, Mar. 23, 1942.
Westinghouse 2:   W. C. Lilliendahl, G. Meister, R. Nagy, D. Wroughton, N. C. Beese,
      and J. W. Marden, Papers Relating to the Production of Uranium by the Electrolytic
      Method, June 21, 1946.

Chapter 5

PHYSICAL PROPERTIES OF URANIUM METAL

1. STRUCTURE AND MECHANICAL PROPERTIES OF SOLID URANIUM

Uranium is a very dense metal, 80 per cent heavier than lead. The freshly polished metal surface is silver-bright but tarnishes in air within a few hours. The low-temperature form of uranium (α uranium) is somewhat malleable; its elasticity is so low that it has been described as "semiplastic." The medium-temperature form (β uranium) is brittle, but the high-temperature form (γ uranium) is plastic. The metal can be forged, drawn, or extruded at high temperature, or it can be cold-worked.

1.1 <u>Crystal Structure of Uranium Metal.</u> The early interpretations of the x-ray diagram of α-uranium metal, by which it was first assigned a body-centered cubic crystal structure (McLennan and McKay, 1930) and later a monoclinic lattice (Wilson, 1933), have proved erroneous. Jacob and Warren (1937) established the crystal lattice of α uranium to be orthorhombic (space group V_h^{17}). This structure can best be interpreted as a distorted hexagonal closest packing. The unit cell contains four atoms in the positions (0 0 0; $\frac{1}{2}$ $\frac{1}{2}$ 0) + (0 y $\frac{1}{4}$; 0 $\bar{y}$ $\frac{3}{4}$) with the parameter y = 0.105 ± 0.005. The values of the lattice parameters in Table 5.1 were found by various observers. For the most precise determination at the Massachusetts Institute of Technology two samples of uranium of highest purity were used. They were quenched from 750 to 800°C, and the structure was determined by means of a symmetrical focusing back-reflection camera.

The exact structure of the β-phase uranium is unknown. It cannot be "frozen" in pure uranium, but, according to observation of the Ames group (MP Ames 1), this can be achieved in uranium-chromium alloys. Study of these alloys indicates that the β phase may be cubic with a giant unit cell ($a_0 = 12.88$ A) containing 58 atoms. At Battelle (Battelle 2) similar quenching experiments were carried out with uranium-molybdenum alloys. At first it was thought that a cubic face-centered phase obtained by furnace-cooling a 5 per cent molybdenum alloy held for 90 min at 650°C was the β phase, but its lattice constant (4.93 A) indicates that it probably is the oxide UO (see Table 11.2 in Chap. 11). Later a 0.6 per cent molybdenum alloy, quenched from 690°C, was found to contain a new, probably orthorhombic phase, which was tentatively interpreted as the β phase. Allen (British 7) confirmed that quenching from the γ and β phases did not retain the structures representative of these ranges. Quenching of certain metal

Table 5.1 — Crystal Lattice Parameters of α Uranium at Room Temperature

Parameters, A			Density,	
a_0	b_0	c_0	g/cc	References
2.852	5.865	4.945	18.97	Jacob, 1937
2.852	5.859	4.944	19.00	Battelle 1
2.8482*	5.8565*	4.9476*	19.050	MIT 1

*±0.01 per cent.

samples from 600°C did show an additional phase which was considered to be associated with an impurity in the metal, probably iron. Attempts to obtain the x-ray diagram of pure β uranium by means of a high-temperature camera have failed because of the weakness of the lines (Battelle 3). Usually only lines of a face-centered cubic phase, probably the oxide UO, are observed in high-temperature photographs (MP Ames 2). More recently it has been reported (Tucker, 1949) that β uranium (stabilized by 2 atom % chromium) has an orthorhombic cell unit of the dimensions $a_0 = 7.51$ A, $b_0 = 15.00$ A, $c_0 = 5.59$ A, containing 30 atoms. It appears probable, despite the uncertainties which at present exist, that β uranium is structurally related to α uranium, that it has a much larger unit cell, and that it is of markedly lower symmetry.

The structure of the γ phase of uranium was determined by the high-temperature camera at 785°C (Battelle 3) and 680°C (MP Ames 3) and found to be body-centered cubic. Like the β phase, this phase cannot be frozen in pure uranium (MP Ames 4). A value of 3.48 A was observed in pure uranium at 785°C (Battelle 4); reduction to room temperature gave $a_0 = 3.43$ A.

The lattice constant of γ uranium at room temperature has been studied by Wilson and Rundle (1949) as a function of molybdenum content, since it has been found that several metals including chromium and molybdenum stabilize the γ-uranium structure at room temperature. Graphical extrapolation to pure uranium gave a lattice constant of 3.467 ± 0.005 A and a room-temperature density of 18.89 ± 0.05 g/cc for γ uranium. The early results of McLennan and McKay (1930) indicate that they obtained the γ form, probably stabilized by some accidental impurity.

1.2 <u>Atomic Dimensions</u>. Most metallic elements, with the exception of such nontypical metals as bismuth or antimony, have crystal structures that deviate only slightly from the closest cubic or hexagonal packing of solid spheres. Uranium is an exception. Probably because of a rather strong atomic binding between a uranium atom in the lattice and its four nearest neighbors, the uranium atoms behave in the α phase as if they were nonspherical. Consequently, the atomic radius of α uranium cannot be defined without ambiguity. Each uranium atom in this phase has two nearest neighbors at a distance of 2.75 A (at room temperature), two second-nearest neighbors at a distance of 2.85 A, four neighbors at a distance of 3.25 A, and four neighbors at a distance of 3.34 A. Thus the uranium atoms can be described roughly as ellipsoids with a small half-axis of 1.4 A and a large half-axis of 1.65 A. Pauling (1947) suggests that α uranium contains straight strings of strongly bonded atoms, similar to the structure he proposes for β tungsten. Tucker (1949) considers α uranium to show a layer structure of corrugated sheets. The binding within the sheets is stronger than between the sheets, similar to the structures found in arsenic, antimony, and bismuth. The binding within the sheets shows a marked dependence on direction. In α uranium the directional properties of the binding appear to arise from hybridization of appropriate atomic orbitals to a trigonal bipyramid arrangement of covalent bonds in which one of the corners is occupied by an electron pair (Tucker, 1949).

The atomic radius of uranium in the cubic γ phase, calculated from the extrapolated lattice constant at room temperature, is 1.485 A.

1.3 <u>Density</u>. Density values calculated from the lattice structure of α uranium are given in Table 5.1. The most reliable value is 19.050 g/cc. If the γ phase existed at room temperature, its density would be 18.7 g/cc. The experimentally determined density of uranium often is somewhat lower than the value calculated from the x-ray structure of the α phase. This is probably due to the presence of impurities, especially carbon. At the Massachusetts Institute of Technology (MIT 2) the density of Metal Hydrides, Inc. (Beverly Plant), uranium metal was found to be a linear function of its carbon content.

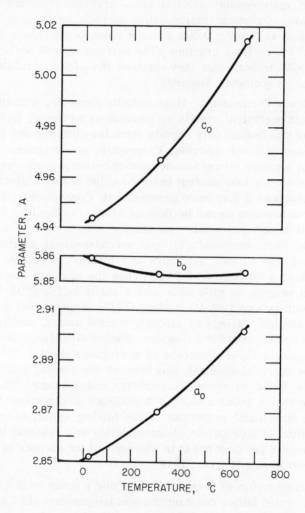

Fig. 5.1—Constants of α uranium as function of temperature.

The density values obtained fitted the equation

$$\rho = 19.05 - 2.14[C] \tag{1}$$

where [C] is the carbon content in weight per cent. The highest measured value was 19.02 g/cc (0.01 per cent C), and the lowest was 18.58

g/cc (0.23 per cent C). Extrapolation to [C] = 0 gave 19.05 g/cc for the density of pure uranium, which agrees well with the value calculated from x-ray data. Measurements made elsewhere (Driggs, 1930; MP Ames 5; MP Chicago 1,2; British 1; Natl. Bur. Standards 1) gave values between 18.7 and 19.08 g/cc, with the lower figures corresponding to carbon contents of 0.05 to 0.06 per cent (Los Alamos 1). The average density of 22 α-rolled bars of commercial metal was 18.882 g/cc (MP Chicago 1,2).

The density of uranium prepared by sintering metal powder may be as low as 15 g/cc without pressing and 17 g/cc after pressing (MP Chicago 3). The bulk density of uranium powder was given by Marden (SAM Columbia 1) as 9.4 g/cc after pressing at 20 tons/in.^{2}, 12.3 g/cc after pressing at 79 tons/in.^{2}, and 15.0 g/cc after pressing at 140 tons/in.^{2}.

1.4 **Thermal Expansion.** The coefficient of linear thermal expansion of uranium in the anisotropic α state is strongly dependent on crystallographic direction (British 11,12,13). Table 5.2 and Fig. 5.1 illustrate this dependence as found by high-temperature x-ray studies

Table 5.2—Coefficient of Linear Thermal Expansion of α
Uranium in the Three Main Crystallographic Directions

Direction parallel to the axis	Average expansion coefficient, 10^6 cm/cm/°C	
	25–300°C	25–650°C
a [1 0 0]	23 ± 3	28 ± 2
b [0 1 0]	−3.5 ± 2	−1.4 ± 1
c [0 0 1]	17 ± 2	22 ± 1

at Battelle (Battelle 5; British 13). The volume expansion coefficient, calculated from the dilatometrically determined linear coefficients, is 44×10^{-6} cc/cc/°C, in the range 25 to 300°C. The x-ray data in Table 5.2 give a volume coefficient of 37×10^{-6}, which is in more or less satisfactory agreement with the dilatometric results.

2. MECHANICAL CONSTANTS OF URANIUM

The mechanical properties of uranium are discussed in more detail in Division IV, Volume 12 A, of the National Nuclear Energy Series. Here we shall give only the most important results.

2.1 **Hardness.** Uranium cast at 1200°C usually has a surface hardness of about 100 on the Rockwell B scale, but hardness declines to 85R$_B$ in 0.05 in. depth (MP Chicago 4). The maximum hardness,

70 to 71 on the Rockwell A scale (corresponding to 115 on the Rockwell B scale), can be obtained by heating uranium at 900°C for 5 hr or more and then quenching in cold water (Los Alamos 2). This hardening effect may be due to the presence of carbide in the metal (British 3). British observers found that the average hardness of cast, refined Widnes metal was 200 to 220 on the Brinell scale (British 3).

Uranium may also be work-hardened by swaging or rolling, e.g., from $90R_B$ to $115R_B$ ($57R_A$ to $71R_A$) (Battelle 8,9). British workers have succeeded in reducing high-purity metal by 80 per cent in thickness without recourse to intermediate annealing or stress-relieving treatment (British 3). Most of the work-hardening is associated with the first 20 per cent reduction in thickness. Work-hardened metal begins to soften at 150°C; as shown in Table 5.3, complete annealing is achieved at 650 to 700°C (Battelle 8; MIT 3,4; Natl. Bur. Standards 2). British observations show that complete recovery from the effect of cold-working requires an annealing temperature of at least 600°C (British 14).

Micro-hardness measurements of uranium by the methods of Knoop and Eberbach were carried out at the National Bureau of Standards (Natl. Bur. Standards 3,4,5,6) and the Massachusetts Institute of Technology (MIT 5). Table 5.4 gives some typical results (Natl. Bur. Standards 3,6).

The decrease in uranium hardness with temperature was measured at Battelle (Battelle 10,11) and in Britain (British 4). Hardness changes very little up to 200°C; it declines rapidly from 252 Brinell at 200°C to 13 Brinell at 650°C. It increases again to 30 or 40 Brinell units in the β state (660 to 700°C). The hardness drops below the range of measurability on the Brinell scale in the plastic γ state (above 770°C).

2.2 Elastic and Inelastic Deformation. Uranium in the α state was called above "semiplastic;" in fact, its stress-strain curves often show curvature even at loads of less than 10×10^3 psi (Natl. Bur. Standards 7; British 15). On the other hand, some stress-strain curves have been obtained which were apparently linear up to 50×10^3 psi (MP Chicago 5), but the elastic modulus calculated from the slope of these linear sections was so low (approximately 5×10^6 psi) that the linearity was probably only an apparent one, caused by insufficient precision of measurements at low stresses. More reliable may be the several stress-strain determinations (MP Chicago 6,7; Los Alamos 1,2,3; Battelle 6,12; British 2,5; MIT 6) which showed linearity up to 10×10^3 to 15×10^3 psi and whose initial slope indicated an elastic modulus of 15×10^6 to 25×10^6 psi. Using a dynamic method, Laquer (1949) found a value for Young's modulus of 29.8×10^6

psi, with a probable accuracy of 1 per cent. Because of anisotropy of the α phase, the elastic modulus probably depends on crystallographic direction. The available measurements are insufficient to establish this influence clearly, but the slightly lower value of the

Table 5.3 —Annealing of Work-hardened Uranium*

Temperature, °C†	Rockwell C hardness
310	36.6
490	24
600	22
720	9
770	0

*MIT 3.
†Samples annealed 1 hr at these temperatures.

Table 5.4 —Hardness of α-rolled Uranium

Scale	Transverse specimen	Longitudinal specimen
Rockwell B	97	98
Rockwell G	80.5	83
Vickers diamond (10- to 15-kg load)	220	240
Vickers diamond (30- to 45-kg load)	...	250
Brinell (calculated from R_B)	240	260
Brinell (calculated from R_G)	250	280
Knoop (200-g load)	260	320
Knoop (500-g load)	250	300
Knoop (1,000-g load)	240	320
Eberbach (263-g load)	270	300
Eberbach (645.6-g load)	260	290

elastic modulus (15×10^6 to 20×10^6 psi) obtained with presumably longitudinal specimens of rolled metal may have been a consequence of crystal orientation due to cold-working (MP Chicago 8,9; MIT 3). The elasticity of uranium, like its hardness, decreases rapidly with temperature in the α phase but increases again in the β phase (MP Chicago 5; Battelle 13) (see Fig. 5.2). Köster (1948) has reported a value of 9×10^6 psi for the elastic modulus of uranium at 20°C, but, since he gives no data on the uranium metal employed, this measurement is of dubious worth.

The shear modulus of uranium was measured by Snyder and Kamm (MP Chicago 8,10). They found values of 6.6×10^6 psi and 6.8×10^6

psi for cold-swaged and annealed wire, respectively. Laquer (1949) gives a value for the shear modulus of 12.10×10^6 psi, about double the value of Snyder and Kamm. Laquer claims a probable accuracy of 0.5 per cent.

A rigidity modulus of 8.5×10^6 psi was observed by Wollan and Stephenson (MP Chicago 11) with a sample whose elasticity modulus was 19×10^6 psi.

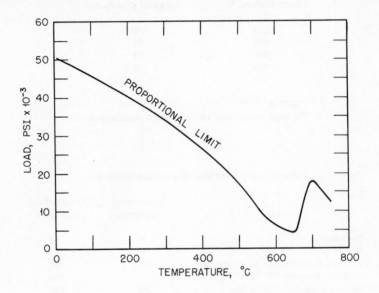

Fig. 5.2 — Elastic change of uranium with temperature.

The measurements of Snyder and Kamm indicated a bulk modulus of 9×10^6 psi for cold-worked metal and 15×10^6 psi for annealed wire. These values correspond respectively to volume compressibilities of 12×10^{-8} and 6.4×10^{-8}/psi. Bridgman (1922) found that the average volume compressibility of probably not very pure uranium at 12,000 atm (about 150×10^3 psi) and 30°C is 9.7×10^{-7} cm^2/kg (about 7×10^{-8}/psi). More recent work by Bridgman (1948) indicates that uranium metal undergoes a volume reduction of 5.9 per cent on the application of a pressure of 100,000 kg/cm^2 compared to values of 12.4 per cent for thorium and 50 per cent for potassium.

Poisson's ratio (contraction normal to stress/elongation parallel to stress) is high, in accordance with the semiplastic nature of uranium. Battelle observers (Battelle 6,12) found ratios of 0.32 to 0.43 for γ-extruded commercial metal, and Foote and coworkers (MP Chicago 8) found 0.42, 0.49, and 0.27 for α-rolled uranium annealed for over 12 hr at 565, 750, and 800°C, respectively. The above-mentioned measurements of Snyder and Kamm lead to a Poisson's ratio of 0.20 for the cold-worked metal and 0.31 for the annealed wire. In good agreement with the results of Snyder and Kamm, Laquer (1949) reports a value of 0.23 for Poisson's ratio. The elasticity and rigidity moduli found by Wollan and Stephenson correspond to an implausibly low value of Poisson's ratio, i.e., about 0.1.

The velocity of sound in α uranium was calculated by Simon to be 1,500 meters/sec (British 1). It probably depends on crystallographic direction.

Being semiplastic, α uranium has no definite yield point. Its yield strength has been determined (MP Chicago 6,7; Los Alamos 1,3; Battelle 7,12,13,14; British 2,3,16) as the stress under which the length of a specimen exceeds by a certain "offset" (usually 0.02, 0.1, or 0.2 per cent) the length which it would have if the deformation were elastic (i.e., if the stress-strain curve were linear). Because of the above-mentioned uncertainty of the initial slope of the stress-strain curve (i.e., of the value of the elastic modulus), the determination of the yield strength by this method is very uncertain. The variation in yield strength reported by British and American workers is great. Most of the values given for the yield strength of uranium at room temperature are of the order of 25×10^3 to 35×10^3 psi for 0.2 per cent offset (MP Chicago 6,7; Los Alamos 1), but some measurements gave results as high as 76×10^3 psi, and some gave values as low as 8×10^3 to 13×10^3 psi (Battelle 12).

The highest yield strength was observed in α-worked metal (MP Chicago 7). Slow cooling after annealing decreases the yield strength (British 16), and quenching, particularly from γ-phase temperatures, increases it. For example, the yield strength (0.2 per cent offset) of a set of α-rolled specimens was increased from 25×10^3 psi for furnace-cooled material to 32×10^3 psi for a specimen quenched from 570°C and to 42×10^3 psi for a specimen quenched from 800°C (MP Chicago 6). In another example the yield strength (0.1 per cent offset) rose from 17×10^3 psi for an "as-rolled" specimen to 57×10^3 psi for a similar specimen quenched from 1000°C (Battelle 14). The yield strength of α uranium decreases rapidly with increasing temperature (Battelle 13; British 17,18), for example, from 43×10^3 psi at room temperature to 8.9×10^3 psi at 600°C. Figure 5.3 shows that β uranium has a definite yield point, 18×10^3 psi at 700°C and 13×10^3 psi at 750°C (MP Chicago 5).

Like the loading curves, the unloading curves of uranium also are nonlinear (Battelle 6), so that the residual deformation after unloading is smaller than the offset measured under stress. The residual deformation remaining after the compression of uranium cylinders was measured in three laboratories (MP Chicago 5; Los Alamos 2; Natl. Bur. Standards 8). Such cylinders can be compressed to barrels with a decrease in height of 35 per cent without shattering.

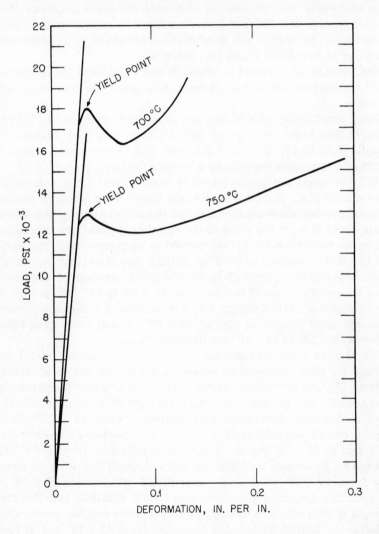

Fig. 5.3—Compression of uranium at 700 and 750°C.

The pressure required for 10 per cent permanent compression decreases with temperature (MP Chicago 5; British 17) from 111×10^3 psi at 25°C to 7×10^3 psi at 650°C but increases again in the β phase to 17×10^3 psi at 700°C and 12×10^3 psi at 750°C (MP Chicago 5).

The offset under stress and the residual deformation after the release of stress both depend on the duration of loading, which fact indicates creep phenomena. Creep measurements on uranium were carried out at the Massachusetts Institute of Technology (MIT 7). Table 5.5 summarizes the results.

2.3 Strength. The ultimate tensile strength of uranium varies between 50,000 and 200,000 psi, depending on cold-working and thermal

Table 5.5 — Creep of Uranium

Temp., °C	Load, 10^{-3}psi	Duration of experiment, hr	Average elongation per 100 hr, %
350	9	260	1.3
356	3.5	263	0.19
356	5.3	934	0.07 (last 300 hr)
355	3.52	~1,000	0.037
355	5.27	1,000	0.089
353	8.80	1,000	0.22

treatment of the specimen (MP Chicago 2,7; British 2; MIT 3,4; Natl. Bur. Standards 3,7,8). Cold-working (α rolling or swaging) gives the highest tensile strength, namely, 170,000 to 200,000 psi (MIT 3,4; Natl. Bur. Standards 3,7). Annealing reduces it, e.g., from 170,000 psi after swaging to 135,000 psi after annealing at 625°C and to 65,000 psi after annealing at 830°C (Natl. Bur. Standards 3). This is confirmed by British tests (British 2). Cast or γ-extruded specimens have almost as low tensile strength as γ-annealed and slowly cooled specimens, roughly 65,000 to 90,000 psi (Los Alamos 3; Battelle 10,15; British 2,3,6). The tensile strength is somewhat improved by quenching, particularly from the high-temperature α region or γ region; values of about 90,000 to 130,000 psi have been obtained for rolled, forged, or swaged specimens quenched from 600 to 1000°C (MP Chicago 2,7; MIT 6; Natl. Bur. Standards 7).

Tensile strength decreases rapidly at elevated temperatures, from 53×10^3 psi at room temperature to 27×10^3 psi at 150°C and 12×10^3 psi at 600°C (Battelle 16).

Seitz (MP Chicago 12) commented on the low value of the ratio of tensile strength to elastic modulus for uranium, $2 \times 10^5/20 \times 10^6 = 0.01$, as compared to a theoretical value of approximately 0.3. He interpreted this as an indication of minute cracks in the metal, which reduce considerably the area to which the stress is applied.

Elongation and area reduction in the tensile-strength test are small in cold-worked uranium metal, i.e., 0 to 3 per cent (British 2,3) and 0 to 8 per cent (MP Chicago 7), respectively. They are a little larger,

5 to 10 per cent and 3 to 15 per cent, respectively, in γ-extruded material (Los Alamos 3; British 6). Much greater ductility (elongation and area reduction values up to 20 per cent) has been found for α-rolled specimens annealed and quenched from the high α region (350 to 600°C). Quenching from the β or γ region (700 to 800°C) reduced the values to 10 per cent or less. Drop-hammer tests showed very little ductility in the low β region (660 to 670°C); the ductility increased somewhat at 705°C (Battelle 17).

The Charpy impact strength of three uranium ingots was given as 13 to 16 ft-lb, corresponding to 1.9 to 2.3 meter-kg (Natl. Bur. Standards 10). British observers gave a value of 5 ft-lb for Widnes cast metal (British 2), which is equal or slightly inferior to extruded metal (British 3).

The fatigue strength of α-extruded bars of uranium was measured on a Krause 10,000-rpm cantilever-beam machine at Battelle (Battelle 18). The endurance limit is in the range of 25,000 psi. At 35,000 psi the specimens failed after 1.2×10^5 to 2.5×10^5 cycles; preliminary treatment at 25,000 psi increased the endurance under higher stresses. The ratio of endurance limit to tensile strength is about 0.2, which is a low value for a metal.

Rupture strength of uranium disks was measured by supporting them in a holding ring and rupturing with a central column (0.5 in. in diameter). The average rupture strength was 3,928 lb for γ-extruded metal and 4,648 for α-extruded material (MP Chicago 13). Annealing for 15 hr at 600 to 800°C decreased the rupture strength slightly, and annealing at temperatures greater than 900°C and quenching increased it considerably (MP Chicago 14).

3. THERMODYNAMIC PROPERTIES OF URANIUM

3.1 Specific Heat, Energy, and Entropy. The specific heats of uranium at low temperatures were measured by Long and coworkers at the U. S. Bureau of Mines at Berkeley (U. S. Bur. Mines 2), and those at high temperatures by Moore (U. S. Bur. Mines 1), Moore and Kelley (1947), and Ginnings and Corruccini (Natl. Bur. Standards 14). The results are given in Tables 5.6 and 5.7 and Figs. 5.4 and 5.5. The Dulong-Petit value is exceeded at a temperature as low as −73°C.

Long and coworkers gave only C_p values; Moore (U. S. Bur. Mines 1) and Moore and Kelley (1947) gave values of only $H_T - H_{298.16}$; and Ginnings and Corruccini gave values of $H_T - H_{273.16}$, C_p, and $S_T - S_{273.16}$. The values of C_v, in Table 5.6, were calculated by Simon (British 1). In Table 5.7 one set of C_p, H_T, and S values was calculated by Simon, and the other set is from Ginnings and Corruccini. In

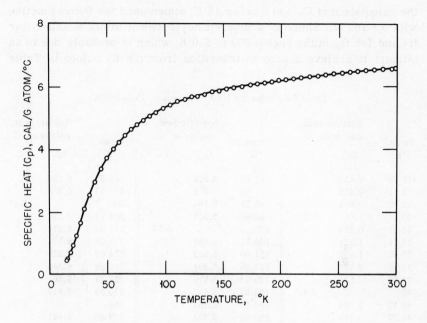

Fig. 5.4—Specific heat of uranium at low temperatures.

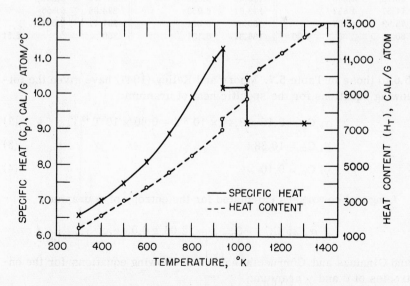

Fig. 5.5—Specific heat of uranium at elevated temperatures.

the calculation of C_v and S below 15°K, Simon used the Debye function with $\theta = 162°K$. There is a discrepancy between Simon's values for H_T and for C_p in the region 300 to 500°K, which is probably due to an attempt to achieve a smooth transition from the C_p values in Table

Table 5.6 — Specific Heat of Uranium, 15 to 300°K

Temp., °K	Specific heat, cal/mole/°C		Temp., °K	Specific heat, cal/mole/°C		Temp., °K	Specific heat, cal/mole/°C	
	C_p	C_v		C_p	C_v		C_p	C_v
15.38	0.432		81.52	4.955		192.57	6.154	
16.61	0.496		87.05	5.072		198.47	6.203	
18.55	0.665		92.96	5.194		200		6.17
20		0.81	98.93	5.325		205.12	6.232	
21.01	0.914		100		5.33	211.07	6.274	
23.91	1.211		104.81	5.435		217.76	6.295	
27.55	1.639		111.00	5.532		224.89	6.335	
31.19	2.080		117.40	5.591		231.88	6.348	
35.32	2.530		123.48	5.639		238.73	6.381	
40		2.94	128.23	5.697		245.59	6.417	
40.10	2.954		134.58	5.754		250		6.39
45.27	3.369		140.90	5.823		252.89	6.441	
50.49	3.731		147.06	5.874		259.36	6.476	
55.53	4.011		150		5.86	266.32	6.486	
60		4.20	153.35	5.924		273.02	6.524	
60.63	4.231		159.76	5.964		280.16	6.544	
65.78	4.446		166.65	6.008		287.50	6.548	
71.57	4.657		173.61	6.053		294.88	6.556	
75.99	4.807		180.04	6.091		297.71	6.574	
80		4.90	186.37	6.131		300		6.51

5.6 to those in Table 5.7. Moore and Kelley (1947) have given the following equations for the specific heat of uranium:

$$\alpha, \; C_p = 3.15 + 8.44 \times 10^{-3}T + 0.80 \times 10^5 T^{-2} \qquad (2)$$

$$\beta, \; C_p = 10.38 \qquad (3)$$

$$\gamma, \; C_p = 9.10 \qquad (4)$$

Long and coworkers calculated for the entropy of α uranium,

$$\alpha, \; 25°C: \qquad S_{298.16} = 12.03 \pm 0.03 \text{ e.u.}$$

and Ginnings and Corruccini gave the following equations for the entropies of β and γ uranium:

β, 941 to 1047°K: $\quad S_T - S_{273.16} = 23.362 \log T - 58.770 \qquad (5)$

γ, 1047 to 1170°K: $\quad S_T - S_{273.16} = 21.062 \log T - 50.745 \qquad (6)$

Table 5.7 — Specific Heat, Heat Content, and Entropy of Uranium, 300 to 1375°K

Temp., °K	C_p, cal/g atom/°C		H_T, cal/g atom		S, e.u.	
	Simon*	Corruccini[†]	Simon*	Corruccini[†]	Simon*	Corruccini[†]
300	6.6	6.649	1,440	1,539.12	12.07	12.052
400	7.0	7.072	2,160	2,202.57	13.97	13.941
500	7.5	7.606	2,940	2,935.42	15.60	15.601
600	8.1	8.227	3,730	3,725.45	17.04	17.056
700	8.9	8.952	4,580	4,582.96	18.35	18.387
800	9.9	9.863	5,535	5,520.47	19.62	19.646
900	11.0	11.107		6,566.80		20.882
935(α)	11.3		6,975		21.29	
941(α)		11.737		7,066.16		21.436
935(β)	10.2[‡]		7,640		22.00	
941(β)		10.147		7,740.37		22.152
1000		10.147		8,336.97		22.760
1045(β)	10.2[‡]		8,770		23.14	
1047(β)		10.147		8,815.97		23.236
1045(γ)	9.2[§]		9,940		24.12	
1047(γ)		9.147		9,947.04		24.316
1100		9.147	10,440	10,430.16	24.55	24.761
1200			11,350		25.38	
1300	9.2[§]		12,260		26.12	
1375			13,000		26.8	

*British 1.
[†]Natl. Bur. Standards 14.
[‡]Value for the range 935(β) − 1045(β).
[§]Value for the range 1045(γ) − 1300.

Moore and Kelley (1947) gave the following equations as representing the experimental results for the three uranium phases:

α, 298 to 935°K: $\quad H_T - H_{298.16} = 3.15T + 4.22 \times 10^{-3}T^2$
$$- 0.80 \times 10^5 T^{-1} - 1046 \quad (\pm 0.2\%) \qquad (7)$$

β, 935 to 1045°K: $\quad H_T - H_{298.16} = 10.38T - 3525 \quad (\pm 0.1\%) \qquad (8)$

γ, 1045 to 1300°K: $\quad H_T - H_{298.16} = 9.10T - 1026 \quad (\pm 0.1\%) \qquad (9)$

Ginnings and Corruccini also gave similar equations for the β- and γ-uranium phases.

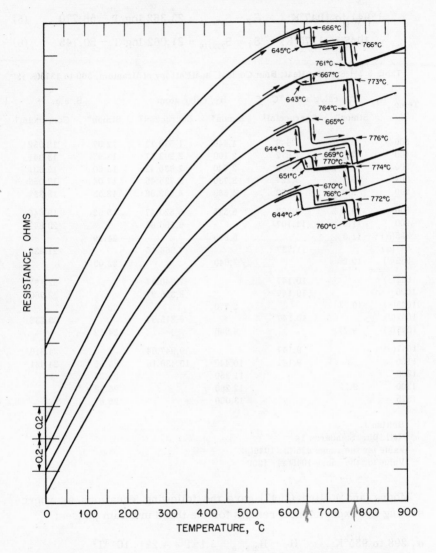

Fig. 5.6—Electrical resistance of uranium (on an arbitrary scale) as a function of temperature.

β, 941 to 1047°K: $H_T - H_{273.16} = -3170.52 + 10.147T$ $(\pm 0.1\%)$ (8a)

γ, 1047 to 1173°K: $H_T - H_{273.16} = -992.80 + 9.147T$ $(\pm 0.1\%)$ (9a)

They stated that the α range from 273 to 941°K cannot be covered by a simple equation.

3.2 <u>Solid Transformations</u>. The transition temperatures of the three uranium phases, α, β, and γ, have been determined by thermal analysis as well as from dilatometric curves and electrical conductivity curves (see Fig. 5.6). The most reliable observed transition temperatures are given in Table 5.8.

Table 5.8 — Transition Temperatures of Uranium

Transition temperature, °C

Material	Heating		Cooling		Method of determination	References
	$\alpha \rightarrow \beta$	$\beta \rightarrow \gamma$	$\gamma \rightarrow \beta$	$\beta \rightarrow \alpha$		
Forged ²/₁₀-in. round bars of cast metal	673	775	770	...	Thermal arrests	MIT 6
Not described	662 ± 3	772 ± 3	...	...	Heat-content measurements	U.S. Bur. Mines 1
U with 0.03% C	...	...	775	645	Thermal arrests	MP Ames 6
U with 0.39% C	...	...	785	652		
U with 1.5% C	...	...	766	633		
Not described	665	776	770	644	Resistivity	Natl. Bur. Stand.9
γ-extruded rod, longitudinal specimen	660.5 ±1.5	...	...	...	Thermal arrests	Battelle 6
Ingots	665−670	766−776	760−770	643−651	Electrical conductivity	Natl. Bur. Stand.11
Extruded uranium	668−676	775−783	767−760	644−639	Dilatometry	British 12
γ-extruded 99.9% uranium	669−671.5	779−785	765	642−643	Dilatometry	British 7
Extrapolated for pure U from results obtained with 98.6−99.9% metal	644 ± 4	765 ± 10	752 ± 3	624 ± 3	Thermal arrests	British 3
Uranium metal (99.9%)	667	772	764	645	Resistivity	Dahl and Van Dusen (1947)

Figure 5.6 illustrates the fact that with α-extruded metal, as well as with cast metal and biscuit metal, the allotropic transitions usually are delayed both on heating and on cooling. In γ-extruded material, on the other hand, the transitions begin without delay. However, even in this material the $\alpha \rightarrow \beta$ transition is not strictly isothermal but extends over a range of several degrees. This may be due to non-uniform internal pressure (Battelle 10). Since at the transition temperature the β phase is less dense than the α phase, increased pressure must raise the transition temperature.

The effect of carbon on the transformation temperatures shown in Table 5.8 is insignificant (Natl. Bur. Standards 11), but other additions may affect these temperatures very much. As mentioned in Sec. 1.1, the γ phase can be preserved in these alloys at room temperature by rapid cooling. The emissivity of uranium changes suddenly at 1048 to 1050°C (Wisconsin 1); this may indicate a third allotropic transformation. A third allotropic change has been reported (British 4) but at temperatures considerably below 1040°C.

The $\alpha \rightarrow \beta$ transformation absorbs 665 cal per gram atom, and the $\beta \rightarrow \gamma$ transformation absorbs 1170 cal per gram atom (see Table 5.7). While the value for the $\beta \rightarrow \gamma$ transition given by Moore and Kelley (1947) agrees with that given here, their value for the $\alpha \rightarrow \beta$ transition, 680 cal per gram atom, is appreciably different. They also reported a temperature of 662°C for the $\alpha \rightarrow \beta$ transition, which is rather lower than the majority of the values reported. The entropy of the first change is 0.71 e.u.; that of the second one, 0.98 e.u. (U. S. Bur. Mines 1).

3.3 Melting Point and Energy of Fusion of Uranium. Uranium was long considered as having a high melting point, a property that seemed consistent with its position in the periodic system at the bottom of the column occupied by metals with increasingly high melting points, namely, chromium, molybdenum, and tungsten. Driggs and Lilliendahl[*] (1930) and Hole and Wright (1939) observed melting points of 1690°C and 1700 ± 25°C, respectively. However, these values were found to be much too high in 1941 to 1942, when work on the preparation of pure metal got under way. The first indication that the melting point was lower appears to be given in a publication from the Metal Hydrides, Inc. (Alexander, 1938). In England (British 8) a melting point of 1150°C was found. More recent determinations gave 1105 to 1116°C for 99.8 and 99.9 per cent pure uranium (British 2,3). In America the first redeterminations gave a value of 1300°C (MP Chicago 15; Natl. Bur. Standards 12), and subsequent, more precise measurements at the National Bureau of Standards (Natl. Bur. Standards 13) led to still lower values, 1080 ± 20°C for 99.1 per cent pure uranium and 1125 ± 25°C for 99.7 per cent pure metal. Values ranging from 1123°C (1.5 per cent carbon) to 1134°C (0.03 per cent carbon) were obtained at Ames (MP Ames 6). A value of 1125°C was extrapolated at the Massachusetts Institute of Technology (MIT 8) from the uranium-aluminum diagram of state.

Probably the most reliable determinations of the melting point were made at the National Bureau of Standards by Dahl and Cleaves (1949). They showed that uranium containing 0.03 to 0.05 per cent

---

[*] These workers attribute the incorrect melting point to an inadvertent choice of a poor method (pyrometric) rather than to impure metal.

carbon first melted at 1125°C, but if the metal was kept just above the melting point for up to 15 hr, its solidification point rose gradually to 1130 to 1132°C. During this time the carbon content decreased to

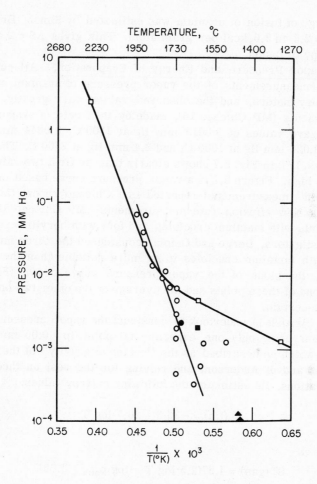

Fig. 5.7—Vapor pressure of uranium at 1300 to 2300°C. Sources: O, Anderson; □, Creutz; ■, Simon; ▲, Derge and Cefola, tantalum crucibles; ●, Derge and Cefola, beryllia crucibles.

1 per cent of its original value (from 500 to 5 ppm), and the remaining carbon was largely segregated as carbide in the crust formed on the metal. It is thus reasonable to attribute the rise in melting point to "self-purification" of the metal. Consequently the most likely value for the melting point of pure uranium is

$$t_F = 1133 \pm 2°C$$

$$T_F = 1406 \pm 2°K$$

The energy of fusion of uranium was estimated by Simon (British 1) as $\Delta H_F = 2.5$ to 3.0 kcal per gram atom. This gives $\Delta S = 2$ e.u. for the entropy of fusion.

3.4 <u>Vapor Pressure and Energy of Evaporation.</u> All currently available measurements of the vapor pressure of uranium are of a preliminary nature, and the observed values vary greatly. Early measurements (MP Chicago 16), made by the "rate of evaporation" method, gave values of 0.0013 mm Hg at 1300°C, 0.0044 mm Hg at 1620°C, 0.026 mm Hg at 1900°C, and 2.4 mm Hg at 2300°C. The curve of log p vs. $1/T$ in Fig. 5.7 shows clearly that the first two values are much too high. Figure 5.7 is a vapor pressure curve based on "rate of effusion" measurements reported in a Chicago report (MP Chicago 17). New effusion rate measurements (MP Chicago 18) gave three points with tantalum crucibles and four with beryllia crucibles. The investigators, Derge and Cefola, considered the three measurements with tantalum crucibles to be more reliable than the others; however, the slope of the vapor pressure curve drawn (Fig. 5.8) through one of these points and the average of the other two obviously is very uncertain.

Simon (British 1) undertook to construct the vapor pressure curve of uranium using only one experimental point (p = 0.05 mm Hg at 2200°K), which he described as the "center of gravity" of the results of Creutz and of Anderson, and relying for the rest on theoretical considerations. He estimated the following entropy values:

$$S° \text{ (liq)} = -36.3 + 20.7 \log T \tag{10}$$

and

$$S° \text{ (gas)} = 4.57(2.5 \log T - \log p_{atm} + 3.66) \tag{11}$$

(The value log g = 0.6 was used for the symmetry factor in the last equation.) With the help of these entropy values and the one above-mentioned empirical point on the vapor pressure curve, he arrived at the equation

$$\log p = -(100 \times 10^3/4.57T) - 2.04 \log T + 15.4 \tag{12}$$

where the pressure is in millimeters of mercury. This gives the pressures in Table 5.9.

Figure 5.8 shows that Simon's curve lies approximately midway between the two curves of Derge and Cefola.

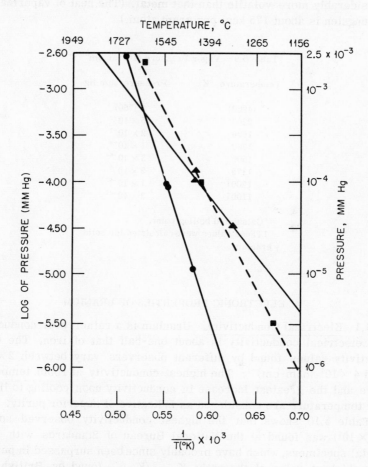

Fig. 5.8 —Vapor pressure of uranium. Sources: ■, Simon, semitheoretical values; ▲, Derge and Cefola, tantalum crucibles; ●, Derge and Cefola, beryllia crucibles.

The energies of vaporization, calculated from the slopes of the log p vs. 1/T curves, are as follows:

	λ, kcal/mole	λ_0, kcal/mole
Simon	93	~100
Derge and Cefola	58 (tantalum crucibles)	...
	137 (beryllia crucibles)	...

About the only thing that can be concluded from these figures is that uranium not only melts 2240°C lower than tungsten but is also considerably more volatile than that metal. (The heat of vaporization of tungsten is about 175 kcal per gram atom.)

Table 5.9 — Vapor Pressure of Uranium

Temperature, °K	Pressure, mm Hg
(4200)*	(760)
2200	5×10^{-2}
1900	2×10^{-3}
1700	1×10^{-4}
1500	3×10^{-6}
1375	2×10^{-7}
1200†	1×10^{-9}
1100†	3×10^{-11}

*Calculated boiling point.
†These values were calculated for solid
γ uranium.

4. ELECTRONIC PROPERTIES OF URANIUM

4.1 Electrical Conductivity. Uranium is a rather poor conductor. Its electrical conductivity is about one-half that of iron. The conductivity values found by different observers vary between 2×10^4 and 4×10^4 (ohm-cm)$^{-1}$. The highest conductivity at room temperature and the greatest increase in conductivity upon cooling to liquid air temperature are considered as indicative of superior purity.

Table 5.10 shows that the highest conductivity observed so far (4×10^4) was found at the National Bureau of Standards with 1943 metal specimens, which have probably since been surpassed in purity. The highest value of the ratio $K_{273°K}/K_{90°K}$, found by British observers, was 2.6; it was obtained with an American specimen having a specific conductivity of 3.5×10^4 at 0°C. This ratio is still considerably below the theoretical value ($K_{273°K}/K_{90°K} = 3.3$ to 4.0) calculated by Simon (British 1) for pure uranium from its Debye temperature. Prolonged annealing in the high α region (at 600°C), which was intended to release working stresses, somewhat improved the room temperature conductivity and particularly the ratio $K_{273°K}/K_{90°K}$, but not enough to approach the theoretical value. According to the National Bureau of Standards (Natl. Bur. Standards 8) the conductivity is decreased by annealing in the β or α region (see also Table 5.11, last column). Ebert and Schulze (1947) found a value of $3.45 \times$

10^4 (ohm-cm)$^{-1}$ for a sample of uranium claimed to be of 99.9 per cent purity. This value, as well as the temperature coefficient of resistivity, 2.18×10^{-3}, is in general agreement with other literature values. This is likewise true for the values reported by Balz (1947).

Table 5.10 — Electrical Conductivity of Uranium

Material	T, °C	Conductivity $(K) \times 10^{-4}$, (ohm-cm)$^{-1}$	Resistivity $(\rho) \times 10^6$, ohm-cm	Temp. coeff. of ρ $(\alpha) \times 10^3$, per °C	$\dfrac{K_{273°K}}{K_{90°K}}$	Reference
Westinghouse metal		3.1	32.1	2.1*	...	Driggs (1930)
Rolled wire 1/32 in.		2.44	41	...	...	MP Chicago 8,9
Cylinder (2.5 in. diameter) from a W ingot, annealed (10 min at 600°C)	25	4.0	25.0	2.78	...	Natl. Bur. Standards 11
Swaged wires (1−5 mm diameter), same material and treatment	25	3.6−3.8	26.2−27.5	2.70−2.82	...	Natl. Bur. Standards 11
99.93% pure American rolled metal, after prolonged annealing (600°C)	0	3.47	28.8	...	2.6	British 3,9
Same, no annealing	0	3.35	29.9	...	2.3	British 3,9
99.86 to 99.90% pure British metal, after prolonged annealing (600°C)	0	3.11−3.38	29.6−32.2	...	2.5	British 3,9
Same, no annealing	0	2.92−3.13	29.8−34.3	...	1.9−2.0	British 3,9
ICI metal, rolled strip 99.8−99.9% pure	23	3.1−3.2	31−32	...	...	British 2
ICI metal, forged, turned	23	3.0	34	2.0	2	British 2

*23 to 140°C.

In another British report (British 10) reference is found to measurements of the conductivity of a uranium sample (containing the following percentages of other elements: Ca, 0.03; Hg, 0.02; Sn, 0.002; Si, 0.14; Cu, 0.19; Fe, 0.60; Al, 0.04; and C, 0.30) from 13.9 to 373°K. No figures are given, but it is stated that the results can be represented, within ±0.5 per cent, by the equation

$$R_T = R_A + (1 - R_A) \frac{T}{273.2} \frac{G(\theta/T)}{G(\theta/273.2)} \tag{13}$$

where R_T is the ratio, $\rho_T / \rho_{273.2}$, of the resistivities at T°K and at 273.2°K, and G is the so-called "Grüneisen function," whose values can be taken from tables. A value of 175°K had to be assumed for θ, which is in satisfactory agreement with the characteristic tempera-

Table 5.11—Effect of Impurities and Heat-treatment on the Resistivity of Rolled Strip

| Analysis, % (Total uranium and major impurities) | | | | Heat-treatment | Resistivity $(\rho) \times 10^6$, ohm-cm | | | | Ratio, $\dfrac{\rho(-183°C)}{\rho(0°C)}$ | Residual resistance ratio, $\dfrac{\rho(-273°C)}{\rho(0°C)}$ | Conductivity (at 20°C) $\times 10^{-4}$, (ohm-cm)$^{-1}$ |
U	Fe	Si	C		20°C (293°K)	0°C (273°K)	-70°C (203°K)	-183°C (90°K)			
99.83	0.086	0.034	0.001	As cold-rolled	32.52	30.90	25.25	15.25	0.49		3.08
				Slow cooled 800–500°C and air-cooled	34.15	32.35	26.10	15.00	0.46		2.93
99.90	0.033	0.022		As cold-rolled	35.95	34.20	27.85	16.95	0.496		2.78
				Slow cooled 800–500°C and air-cooled	33.40	31.65	25.35	14.15	0.45		3.00
				Annealed 8 hr at 600°C and furnace-cooled	34.15	32.15	25.00	12.75	0.397		2.93
99.83				Slow cooled 800–500°C and air-cooled	35.50	33.60	26.95	15.75	0.47		2.82
99.87	0.038	0.019		As cold-rolled	33.65	32.00	26.10	15.80	0.494		2.97
				Slow cooled 800–500°C and air-cooled	35.55	33.75	27.35	15.35	0.455		2.82
				Annealed 8 hr at 600°C and furnace-cooled	31.45	29.60	23.10	11.95	0.404		3.18
99.86	0.059	0.052	0.004	As cold-rolled	34.50	32.90	27.25	17.30	0.526		2.90
				Annealed 8 hr at 600°C and furnace-cooled	31.60	29.75	23.25	12.00	0.403		3.17
99.79	0.054	0.053		As cold-rolled	32.93	31.50	26.30	17.00	0.54		3.04
				Annealed 8 hr at 600°C and furnace-cooled	28.70	27.20	21.75	11.30	0.416		3.49
99.86	0.052	0.0005	0.027	As cold-rolled	31.30	29.75	24.10	13.56	0.455	0.221	3.20
				Annealed 8 hr at 600°C and furnace-cooled	29.90	28.30	22.50	11.45	0.405	0.092	3.34
				(Annealed ¾ hr at 800°C, cooled to 600°C in 2½ hr, held 8 hr at 600°C, and furnace-cooled)	29.70	28.20	22.40	11.55	0.410	0.140	3.37
99.79	0.057	0.023	...	As cold-rolled						0.244	
99.74	0.14	0.018	0.015	As extruded						0.194	

ture of 162°K derived from specific-heat measurements (see Sec. 3.1). The quantity R_A can be represented by

$$R_A = 0.194 + 0.000511T \qquad (14)$$

A "residual resistivity" of 0.194 $\rho_{273.2}$ is much larger than could be expected from the presence of 1.0 per cent impurities listed above. Thus the metal used must have contained either marked quantities of hydrogen or oxygen (for which no analysis was available) or numerous physical disturbance centers.

In renewed British investigations (British 6) the value of residual resistivity was redetermined for a uranium sample 99.86 per cent pure, which was annealed for 4 hr at 595°C, furnace-cooled, and cold-rolled to 75 per cent reduction. Without annealing, $\rho_0/\rho_{273.2}$ was 0.221. After annealing for ¾ hr at 800°C, cooling to 600°C in 2½ hr, subsequent annealing for 8 hr at 600°C, and furnace-cooling, it dropped as low as 0.140. The resistivity values given in Table 5.11 were obtained from the latest available British report (British 3).

The change of resistivity at elevated temperature is illustrated in Fig. 5.6, taken from measurements made at the National Bureau of Standards (Natl. Bur. Standards 9). The sudden changes at 640 to 670°C* and 760 to 780°C* correspond to the $\alpha \rightarrow \beta$ and the $\beta \rightarrow \gamma$ transitions, respectively. All curves show hysteresis, i.e., overheating on the ascending branch or undercooling on the descending branch; this may be due to one or the other, or both. The temperature coefficient of resistivity declines in the α phase from about 3×10^{-3} at room temperature to about 0.5×10^{-3} at 650°C and drops to 0.14×10^{-3} in the β phase (Natl. Bur. Standards 9,11). In later work from the National Bureau of Standards, the resistance temperature for five samples of quite pure uranium annealed at 600°C was found to be 2.76×10^{-3} in the range 0 to 100°C; for uranium annealed at 910°C, the mean temperature coefficient for the same temperature range is 2.61×10^{-3} (Dahl and Van Dusen, 1947). These are probably the best values for this constant.

Under high pressure (up to 12,000 atm) the resistance of uranium was found by Bridgman (1923) to decrease at an average rate of 4.36×10^{-4} per cent per atmosphere. However, the sample used by Bridgman for this determination was not very pure, as shown by its high resistivity at 0°C (76×10^{-6} ohm-cm).

---

*The lower value of each "range" refers to cooling conditions, $\beta \rightarrow \alpha$ and $\gamma \rightarrow \beta$ transformations, and the other values to transitions on heating.

It has been reported by Aschermann and Justi (1942) that uranium becomes superconducting below 1.3°K. Alekseevskiĭ and Migunov (1947) found this to be the case for some specimens of metal but found that superconductivity failed to appear in less pure samples. Shoenberg (1947) at first attributed the superconductivity exhibited by some samples of metal to traces of superconducting impurities, but in more recent work Goodman and Shoenberg (1950) conclude that uranium is

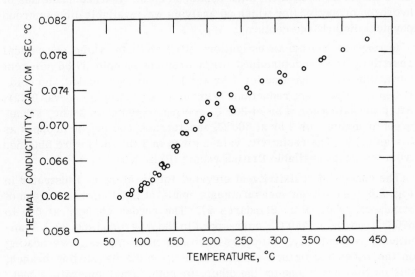

Fig. 5.9 — Thermal conductivity of uranium.

indeed a superconductor which has different transition temperatures according to the purity of the metal. Temperatures as low as 0.75°K may be required for the appearance of superconduction.

4.2 Thermal Conductivity. With the exception of some early measurements on sintered Metal Hydrides, Inc., metal (MP Chicago 3,8,11,19), which gave thermal conductivities of the order of 50×10^{-3} cal/cm sec °C or less, all other observations at Chicago (MP Chicago 11,20), Battelle (Battelle 19), and England (British 2) lead to values of about 60 to 65×10^{-3} cal/cm sec °C for the thermal conductivity of uranium at room temperature and slightly above (30 to 60°C). The average temperature coefficient of thermal conductivity is, according to Fig. 5.9, about 1.5×10^{-3} per degree centigrade in the range 100 to 225°C and 0.4×10^{-3} per degree centigrade in the range 225 to 450°C. The rather sharp break in the curve at 225°C has not been explained.

Battelle measurements on the same kind of material (γ-extruded metal) gave a linear increase in conductivity between 125 and 300°C, with a temperature coefficient of about 1.3×10^{-3} per degree centigrade.

The thermal conductivity of α uranium probably depends on crystallographic direction, and the conductivity of polycrystalline material with preferred orientation of crystals (rolled or extruded metal)

Table 5.12 — Thermoelectric Potential of Uranium
against Platinum

Temp., °C	Potential, mv	Temp., °C	Potential, mv
0	0.00	500	10.54
100	1.19	600	13.90
200	2.87	700	17.51
300	5.03	800	21.35
400	7.59	900	25.50

may therefore differ in transverse and longitudinal specimens. However, Raeth and King's measurements (MP Chicago 20) of radial conductivity (conductivity transverse to extrusion stress) of γ-extruded uranium gave values not significantly different from those obtained with other specimens at 84 to 241°C, namely, 62×10^{-3} to 69×10^{-3} cal/cm sec °C.

Assuming that electrical conductivity of pure uranium at room temperature is 4×10^4 (ohm-cm)$^{-1}$ and using a Wiedemann-Franz ratio of 5.8×10^{-9}, a thermal conductivity of 70×10^{-3} cal/cm sec °C is calculated. This is slightly larger than the highest measured value of 66×10^{-3}.

4.3 Thermoelectric Potential. Table 5.12 gives the values found at the National Bureau of Standards (Natl. Bur. Standards 10) for the thermoelectric potential of 99.9 per cent pure swaged uranium wire annealed for 10 min at 900°C. Dahl and Van Dusen (1947) found that relatively small variations in the purity of the uranium produce relatively large effects on the thermal emf; differences of 0.07 per cent iron and 0.02 per cent oxygen between two samples resulted in a difference of 1.92 millivolts at 800°C. Ebert and Schulze (1947) found a value of dE/dt against copper of about 5 microvolts/°C. This value places uranium in the same position as molybdenum in the thermoelectric series.

The emf vs. temperature curve shows only very slight discontinuities at the transformation points.

4.4 Electron Emission. The available determinations of the work function of uranium were all made before 1940 with metal of questionable purity. Dushman (1923) had calculated from the emission of a tungsten wire containing uranium an upper limit of 3.28 volts for the thermionic work function of uranium. Hole and Wright (1939) measured the thermionic emission of a Mackay strip and a Westinghouse rod of uranium at 680 to 1030°C and found values of the thermionic work function between 3.60 and 3.15 volts, generally decreasing with progressive outgassing. They estimated 3.27 ± 0.05 volts as the most probable value for pure metal.

From the threshold of the photoelectric effect, Rentschler, Henry, and Smith (1932) found 3.63 volts as the value of the photoelectric work function of uranium and pointed out that the difference between this value and the value of 3.28 volts found by Dushman for the thermionic work function is outside the limits of experimental error.

4.5 Magnetic Susceptibility. Uranium is weakly paramagnetic. The available measurements (Honda, 1910, 1912; Owen, 1912) are rather old, and the metal used probably contained considerable impurities, particularly iron. This may be the reason that the specific susceptibility ($X = 2.6 \times 10^{-6}$ at 18°C, according to Owen) was found to decrease somewhat with temperature.

4.6 Optical Emissivity. Hole and Wright (1939) found an average emissivity of 0.51 at 670 mμ. By comparison of the apparent temperature of a uranium surface with its true temperature determined from the emission of a hole in the metal that served as a black body, Wahlin (Wisconsin 1) calculated emissivity values of 0.453 at 97 to 1050°C and 0.415 at 1052 to 1097°C. The sudden change at 1050 to 1052°C seems to indicate an allotropic transformation. Since both emissivity coefficients are low, it is unlikely that one of them belongs to an oxide.

REFERENCES

1910 K. Honda, Ann. Physik, (4) 32: 1047.

1912 K. Honda, Science Repts. Tôhoku Imp. Univ., 1 (I): 21.

1912 M. Owen, Proc. Acad. Sci. Amsterdam, 14: 637; Ann. Physik, (4) 37: 667, 668.

1922 P. W. Bridgman, Proc. Natl. Acad. Sci. U. S., 8: 361.

1923 P. W. Bridgman, Proc. Am. Acad. Arts Sci., 58: 158.

1923 S. Dushman, Phys. Rev., 21: 623.

1930 F. H. Driggs and W. C. Lilliendahl, Ind. Eng. Chem., 22: 516.

1930 J. C. McLennan and R. W. McKay, Trans. Roy. Soc. Can., (3) 24, III: 1.

1932 H. C. Rentschler, D. E. Henry, and K. O. Smith, Rev. Sci. Instruments, 3: 794.

1933 T. A. Wilson, Physics, 4: 148; Phys. Rev., 43: 781.

1937 C. W. Jacob and B. E. Warren, J. Am. Chem. Soc., 59: 2588.

1938 P. P. Alexander, Metals & Alloys, 9: 270-274.

1939 W. L. Hole and R. W. Wright, Phys. Rev., 56: 785.
1942 G. Aschermann and E. Justi, Physik. Z., 43: 207.
1947 E. Alekseevskiĭ and L. Migunov, J. Phys. U.S.S.R., 11: 95.
1947 G. Balz, Metallforschung, 2: 144-146.
1947 A. I. Dahl and M. S. Van Dusen, J. Research Natl. Bur. Standards, 39 (1): 53-58.
1947 H. Ebert and A. Schulze, Metallforschung, 2: 46-49.
1947 G. E. Moore and K. K. Kelley, J. Am. Chem. Soc., 69: 2105-2107.
1947 L. Pauling, J. Am. Chem. Soc., 69: 542-553.
1947 D. Shoenberg, Nature, 159: 303.
1948 P. W. Bridgman, Proc. Am. Acad. Arts Sci., 76: 55-70.
1948 W. Köster, Z. Metallkunde, 39: 1-9.
1949 A. I. Dahl and H. E. Cleaves, J. Research Natl. Bur. Standards, 43: 513.
1949 H. Laquer, Report AECD-2606.
1949 C. W. Tucker, Jr., Report AECD-2716.
1949 A. S. Wilson and R. E. Rundle, Acta Crystallographica, 2: 126.
1950 B. B. Goodman and D. Shoenberg, Nature, 165: 441.

Project Literature

Battelle 1: Battelle Memorial Institute, Report CT-2144, Sept. 1, 1944, p. 218.

Battelle 2: Battelle Memorial Institute, Reports CT-2374, November, 1944, p. 249, and CT-2700, February, 1945, p. 33.

Battelle 3: Battelle Memorial Institute, Reports CT-1795, June 1, 1944, p. 143, and CT-2483, Dec. 1, 1944, p. 282.

Battelle 4: Battelle Memorial Institute, Report CT-2374, Nov. 1, 1944, p. 252.

Battelle 5: Battelle Memorial Institute, Reports CT-2002, Aug. 1, 1944, p. 194, and CT-2144, Sept. 1, 1944, p. 216.

Battelle 6: Battelle Memorial Institute, Report CT-1571, Apr. 1, 1944.

Battelle 7: Battelle Memorial Institute, Report CT-1937, July 1, 1944, pp. 150, 165.

Battelle 8: Battelle Memorial Institute, Report CT-393, Dec. 15, 1942, p. 2.

Battelle 9: Battelle Memorial Institute, Report CT-428, Jan. 1, 1943, p. 10.

Battelle 10: Battelle Memorial Institute, Report CT-1697, May 1, 1944, pp. 91, 102.

Battelle 11: Battelle Memorial Institute, Report CT-688, May 10, 1943.

Battelle 12: Battelle Memorial Institute, Report CT-1697, May 1, 1944, p. 113.

Battelle 13: Battelle Memorial Institute, Report CT-468, Feb. 1, 1943, pp. 56, 57.

Battelle 14: Battelle Memorial Institute, Reports CT-893, Aug. 10, 1943, p. 279, and CT-956, Sept. 10, 1943, p. 321.

Battelle 15: Battelle Memorial Institute, Report CT-893, Aug. 10, 1943.

Battelle 16: Battelle Memorial Institute, Report CT-753, June 10, 1943, p. 197.

Battelle 17: Battelle Memorial Institute, Report CT-611, Apr. 30, 1943, p. 122.

Battelle 18: Battelle Memorial Institute, Report CT-1795, June 1, 1944, p. 160.

Battelle 19: Battelle Memorial Institute, Report CT-2700, Feb. 1, 1945, p. 31.

British 1: F. E. Simon, Report BR-280, Aug. 16, 1943.

British 2: Directorate of Tube Alloys, Report BR-403, Mar. 1, 1944.

British 3: Imperial Chemical Industries, Report BR-658, Sept. 21, 1945.

British 4: C. Sykes, Report BR-203, Apr. 19, 1943.

British 5: H. Greenwood, Report BR-78, Dec. 18, 1942.

British 6: Directorate of Tube Alloys, Report [B]LRG-42, March, 1945.

British 7: N. Allen, Reports BR-579, Feb. 26, 1945, and BR-592, Apr. 11, 1945.

British 8: J. Ferguson, Report B-37, January, 1942.

British 9: Directorate of Tube Alloys, Report [B]LRG-41, February, 1945.

British 10: W. H. Denton, Report [B]LRG-39, December, 1944.

British 11: N. Allen, Report BR-679, 1946.

British 12: N. Allen, Report BR-703, 1946.
British 13: N. Allen, Report BR-717, 1946.
British 14: W. O. Alexander and K. Forrest, Report BR-727, 1946.
British 15: E. Orowan, Report BR-685, 1946.
British 16: E. W. Colbeck and R. P. Garner, Report BR-707, April, 1946.
British 17: Morrison and Parker, Report BR-735, 1946.
British 18: N. Allen, Report BR-718, 1946.

SAM Columbia 1: J. W. Marden, Report A-605, Mar. 23, 1943.

MIT 1: P. Gordon, Report CT-2780, Mar. 3, 1945.
MIT 2: A. R. Kaufmann, Report CT-685, May 29, 1943, p. 15.
MIT 3: A. R. Kaufmann, Report CT-422, Jan. 15, 1943, Part D.
MIT 4: A. R. Kaufmann, Report CT-539, Mar. 27, 1943, Part G, p. 7.
MIT 5: P. Gordon, private communication to F. Foote, Dec. 7, 1944.
MIT 6: A. R. Kaufmann, Report CE-345, Nov. 15, 1942, Part B, p. 5.
MIT 7: R. N. Palmer, Reports CT-2606, Apr. 7, 1945, p. 12; CT-3060, May, 1945, p. 13; CT-3122, Aug. 9, 1945, p. 9; CT-3194, July 10, 1945, p. 8; and CT-3459 Mar. 14, 1946, p. 8.
MIT 8: A. R. Kaufmann, P. Gordon, and R. N. Palmer, Report CT-953, Sept. 25, 1943, p. 18.

MP Ames 1: Metallurgical Project Information Meeting, Report CS-2745, Feb. 21, 1945, p. 6.
MP Ames 2: A. S. Wilson, Reports CN-1495, Apr. 10, 1944, p. 24, and CT-1501, May 10, 1944, p. 14.
MP Ames 3: A. S. Wilson, Report CT-1985, Nov. 10, 1944, p. 20.
MP Ames 4: A. S. Wilson, Report CT-1775, June 10, 1944, p. 19.
MP Ames 5: D. Peterson, Report CC-682, May 15, 1943, Part V, p. 7.
MP Ames 6: J. H. Carter, Report CT-609, Apr. 24, 1943, Part A, p. 5.

MP Chicago 1: P. A. Lauletta and N. E. Hamilton, Report N-1611a, July 3, 1944.
MP Chicago 2: A. Van Echo, Report N-1611c, July 8, 1944.
MP Chicago 3: R. F. Plott and C. H. Raeth, Report CP-228, Aug. 14, 1942.
MP Chicago 4: E. C. Creutz and J. Simmons, Report CP-322, Oct. 29, 1942, p. 4.
MP Chicago 5: A. Van Echo, Report CT-2668, January, 1945, pp. 10-15.
MP Chicago 6: N. E. Hamilton, P. A. Lauletta, and A. Van Echo, Report N-1611c, July 8, 1944.
MP Chicago 7: A. Van Echo, Report CT-2743, February, 1945, pp. 5, 6.
MP Chicago 8: T. M. Snyder and R. L. Kamm, Report CP-124, June 13, 1942, p. 3.
MP Chicago 9: T. M. Snyder and R. L. Kamm, Report CT-192, p. 38.
MP Chicago 10: T. M. Snyder and R. L. Kamm, Report CT-96, Part E, p. 5.
MP Chicago 11: E. O. Wollan and R. L. Stephenson, Report CP-76, May 16, 1942.
MP Chicago 12: F. Seitz, Report CP-1598, Apr. 21, 1944, p. 4.
MP Chicago 13: E. C. Creutz, Report CP-1507, Mar. 27, 1944, p. 34.
MP Chicago 14: E. C. Creutz, Report CP-1576, Apr. 24, 1944.
MP Chicago 15: L. Szilard, Report A-24, Aug. 16, 1941.
MP Chicago 16: E. C. Creutz, Report A-95.
MP Chicago 17: E. C. Creutz, Report CP-255, Sept. 15, 1942.
MP Chicago 18: G. Derge and M. Cefola, Report CT-2277, Oct. 26, 1944.
MP Chicago 19: R. F. Plott and C. H. Raeth, Report CE-236, Aug. 15, 1942.

MP Chicago 20: H. R. Kratz and C. H. Raeth, Report CT-539, Mar. 27, 1943, Part B, p. 8; H. R. Kratz, Report CT-861, July 20, 1943, p. 11; E. C. Creutz, Report CT-890, Aug. 28, 1943, p. 7; H. R. Kratz, Report CT-953, Oct. 2, 1943; C. H. Raeth and E. King, Report CP-1087, Nov. 27, 1943, p. 18; H. R. Kratz, Report CP-1728, May 25, 1944, p. 30; C. H. Raeth, Reports CP-2332, Dec. 7, 1944; MUC-RS-4 (N-1880), Jan. 20, 1945; and MUC-RS-9 (N-1568), Feb. 21, 1945; H. R. Kratz and C. H. Raeth, Report CP-2315, Jan. 27, 1945; C. H. Raeth, Report MUC-RJM-2 (N-1936), Mar. 9, 1945.

Los Alamos 1: A. U. Seybolt, L. B. Stark, W. F. Arnold, and F. J. Schnettler, Report LA-68, Feb. 15, 1944, p. 6.
Los Alamos 2: A. U. Seybolt, L. B. Stark, and W. F. Arnold, Report LA-55, Jan. 14, 1944, p. 7.
Los Alamos 3: A. U. Seybolt, Report LA-180, Dec. 6, 1944, p. 25.

Natl. Bur. Standards 1: J. G. Thompson, Report CT-539, Mar. 27, 1943, Part F, p. 1.
Natl. Bur. Standards 2: H. E. Cleaves, Report CT-2375, October, 1944, p. 1.
Natl. Bur. Standards 3: J. G. Thompson, Report CT-2478, November, 1944, p. 2.
Natl. Bur. Standards 4: J. G. Thompson, Report CT-2252, September, 1944, p. 32.
Natl. Bur. Standards 5: J. G. Thompson, Report MUC-FF-135B (N-1404), July 10, 1944.
Natl. Bur. Standards 6: J. G. Thompson, Report CT-2692, January, 1945.
Natl. Bur. Standards 7: J. G. Thompson, Report CT-750, June 26, 1943, p. 22.
Natl. Bur. Standards 8: H. E. Cleaves, Report CT-1179, Jan. 1, 1944, pp. 7, 8.
Natl. Bur. Standards 9: J. G. Thompson, Report CT-685, May 29, 1943, p. 25.
Natl. Bur. Standards 10: J. G. Thompson, Report CT-890, Aug. 28, 1943, pp. 16, 18.
Natl. Bur. Standards 11: J. G. Thompson, Report CT-815, July 24, 1943.
Natl. Bur. Standards 12: C. J. Rodden, Report A-36, p. 4.
Natl. Bur. Standards 13: H. T. Wensel and W. F. Roeser, Report A-67.
Natl. Bur. Standards 14: R. J. Corruccini and D. C. Ginnings, Report A-3947, undated, received July 25, 1946.

U. S. Bur. Mines 1: G. E. Moore and E. A. Long, Report A-502, Dec. 31, 1942; and E. A. Long, K. K. Kelley, and G. E. Moore, Report CT-385, Dec. 24, 1942.
U. S. Bur. Mines 2: E. A. Long, W. M. Jones, and J. Gordon, Report A-329, October, 1942.

Wisconsin 1: H. B. Wahlin, Report CT-2149, Sept. 12, 1944.

Chapter 6

CHEMICAL PROPERTIES OF URANIUM METAL

In this chapter a summary of the chemical reactions of uranium metal is given. Little more than an enumeration of the various reactions observed can be included because practically no quantitative data bearing on the kinetics of these reactions or their mechanisms are available. To avoid duplication, cross references are given to other chapters of this volume where more detailed information on the individual reactions can be found.

Uranium metal is highly reactive. It reacts readily with all the nonmetallic elements and also forms numerous intermetallic compounds with Hg, Sn, Cu, Pb, Al, Bi, Fe, Ni, Mn, Co, Zn, and Be. These compounds are discussed in Chap. 7. The general chemical character of uranium is that of a strong reducing agent, particularly in aqueous systems. The position of uranium in the electromotive series is not known exactly but appears to be close to that of beryllium.

The rates of reactions given below usually refer to massive cast uranium metal (99.9 per cent uranium) cleaned of oxide with dilute nitric acid. Finely divided uranium metal (such as that obtained by the decomposition of uranium hydride) often reacts much more rapidly than massive metal. Many reactions ascribed to uranium hydride (see Chap. 8), particularly those occurring above 400 to 500°C, are actually reactions of finely divided uranium metal.

1. REACTIONS WITH NONMETALLIC ELEMENTS

The elements are discussed below in the order in which they occur in the periodic table.

1.1 Hydrogen and Deuterium. (See Chap. 8 where this reaction is discussed in considerable detail.) Uranium turnings or lumps are converted to uranium hydride by gaseous hydrogen at or above 250°C.

$$U + \tfrac{3}{2}H_2 \rightarrow UH_3 \tag{1}$$

1.2 Boron. (See Chap. 9.) Finely divided uranium reacts with amorphous boron at the temperature of the electric furnace to give uranium boride (Wedekind, 1913).

1.3 Carbon. (See Chap. 9.) By heating an intimate mixture of powdered uranium with the appropriate amount of powdered carbon to 800 to 1200°C, either of the two known uranium carbides, UC and UC_2, can be obtained.

When uranium is melted in a graphite crucible, the crucible is protected by a film of uranium carbide formed at the interface. Therefore, the attack does not become serious until temperatures of 1500 to 1650°C are reached.

1.4 Silicon. (See Chap. 9.) Uranium and silicon form alloys when the powdered reactants are melted together. The phase diagram, which is very complex, indicates the existence of at least five uranium-silicon compounds.

1.5 Nitrogen. (See Chap. 10.) At atmospheric pressure, uranium turnings react slowly with nitrogen at 450°C. At 700°C the reaction is rapid, and the compound $UN_{1.75}$ is formed. Reaction with powdered uranium is rapid even at 520°C. With higher nitrogen pressures, nitrides of the composition UN_2 can be prepared. If the reaction temperature is raised above 1300°C the mononitride UN is produced, since in this temperature region all the higher nitrides are unstable with respect to UN.

1.6 Phosphorus. (See Chap. 10.) Finely divided uranium when heated with powdered phosphorus to 600 to 1000°C forms the phosphide U_3P_4 (Driggs, 1929).

1.7 Arsenic. (See Chap. 10.) At least two compounds, U_2As and UAs, have been identified as products of the reaction of arsenic and uranium.

1.8 Oxygen. (See Chap. 11.[*]) Uranium as turnings or small lumps burns brilliantly in oxygen at 700 to 1000°C with the emission of white light.

$$3U + 4O_2 \rightarrow U_3O_8 \tag{2}$$

At very low partial pressures of oxygen (less than 10^{-4} atm), films of uranium monoxide, UO, are formed on uranium metal.

---

[*]See Gmelin (1936) for references to older observations on impure metal.

In air at room temperature massive uranium metal oxidizes slowly (Moore, 1923; Lely, 1914). It first assumes a golden-yellow color; as the oxidation proceeds the film becomes darker, and at the end of 3 or 4 days the metal appears black. The oxide films that form on uranium in air do not protect the metal from further attack.

Uranium turnings oxidize with moderate rapidity at 125°C. When ignited in air they burn without a flame. Massive metal oxidizes slowly at 500 to 700°C, but complete oxidation is achieved within 1 hr when it is ignited at 700 to 1000°C.

Powdered uranium metal is usually pyrophoric and burns with a bright glow. Spectacular displays of sparks occur when metallic uranium is filed or held to a grindstone.

According to British observers (British 1), below 100°C only UO_2 is formed by the oxidation of uranium in air, while between 100 and 200°C both UO_2 and U_3O_8 are produced.

1.9 Sulfur and Selenium. (Péligot, 1842; Moissan, 1896; Zimmermann, 1882. See Chap. 11.) Uranium reacts slowly with molten sulfur at 250 to 300°C. Uranium burns at 500°C in sulfur vapor. Depending on the exact conditions, the disulfide, US_2, the sesquisulfide, U_2S_3, or mixtures of the two are obtained. Selenium reacts in an analogous fashion.

1.10 Fluorine. (See Chaps. 12 and 13.) Fluorine reacts vigorously with metallic uranium at room temperature with the formation of uranium hexafluoride. The metal may easily become incandescent if it is finely divided.

1.11 Chlorine. (See Chap. 14.) Chlorine reacts with massive uranium metal at a moderate rate at 500 to 600°C. Finely divided metal burns in chlorine at 150 to 180°C. The reaction products consist of UCl_4, UCl_5, and UCl_6. The chlorides sublime and collect in the cooler parts of the apparatus.

1.12 Bromine. (See Chap. 15, Sec. 2.1.) At 650°C bromine reacts smoothly with uranium turnings to form uranium tetrabromide, which distills from the reaction zone.

$$U + 2Br_2 \rightarrow UBr_4 \tag{3}$$

When less bromine is used, uranium tribromide is obtained (see Chap. 15, Sec. 1.1).

1.13 Iodine. (See Chap. 15.) Uranium metal is attacked by iodine vapor at 350°C. Either UI_3 or UI_4 may be obtained, depending on the experimental conditions chosen.

1.14 Noble Gases. (See Chap. 8.) Because of its tendency to react with the usual impurities present in these gases, uranium metal can be used to prepare very pure helium, argon, or other noble gases (MP Ames 1).

2. REACTIONS WITH COMPOUNDS OF THE NONMETALLIC ELEMENTS

2.1 Water. The behavior of uranium metal with water and steam has been very thoroughly studied (see Division IV, Volume 6 A of the National Nuclear Energy Series), but the data are too extensive to permit detailed consideration here. Boiling water attacks massive uranium slowly.

$$U + 2H_2O \rightarrow UO_2 + 2H_2 \qquad (4)$$

Hydrogen accelerates the corrosion of uranium because of hydride formation. In aerated distilled water the rate of reaction is at first less than in hydrogen-saturated water, a phenomenon which is probably contingent on the formation of protective oxide films. Eventually the rate of reaction increases and approaches the value in hydrogen-saturated water (MP Chicago 1,2,3).

Steam reacts with uranium at 150 to 250°C. A difference of opinion exists as to the products formed. American workers postulated the reaction

$$7U + 6H_2O \text{ (gas)} \xrightarrow{250°C} 3UO_2 + 4UH_3 \qquad (5)$$

At 600 to 700°C the products were reported to be pure UO_2 and hydrogen; no U_3O_8 was formed even at 1000°C (MP Ames 2,3). British workers, however, claimed that U_3O_8 was the principal product of the reaction of steam and uranium at temperatures above 300°C (British 1). No explanation is available for this discrepancy. All observers are agreed that attack by steam is much more vigorous than attack by oxygen (British 2; MP Chicago 4).

2.2 Hydrogen Fluoride. (See Chap. 12.) Powdered uranium metal reacts with anhydrous hydrogen fluoride at elevated temperatures to form uranium tetrafluoride.

$$U + 4HF \rightarrow UF_4 + 2H_2 \qquad (6)$$

(With uranium hydride and hydrogen fluoride, production of UF_4 occurs in the temperature range 200 to 400°C.) By using a mixture of hydrogen and hydrogen fluoride, reaction with massive metal can be initiated at 250°C. Since hydrogen is produced in the reaction, the external source of hydrogen may be removed once the reaction has started, and the reaction will then proceed until all the metal is consumed.

2.3 Hydrogen Chloride. (See Chap. 14.) With finely divided uranium, hydrogen chloride forms uranium trichloride.

$$U + 3HCl \rightarrow UCl_3 + \tfrac{3}{2}H_2 \tag{7}$$

This reaction is incomplete with massive metal.

2.4 Hydrogen Bromide and Hydrogen Iodide. (See Chap. 8.) The reactions of these compounds with uranium metal have not been studied in detail. Uranium tribromide is formed from uranium hydride by treatment with hydrogen bromide. With hydrogen iodide and uranium hydride, mixtures of UI_4 and UI_3 appear to be formed.

2.5 Carbon Monoxide. (See Chap. 9.) There is little reaction between powdered uranium metal and carbon monoxide below 400°C. Reaction definitely occurs with turnings at 750°C and yields a mixture of uranium oxide and carbide (MP Ames 4,5).

2.6 Carbon Dioxide. (See Chap. 9.) At 750°C, reaction between uranium and carbon dioxide is quite rapid and leads to the formation of uranium oxides and carbides. Finely powdered metal was occasionally observed to ignite spontaneously in carbon dioxide (MP Ames 5).

2.7 Ammonia. (See Chap. 10.) Uranium powder rapidly reacts with ammonia at 400°C (uranium turnings react at 700°C) to form the nitride $UN_{1.75}$.

2.8 Nitric Oxide. Uranium in the form of turnings burns in NO at 400 to 500°C to form U_3O_8 and nitrogen (Emich, 1894).

2.9 Methane. Finely divided uranium reacts with methane at 900°C to give the monocarbide, UC (Litz, Garrett, and Croxton, 1948).

3. REACTIONS WITH AQUEOUS ACID SOLUTIONS

3.1 Hydrofluoric Acid. Massive uranium metal is attacked only slowly by concentrated hydrofluoric acid even at 80 to 90°C, presumably because of the formation of an insoluble coating of uranium tetrafluoride. The addition of oxidizing agents, such as hydrogen peroxide, does not appear to accelerate the reaction appreciably.

3.2 Hydrochloric Acid. Uranium metal is attacked by concentrated hydrochloric acid with remarkable rapidity. The rate is much slower in 1N acid than in 6N acid. The reaction appears to be complex, since variable amounts of the metal are converted into an insoluble black material. It has been suggested that this product is a hydrated uranium oxide (MP Ames 6). With a large excess of acid only small amounts of the black material are formed; if smaller amounts of acid are used as much as 20 per cent of the metal may be converted to this product (MP Ames 7). The ratio of U(III) to U(IV) in the final solution also varies. In 12N acid practically all of the metal is oxidized to the quadrivalent state, but in 6N acid the average oxidation state of the final product is between 3.2 and 3.4. The degree of oxidation depends upon acid strength, ratio of acid to metal, temperature, time, and probably other as yet unrecognized factors (MP Ames 8).

Mixtures of hydrochloric acid and oxidizing agents may be used to effect complete solution of uranium metal. Hydrogen peroxide, nitric acid, bromine water, ammonium persulfate, potassium chlorate, or perchloric acid may also be used. The presence of 0.05M fluosilicic acid also permits complete solution of uranium metal in concentrated hydrochloric acid without the formation of a black residue (MP Ames 9). No trivalent uranium is present in such a solution. Methanolic hydrogen chloride dissolves uranium at a moderate rate, but about 28 per cent of black residue remains. A violet solution stable for some time is formed (MP Ames 6).

3.3 **Hydrobromic Acid and Hydriodic Acid.** Hydrobromic acid resembles hydrochloric acid in its effect on uranium, except that the reaction is slower. A black precipitate forms in this case also. Hydriodic acid reacts even more slowly.

3.4 **Nitric Acid.** Massive uranium is dissolved with only moderate rapidity by dilute or concentrated nitric acid to form uranyl nitrate. Since nitric acid vapors or nitrogen dioxide can react with uranium with explosive violence, it is necessary to add finely divided uranium to the nitric acid in small portions in order to avoid accidents (MP Chicago 5,6,7).

3.5 **Sulfuric Acid.** Dilute (6N) sulfuric acid does not attack uranium; at the boiling point the action is about that of boiling water alone. With hot concentrated sulfuric acid a slow reaction occurs with the formation of U(IV) acid sulfate, SO_2, S, H_2S, and other products. In conjunction with oxidizing agents such as hydrogen peroxide or nitric acid, dilute sulfuric acid will dissolve uranium. Electrolytic oxidation in sulfuric acid solution can also be used to achieve solution (MP Chicago 8).

3.6 **Phosphoric Acid.** Uranium is attacked slowly by cold 85 per cent phosphoric acid. On heating, the rate is at first only slightly increased. On further heating, sufficient water is driven off to raise the concentration of the phosphoric acid to the point where a rapid exothermic reaction takes place, and a green solution of uranium(IV) acid phosphate is formed. Prolonged heating of a phosphoric acid solution of uranium(IV) phosphate may result in the formation of glasses that are extremely resistant to further chemical action.

3.7 Perchloric Acid. Dilute perchloric acid is rather inert as far as action on uranium is concerned. As water is boiled off, a vigorous reaction ensues when the concentration of perchloric acid reaches 90 per cent. The oxidation is very vigorous; the use of this reagent with large quantities of metal is not recommended. Dilute perchloric acid dissolves uranium smoothly with the aid of oxidizing agents.

3.8 Organic Acids. Although formic, acetic, propionic, or butyric acid (dilute or anhydrous) does not react with metallic uranium, rapid exothermic reactions occur in the presence of hydrogen chloride or hydrochloric acid which result in the formation of the corresponding uranium(IV) salts (MP Ames 10). Uranium acetate can also be obtained by reaction of acetic anhydride or acetyl chloride on the metal. Uranium reacts with benzoic acid in ether solution to form uranium(IV) benzoate.

4. REACTIONS WITH AQUEOUS ALKALI SOLUTIONS

Solutions of alkali metal hydroxides have little effect on uranium metal (MP Chicago 7; Zimmermann, 1882; Lely, 1914). Sodium hydroxide solutions containing hydrogen peroxide (or sodium peroxide − water mixtures) dissolve uranium. Soluble sodium peruranates are formed (MP Ames 6). For a similar reaction with uranium oxide, see Chap. 11.

5. REACTIONS WITH HEAVY METAL SALT SOLUTIONS

Uranium metal is a sufficiently powerful reducing agent to displace many metals from solutions of their salts. Solutions of $Hg(NO_3)_2$, $AgNO_3$, $CuSO_4$, $SnCl_2$, $PtCl_4$, and $AuCl_3$ yield precipitates of the corresponding metals when treated with metallic uranium (Zimmermann, 1882). This reaction has been studied in connection with the problem of devising a method for the determination of uranium in metallic materials. Procedures based on solution of uranium in hydrochloric acid and measurement of the hydrogen evolved have been proposed, but the complexity of the reaction (see Sec. 3.2) probably renders it unsuitable. Therefore efforts have been made to determine uranium by measuring the amount of another metal displaced from solution; so far these attempts have met with only indifferent success.

5.1 Silver Salts. (MP Ames 6,11.) Solutions of silver sulfate react slowly with uranium. The surface of the uranium appears to become coated with silver, which practically prevents further reaction.

Solutions of silver perchlorate react much more vigorously. The reaction is complex, since not only is silver produced but also silver chloride, owing to the simultaneous reduction of the perchlorate to chloride.

$$U + 6AgClO_4 + 2H_2O \rightarrow 6Ag + UO_2(ClO_4)_2 + 4HClO_4 \tag{8}$$

$$U + 4AgClO_4 \rightarrow U(ClO_4)_4 + 4Ag \tag{9}$$

$$2U + 5AgClO_4 \rightarrow 4Ag + AgCl + 2UO_2(ClO_4)_2 \tag{10}$$

About 20 per cent of the uranium reacts according to Eq. 10, whereas less than 0.1 per cent reacts according to Eq. 9. The simultaneous formation of silver and silver chloride in varying ratios makes it impossible to use this reaction for determining uranium. In toluene (in which silver perchlorate is very soluble) the reaction proceeds in the same way with the formation of both silver and silver chloride.

5.2 <u>Copper Salts</u>. (Riott, 1941; Willems, 1941; MP Chicago 9, 10,11.) Although uranium seems to be only slightly attacked by cupric sulfate (cf., however, Zimmermann, 1882), it dissolves readily in solutions of copper ammonium chloride. Copper first precipitates but is redissolved on shaking by the excess ammonium chloride present in the solution. Since uranium carbides and oxides as well as uranium metal are soluble in this reagent to an appreciable extent, the method has little analytical significance.

6. MISCELLANEOUS REACTIONS

6.1 <u>Uranium Tetrafluoride</u>. (See Chap. 12.) Powdered uranium reduces uranium tetrafluoride to the trifluoride at 1100°C.

$$3UF_4 + U \rightarrow 4UF_3 \tag{11}$$

6.2 <u>Uranium Dioxide</u>. (See Chap. 11.) At 2400°C uranium metal reduces UO_2 to the monoxide, UO.

$$UO_2 + U \rightarrow 2UO \tag{12}$$

6.3 <u>Substances Containing Silica</u>. Glass, porcelain, and silica ware are attacked by finely divided uranium metal at 700 to 800°C with the formation of mixtures of uranium oxide and silicide.

6.4 <u>Boric Acid</u>. (MP Ames 12.) Uranium metal is substantially inert to fusion with boric acid.

6.5 <u>Solutions of Potassium and Ammonium Persulfate</u>. (Levi, 1908.) Vigorous reaction is reported to occur when metallic uranium is treated with an aqueous solution of potassium persulfate. With ammonium persulfate the reaction is more sluggish.

6.6 <u>Methanol</u>. (MP Ames 13.) It has been reported that powdered uranium does not react with anhydrous methanol. This observation is in need of confirmation.

6.7 <u>Chlorinated Hydrocarbons</u>. At temperatures above 150 to 200°C the vapors of carbon tetrachloride, chloroform, and trichloroethylene react slowly with metallic uranium. Above 1000°C the reaction with carbon tetrachloride is rapid. Caution is necessary since finely divided uranium may react violently with liquid halogenated hydrocarbons, especially if the metal is prepared by hydride decomposition and still contains some hydride.

REFERENCES

1842 E. Péligot, Ann. chim. et phys., (3) 5: 5.
1882 C. Zimmermann, Ber., 15: 849.
1894 F. Emich, Monatsh., 15: 375.
1896 H. Moissan, Compt. rend., 122: 1092.
1908 M. G. Levi, E. Miglorini, and G. Ergolini, Gazz. chim. ital., 38 (I): 599.
1913 E. Wedekind and O. Jochem, Ber., 46: 1204.
1914 D. Lely and L. Hamburger, Z. anorg. Chem., 87: 220.
1923 R. W. Moore, Trans. Am. Electrochem. Soc., 43: 223.
1929 F. H. Driggs and W. C. Lilliendahl, U. S. Patent No. 1893296.
1936 "Gmelins Handbuch der anorganischen Chemie," System No. 55, p. 65, Verlag Chemie, Berlin.
1941 J. P. Riott, Ind. Eng. Chem. Anal. Ed., 13: 546.
1941 F. Willems, Z. anorg. u. allgem. Chem., 246: 46.
1948 L. M. Litz, A. B. Garrett, and F. C. Croxton, J. Am. Chem. Soc., 70: 1718.

Project Literature

British 1: T. Wathen, Report BR-223, May 13, 1943.
British 2: Report [B] LRG-14, November, 1942.

MP Ames 1: A. S. Newton, The Use of Uranium and Uranium Compounds in Purifying Gases, in National Nuclear Energy Series, Division VIII, Volume 6.
MP Ames 2: J. G. Feibig and J. C. Warf, Report CC-1524, Mar. 10, 1944.
MP Ames 3: A. S. Newton, Report CC-695, May 26, 1943.
MP Ames 4: J. C. Warf, Reports CC-580, Apr. 15, 1943, and CC-587, Apr. 19, 1943.
MP Ames 5: H. A. Wilhelm and R. Hoxeng, Report CC-238, Aug. 15, 1942.
MP Ames 6: R. W. Fisher, J. Powell, and J. C. Warf, Report CC-1194, Dec. 9, 1943.
MP Ames 7: D. W. Peterson, Report CC-1061, Oct. 8, 1943.
MP Ames 8: R. Fulmer, Report CC-1194, Dec. 9, 1943.
MP Ames 9: C. V. Banks, W. K. Noyce, J. H. Patterson, and J. C. Warf, Report CC-2942, July 18, 1945.
MP Ames 10: J. G. Feibig, Report CC-1504, June 10, 1944.
MP Ames 11: R. W. Fisher, Report CC-1057, Nov. 6, 1943.
MP Ames 12: R. Tevebaugh, Report CC-1194, Dec. 9, 1943.
MP Ames 13: H. D. Brown, Report CC-1524, Mar. 10, 1944.

MP Chicago 1: W. A. Mollison, G. C. English, and F. Nelson, Report CT-3055, Aug. 8, 1945.

MP Chicago 2: N. Bensen, R. P. Straetz, and J. E. Draley, Report CT-3043, June 4, 1945.

MP Chicago 3: J. M. Hopkins, F. Nelson, and W. W. Binger, Report CT-3031, May 31, 1945.

MP Chicago 4: G. Rosner, Report CT-2548, Jan. 5, 1945.

MP Chicago 5: J. B. Sutton, Report CN-566, Apr. 12, 1943.

MP Chicago 6: T. R. Cunningham, Report N-42, undated.

MP Chicago 7: A. C. Hyde, Report CN-1751, May 15, 1944.

MP Chicago 8: L. Safranski, R. P. Straetz, and R. Spence, Report CC-934, Sept. 11, 1943.

MP Chicago 9: L. Safranski, Report CC-1047, Nov. 6, 1943.

MP Chicago 10: L. Safranski and H. A. Potratz, Report CK-1064, Nov. 6, 1943.

MP Chicago 11: R. E. Fryxell, Report CC-1448, Mar. 14, 1944.

Chapter 7

INTERMETALLIC COMPOUNDS AND ALLOY SYSTEMS
OF URANIUM

A detailed description of the various uranium alloy systems is found in Division IV, Volume 12 of this Series. Here space permits only the discussion of certain aspects of the subject that have a particular chemical interest. These will include a brief survey of the mutual solubility of uranium and various metals and a description of some intermetallic compounds that have been recognized during the study of the phase relations of the intermetallic systems.

1. PREPARATION OF ALLOYS AND INTERMETALLIC
COMPOUNDS

Almost invariably fusion of the component metals in a vacuum or in an inert atmosphere of argon or helium is required. Refractory crucibles of beryllia, zirconia, or thoria are now often employed; alumina ones have also been used on occasion. A very good vacuum is required to prevent oxidation. Where one of the metals is particularly volatile, highly purified argon may be used to minimize losses by distillation. Heating is best carried out by induction; this is particularly desirable when metals of very different densities are to be alloyed, because of the stirring action obtained. For low-melting metals, such as lead or bismuth, electrolytic procedures have been employed; uranium tetrachloride was dissolved in a molten mixture of sodium and calcium chlorides (m.p. 750°C) and electrolyzed with a pool of lead or bismuth covering the steel cathode (British 1). For preparing mercury amalgams it is essential to utilize very pure uranium metal prepared by decomposition of the hydride. Some alloys have been prepared accidentally by the simultaneous reduction of uranium tetrafluoride and another metal fluoride. This procedure is not to be recommended for systematic studies, since it is difficult to control the final composition and structure of the alloy.

2. MUTUAL SOLUBILITY OF URANIUM AND VARIOUS METALS

A large amount of information is available on the solubility of uranium in various metals and the solubility of various metals in uranium. For the most part the data are the results of the application of both x-ray and metallographic techniques. As was pointed out in

Table 7.1 — Mutual Solubility of Uranium and Some Metals

Atomic no.	Metal	Solubility of uranium in metal, atom %	Solubility of metal in uranium, atom %			Reference report no.
			α	β	γ	
13	Aluminum	...	Nil	Nil	4−5	CT-2721
23	Vanadium	...	Nil	...	...	N-1779
24	Chromium	Nil	Nil	<2.5	<4	CT-3335
25	Manganese	?	...	...	3−4	CT-2945
26	Iron	Nil in γ Fe	Nil	Nil	1.5−2.0	CT-2945
28	Nickel	...	...	1	2	CT-3013
29	Copper	Nil	Nil	Nil	Nil	CT-1784
41	Columbium	...	<0.25	<0.25	3.6 (656°C)	CT-3071
					85 (1350°C)	
42	Molybdenum	1 (900°C)	Nil	...	35.8	CT-2946
47	Silver	...	Nil	...	...	CN-1048
50	Tin	<0.02 (600°C)	Nil	...	...	CN-1784
58	Cerium	...	Nil	...	Nil	CT-2619
73	Tantalum	...	Nil	...	...	N-1779
74	Tungsten	1	<1	...	...	CT-3375
79	Gold	...	2	...	...	CT-2483
80	Mercury	0.001−0.01 (25°C)	Nil	...	...	CT-2960
		1.25 (350°C)				
83	Bismuth	Nil	Nil	...	...	CT-2961
90	Thorium	<2	Nil	...	...	CT-2717

Chap. 5, α uranium possesses an unusually complex crystal structure. This makes it difficult to apply the usual crystallographic chemical criteria of solid solubility (Hume-Rothery, 1933). As far as could be ascertained, no extensive solubility of any metal in α uranium has been observed. However, a number of metals possess considerable solubility in γ uranium, which is stable above 770°C. In the case of molybdenum, quenching to room temperature leads to the formation of solid solutions in which the γ structure is retained. Table 7.1 summarizes the available solubility data.

Table 7.2 — Crystal Structures of Some Uranium Intermetallic Compounds*

Compound	Unit cell	Lattice dimensions, A a_1	a_3	Molecules per unit cell	Density, g/cc	Space group or structure type	Related structures
UCo	Body-centered cubic	6.3557 ± 0.0004		8	15.37	T^5	Related to the UCu_5 structure
UAl₂	Face-centered cubic	7.795		8	8.138	O_h^7 or O^4	
UAl₃	Cubic	4.278		1	6.70	L12	USn_3
UCu₅	Face-centered cubic	7.033 ± 0.002		4	10.61	T_d^2 or T^2	UAl_4, $AuBe_5$, $PdBe_5$
UNi₅	Face-centered cubic	6.7830 ± 0.0005		4	11.31	T_d^2 or T^2	UCu_5, $PdBe_5$
UHg₂	Hexagonal	4.976 ± 0.01	3.218 ± 0.005	1	15.29		
UHg₃	Hexagonal	3.320 ± 0.005	4.878 ± 0.005	½	14.88		
UHg₄	Pseudo unit is body-centered cubic	3.62			14.5	True unit and structure are complex	
USn₃	Simple cubic	4.62		1	9.95	L12	UAl_3
UMn₂	Face-centered cubic	7.1628 ± 0.0014		8	12.57	C15	UAl_2
UFe₂	Face-centered cubic	7.058		8	13.21	C15	UAl_2
UCo₂	Face-centered cubic	6.9924 ± 0.0004		8	13.83	C15	UAl_2
UNi₂	Hexagonal	4.966 ± 0.005	8.252 ± 0.009	4	13.46	C14	$MgZn_2$
U₆Mn	Tetragonal body-centered	10.29 ± 0.01	5.24 ± 0.02	4	17.8	D_4^9 or D_{2d}^{10}	
U₆Fe	Tetragonal body-centered	10.31 ± 0.04	5.24 ± 0.02	4	17.7	D_4^9 or D_{2d}^{10}	
U₆Co	Tetragonal body-centered	10.36 ± 0.02	5.21 ± 0.02	4	17.7	D_4^9 or D_{2d}^{10}	
U₆Ni	Tetragonal body-centered	10.37 ± 0.04	5.21 ± 0.02	4	17.6	D_4^9 or D_{2d}^{10}	
UBi†	Cubic	6.364 ± 0.004		1	11.52	Sodium chloride structure	

*This table is compiled, unless otherwise noted, from the results of Rundle and Wilson (1949) and Baenziger, Rundle, Snow, and Wilson (1950). These papers should be consulted for detailed discussions of the structures.

†Brewer, Edwards, and Templeton (UCRL 1).

The solubility of uranium in mercury deserves special mention (MP Ames 1). Prior to 1941 all investigators agreed that the solubility of uranium in liquid mercury at 25°C was negligibly small; values of the order of 0.00001 per cent were quoted for it. Recent work has shown that this result must be attributed to the use of impure or oxide-coated uranium. It was found that pure, finely divided uranium metal prepared in an oxide-free state by decomposition of uranium hydride readily amalgamated with mercury to form a silvery pasty mass quite similar in appearance to other metal-mercury amalgams. Amalgams containing up to 1 per cent uranium are liquid and fairly stable to air. Those having between 1 and 15 per cent uranium are gray pyrophoric solids. A number of intermetallic uranium-mercury compounds have been isolated and are described below. The solubility of uranium in mercury was determined by separating the solid phase from the liquid phase in the amalgam by centrifugation at various temperatures and measuring the uranium content of the liquid. The results are not very precise but indicate a solubility range from 0.001 to 0.01 per cent at 25°C to 1.06 per cent at 350°C. Very little is known about the chemistry of these amalgams; even less is known about their physical properties. The stability of these amalgams is indicated by the observation that heat is evolved when finely divided uranium is wetted by mercury.

The solubility of uranium in other solid metals appears to be limited. The same seems to be true of liquid metals of low melting point. Thus massive uranium is not appreciably attacked by liquid sodium at 500°C even after several days of exposure (MP Chicago 1). Uranium also appears to be insoluble in liquid sodium-potassium alloys. No alloys or intermetallic compounds of uranium with any of the members of the alkali or alkaline earth metals are known (MP Ames 2), with the exception of beryllium for which an intermetallic compound with 93 atom % beryllium (UBe_{12-13}) appears probable (Montreal 1). The case of magnesium is of interest. It was shown at Ames that it is extremely difficult to prepare uranium containing more than a few parts per million of magnesium.

3. INTERMETALLIC COMPOUNDS

A considerable number of intermetallic compounds of uranium have thus far been prepared and identified (chiefly by x-ray methods). In Table 7.2 are listed all the intermetallic compounds for which x-ray structures are available. In addition to these, the existence of the compounds UAl_5, UNi, $UFeNi$, UBi_2, U_5Sn_4, and U_3Sn_5 has been indicated by either chemical analysis or microscopic examination, but they have not been definitely identified as pure phases. Table 7.3 shows all the melting or decomposition point data available at the time of this writing.

The intermetallic compounds vary considerably in physical properties and chemical reactivity. Compounds of the type U_6M are found to be exceedingly brittle. They can be readily fractured by a hammer blow but are not quite hard enough to scratch glass. Compounds of

Table 7.3 — Melting or Decomposition Points of Intermetallic
Compounds of Uranium

Compound	Melting or decomposition temperature, °C	Compound	Melting or decomposition temperature, °C
UAl_2	1590	U_6Fe	815*
UAl_3	1350*	UMn_2	1120
UAl_5	730*	U_6Mn	726*
UCu_5	1052*	UHg_4	360*
UNi_2	810*	UHg_3	390*
UNi_5	1295	UHg_2	450*
U_6Ni	754*	U_5Sn_4	1500
UFe_2	1235	USn_3	1350*

*Decomposes by peritectic reaction.

the type U_6M can be isolated conveniently by utilizing their inertness to nitric acid. Treatment with either dilute or concentrated nitric acid of iron, cobalt, or manganese alloys of the approximate composition U_6M results in the solution of any excess uranium present. The uranium-mercury, uranium-tin, uranium-lead, and uranium-bismuth systems are particularly noteworthy for their chemical reactivity. Amalgams containing up to 15 per cent uranium oxidize readily in air to form a black powder which contains both uranium and mercury; those containing more than 15 per cent uranium are spontaneously inflammable in air. In the uranium-tin system pyrophoric alloys are also encountered; the 50 per cent alloy is very easily inflammable in air. The same is true of the lead system. In the case of bismuth, both UBi and UBi_2 are highly reactive. They heat up in a few minutes in air and react with water, alcohols, mineral oil, kerosene, benzene, and carbon tetrachloride. No systematic investigation of the reactions of any of these intermetallic compounds has been conducted, and, as can be seen from the above, the available information is very fragmentary.

A great number of alloy systems have been studied for resistance to corrosion by water and air. More or less detailed studies have been made of the systems of uranium with the following elements: Na, K, Cu, Ag, Au, Be, Mg, Zn, Cd, Hg, Al, Ga, In, Ce, La, Nd, Ti, Ge, Zr, Sn, Th, V, Cb, Ta, Bi, Cr, Mo, W, Mn, Re, Fe, Co, Ni, Ru, Rh, Pd, Os, Ir, and Pt. In most cases the complete phase diagram has not as yet been worked out. Descriptions of the uranium-aluminum and uranium-iron systems have recently been published by Gordon and Kaufmann (1949); the uranium-tungsten and uranium-tantalum systems have been discussed by Schramm, Gordon, and Kaufmann (1949), and the uranium-manganese and uranium-copper systems by Wilhelm and Carlson (1949).

REFERENCES

1933   W. Hume-Rothery, G. W. Mabbot, and K. M. C. Evans, Trans. Roy. Soc. London, 233: 1-98.

1949   P. Gordon and A. R. Kaufmann, Report AECD-2683.

1949   R. E. Rundle and A. S. Wilson, Acta Crystallographica, 2: 148.

1949   C. H. Schramm, P. Gordon, and A. R. Kaufmann, Report AECD-2686.

1949   H. A. Wilhelm and O. N. Carlson, Report AECD-2717.

1950   N. C. Baenziger, R. E. Rundle, A. I. Snow, and A. S. Wilson, Acta Crystallographica, 3: 34.

Project Literature

British 1:   J. Ferguson, Report B-38, February, 1942.

MP Ames 1:   D. H. Ahmann, R. R. Baldwin, and A. S. Wilson, Report CT-2960, Dec. 21, 1945.

MP Ames 2:   D. H. Ahmann, Report CT-2959, Dec. 5, 1945.

MP Chicago 1:   F. Foote, Report CT-2857, Mar. 24-31, 1945.

Montreal 1:   R. L. Cunningham, Report MX-180, Oct. 16, 1945, p. 2.

UCRL 1:   L. Brewer, R. K. Edwards, and D. H. Templeton, Report UCRL-433 (revised) Nov. 15, 1949.

Part 3

BINARY COMPOUNDS OF URANIUM OTHER THAN HALIDES

Chapter 8

THE URANIUM-HYDROGEN SYSTEM

1. SOLUBILITY OF HYDROGEN IN SOLID AND LIQUID
URANIUM METAL

Formation of uranium hydride (see Sec. 2 of this chapter) prevents study of the solubility of hydrogen in uranium at low temperatures and high pressures. Under a partial pressure of 1 atm of hydrogen, for example, the solubility can be measured only above 435°C, which is the decomposition temperature of the hydride under atmospheric pressure. The capacity of uranium to dissolve hydrogen is small compared to that of metals such as iron. However, it increases markedly with temperature, particularly in the liquid state. Consequently, considerable quantities of hydrogen that were dissolved in the melt may remain occluded in castings. The solubility, occlusion, and liberation of hydrogen from uranium have been studied in considerable detail by the Battelle group (Battelle 1).

In the earlier investigations the hydrogen content of uranium samples was determined by pumping off and collecting the gas that was liberated by heating metal specimens to about 750°C in a vacuum. It was later found, however, that some hydrogen was formed in this procedure by the reduction of moisture present in the system. More recent determinations of occluded hydrogen therefore were carried out by heating the metal in a closed vessel and measuring the increase in pressure. Comparable results are obtained by burning the metal in oxygen and weighing the liberated water after absorption in a desiccant. A correction had to be applied for the solubility of hydrogen at the temperature of degassing, as determined by the first-named method. The analytical methods are described in more detail in Division VIII, Volume 2, of this Series.

The solution isobar for hydrogen pressure of 1 atm is reproduced in Fig. 8.1. It shows that the equilibrium content of hydrogen in the

α-phase uranium is about 2 ppm and changes but little with temperature. This corresponds to about 0.4 cc of hydrogen per cubic centimeter of metal or 1 atom of hydrogen per 2,000 atoms of uranium. The $\alpha \rightarrow \beta$ transformation, at 660°C, causes the solubility to increase

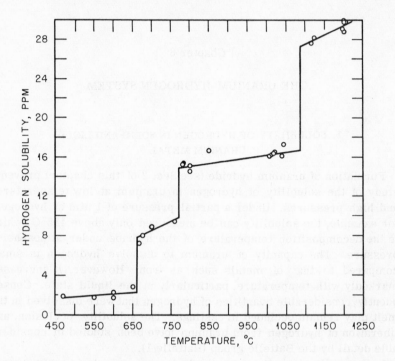

Fig. 8.1—Solubility of hydrogen in uranium metal in equilibrium with hydrogen at atmospheric pressure.

from 2 to about 8 ppm, and the $\beta \rightarrow \gamma$ transformation, at about 770°C, leads to another increase to 15 ppm. Melting, which occurs at 1133°C, increases the solubility from 17 to 28 ppm, and the hydrogen content of the liquid metal continues to increase rapidly with temperature, reaching 30 ppm at 1250°C.

From experiments on the rate of degassing rods of different sizes, the values of the diffusion coefficient of hydrogen in α uranium given in Table 8.1 have been derived.

As mentioned before, the solubility of hydrogen in uranium is proportional to the square root of the partial pressure of hydrogen. This is illustrated by Fig. 8.2 which shows that solution isotherms are straight lines if the square root of pressure is used as the abscissa.

The proportionality of solubility to the square root of pressure indicates that hydrogen is dissolved in uranium in the form of free atoms. The rule holds for all three solid uranium phases, α, β, and γ, since each of them is represented by a straight-line isotherm in Fig. 8.2.

Table 8.1—Diffusion of Hydrogen in Uranium

Temperature, °C	Diffusion coefficient, in.²/hr
566	0.0040
593	0.0064
640	0.0125*

*Extrapolated.

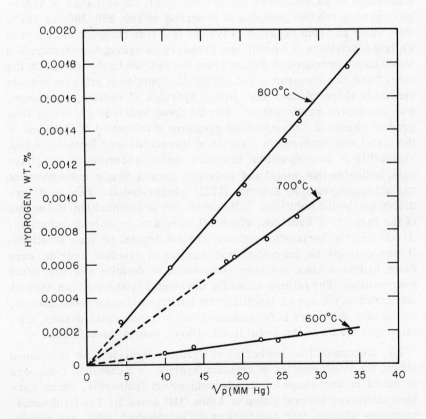

Fig. 8.2—Solubility of hydrogen in uranium metal as a function of hydrogen pressure for 600, 700, and 800°C.

2. URANIUM HYDRIDE

The existence of uranium hydride was discovered by Driggs (1929). He obtained the hydride by heating uranium powder in a hydrogen atmosphere to 225°C and observed that it decomposed at 350 to 400°C under 1 atm partial pressure of hydrogen. A detailed study of uranium hydride has been carried out since 1943 (MP Ames 1; Battelle 2; Los Alamos 1; MP Clinton 1; UCRL 1; Burke and Smith, 1947; Gueron and Yaffe, 1947; Newton, 1948).

Judged by the criteria of simple stoichiometric composition and constant decomposition pressure at a given temperature, uranium hydride is a true chemical compound. If two bulbs containing uranium powder and uranium hydride powder, respectively, are kept in communication in an evacuated system for 10 hr, no exchange of hydrogen between the two samples is observed at 269, 288, 300, or 325°C (MP Ames 2). This behavior is typical of a true compound since even a slight dependence of equilibrium pressure on hydrogen concentration would cause hydrogen to diffuse from the hydride to the metal. On the other hand, the appearance and physical properties of uranium hydride resemble those of the "metallike" hydrides of cerium, lanthanum, and other rare earth metals. For all these hydrides a more or less gradual change of decomposition pressure with content of hydrogen in the metal was observed in a series of investigations. Because of this variability of decomposition pressure, which, according to the phase rule, indicates that metal and hydrogen form a single solid phase of variable composition, Sieverts (1925) interpreted the rare earth hydrides as "solid solutions." However, the decomposition isotherms of the rare earth hydrides, which all show a more or less extensive, if not strictly horizontal, plateau, are not typical of true solutions. It may perhaps be suggested that, similar to uranium hydride, rare earth hydrides also are true compounds of definite stoichiometric composition. The failure to obtain horizontal isotherms can then be attributed to the use of insufficiently pure and homogeneous metals; the metals used may have contained two or more solid phases, e.g., allotropic forms of the metal itself, alloys, oxides, or carbides.

2.1 Decomposition Isotherms of Uranium Hydride. As mentioned above, the interpretation of uranium hydride as a chemical compound is based on the shape of its decomposition isotherms, which have been measured several times at Ames (MP Ames 3). The first measurements showed that equilibrium is established only very slowly. The 357°C isotherm, obtained by pumping hydrogen away from the hydride and waiting for 30 min, had a horizontal plateau at about 120 mm Hg pressure; however, a similar isotherm, obtained by admitting hydrogen to the metal and waiting for the same length of time, had a plateau at about 170 mm Hg. This isotherm was redetermined later, waiting at each point until no pressure changes occurred in 24 hr; two days was usually required to reach this degree of constancy. The results are shown in Fig. 8.3. If the equilibrium is approached "from below," i.e., by the decomposition of hydride, the curve shows

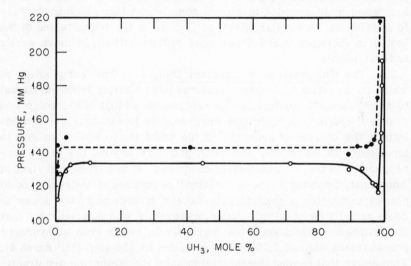

Fig. 8.3—Pressure-composition isotherm for the uranium−uranium hydride−hydrogen system at 357°C. ○, apparent equilibrium upon decomposition; ●, apparent equilibrium upon formation.

a plateau at 134 mm Hg which stretches from 5 to 90 per cent decomposition. A small but reproducible dip at 90 to 98 per cent hydrogenation, with a minimum at about 98 per cent UH_3, is the most remarkable feature of this curve. The points at the bottom of the dip are not changed even by 14 days of waiting. In addition to the isotherm obtained at the temperature of boiling mercury (357°C), the dip could be observed also on dehydrogenation isotherms obtained at 440°C (boiling sulfur) and 307°C (boiling benzophenone).

The hydrogenation isotherm (dotted line) shows no dip, and its plateau extends from 5 to 95 per cent hydrogenation. Despite long waiting, this isotherm is still several millimeters of mercury above the dehydrogenation isotherm, showing that perfect thermodynamic equilibrium has not been realized. The "pseudoequilibrium" field is particularly wide in the region of the dip. Experiments in which the uranium—uranium hydride mixture was carefully protected from mercury vapor showed that the dip was not due to a reaction with mercury. In the region of the dip the x-ray diagram showed only lines of uranium metal and of UH_3.

The origin of the dip remains unexplained. Assuming that the hydrogenation isotherm corresponds to true equilibrium while the dehydrogenation isotherm runs "too low" because of slow approach to the equilibrium, the dip seems to indicate that delay is particularly long when hydrogenation is almost complete, a fact which is not easy to understand. A similar effect, noticeable at the opposite end of the isotherm (between 0 and 5 per cent hydrogenation), is more easily understandable.

2.2 The Composition of Uranium Hydride. The composition at which the uranium-hydrogen isotherms bend sharply from horizontal to approximately vertical can be considered as that of the compound uranium hydride. If hydrogen were soluble in the uranium hydride lattice, the content of hydrogen in the solid phase would continue to increase with increasing hydrogen pressure beyond the value corresponding to the stoichiometric compound. It was thought at first at Ames that, by using "supersaturating" pressures of hydrogen, solid phases containing a considerable excess of hydrogen (MP Ames 4) could actually be obtained. Later, however, it was demonstrated that the hydrogen content does not markedly increase even at hydrogen pressures as high as 2,000 psi (equivalent to 136 atm) (MP Ames 5). This shows that beyond the saturation point the isotherms are practically vertical. The only visible result of high pressure on the hydride structure is crystal growth, which could be observed at temperatures above 600°C. The x-ray diffraction lines were sharper, and fibrous crystals could be noticed in the hydride powder obtained under high hydrogen pressure.

Since uranium hydride is a single compound and does not markedly dissolve hydrogen, its analysis gives the same composition regardless of the temperature and of the hydrogen pressure under which it was prepared. The first determinations at Ames gave compositions between $UH_{3.85}$ (cooled in vacuum) and $UH_{4.15}$ (cooled in hydrogen), so that a composition UH_4 was postulated (MP Ames 6). It was, however, soon found at Battelle (Battelle 2) as well as at Ames (MP Ames 7) that these results were incorrect and that the true composition was close to UH_3. Thus Battelle measurements gave $UH_{2.91}$ for material made at 300°C and at a series of hydrogen pressures from 253 to 1,265 mm Hg. At Ames, compositions between $UH_{2.91}$ and $UH_{3.10}$ (average $UH_{2.99}$) were found by measuring the gain in weight of uranium turnings in hydrogen at 250°C and of uranium powder obtained by hydride decomposition at room temperature, as well as by determining the volume of gas absorbed in hydrogenation or liberated in the decomposition of the hydride.

Six combustions of hydride prepared at 250°C, in which the weight of water formed from hydrogen produced by hydride decomposition was determined (MP Ames 8), gave compositions between $UH_{2.94}$ and $UH_{2.96}$. The product prepared at 420°C had the same composition. The deviations from the formula UH_3 found in this work, although small, were beyond the limit of experimental error. However, they can be

Table 8.2 — Hydrogen/Uranium Ratios
in Uranium Hydride

Temp., °C	H/U	Temp., °C	H/U
150	3.03	250	3.03
175	2.99	275	3.03
200	2.96	300	2.99
225	3.04	350	2.92

explained by the presence of oxides, carbides, or other impurities in uranium metal. For example, the presence of 0.14 per cent carbon as UC in the metal powder would reduce the hydrogen/uranium ratio to 2.91 if all UC remained unreduced by hydrogen (Battelle 2). At Ames (MP Ames 9), impurities in a particular uranium sample were calculated to account very nearly for the hydrogen/uranium ratio of 2.97 observed in the formation of the hydride. At the University of California Radiation Laboratory (UCRL 1) a composition of $UH_{3.06}$ was calculated from reaction of a known amount of uranium with hydrogen by correcting the result for the known content of oxide in the metal (which was as high as 11.7 per cent). At Los Alamos (Los Alamos 1), hydrogen/uranium ratios, as given in Table 8.2, were found at different temperatures. Thus the formula UH_3 is well established.

2.3 Hydrogen-pressure Curve and Thermodynamics of Formation of UH_3. Since UH_3 behaves as a true chemical compound (i.e., has a constant hydrogen pressure over a wide range of average hydrogen contents of the solid phase), a unique decomposition-pressure curve can be constructed, and the heat of formation from the elements can be calculated from the slope of this curve. The first measurements were carried out at Ames (MP Ames 9) at 252 to 438°C; they gave pressures from 4 to 808 mm Hg; the pressure of 1 atm was reached at 436°C. However, no exact thermodynamic equilibrium was reached in these measurements, as shown by considerable deviation between heating and cooling curves. According to a later review (MP Ames 9,10) the heating curve can be represented by the equation

$$\log p_{mm} = -\frac{4500}{T} + 9.23 \qquad (1)$$

corresponding to $\Delta H = 31.0$ kcal per mole for the heat of decomposition. The results of similar measurements carried out at Battelle

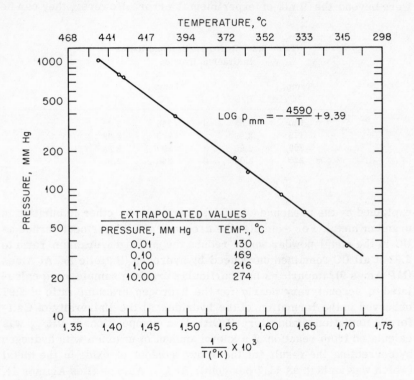

Fig. 8.4 — Equilibrium pressure as a function of temperature for the system uranium hydride $\rightleftharpoons$ uranium + hydrogen.

(Battelle 2) are shown in Fig. 8.4. The equation of the straight line in Fig. 8.4 is

$$\log p_{mm} = -\frac{4590}{T} + 9.39 \qquad (2)$$

The corresponding ΔH value is 31.5 kcal per mole. A rather rough direct calorimetric determination of ΔH gave 30.5 ± 0.5 kcal per mole, which is in satisfactory agreement with the values calculated from the decomposition curves (MP Ames 11).

According to the Ames data (MP Ames 10) the decomposition pressures used in the above calculations are probably still too low because of incomplete equilibrium. Three points on the decomposition curve, determined more carefully by waiting several days at constant temperature, were markedly higher than those obtained in more rapid experiments, but they gave no significant change in the slope of the pressure-temperature curve and thus in the calculated value of ΔH. The equation of the straight line passing through these three points was

$$\log p_{mm} = -\frac{4500}{T} + 9.28 \tag{3}$$

Attempts to approach equilibrium from the side of the higher temperatures, i.e., by hydrogenation of uranium rather than dehydrogenation of UH_3, gave less satisfactory results. The final values were not reached even after waiting for weeks.

At the University of California Radiation Laboratory (UCRL 1,2) the pressures given in Table 8.3 were obtained by approaching the

Table 8.3—Decomposition Pressure of Uranium Hydride*

Temp., °C	H_2 pressure, mm Hg	Temp., °C	H_2 pressure, mm Hg
200	0.6	350	103.0
250	4.5	400	345
300	24.8	436	760†

*UCRL 2.
†Extrapolated.

equilibrium from both above and below. These results can be represented by the equation

$$\log p_{mm} = -\frac{4480}{T} + 9.20 \tag{4}$$

corresponding to a heat of decomposition of 30.7 kcal per mole.

The calculation of ΔH_0, ΔF, and ΔS for the formation of UH_3 from the elements from this set of decomposition pressures was carried out at University of California Radiation Laboratory (UCRL 2) and later revised (UCRL 3). It is based on an approximate value, $\Delta C_p = 1$ cal per mole, for the reaction

$$UH_3 \rightarrow U + \tfrac{3}{2}H_2 \qquad (5)$$

This value was derived from Kopp's rule and gives for the heat of decomposition

$$\Delta H = \Delta H_0 + 10^{-3} \times T \text{ kcal per mole}$$
$$\Delta H_0 = 30.1 \text{ kcal per mole} \qquad (6)$$
$$\Delta H_{298} = 30.4 \text{ kcal per mole}$$

Using the equation

$$R \ln K = R \ln p^{3/2} = \Delta C_p \ln T - \frac{\Delta H_0}{T} + I \qquad (7)$$

with $\Delta C_p = 1$, the following is obtained:

$$\frac{3R}{2} \ln p = \ln T - \frac{\Delta H_0}{T} - I \qquad (8)$$

MacWood (UCRL 2) obtained, from 17 individual measurements of decomposition pressure between 530 and 652°K, I values between -35.5 and -36.4, with an average of $I = -35.85$. This gives, for the standard free energy of decomposition of UH_3,

$$\Delta F° = 30,100 - 2.303T \log T - 35.85T \text{ cal per mole} \qquad (9)$$

and thus

$$\Delta F°_{298} = 17.7 \text{ kcal per mole}$$

Combining the last value with $\Delta H_{298} = -30.4$ kcal per mole, Eq. 10 is obtained for the entropy of decomposition.

$$\Delta S°_{298} = 42.6 \text{ e.u.} \qquad (10)$$

The corresponding values for the formation of UH_3 from the elements are

$$\Delta H_{298} = -30.4 \text{ kcal per mole}$$

$$\Delta F^{\circ}_{298} = -17.7 \text{ kcal per mole}$$

$$\Delta S^{\circ}_{298} = -42.6 \text{ e.u.}$$

2.4 Kinetics of Formation and Decomposition of Uranium Hydride.

Driggs (1929) first obtained uranium hydride by heating uranium powder to 225°C or higher in a hydrogen atmosphere. Early experiments at Ames (MP Ames 12) confirmed that a temperature of 225 to 250°C is best for rapid hydrogenation. As an example, 100 g of uranium turnings could be completely hydrogenated in 30 min at 250°C, and a 100-g lump of uranium metal could be hydrogenated in 2 hr (MP Ames 12; MP Clinton 2; Los Alamos 1).

(a) Effect of Temperature. Rapid decline in the rate of hydrogenation at temperatures above 300°C is characteristic of the hydrogenation process, independently of whether powder or solid metal is used for the experiment. For example, under 1 atm partial pressure, uranium powder took up no hydrogen in two days at 440°C; three days

Table 8.4— Penetration Rate of Hydrogen
into Solid Uranium

Temperature, °C	Rate of penetration, cm/hr
200	0.27
250	0.30
300	0.19

was required for complete hydrogenation at 306 to 420°C (MP Ames 2) and less than 1 hr at 250°C. Similar results were obtained at Clinton (MP Clinton 2) for the rate of penetration of hydrogen into solid uranium (cf. Table 8.4). Figure 8.5 shows the temperature dependence of the rate of hydrogenation of uranium wire (0.0625 in. in diameter) (Los Alamos 1). The values plotted were the approximately constant rates prevailing after the termination of the induction period (see below) and before the slowing down of the process toward the end of hydrogenation. This curve also shows a maximum at 225°C.

The decline in reaction velocity at temperatures below 200°C depends on the state of the metal and the pressure of hydrogen. With turnings the reaction first becomes noticeable at 200°C under 1 atm hydrogen pressure, at 150°C under 1,000 psi, and at 130°C under

1,800 psi. With finely powdered uranium obtained by the decomposition of the hydride, the hydrogen pickup remains very rapid even at 0°C and is marked but slow at −80°C; no reaction could be observed at the temperature of liquid air.

(b) Effect of Pressure. The decline of the rate of hydrogenation at temperatures above 250°C is associated with the fact that the rate becomes zero when the decomposition pressure reaches the external

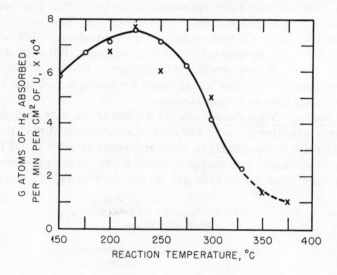

Fig. 8.5—Effect of temperature on reaction rate. ○, purified hydrogen; ×, tank hydrogen.

pressure of hydrogen. Thus the rate of hydrogenation must be a function of the excess hydrogen pressure, i.e., of the difference between the external hydrogen pressure, p, and the decomposition pressure of the hydride, p_0. Rate measurements at constant temperature (357°C) (MP Ames 14) have shown that the initial rate of hydrogenation of uranium powder (see Fig. 8.6) can be represented by the equation

$$v_0 = -\left(\frac{dp}{dt}\right)_{t=0} = K(p - p_0)^{5/2} \tag{11}$$

where $p_0 = 134$ mm Hg. Integration of Eq. 9 indicates that the graph of $(p_t - p_0)^{-3/2}$ against t must be linear. The quantity p_t is the residual hydrogen pressure after the hydrogenation has proceeded for the time t. This conclusion is confirmed by experiments (see Fig. 8.7).

(c) Effect of Surface. The results of measurements (Los Alamos 1) on the slowing down of hydrogen consumption, which always occurs after hydrogenation has proceeded for a while, can be accounted for by the assumption that the rate is proportional to the active uranium surface. In the calculation of the active surface it was assumed that the hydrogenation of a wire progresses from the surface to the axis with the surface retaining its cylindrical shape.

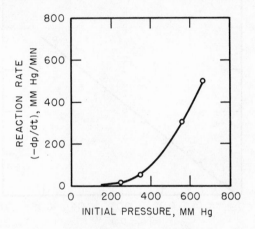

Fig. 8.6 — Effect of initial hydrogen pressure on reaction rate of hydrogen and uranium at 357°C.

This assumption obviously cannot be exactly correct, but no other quantitative data on the relation between surface and hydrogenation velocity are available, apart from the trivial observation that fine powder is hydrogenated much more rapidly than solid metal.

(d) Induction Period. It was suggested above that the decline of hydrogen consumption with progressing hydrogenation could be accounted for by calculating the rate for unit active surface. Figure 8.8 shows that the specific rate calculated in this way remains constant throughout the process, except for a slight dip toward the very end of hydrogenation (which is probably due to errors of method) and an "induction period" at the beginning of the process.

The duration of the induction period depends on the purity of the hydrogen; at 250°C it is as long as 28 min in tank hydrogen but drops to practically zero in hydrogen purified by hot uranium. At 150°C induction is noticeable even in most thoroughly purified hydrogen. Induction must be attributed to the formation of protective surface layers of oxide or nitride. It is noteworthy, however, that preliminary exposure of the metal to nitrogen or oxygen does not lead to an induction period in subsequent treatment by pure hydrogen.

After the induction period is over, the hydrogenation rate is about the same in tank hydrogen as in pure hydrogen (Los Alamos 1).

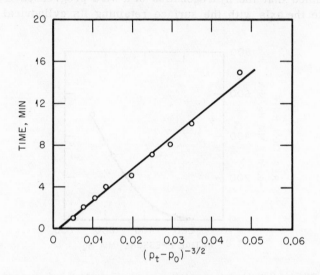

Fig. 8.7—Rate of hydrogenation of uranium powder at 357°C.

An attempt was made at Ames (MP Ames 3) to prepare uranium hydride under an inert solvent (mineral oil); however, the rate of hydrogenation at 250°C was only 1 per cent of the value obtained in hydrogen gas.

(e) <u>Mechanism of Hydride Formation</u>. The formation and decomposition of uranium hydride provide an interesting example for the study of kinetics of heterogeneous reactions. The occurrence of the difference $(p - p_0)$ in the rate vs. pressure function seems to indicate that the rate is limited by a diffusion process, but no offhand explanation can be given for the $5/2$ power with which this factor occurs in Eq. 11. The same factor indirectly accounts for the drop in the rate of hydrogenation at high temperatures, where p_0 approaches p.

(f) <u>Kinetics of Decomposition</u>. The rate of decomposition of the hydride also was studied at Ames (MP Ames 15). At 250°C about 50 per cent of a 6- to 7-g sample of hydride could be decomposed in 1 hr by evacuation at a pressure of approximately 0.1 mm Hg. Systematic measurements were carried out at 294, 367, and 405°C (MP Ames 13).

However, the temperature dropped by as much as 100°C immediately after the beginning of pumping and returned only gradually to the initial level. Furthermore, the true pressure of hydrogen at the surface of the hydride is not known and may vary for different parts of the sample. Thus the decomposition curves (Fig. 8.9) cannot be considered as having any general significance. These curves indicate an approximately first order reaction and a rather slow increase in decomposition rate with temperature. For example, 50 per cent decomposition is reached in 30 min at 294°C and in 3 min at 405°C. This

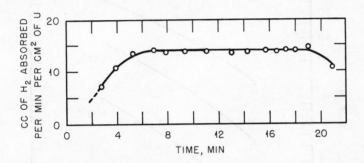

Fig. 8.8—Rate of hydrogen absorption corrected for surface area of uranium specimen. Reaction temperature, 225°C.

slow increase may be due to the fact that the hydrogen pressure above the surface of the hydride becomes higher with increasing temperature because of the limited rate of removal of the gas.

2.5 **Hydride Formation from Uranium Compounds or from Uranium and Water.** (a) Formation of Hydride from Uranium Compounds. According to the Ames group (MP Ames 12) no hydride is formed by heating uranium carbide or oxide in hydrogen. Uranium alloy with 20 per cent aluminum reacts with hydrogen to give a powder that apparently contains metallic uranium and a uranium-aluminum compound. Sodium hydroxide dissolves aluminum out of it, at least partly. Uranium amalgam does not react with hydrogen (MP Ames 2).

(b) **Hydride Formation from Uranium and Water Vapor.** Water can be decomposed by uranium at 600 to 700°C, uranium hydride being formed by reaction of hydrogen with uranium at 250°C in the same piece of apparatus (MP Ames 52). Considerable amounts of the hydride were also obtained by direct reaction of uranium with steam at 250°C (MP Ames 57).

(c) **Hydride Formation from Uranium and Decalin or Tetralin.** Small amounts of UH_3 are formed by reaction of uranium metal with decalin or tetralin at 210 to 270°C.

2.6 Physical Properties of Uranium Hydride. The hydride is described as a brownish-black or brownish-gray pyrophoric powder. It passes easily through 400-mesh sieve or silk bolting cloth. Prepared from metal, it contains shiny particles, probably oxide inclusions from the original metal. Screened, decomposed in vacuum, and resynthesized, the hydride appears more homogeneous and more grayish in color (MP Ames 10,12).

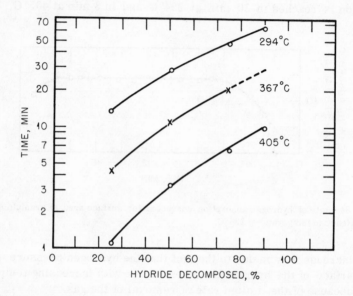

Fig. 8.9—Decomposition rate of uranium hydride with temperature. Values given beside curves are initial temperatures.

(a) **Crystal Structure and Density.** The first x-ray analysis of uranium hydride at Ames (MP Ames 16) showed that it has a simple cubic structure quite different from the three structures of uranium metal. The determination was repeated later with especially pure samples (MP Ames 17,18). The samples prepared at 200 to 300°C and atmospheric pressure of hydrogen gave satisfactory but not very sharp reflections in the back reflection region of a symmetrical self-focusing camera of 5 cm radius with CuKα radiation. The reason for the diffuseness is the small size of the hydride particles. Hydride prepared under 1,800 psi at 500 to 600°C consists of much larger crystals and gives very sharp maxima. The best value of the lattice spacing

$$a_0 = 6.6310 \pm 0.0008 \text{ A}$$

has been obtained with such high-pressure preparations. The "low pressure" hydride gave the value

$$a_0 = 6.632 \pm 0.001 \text{ A}$$

At Chicago (MP Chicago 1) a lattice constant

$$a_0 = 6.634 \pm 0.002 \text{ A}$$

was found, in satisfactory agreement with the Ames result.

The missing reflections in the diffraction diagram lead to the identification of O_h^3, O^2, and T_d^4 as the three possible space groups in the lattice of uranium hydride; final selection of the space group is impossible without knowledge of the positions of the hydrogen atoms. The elementary cell contains eight molecules. The positions of the eight uranium atoms can be divided into two groups: (I) two equivalent positions at 0 0 0 and $\frac{1}{2}$ $\frac{1}{2}$ $\frac{1}{2}$, and (II) six equivalent positions at $\frac{1}{4}$ $\frac{1}{2}$ 0; 0 $\frac{1}{4}$ $\frac{1}{2}$; $\frac{1}{2}$ 0 $\frac{1}{4}$; $\frac{3}{4}$ $\frac{1}{2}$ 0; $\frac{1}{2}$ 0 $\frac{3}{4}$; and 0 $\frac{3}{4}$ $\frac{1}{2}$. This arrangement is represented in Fig. 8.10. Each of the atoms of group I has twelve atoms of group II as its nearest neighbors at a distance of 3.707 A; each atom of group II has two nearest neighbors of the same group at 3.316 A distance, four tetragonally arranged neighbors of group I at 3.707 A distance, and eight atoms of group II at a distance of 4.06 A.

The structure of uranium hydride has recently been discussed by Rundle (1947) and Pauling and Ewing (1948). Rundle has advanced a theory of electron-deficient structures which appears to interpret adequately the structure of UH_3 and to account qualitatively for the physical properties of the compound. The crystal structure of UH_3 is fundamentally different from any of the forms of metallic uranium; from the metal-metal distances in UH_3 it is clear that metal-metal bonds are practically nonexistent in the hydride. Nevertheless UH_3 shows distinctly metallic properties, such as a high electrical conductivity and metallic luster. The high melting point, brittleness, and hardness are also inconsistent with very weak metal-metal bonds and indicate rather a continuous-type structure held together by covalent bonds. Rundle therefore proposes a structure in which each uranium atom of type I is bonded to twelve atoms of type II by hydrogen bridges, U–H–U; each uranium atom of type II is bonded directly to two other type II atoms and by hydrogen bridges to four uranium atoms of type I. Each hydrogen bridge U–H–U is interpreted in terms of electron-deficient "half-bonds," in which the bridge contains only one electron pair. The resonating system of half-bonds accounts for the polarizability, metallic luster, and high electrical conductivity; provides a satisfactory explanation for the uranium-

uranium distances in UH$_3$; and provides satisfactory positions for the correct number of hydrogen atoms. According to Pauling and Ewing, Rundle's conclusions are substantiated by considerations of inter-atomic distances, and they suggest that in UH$_3$ there occurs a new low-valence form of uranium similar to the low-valence forms of chromium and manganese previously described by Pauling (1947).

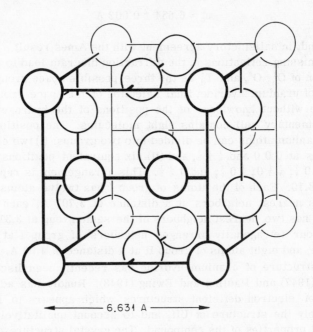

6.631 A

Fig. 8.10 — The structure of uranium hydride.

The density of UH$_3$ calculated from the x-ray structure is 10.92 g/cc. This result is in excellent agreement with the experimental density determination by the helium-displacement method (MP Ames 10,19), which gave 10.95 g/cc. Earlier, less precise density determinations (under hexane) gave a value of 11.4 g/cc. The bulk density of dry UH$_3$ powder is 3.4 g/cc; when wetted with hexane, the powder settles to a density of 3.5 g/cc. Centrifugation increases the density to 4.0, pressure of 28 tons/in.^{2} increases it to 7.3, and pressure of 160 tons/in.^{2} increases it to 8.4 g/cc (MP Ames 10). Uranium hydride powder prepared at 700 to 800°C under 1,800 psi hydrogen pressure can be pressed to a bulk density of more than 9.0 g/cc (MP Ames 5).

The low bulk density of uranium hydride causes a large increase of the volume of the metal during hydrogenation. This swelling must be taken into account in the construction of apparatus for hydrogenation.

In partly decomposed or incompletely hydrogenated hydride samples, the density is within ±1 per cent a linear function of the hydrogen content (MP Chicago 1). The lattice constant of the hydride was unchanged by heating a mixture of hydride with uranium metal; in a heated mixture consisting of 1 part of UH_3 and 1 part of uranium, for example, the spacing was a_0 = 6.630 ± 0.002 A. These results are typical of a mechanical mixture of two solid phases without mutual solubility. As mentioned before, a hydride sample made under 1,800 psi of hydrogen showed no significant displacement of the x-ray diffraction maxima compared to their position in the diagram of hydride prepared under low hydrogen pressure. In other words, there was no indication of lattice distortion by dissolved or occluded hydrogen. The hydride separated from the uranium hydride amalgam by aeration (see Sec. 2.8) had the same lattice constant as ordinary hydride.

(b) Electrical Conductivity. Hydride powder tapped between electrodes showed metallic conductivity. The specific resistance was 0.47 ohm-cm, not very different from that of uranium metal powder obtained by the decomposition of the hydride (0.68 ohm-cm) (MP Ames 6). Pressed material conducted better than loose powder (MP Ames 10).

2.7 Chemical Reactions of Uranium Hydride. Uranium hydride is a very reactive substance and can be used for the preparation of numerous uranium compounds. In many cases the reaction probably proceeds through the intermediary of free uranium metal which is formed by the decomposition of the hydride into a very finely divided and chemically active state. The reactions of the hydride have been studied in a series of investigations at Ames (MP Ames 6,10,20,21). A few experiments also were performed at Clinton Laboratories (MP Clinton 3) and at the Massachusetts Institute of Technology (MIT 1).

(a) Reactions with Gases. Table 8.5 is a summary of the reactions of uranium hydride with various gases (MP Ames 10,22). The hydride is often pyrophoric (MP Ames 12) and must be handled with care. Nonpyrophoric samples probably are protected by surface layers of oxide and can be obtained by first exposing the hydride to air at the temperature of dry ice. Carbon dioxide and nitrogen are safe atmospheres up to 200 to 225°C; however, these gases also react with the hydride at higher temperatures. Once reaction with carbon dioxide or nitrogen has started it cannot be stopped, but there is no risk of explosion as in the presence of oxygen. In open air the hydride burns smoothly to water and U_3O_8; UO_2 may be formed if the air supply is insufficient.

The rate of oxidation of UH_3 in air can be measured if ignition is avoided by careful initial exposure (MP Ames 23). The initial rate of oxygen pickup of a stable UH_3 sample was 0.25 mg per day per gram of hydride; after 45 days the rate still was about half as large. So-called "quasi-hydride" (see Sec. 2.8) gained 0.83 mg per gram per day initially and about one-fifth of this after 45 days.

Table 8.5 shows that the reactions of uranium hydride with hydrogen chloride, hydrogen bromide, phosphine, and ammonia give compounds of trivalent uranium, and those with chlorine, bromine, water, hydrogen fluoride, phosgene, and probably hydrogen iodide, lead to quadrivalent uranium compounds.

The reaction of UH_3 with hydrogen fluoride was studied in some detail at Clinton Laboratories (MP Clinton 3). When UH_3 is treated with hydrogen fluoride at 270°C a surface layer of UF_4 is formed which prevents further penetration of the hydride by hydrogen fluoride. No such caking occurs at 500°C, and all hydride is converted to UF_4. According to studies at Ames (MP Ames 29) a reaction of uranium with a mixture of equal volumes of hydrogen fluoride and hydrogen H_2 at 250°C gives a smooth conversion to UF_4; uranium hydride undoubtedly is an intermediate compound in this reaction.

(b) Reactions with Water and Aqueous Acids and Bases. A small quantity of UH_3 powder disperses in water without reaction. Sometimes, however, ignition occurs, although once the powder is wetted, this danger is over (MP Ames 30). Large quantities always ignite unless water is added very slowly (MP Ames 31). UH_3 reacts slowly with water vapor at 200 to 300°C; the oxide formed is probably UO_2 (MP Ames 24).

Uranium hydride dissolves much more slowly than uranium metal in 6N or 12N hydrochloric acid, giving a pale-green solution (MP Ames 30,32). The hydride is easily dissolved in chloric acid ($KClO_3$ + H_2SO_4) (MP Ames 33). Dilute perchloric acid dissolves the hydride slowly if at all, but concentrated acid reacts upon heating to give uranyl perchlorate, $UO_2(ClO_4)_2$. This reaction is safe if performed on a small scale but becomes dangerous with quantities exceeding 1 or 2 g (MP Ames 30). Dilute phosphoric acid does not react, although concentrated acid dissolves the hydride to form uranous phosphate (MP Ames 30). Cold 8N acetic acid does not react, but a brown suspension is formed on boiling; this may be merely peptized hydride, or it may be a product of chemical transformation. Glacial acetic acid does not affect the hydride (MP Ames 30). Uranium hydride readily reduces 6N or concentrated nitric acid to NO_2, while uranium,

Table 8.5—Reaction of Uranium Hydride with Gases

Gas	Temp., °C	Product	Remarks
		Elementary Gases	
N_2	225	$UN_{1.6}$	Reaction slow; incomplete
	250		Pyrophoric nitride rapidly formed (MP Ames 24)
	900	UN (?)	Chicago work showed 3.6% N (calculated for UN 5.5%) (MIT 1)
O_2	Ignites	U_3O_8	UO_2 can be formed if air supply is insufficient
Cl_2	200	UCl_4	Product melts during reaction
Br_2	200−300	UBr_4	Brown, partly fused powder
I_2	400	$UI_{(4)}$(?)	Product unstable, analysis uncertain (MP Ames 25)
		Hydrides	
H_2O	350	UO_2	Black oxide; by weight gain, $UO_{2.14}$ (MP Ames 6) (excess over $UO_{2.0}$ probably due to air leakage)
	500	UO_2	Reaction mass glows
H_2S	400−500	US_2	Black powder; U_2S_3 as intermediate can be isolated; reaction slow
NH_3*	100	UN (?)	"Low temperature" nitride; slow reaction
	250	$UN_{1.5}$−UN_2	Pyrophoric (?); faster reaction
PH_3	400	U_2P_3	Black powder
HF	20−400	UF_4	Green dry powder; typical "green salt"
HCl	250−300	UCl_3	Dry powder; olive green at 25°C; reddish brown at higher temperatures
HBr	200	UBr_3	Reddish-brown dry powder; best results obtained by conducting HBr through UH_3 from below at 250 to 275°C (MP Ames 26)
HI	300−400	UI_4 (?)	Product unstable; analysis UI_{3-4}
		Oxides	
CO	Up to 400	U	No volatile carbonyl formed
CO_2	200	UO_2	Impure product (carbide); UH_3 burns once reaction started
		Organic Gases	
C_2H_4	500	UC (?)	Product not known; contains 1.5% C; reaction probably incomplete
HCN	400	Carbide, nitride	Black powder; pyrophoric; contains no CN
$COCl_2$	250	UCl_4	Yellow-green dry powder; slightly contaminated by carbon
CH_3I	275−300	UI_3 (?)	Uncertain; reaction moderately fast
CCl_4	250	UCl_4 + C (?)	CH_4 and C_2H_6 probable by-products (MP Ames 27,28)

*For clarification of reaction of UH_3 with nitrogen and ammonia see the discussion of uranium nitrides in Chap. 10.

after passing through an intermediate green stage, forms yellow uranyl nitrate (MP Ames 30). Dilute sulfuric acid reacts very slowly, but hot concentrated acid is rapidly reduced to sulfur and hydrogen sulfide, while uranium is dissolved (MP Ames 30) as uranium(IV) sulfate, $U(SO_4)_2$.

Hot or cold solutions of potassium hydroxide, sodium hydroxide, ammonium hydroxide, or sodium cyanide do not react with the hydride. Nor is the hydride dissolved by liquid ammonia (MP Ames 30).

(c) Reactions with Organic Solvents. Organic solvents containing no halogen (e.g., benzene, toluene, hexane, dioxane, alcohol, acetone, ethyl acetate) do not dissolve the hydride and do not react with it (MP Ames 30). Halogenated solvents, on the other hand, are dangerous. Upon addition of hydride to carbon tetrachloride, for example, a violent explosion may ensue, producing UCl_4, hydrogen chloride, hydrogen, and carbon. Carbon tetrachloride vapor reacts with uranium hydride only above 200°C, producing UCl_3, UCl_4, and carbon (MP Ames 28).

(d) Oxidation of Hydride by Weak Inorganic Oxidants. Many weak inorganic oxidants oxidize UH_3 to $U(IV)$ or $U(VI)$ salts and liberate hydrogen. Particularly efficient are silver salts, as in the following reactions (MP Ames 34):

$$2UH_3 + 8AgF \rightarrow 2UF_4 + 8Ag + 3H_2 \tag{12}$$

This is a vigorous reaction.

$$2UH_3 + 12AgNO_3 + 4H_2O \rightarrow 12Ag + 2UO_2(NO_3)_2 + 8HNO_3 + 3H_2 \tag{13}$$

This is a very rapid reaction; the solution heats up to boiling, and the color of the resulting solution indicates complex formation.

$$2UH_3 + 6Ag_2SO_4 + 4H_2O \rightarrow 2UO_2SO_4 + 12Ag + 4H_2SO_4 + 3H_2 \tag{14}$$

This reaction is slower than the two preceding ones.

$$2UH_3 + 12Ag(CH_3COO) + 4H_2O \rightarrow 2UO_2(CH_3COO)_2 + 12\ Ag + 8CH_3COOH + 3H_2 \tag{15}$$

This reaction is slow.

$$2UH_3 + 6Ag_2(OOCCHOHCHOHCOO) + 4H_2O$$
$$\rightarrow 2UO_2(OOCCHOHCHOHCOO) + 12Ag + 4(CHOH)_2(COOH)_2 + 3H_2 \tag{16}$$

This reaction is still slower.

Three parallel reactions occur with silver perchlorate:

$$2UH_3 + 12AgClO_4 + 4H_2O \rightarrow 12Ag + 2UO_2(ClO_4)_2 + 8HClO_4 \\ + 3H_2 \quad (80\%) \quad (17)$$

$$2UH_3 + 8AgClO_4 \rightarrow 8Ag + 2U(ClO_4)_4 + 3H_2 \quad (<0.1\%) \quad (18)$$

$$2UH_3 + 5AgClO_4 \rightarrow 4Ag + AgCl + 2UO_2(ClO_4)_2 + 3H_2 \quad (20\%) \quad (19)$$

In toluene the reaction of UH_3 with silver perchlorate gives silver and silver chloride, perhaps according to the equation (MP Ames 35)

$$2UH_3 + 6AgClO_4 \rightarrow 2Ag + 2AgCl + Ag_2O + 2UO_2(ClO_4)_2 + 3H_2O \quad (20)$$

Similar reactions occur with mercuric salts. Mercuric chloride reacts vigorously, giving UCl_4 (15 per cent) and UO_2Cl_2 (35 per cent); the rest of the uranium is found in the gray $Hg + Hg_2Cl_2$ precipitate. Mercuric nitrate also reacts rapidly, giving uranyl nitrate and mercury. Antimony(III) chloride in 3N HCl oxidizes UH_3 to UCl_4 and liberates H_2

$$8SbCl_3 + 6UH_3 \rightarrow 8Sb + 6UCl_4 + 9H_2 \quad (21)$$

Cupric sulfate does not react at room temperature but is reduced slowly at 80 to 90°C, giving copper and $U(SO_4)_2$. Cupric chloride and NH_4Cl evolve hydrogen rapidly and give a green solution (MP Ames 23).

Ferric sulfate oxidizes UH_3 to UO_2^{++} upon heating (MP Ames 34). No reaction is observed with bismuth or lead salts (MP Ames 34).

(e) <u>Oxidation of Hydride by Strong Oxidants</u>. The effect of saturated sodium hypochlorite (NaOCl) solution on UH_3 is slight (MP Ames 33). A 30 per cent solution of hydrogen peroxide reacts vigorously with sparking and formation of black oxide (MP Ames 36). The reaction of hydrogen peroxide with suspension of UH_3 in hydrochloric acid gives UO_2Cl_2. This reaction is slower in sulfuric acid. In the presence of organic acids, hydrogen peroxide produces the corresponding uranyl salts; with excess peroxide peruranic acid may be formed. Ceric salts oxidize the uranium in UH_3 to the sexivalent state and release hydrogen. Dichromate reacts only upon heating, giving uranyl salt. Cold permanganate does not react without acid but reacts rapidly in the presence of acid giving uranyl salt. The same is true of bromate

(MP Ames 10). Reactions with perchloric, nitric, and sulfuric acids were mentioned in Sec. 2.7b.

2.8 The Uranium Hydride Amalgam and Similar Systems and the "Quasi-hydride." Uranium hydride can be dispersed in mercury (MP Ames 10,37) without marked heat liberation or hydrogen evolution. The resulting "amalgam" is fluid up to 10 per cent UH_2, pasty at 30 per cent, and semisolid at 40 per cent. Above 60 per cent, UH_3 takes up mercury as sawdust takes up oil. The 90 per cent amalgam is a gray pyrophoric powder.

Filtration through alundum or sintered glass separates the amalgam into practically pure mercury and a residue that looks like more concentrated amalgam. Pure mercury can also be squeezed out from 35 per cent amalgam by pressure of 40 tons/in.^{2} (MP Ames 38). These results indicate that the amalgam is a colloidal system rather than a true molecular solution or a chemical compound.

The behavior of the amalgam in air supports the interpretation of its being a capillary system. Uranium hydride that has been exposed to air ceases to be wetted by mercury. Aeration of the amalgam prepared under exclusion of air causes a precipitation of the hydride. In the course of this precipitation partial oxidation takes place, causing the precipitate to be somewhat different from the original hydride and leading to its designation as "quasi-hydride" (MP Ames 39). Its x-ray diffraction picture is the same as that of the true hydride, but it is brown in color, and a typical analysis shows 97.4 per cent uranium and 1.15 per cent hydrogen, the rest probably being oxygen. This corresponds to the composition $UH_{2.82}O_{0.2}$; therefore the quasi-hydride is probably an intimate mixture of UH_3 and uranium oxide. The quasi-hydride is sometimes stable and sometimes pyrophoric.

Because of the partial oxidation by air of the hydride during its precipitation from mercury, some water is formed in this process (MP Ames 40); the absorption of oxygen and formation of water lead to liberation of heat. If the amalgam is exposed to quiet or slowly circulating air in an open vessel, the pickup of oxygen and the formation of water are rapid the first day and continue at a diminishing rate for months or even years (MP Ames 38).

Freshly precipitated, the quasi-hydride carries a large proportion of finely dispersed mercury. It can be freed from mercury by suspension in hexane or petroleum ether and filtration through sintered glass (MP Ames 41). Amalgamation of UH_3 by mercury is not affected by the presence of magnesium in mercury (MP Ames 42).

The hydride disperses not only in mercury but also in other low-melting metals and alloys. For example, 2 g of UH_3 could be "dissolved" in 20 ml of liquid sodium-potassium alloy (MP Ames 43). In this case, also, the heat effect is negligible. Treatment of uranium hydride with liquid Wood's metal or pure tin leads to the formation of a spongy mass. Hydrogen is liberated, and intermetallic compounds are formed in this reaction (MP Ames 38). Some experiments were made with a hydride formed from uranium-iron alloy (MP Ames 42). The product probably contained UH_3 together with an unchanged uranium-iron compound.

2.9 Uses of Uranium Hydride. The first and most obvious use of uranium hydride is the preparation of pure, finely dispersed uranium metal (MP Ames 12). This metal is reactive and can be used for the preparation of alloys and of many other compounds, such as uranium carbide (MP Ames 44). Pure hydrogen and deuterium also can be obtained by synthesis and decomposition of uranium hydride, a method in practical use at Ames (MP Ames 10). Intermetallic or related compounds of uranium do not react with hydrogen; therefore inclusions can be separated from uranium metal by hydrogenation and sifting. This method has been applied to uranium-aluminum alloys (MP Ames 45). Uranium alloys can be etched by hydride formation (MP Ames 46).

The preparation of various trivalent and quadrivalent compounds of uranium from uranium hydride has been suggested (MP Ames 47), and many of these procedures, based on results summarized in Table 8.5, have actually been utilized. One of these methods is the preparation of UF_4 by reaction between UH_3 and hydrogen fluoride (MP Ames 48) or by simultaneous action of hydrogen and hydrogen fluoride on uranium (MP Ames 29,49). Preparation of certain uranyl salts also can be achieved by using uranium hydride as starting material (MP Ames 10), as, for example, uranyl fluosilicate and lactate

$$2Ag_2SiF_6 + 2UH_3 + 4Ag_2O \rightarrow 2UO_2SiF_6 + 12Ag + 3H_2 \qquad (22)$$

$$4(CH_3CHOHCOO)Ag + 2UH_3 + 4Ag_2O \rightarrow 2(CH_3CHOHCOO)_2UO_2$$
$$+ 12\,Ag + 3H_2 \quad (23)$$

Attempts to use uranium as a hydrogenation catalyst (with naphthalene and maleic acid as hydrogen acceptors) were unsuccessful (MP Ames 50).

The formation and decomposition of the hydride and volumetric determination of hydrogen formed has been suggested as an analytical method for the determination of uranium (MP Ames 51). A similar procedure has been used for the conversion of water and heavy water to hydrogen and deuterium in the analysis of heavy water (MP Ames 10).

3. URANIUM DEUTERIDE

The uranium deuteride UD_3 has been prepared at Ames and studied especially from the point of view of its possible use for the separation of deuterium from hydrogen.

Uranium deuteride can be prepared by decomposing heavy water with uranium at 600 to 700°C and reacting the deuterium produced in this way with uranium at 250°C.

$$U + 2D_2O \xrightarrow{600-700°C} UO_2 + 2D_2 \tag{24}$$

$$U + \tfrac{3}{2}D_2 \xrightarrow{250°C} UD_3 \tag{25}$$

The two operations can be carried out in the same piece of apparatus (MP Ames 52).

The decomposition pressure of UH_3-UD_3 mixtures between 20 and 80 per cent UH_3 was measured at 357°C (MP Ames 53). The results (see Fig. 8.11) indicate that this system behaves as a perfect solution, i.e., the pressure is a linear function of composition. The decomposition pressure of pure UD_3 at 357°C is 185 mm Hg and thus is considerably higher than that of UH_3 (134 mm Hg). The ratio P_{UD_3}/P_{UH_3} is constant (about 1.4) in the temperature range 250 to 430°C (MP Ames 54). The decomposition pressure equation is

$$\log p_{mm} = -\frac{4500}{T} + 9.43 \tag{26}$$

(cf. Eq. 3).

The heat of formation of UD_3 is 31 kcal per mole, the same as that of UH_3.

The rate of formation of UD_3 from uranium and deuterium at 356°C is much slower than that of UH_3 (Fig. 8.12) (MP Ames 14), a difference which can be explained by the larger zero-point energy of the hydrogen molecule as compared to the deuterium molecule. However, contrary to expectation, this difference is not increased but rather diminished if the synthesis is carried out at low temperature. Measurements at −76°C were rather erratic (MP Ames 55), but the rate of consumption of both hydrogen and deuterium was about the same, 7 cc of gas per hour per gram of uranium (compared to an initial rate of about 30 cc of hydrogen per minute at 357°C).

The slower formation of the deuteride and its higher decomposition pressure appeared to offer a way for the separation of hydrogen and deuterium. However, experiments on decomposition of UH_3 + DH_3

mixtures at Ames (MP Ames 56) lead only to a slight enrichment of deuterium in the gas phase. For example, with a solid material containing 19.8 mole % deuterium in total hydrogen, the ratio of (D_2/H_2) to (UD_3/UH_3) was equal to 1.2. The first gaseous fraction (out of ten) contained 23.2 per cent deuterium and the tenth and last one 16.8 per cent deuterium. The difference in the rate of formation of UH_3 and UD_3 cannot be utilized for separation because the isotopic exchange

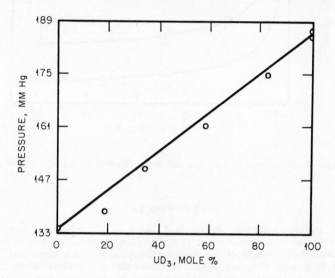

Fig. 8.11 — Equilibrium pressures of hydrogen and deuterium over UH_3-UD_3 mixtures at 357°C.

between solid hydride and gas is very rapid, the exchange being complete after 5 min at 300°C (MP Ames 56).

Some separation experiments were carried out with the hydride and deuteride amalgams. In this case the isotopic exchange between gas phase and liquid phase was found to be slow. For example, when hydrogen was heated with a 25 per cent DH_3 and 75 per cent UH_3 amalgam, the concentration of deuterium in the gas was only 2.7 per cent after 15 min and 10.3 per cent after 2 hr. The reverse exchange, i.e., replacement of hydrogen in UH_3 by deuterium from the gas phase, was even slower. This slow exchange prevents the use of an amalgam column for effective separation of deuterium and hydrogen.

The lattice constant of uranium deuteride (MP Ames 39) is $a_0 = 6.625$ A, i.e., 0.006 A smaller than the lattice constant of the hydride. The calculated density is 11.16 g/cc compared with 10.91 g/cc for UH_3.

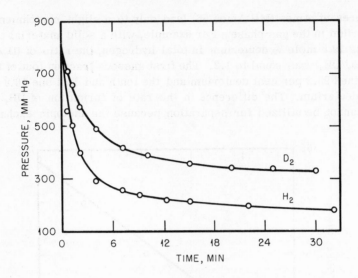

Fig. 8.12—Reaction rates of hydrogen and deuterium with powdered uranium at 357°C.

REFERENCES

1925 A. Sieverts, numerous articles in Z. anorg. u. allgem. Chem., from 1925 to
      1935. See H. J. Emeléus and J. S. Anderson, "Modern Aspects of Inorganic
      Chemistry," Chap. 6, D. Van Nostrand Company, Inc., New York, 1942.
1929 F. H. Driggs, U. S. Patent Nos. 1816830 (1929) and 1835024 (1929); Canadian
      Patent No. 325501 (1930) transferred to Canadian Westinghouse Co., Ltd.
1947 J. E. Burke and C. S. Smith, J. Am. Chem. Soc., 69: 2500-2502.
1947 J. Gueron and L. Yaffe, Nature, 160: 575.
1947 L. Pauling, J. Am. Chem. Soc., 69: 542.
1947 R. E. Rundle, J. Am. Chem. Soc., 69: 1719.
1948 A. S. Newton, U. S. Patent No. 2446780.
1948 L. Pauling and F. J. Ewing, J. Am. Chem. Soc., 70: 1660.

Project Literature

Battelle 1: Battelle Memorial Institute, Reports CT-611, Apr. 10, 1943, pp. 131-134;
      CT-688, May 10, 1943, pp. 144-154; CT-753, June 10, 1943, pp. 176-186; CT-818,
      July 10, 1943, pp. 222-229; CT-893, Aug. 10, 1943, pp. 239-261; CT-956, Sept. 10,
      1943, pp. 314-319; CT-1009, Oct. 10, 1943, pp. 345-350; CT-1388, Feb. 10, 1944, pp.
      3-6; CT-2374, Nov. 1, 1944, pp. 252-255; and CT-2483, Dec. 1, 1944, pp. 283-284.
Battelle 2: Battelle Memorial Institute, Report CT-818, July 10, 1943.

Los Alamos 1: J. E. Burke and C. S. Smith, Report LA-37, Nov. 13, 1943.

MIT 1: L. S. Foster, Report CT-2106, Sept. 14, 1944, p. 29.

MP Ames 1: J. C. Warf, A. S. Newton, T. A. Butler, and J. A. Ayres, Report CC-580,
      Apr. 15, 1943; F. H. Spedding and I. B. Johns, Report CC-587, Apr. 19, 1943; A. S.
      Newton, J. C. Warf, O. Johnson, and R. W. Nottorf, Report CC-1201, Jan. 1, 1944;

J. C. Warf, Iowa State College, Report ISC-50, Sept. 29, 1949. In National Nuclear Energy Series, Division VIII, Volume 6: F. H. Spedding, A. S. Newton, I. B. Johns, O. Johnson, A. H. Daane, R. W. Nottorf, and J. C. Warf, Preparation and Physical Properties of Uranium Hydride; F. H. Spedding, J. C. Warf, A. S. Newton, O. Johnson, I. B. Johns, J. A. Ayres, T. A. Butler, R. W. Fisher, and R. W. Nottorf, The Chemical Properties of Uranium Hydride; R. E. Rundle, A. S. Wilson, and R. A. McDonald, The X-ray Investigation of the Uranium-Hydrogen System; The Structure of UH_3. The substance of the above papers has appeared recently in the open literature: F. H. Spedding, A. S. Newton, J. C. Warf, O. Johnson, R. W. Nottorf, I. B. Johns, and A. H. Daane, Nucleonics, 4 (1): 4-15 (1949); A. S. Newton, J. C. Warf, F. H. Spedding, O. Johnson, I. B. Johns, R. W. Nottorf, J. A. Ayres, R. W. Fisher, and A. Kant, Nucleonics, 4 (2): 17-25 (1949); J. C. Warf, A. S. Newton, T. A. Butler, and F. H. Spedding, Nucleonics, 4 (3): 43-47 (1949).

MP Ames 2: O. Johnson, Report CC-1059, Oct. 9, 1943.

MP Ames 3: F. H. Spedding and I. B. Johns, Report CC-803, July 15, 1943; A. S. Newton, A. H. Daane, O. Johnson, and R. W. Nottorf, Report CC-1059, Oct. 19, 1943; A. S. Newton, J. C. Warf, O. Johnson, and R. W. Nottorf, Report CC-1201, Feb. 8, 1944; R. W. Nottorf, Report CC-1212, Dec. 12, 1943; F. H. Spedding et al., Preparation and Physical Properties of Uranium Hydride, in National Nuclear Energy Series, Division VIII, Volume 6.

MP Ames 4: J. C. Warf, Report CC-862, Aug. 8, 1943.

MP Ames 5: W. Tucker and P. Figard, Reports CC-1781, Oct. 18, 1944, and CC-1975, Oct. 24, 1944.

MP Ames 6: J. C. Warf, A. S. Newton, T. A. Butler, J. A. Ayres, and I. B. Johns, Reports CC-580, Apr. 15, 1943, and CC-587, Apr. 19, 1943.

MP Ames 7: A. S. Newton, R. W. Nottorf, A. H. Daane, and O. Johnson, Report CC-858, Aug. 7, 1943.

MP Ames 8: J. C. Warf and O. Johnson, Reports CC-1059, October, 1943, and CC-1061, October, 1943.

MP Ames 9: F. H. Spedding et al., Preparation and Physical Properties of Uranium Hydride, in National Nuclear Energy Series, Division VIII, Volume 6.

MP Ames 10: A. S. Newton, J. C. Warf, O. Johnson, and R. W. Nottorf, Report CC-1201, Feb. 8, 1944.

MP Ames 11: A. S. Newton, Report CC-1212, Dec. 10, 1943.

MP Ames 12: J. C. Warf, A. S. Newton, T. A. Butler, J. A. Ayres, and I. B. Johns, Report CC-580, Apr. 15, 1943.

MP Ames 13: A. S. Newton, O. Johnson, A. H. Daane, R. W. Nottorf, and J. C. Warf, Report CC-803, July 15, 1943.

MP Ames 14: O. Johnson and A. S. Newton, Report CC-1063, Nov. 6, 1943.

MP Ames 15: J. C. Warf, Report CC-1061, Oct. 8, 1943.

MP Ames 16: R. E. Rundle, Report CT-609, Apr. 24, 1943.

MP Ames 17: R. E. Rundle, A. S. Wilson, and R. A. McDonald, Report CC-1131, Dec. 18, 1943; R. E. Rundle, N. C. Baenziger, A. S. Wilson, and R. A. McDonald, Report CC-2397, Apr. 2, 1945, pp. 6-10.

MP Ames 18: R. E. Rundle, A. S. Wilson, and R. A. McDonald, The X-ray Investigation of the Uranium-Hydrogen System; The Structure of UH_3, in National Nuclear Energy Series, Division VIII, Volume 6.

MP Ames 19: R. W. Nottorf, Reports CC-1063, Nov. 6, 1943, and CC-1212, Dec. 10, 1943.

MP Ames 20: T. A. Butler, Report CC-664, May 14, 1943; A. S. Newton, O. Johnson, A. Kant, and R. W. Nottorf, Reports CC-705, June 7, 1943, and CC-725, June 15, 1943; O. Johnson, A. Kant, R. W. Nottorf, and J. C. Warf, Report CC-803, July 15, 1943; A. H. Daane, O. Johnson, J. C. Warf, and R. W. Nottorf, Reports CN-852, Aug. 8, 1943, and CC-1059, Oct. 9, 1943; R. W. Fisher and J. C. Warf, Report CC-1091,

Jan. 7, 1944; J. C. Warf, M. Goldblatt, R. D. Tevebaugh, and J. Powell, Report CC-1194, Dec. 9, 1943; J. A. Ayres, Report CN-1243, Jan. 8, 1944; W. Lyon, J. Iliff, and H. Lipkind, Report CK-1494, Apr. 29, 1944.

MP Ames 21: F. H. Spedding et al., The Chemical Properties of Uranium Hydride, in National Nuclear Energy Series, Division VIII, Volume 6.

MP Ames 22: O. Johnson, A. Kant, and R. W. Nottorf, Report CC-705, June 7, 1943.

MP Ames 23: R. W. Fisher, Report CC-1091, Feb. 14, 1944.

MP Ames 24: R. W. Nottorf, Report CC-803, July 15, 1943.

MP Ames 25: A. Kant, Report CC-803, July 15, 1943.

MP Ames 26: W. Lyon, J. Iliff, and H. Lipkind, Report CK-1494, Apr. 29, 1944.

MP Ames 27: J. A. Ayres, Report CN-1243, Jan. 8, 1944.

MP Ames 28: A. H. Daane, O. Johnson, and R. W. Nottorf, Report CC-1059, Oct. 9, 1943.

MP Ames 29: I. B. Johns and R. D. Tevebaugh, Report CC-1059, Oct. 9, 1943.

MP Ames 30: J. C. Warf, A. S. Newton, T. A. Butler, and J. A. Ayres, Report CC-580, Apr. 15, 1943.

MP Ames 31: R. D. Tevebaugh, Report CC-1194, Dec. 9, 1943.

MP Ames 32: A. F. Voigt, F. J. Walter, J. A. Ayres, and R. E. Hein, Report CN-578, Apr. 15, 1943.

MP Ames 33: M. Goldblatt, Report CC-1194, Dec. 9, 1943.

MP Ames 34: J. A. Ayres, Reports CN-853, Aug. 8, 1943; CN-858, Aug. 7, 1943; and CN-925, Sept. 8, 1943.

MP Ames 35: J. C. Warf, Report CC-1194, Dec. 9, 1943.

MP Ames 36: J. C. Warf, A. S. Newton, T. A. Butler, J. A. Ayres, and I. B. Johns, Report CC-580, Apr. 15, 1943; M. Goldblatt, Report CC-682, May 15, 1943.

MP Ames 37: T. A. Butler, A. F. Voigt, F. J. Walter, and J. A. Ayres, Report CN-925, Sept. 8, 1943.

MP Ames 38: J. C. Warf, Report CC-1059, Oct. 9, 1943.

MP Ames 39: R. E. Rundle, A. S. Wilson, and R. A. McDonald, Report CC-1131, Dec. 18, 1943.

MP Ames 40: J. C. Warf, Report CC-1091, Oct. 9, 1943.

MP Ames 41: R. W. Fisher, Report CC-1059, Oct. 9, 1943.

MP Ames 42: J. Powell, Report CC-1194, Dec. 9, 1943.

MP Ames 43: A. H. Daane, Report CC-1059, Oct. 9, 1943.

MP Ames 44: A. H. Daane, Report CT-686, May 22, 1943.

MP Ames 45: J. H. Carter, Report CC-664, May 15, 1943.

MP Ames 46: T. A. Butler, Report CC-725, June 15, 1943.

MP Ames 47: A. S. Newton, O. Johnson, A. Kant, and R. W. Nottorf, Report CN-725, June 15, 1943.

MP Ames 48: A. S. Newton, O. Johnson, and F. H. Spedding, Report CN-717, June 17, 1943.

MP Ames 49: R. D. Tevebaugh, K. A. Walsh, J. Iliff, W. H. Keller, and I. B. Johns, Report CC-1063, Nov. 6, 1943.

MP Ames 50: J. C. Warf and R. W. Nottorf, Report CC-803, July 15, 1943.

MP Ames 51: J. C. Warf, Report CC-1782, Aug. 10, 1944.

MP Ames 52: A. S. Newton, Report CC-695, May 27, 1943.

MP Ames 53: O. Johnson, Report CC-1063, Nov. 6, 1943.

MP Ames 54: A. S. Newton, O. Johnson, A. H. Daane, and R. W. Nottorf, Report CC-803, July 15, 1943.

MP Ames 55: O. Johnson, Report CC-1212, Dec. 10, 1943.

MP Ames 56: A. S. Newton, A. H. Daane, O. Johnson, and R. W. Nottorf, Report CC-1059, Oct. 9, 1943.

MP Ames 57: J. C. Warf, Some Reactions of Uranium Metal, in National Nuclear Energy Series, Division VIII, Volume 6.

MP Chicago 1: W. H. Zachariasen and R. C. L. Mooney, Report CK-1096, Nov. 27, 1943.

MP Clinton 1: I. Perlman, Report CN-1025, Nov. 8, 1943.
MP Clinton 2: H. H. Hubble and F. C. McCullough, Report CN-1025, Nov. 8, 1943.
MP Clinton 3: M. Lindner, Report CN-1025, Nov. 8, 1943.

UCRL 1: G. E. MacWood, Report RL-4.6.234, Dec. 22, 1943.
UCRL 2: G. E. MacWood and D. Altman, Report RL-4.7.600, Oct. 24, 1944.
UCRL 3: G. E. MacWood, private communication, May 22, 1945.

Chapter 9

URANIUM BORIDES, CARBIDES, AND SILICIDES

1. URANIUM-BORON SYSTEM

Three uranium borides have been described in the literature. Wedekind and Jochem (1913) obtained uranium diboride, UB_2, by forming electrodes from uranium powder pressed with amorphous boron under 200 atm pressure at 1000°C and melting these electrodes in an electric arc. The silver-gray product obtained was washed with dilute acid. The diboride proved resistant to aqueous alkalis and acids with the exception of nitric acid and hydrofluoric acid. It dissolved in melted alkali with liberation of hydrogen.

Andrieux (1929, 1948) electrolyzed at about 1000°C a melt of U_3O_8 and B_2O_3, to which alkaline earth oxides and fluorides were added as flux, and obtained crystals with metallic luster, the composition of which he determined as UB_4. This uranium tetraboride dissolves in cold hydrofluoric or hydrochloric acid, reduces concentrated sulfuric acid, and dissolves easily in nitric acid and concentrated hydrogen peroxide. It is rapidly decomposed by melting with hydroxides and carbonates and reacts vigorously with peroxides.

Bertaut and Blum (1949) have studied two uranium borides, UB_4 and UB_{12}, by x-ray methods. According to them, UB_4 is tetragonal, with $a_1 = 7.066$ A and $a_3 = 3.97$ A. There are four molecules per unit cell, and the symmetry group P4/mbm is suggested for this structure. The intensities fix the positions of the uranium atoms in 4(g), with $x = 0.1875$; the positions of the boron are as yet undetermined.

The boride UB_{12}, recently described by Andrieux and Blum (1949), is cubic, with $a = 7.473$ A. There are four molecules per unit cell, and the calculated density is 5.825, apparently confirmed by a direct density determination. The probable space group is O_h^5-Fm3m, with the uranium atoms in 4(a), the boron in 48(i), and $x = \frac{1}{6}$. Each boron atom has five neighboring boron atoms at a distance B-B = 1.76 A, and two neighboring uranium atoms at a distance U-B = 2.79 A. Each uranium atom is surrounded by 24 boron atoms arranged in a regular cuboctahedron. The uranium radius, 1.91 A, is unusually high; this is attributed to the high coordination, and it is pointed out that a similar enlargement of atomic radius is observed in the case of thorium in the compound ThB_6.

A uranium boride of ill-defined composition is obtained by the thermal decomposition of uranium(IV) borohydride, $U(BH_4)_4$ (see Chap. 15).

2. URANIUM-CARBON SYSTEM

The existence of a uranium carbide was discovered by Moissan (1896), who obtained it by heating U_3O_8 and carbon in an electric furnace. Moissan, and later Rideal (1913), ascribed to this compound the formula U_2C_3, but Lebeau (1911) and Heusler (1925) favored the formula UC_2. Later investigations proved the existence of a uranium monocarbide, UC, in addition to confirming that of the uranium dicarbide, UC_2. The monocarbide is thermodynamically stable at room temperature, the dicarbide probably only at high temperatures. The existence at still higher temperatures (above 2000°C) of a true sesquicarbide, U_2C_3, has been postulated but is as yet quite uncertain (see Division IV, Volume 12 A, of the National Nuclear Energy Series).

2.1 Phase Relations in the Uranium-Carbon System. The solubility of pure carbon (graphite) in liquid uranium was determined at Ames by heating the metal in graphite crucibles and maintaining constant temperature long enough to attain equilibrium between the melt and the graphite. The values obtained (MP Ames 1,2) are given in Table 9.1.

The uranium-carbon phase diagram between UC (4.8 per cent carbon) and UC_2 (9.16 per cent carbon) also was studied at Ames (MP Ames 4) (see Fig. 9.1). Evidence for the existence of a high-temperature compound, U_2C_3, is slight. Widmanstätten patterns of alloys between U_2C_3 and UC_2 have been interpreted as indicating the existence of solid solutions of δ U_2C_3 and ϵ UC_2. No carbides with more than two carbon atoms per uranium atom seem to exist. There is some x-ray evidence, however, for solubility of carbon in the UC_2 lattice, particularly at temperatures near the melting point (2375°C).

2.2 Physical Properties of Uranium Monocarbide. (a) Melting Point. Evidence of melting of UC was observed at Ames at 2250°C (MP Ames 4,9).

(b) Crystal Structure of Uranium Monocarbide. The monocarbide is face-centered cubic with a = 4.951 ± 0.001 A, according to Rundle, Baenziger, Wilson, and McDonald (1948), and a = 4.955 A, according to Litz, Garrett, and Croxton (1948). Experimentally determined lattice constants are listed in Table 9.2. There are four molecules per unit cell, and the x-ray density is 13.63 g/cc (Rundle et al., 1948). Uranium monocarbide is isomorphous with UN and UO, and, according

Table 9.1—Solubility of Carbon (or UC) in Liquid Uranium

Temperature, °C	Heating time, min	Carbon by wt., %	Atom C/atom U
1350 – 1400		0.10*	0.02
1450 – 1500		0.17*	0.034
1550 – 1600		0.36*	0.072
1650 – 1700		0.42*	0.084
1750 – 1800		0.51*	0.102
1850 – 1900		1.60*	0.325
1950 – 2000		1.74*	0.354
2080 – 2130		2.92*	0.592
Max. 1375	40	0.137†	
Max. 1400	20	0.168†	
Max. 1500	60	0.385†	
Max. 1600	20	0.626†	
Max. 1800	30	0.775†	
Max. 1900	30	1.20†	
Max. 2000	30	1.50†	

*MP Ames 1.
†MP Ames 2.

Table 9.2—Lattice Constant and X-ray Density of UC

a_1, A	Density, g/cc	Reference
4.98	...	MP Ames 6
4.966 ± 0.005	13.56	MP Ames 7
4.948 ± 0.001	13.66	MP Ames 8
4.948 to 4.951 ± 0.001*	13.63†	MP Ames 9
4.94	...	British 1
4.995	...	Litz, Garrett, and Croxton (1948)

*For details, see Rundle, Baenziger, Wilson, and McDonald (1948).
†For a_1 = 4.951 A.

to Ames observations, carbon in UC can be replaced by oxygen or nitrogen with a slight decrease in lattice constant (to as low as 4.936 A when UO_2 is also found). It is likely that UC will exhibit complete solid miscibility with UN and UO. Since replacement of carbon by oxygen or nitrogen decreases the lattice parameter, the highest value, 4.955 A, would be expected to be the most reliable. The lattice constant of UC does not depend markedly upon the presence of excess uranium or carbon, thus indicating low solubility of both components in the carbide at room temperature.

Since the x-ray scattering power of carbon relative to uranium is very slight, it has been somewhat of a problem as to which of the two

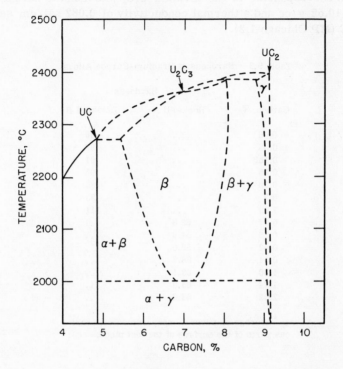

Fig. 9.1—Uranium-carbon system, 4.8 to 9.16 per cent carbon.

possible structures, the sodium chloride or zinc blende type, is correct. On the basis of intensity measurements, it appears likely that UC is of the sodium chloride type (O_h^5-Fm3m) with the atoms in the following positions:

$$(0\ 0\ 0;\ 0\ \tfrac{1}{2}\ \tfrac{1}{2};\ \tfrac{1}{2}\ 0\ \tfrac{1}{2};\ \tfrac{1}{2}\ \tfrac{1}{2}\ 0)+$$

$$U\ 4(a)\ 0\ 0\ 0$$

$$C\ 4(b)\ \tfrac{1}{2}\ \tfrac{1}{2}\ \tfrac{1}{2}$$

(c) $\underline{U + UC}$. Uranium-carbon alloys with less than 4.8 per cent carbon are mixtures of uranium (containing less than 0.01 per cent carbon) with UC. Because of this inhomogeneous character, determinations of physical constants of these alloys give variable results. The hardness figures given in Table 9.3 were obtained at Ames (MP Ames 1,5).

At the Metallurgical Laboratory a carbide preparation with a composition corresponding to the formula U_3C_2 was found to have a density of 10.52 g/cc and a thermal conductivity of 0.082 cal/cm sec °C at 56°C (MP Chicago 1,2).

Table 9.3 — Hardness of Uranium-Carbon Alloys

Carbon, %	Hardness	
	Rockwell A	Rockwell B
0.06	...	88
0.08	...	88
0.168	...	91
0.25	...	95
0.30	...	95
0.7	...	93
1.99	...	111
0.17	52.8*	...
0.36	54.8	...
0.42	56.0	...
0.51	58.7	...
1.60	63.1	...
1.74	63.8	...
2.92	63.6	...

*Average of several measurements in various parts of the surface of the test piece.

2.3 Uranium Sesquicarbide. X-ray studies of carbide samples with the composition U/C = 2/3 showed that they always consisted of UC and UC_2 (MP Ames 10). Evidence in favor of the existence of a compound U_2C_3 at high temperatures was derived from the fact that apparently single cubic crystals are formed by quenching an alloy corresponding to U_2C_3 from a high temperature, even though x-rays prove these single crystals to be diphasic. Microscopic examination of the single crystals revealed a Widmanstätten structure. Therefore, it appears that U_2C_3 is a cubic phase stable somewhere above 2400°C which has not yet been successfully quenched. A sample with the composition U_2C_3 was brittle, and its hardness could be determined only approximately as Rockwell 396 (MP Ames 11). The melting point of a carbide sample with 6.95 per cent carbon (U_2C_3 contains

7.04 per cent carbon) was found at Ames to be the same as that of UC_2, 2350 to 2400°C (MP Ames 9).

2.4 Physical Properties of Uranium Dicarbide. (a) Melting and Boiling Points. A uranium-carbon sample containing 9.18 per cent carbon (UC_2 contains 9.16 per cent carbon) showed signs of melting at 2350 to 2400°C (MP Ames 4,9). An estimate of the boiling point as 4370°C at 760 mm Hg was made by Mott (1918). Ruff and co-workers (1911, 1919) gave 2425°C for the melting point of UC_2. Tiede and Birnbräuer (1914a) found the melting point to be 2200°C in an electric vacuum furnace and 2260°C in a cathode-ray furnace.

(b) Crystal Structure. (Rundle et al., 1948.[*]) Moissan (1897) described uranium dicarbide, UC_2, as metallic, dense, and finely crystalline with a density of 11.28 g/cc. The crystal structure of UC_2 was first studied by Hägg (1931), who found it to be face-centered tetragonal. This was confirmed at Ames (MP Ames 12). The structure is similar to that of LaC_2 with two molecules per unit cell. The lattice constants are $a_1 = 3.517 \pm 0.001$ A and $a_3 = 5.987 \pm 0.001$ A. Dissolution of carbon in the UC_2 lattice seems to decrease the lattice constant to about $a_1 = 3.505$ A and $a_3 = 5.951$ A. The Ames investigators consider this decrease an indication that UC_2 is not an "interstitial solid solution" of carbon in uranium. They suggest that UC_2 is built of small positive U ions and large negative C_2 ions, with uranium in the interstices of the anion lattice, and that in alloys with $UC_{>2}$ some uranium ions are missing. As discussed in Chap. 11 in connection with a similar interpretation of the decrease in lattice dimensions from $UO_{2.0}$ to $UO_{2.3}$, this interpretation requires a marked decrease in density with increasing carbon content. This decrease was apparently not found in the case of the oxides; no systematic density measurements are available for carbides. The calculated x-ray density is 11.68 g/cc. At the Metallurgical Laboratory a density of only 9.97 g/cc ± 2 per cent was found by Plott and Raeth (MP Chicago 1,2).

(c) Specific Heat. At Berkeley (UCRL 1) the specific heat of UC_2 was estimated as

$$C_p = 8.92 + 3.95 \times 10^{-3}T \quad \text{cal per mole} \tag{1}$$

At Chicago (MP Chicago 3) an estimate of C_p was made by extrapolation from the known values for other carbides, particularly TaC.

---

[*]For an alternative, and generally less satisfactory, treatment of the uranium-carbon system, see Esch and Schneider, Z. anorg. Chem., 527: 254-256 (1948).

The latter is 75 per cent of the Dulong-Petit value at 21.3°C. Assuming that in the case of UC_2 the corresponding value is 80 per cent, the following value is obtained:

$$C_p \ (UC_2) = 15 \text{ cal per mole at } 1940°K$$

(d) <u>Thermal Conductivity</u>. At the Metallurgical Laboratory (MP Chicago 2) a thermal conductivity of 0.082 cal/cm sec °C was found for a UC_2 sample with a density of 10 g/cc (see Sec. 2.4b).

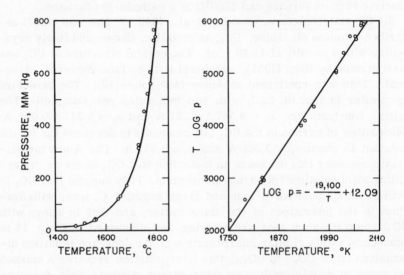

Fig. 9.2—Carbon monoxide equilibrium pressure over uranium dioxide—carbon mixture [from O. Heusler, Z. anorg. u. allgem. Chem., 154: 364 (1926)].

2.5 <u>Preparation of Uranium Carbides</u>. (a) <u>Thermodynamics of Uranium Monocarbide Formation</u>. At Berkeley the heat of formation of UC from the elements was estimated at −30 kcal per mole. This estimate was based on the observation that UC_2 decomposes into U and UC at low temperatures (MP Berkeley 1).

(b) <u>Thermodynamics of Uranium Dicarbide Formation</u>. At Chicago (MP Chicago 3) estimates were made of the energy and free energy of the reaction

$$UO_2 + 4C \text{ (graphite)} \rightleftarrows UC_2 + 2CO \qquad (2)$$

The required experimental data were taken from Heusler's equilibrium measurements of the reaction given in Eq. 2. Heusler (1926a) measured the CO equilibrium pressure over a UO_2-carbon mixture at

1480 to 1801°C. Figure 9.2 shows that the pressures ranged from 18 mm Hg to 1 atm and that the graph of T log p vs. T was approximately linear. The equation

$$\log p = -\frac{19,100}{T} + 12.09 \tag{3}$$

represents the results approximately but not exactly; ΔH apparently changes markedly with temperature. According to Eq. 3, the average value of $-\Delta H$ in the region studied is 174.6 kcal per gram atom of uranium. At Chicago (MP Chicago 3) an estimate also was made of ΔC_p for the reaction given in Eq. 2:

$$\Delta C_p = 0.8 - 8.1 \times 10^{-3}T + 4.68 \times 10^5 T^{-2} \tag{4}$$

Combining this equation with the results of Heusler, the following thermodynamic equations were obtained for the reaction given in Eq. 2:

$$\Delta H = 188,540 + 0.8T - 4.05 \times 10^{-3}T^2$$
$$- 4.68 \times 10^5 T^{-1} \quad \text{cal/g atom of U} \tag{5}$$

$$\Delta F° = 188,540 - 1.842T \log T + 4.05 \times 10^{-3}T^2 - 2.34 \times 10^5 T^{-1}$$
$$- 93.4T \quad \text{cal/g atom of U} \tag{6}$$

For the entropy change in the same reaction the following may be obtained:

$$\Delta S°_{298} = 93.8 \text{ e.u.}$$

At Berkeley similar estimates were made, using Eq. 1 to calculate C_p for UC_2 and the equilibrium data of Heusler (UCRL 1). The resulting thermodynamic equations for the reaction given in Eq. 2 were

$$\Delta F° = 202,500 + 16.167T \log T + 2.775 \times 10^{-3}T^2 - 3.495 \times 10^5 T^{-1}$$
$$-156.74T \quad \text{cal/g atom of U} \tag{7}$$

$$\Delta H°_{298} = 199.000 \text{ kcal/g atom of U} \tag{8}$$

$$\Delta F°_{298} = 166.790 \text{ kcal/g atom of U}$$

$$\Delta S°_{298} = 108.05 \text{ e.u.}$$

Combining the thermodynamic data for the reaction in Eq. 2 with the known thermodynamic constants for UO_2, C, and CO, the heat, free energy, and entropy of formation of UC_2 from the elements can be calculated. The values in Table 9.4 were obtained in this way.

Table 9.4 — Thermodynamics of Formation of UC_2

$$U + 2C \rightarrow UC_2$$

$-\Delta F^\circ$, cal/g atom of U	$-\Delta F^\circ_{298}$, kcal/g atom of U	$-\Delta H^\circ_{298}$, kcal/g atom of U	ΔS°_{298}, e.u.	Reference
	...	...	23.6	MP Chicago 3
	8.8	...	15.1	MP Berkeley 2
$3925 + 6.3T \log T - 3.48$ $\times 10^{-3}T^2 + 7.545 \times 10^4 T^{-1}$ $+ 19.85T$	9.83	3.92	19.8	UCRL 1

The dicarbide is unstable at room temperature (with respect to UC + C); it apparently decomposes slowly into UC and C below about 2400°C. Quenching of the dicarbide from above 2400°C produces pure UC_2, and quenching from 1000°C or slow cooling results in UC_2 containing both UC and C, independently of which component is in excess in the melt.

(c) Preparation of Uranium Carbide from Uranium Metal and Carbon. At Ames (MP Ames 11,13) it was noticed that when uranium metal was heated to 1800 to 1900°C in graphite crucibles, the crucible walls were attacked and a hard, brittle carbide was formed. The formation of the carbide seems to occur by penetration of uranium into graphite rather than by penetration of graphite into uranium. Therefore the reaction takes place in the presence of excess carbon and leads to the carbon-rich compound UC_2 (MP Ames 3). Alternatively, graphite powder can be dissolved in molten uranium or can be pressed with uranium turnings before melting (MP Ames 5).

Sintering of powdered metal mixed with excess carbon was recommended at Ames for the preparation of uranium-carbon alloys with high carbon content (MP Ames 6). This method (pressing at 50 tons, sintering at 1800 to 2000°C in vacuum) was used at Ames to prepare alloys with carbon content between 0.174 per cent and 8.54 per cent by weight (MP Ames 10). In England uranium carbide was prepared by heating uranium turnings with pure carbon to 1200 to 1600°C (British 1).

All uranium-carbon alloys obtained in this way are inhomogeneous; they contain dendritic carbide inclusions in a matrix of a saturated carbon solution in uranium. The inclusions can be separated from the matrix by dissolving the latter in 3N HCl containing some peroxide.

The residue is found both by x-ray and chemical analysis (4.70 per cent C) to be practically pure monocarbide, UC. Uranium monocarbide can also be obtained by heating uranium and UC_2 above 1800°C.

According to Litz, Garrett, and Croxton (1948), either UC or UC_2 can be obtained from metal and carbon, depending on the temperature.

$$U + C \xrightarrow{\text{2100°C}} UC \tag{9}$$

$$U + 2C \xrightarrow{\text{2400°C}} UC_2 \tag{10}$$

They imply, but do not state, that the products are homogeneous.

(d) Preparation of Uranium Monocarbide from Uranium Metal and Methane. Litz, Garrett, and Croxton (1948) report that UC can be readily prepared by the reaction of methane on metallic uranium.

$$U + CH_4 \xrightarrow{\text{625-900°C}} UC + 2H_2 \tag{11}$$

This is probably the most convenient method for preparing the pure monocarbide; the uranium metal should first be converted to a finely divided form through the hydride, UH_3.

(e) Preparation of Uranium Carbide from Uranium Metal and Carbon Monoxide. At Battelle, carbon monoxide was admitted to an outgassed uranium slug, and the latter was heated to 1200°C. Reaction was rapid at first; then it slowed down. After 45 min, while the liquid metal was still slowly taking up carbon monoxide, it was cooled and examined. A black flaky crust was found on the surface; x-ray examination showed it to consist of UO_2 and UC (Battelle 1).

(f) Preparation of Uranium Carbide from Uranium Oxides. Moissan (1897) mixed U_3O_8 (50 g) and sugar carbon (50 g) in a graphite crucible and heated the mixture 8 to 10 min in an electric furnace (45,000 joules). According to Lebeau (1911) the product obtained in this way always contains graphite. Ruff and Heinzelmann (1911) heated UO_2 with sugar carbon in a vacuum furnace slowly to 2450°C. The reaction is less violent than with U_3O_8, and the product is almost free from graphite. Heusler (1926b) carried out the reaction of UO_2 with excess carbon in a vacuum furnace at 1800°C; the compound would not melt; liberated carbon monoxide was pumped away.

At Ames mixtures of 1 part of Acheson graphite and 5 to 7 parts of U_3O_8 were heated in graphite crucibles in a hydrogen atmosphere to produce carbide (MP Ames 14). Later (MP Ames 11,15) it was determined that with UO_2/C weight ratio between 5 and 6.4 the product was UC_2 (9.15 per cent carbon compared to 9.16 per cent in UC_2). With UO_2/C ratios of 7 and 8 the product had a composition close to U_2C_3 (7.21 per cent carbon compared to 7.04 per cent in U_2C_3).

No pure UC could be obtained from UO_2 and carbon. Attempts made at Ames to prepare uranium carbide by reduction of mixtures of UF_4 and carbon with calcium were unsuccessful. Heating of UO_2 with UC_2 gave the monocarbide, UC, but the product obtained in this way was not very pure.

Litz, Garrett, and Croxton (1948) state that UC is readily prepared by heating U_3O_8 with the stoichiometric amount of graphite to 1800°C. Higher temperatures yield UC_2.

$$U_3O_8 + 11C \xrightarrow{1800°C} 3UC + 8CO \qquad (12)$$

$$U_3O_8 + 14C \xrightarrow{2400°C} 3UC_2 + 8CO \qquad (13)$$

2.6 Chemical Properties of Uranium Carbides. The data on the chemical behavior of uranium carbide refer to products whose composition is given either as UC_2 or as U_2C_3. It was repeatedly stated above that the existence of a homogeneous phase of the latter composition is highly doubtful, except perhaps at high temperatures.

(a) $\underline{UC_2 \text{ and } O_2}$. Uranium carbide is pyrophoric. Dropping it on a hard surface or striking it with a hammer produces a spray of sparks (Tiede and Birnbräuer, 1914b; Moissan, 1897). Crushing in an agate mortar may cause ignition (Moissan, 1897). According to Rideal (1913), the carbide ignites in air at 400°C. Moissan (1897) found that it ignites in oxygen at 370°C and burns to U_3O_8 and CO_2. According to Ames observers, UC_2 decomposes completely within a week in air, in all probability by primary reaction with water vapor (MP Ames 13).

At Ames it was observed that UC_2 did not oxidize at 300°C but oxidized completely within 4 hr in an air stream at 400 and 500°C. A sample with the composition U_2C_3 was merely covered with a black film after the same treatment. After 5½ hr at 600°C the U_2C_3 sample was found to have disintegrated into granules still containing metallic-appearing material inside.

(b) $\underline{UC_2 \text{ and } N_2}$. According to Moissan (1897) nitrogen reacts with uranium carbide at 1100°C. Heusler (1926b) found that the reaction was rapid at 1180°C. At this temperature the carbide takes up nitrogen until saturation is reached. After 12 hr all carbide is converted to nitride.

(c) $\underline{UC_2 \text{ and } F_2}$. Moissan (1897) reported no reaction at room temperature, but slight heating was enough to cause explosive reaction.

(d) $\underline{UC_2 \text{ and } Cl_2}$. Moissan (1897) found that the carbide ignites with chlorine at 350°C giving a volatile chloride. At Ames it was observed that $UO_2 + UC_2$ mixtures react with chlorine at 600°C to form UCl_4, but a large residue is left (MP Ames 16). At 800 and 1000°C higher uranium chlorides are produced.

(e) UC_2 and Br_2. Moissan (1897) found that uranium carbide ignites in bromine vapor at 390°C. At Ames UC_2 was found to react with bromine above 300°C; at 900°C the reaction gave UBr_4 (MP Ames 17). Later, the same reaction,

$$UC_2 + 2Br_2 \rightarrow UBr_4 + 2C \tag{14}$$

was observed at 800°C (MP Ames 18). The carbon produced is finely dispersed and difficult to separate. Most construction materials are attacked in this reaction.

(f) UC_2 and I_2. According to Moissan (1897) uranium carbide reacts with iodine below red heat without ignition. At Ames a carbide with the composition U_2C_3 was observed to react with iodine at 600°C giving uranium iodide (MP Ames 17). At the University of California Radiation Laboratory it was observed that, when iodine vapor at a partial pressure of 100 mm Hg is passed over UC_2 at 500°C, UI_4 is formed (UCRL 1). It was suggested that the carbide may be a satisfactory material for the preparation of UI_3 or UI_4.

(g) UC_2 and S or Se. Moissan (1897) found that the uranium carbide burns in sulfur vapor, giving uranium sulfide and carbon disulfide. Similar reaction occurs with selenium.

(h) UC_2 and H_2O. According to Moissan (1897) uranium carbide decomposes water slowly at room temperature and rapidly when heated. In the absence of air a green hydroxide is produced, whereas in the presence of air the product is grayish black. The carbon content is converted to hydrocarbons, partly gaseous (one-third) and partly liquid and solid (two-thirds). [See Lebeau and Damiens (1913, 1914, 1917) for the composition of this mixture.] The mechanism of formation of hydrocarbons other than acetylene (which could be expected as the main product because the crystal structure of UC_2 is analogous to that of CaC_2) was discussed by Schmidt (1934). He attributed the multiplicity of products to the changes in uranium oxidation state during decomposition.

The carbide reacts with water vapor at dark-red heat with ignition to form a black oxide (Moissan, 1897). According to observations at Ames, uranium carbide decomposes completely in moist air within a week (MP Ames 13).

(i) UC_2 and NH_3. Moissan (1897) observed a partial decomposition of UC_2 by ammonia at red heat.

(j) UC_2 and H_2S. Moissan (1897) found that UC_2 ignited in hydrogen sulfide at 600°C and produced a sulfide.

(k) UC_2 and HCl. Moissan (1897) observed ignition of UC_2 in HCl at 600°C and the formation of a chloride which was decomposed by water.

(l) UC_2 and Acids. According to Moissan (1897), dilute hydrochloric, nitric, and sulfuric acids decompose UC_2 similarly to water, giving yellow uranyl salt solutions. Concentrated acids (except nitric) react only slowly at room temperature but very vigorously when heated. At Ames, UC_2 and U_2C_3 were treated with 85 per cent H_3PO_4; reaction was slow at room temperature but vigorous when heated, giving a mixture of gaseous, liquid, and solid hydrocarbons (MP Ames 10).

(m) UC_2 and Alkalis. According to Rideal (1913) the carbide is readily decomposed by alkalis.

(n) UC_2 and Salts. Melting of UC_2 with potassium nitrate or potassium perchlorate causes ignition and produces potassium uranate. At Ames, $Na_2UO_2Cl_4$ was obtained by heating a mixture of NaCl, UC_2, and UO_2 to 1000°C (MP Ames 16).

3. URANIUM-SILICON SYSTEM

The uranium-silicon system was investigated at the Massachusetts Institute of Technology by thermal, microscopic, and x-ray methods and was found to be extremely complex (MIT 1).

3.1 Phase Diagram. The phase diagram shown in Fig. 9.3 shows five compounds to which were originally attributed the compositions U_5Si_3, USi, U_2Si_3, USi_2, and USi_3. A sixth, formed by peritectoid reaction between carbon-free γ uranium and U_5Si_3 at 940°C, was found at the Massachusetts Institute of Technology to exist between 4 and 28 atom % silicon and was given the formula $U_{10}Si_3$ (MIT 2). The uranium-silicon system thus is found to have six compounds, three eutectoid and three peritectoid. Recently, Zachariasen (1949) has examined the uranium-silicon system by x-ray methods, and he has shown that some of the original formula attributions were in error. Thus, he has shown that U_5Si_3 is actually U_3Si_2, that U_2Si_3 is a crystal modification of USi_2, and that $U_{10}Si_3$ is U_3Si. The identified compounds of the uranium-silicon system are then U_3Si_2, USi, β USi_2, α USi_2, USi_3, and U_3Si.

The eutectic at 1570°C, between U_3Si_2 and USi, was established by microstructural analysis and is located at 47 atom % silicon. The eutectic between γ uranium and U_3Si_2 has been found at 9 atom % silicon at 985°C. The maximum solubility of silicon in γ uranium is about 1.75 atom % at 980°C; the solubility in β uranium is less than 1 atom %. The $\beta \rightarrow \gamma$ transformation temperature of uranium is

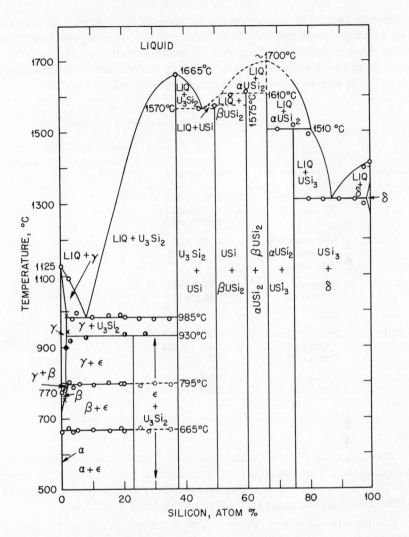

Fig. 9.3 — Uranium-silicon phase diagram. O, thermal arrest obtained on heating; ×, two-phase alloy by microscopic examination; ●, one-phase alloy by microscopic examination; ◑, epsilon peritectoid temperature.

somewhat increased by silicon from 770 to about 795°C; the $\alpha \rightarrow \beta$ transformation temperature appears unchanged at 660°C.

The compound USi_2, uranium disilicide, was first described by Defacqz (1908), who prepared it by aluminothermy, in the form of a light-gray metallic microscopically crystalline (cubic) powder. It is insoluble in cold or hot concentrated hydrochloric, nitric, or sulfuric acid or aqua regia but soluble in concentrated hydrofluoric acid. It is converted to silicate and uranate by molten alkalis or alkali carbonates at red heat and is slowly attacked by molten potassium bisulfate. It burns in oxygen at 800°C and reacts with chlorine at 500°C.

Table 9.5 — Crystal Structure Data for Uranium Silicides*

Compound	Symmetry	Unit cell dimensions, A	No. of molecules per unit cell	Density, g/cc
USi	Orthorhombic Pbnm	$a_1 = 5.65 \pm 0.01$ $a_2 = 7.65 \pm 0.01$ $a_3 = 3.90 \pm 0.01$	4	10.40
α USi$_2$	Body-centered tetragonal I4/amd	$a_1 = 3.97 \pm 0.03$ $a_3 = 13.71 \pm 0.08$	4	8.98
β USi$_2$	Hexagonal C6/mmm	$a_1 = 3.85 \pm 0.01$ $a_3 = 4.06 \pm 0.01$	1	9.25
USi$_3$	Cubic	$a = 4.03$		
U$_3$Si	Body-centered tetragonal I4/mcm	$a_1 = 6.017 \pm 0.002$ $a_3 = 8.679 \pm 0.003$	4	15.58
U$_3$Si$_2$	Tetragonal P4/mbm	$a_1 = 7.3151 \pm 0.0004$ $a_3 = 3.8925 \pm 0.0005$	2	12.20

*Zachariasen (1949).

The compound USi$_3$, uranium trisilicide, was identified by microscopic and x-ray analysis; it decomposes peritectically at about 1515°C. A eutectic is formed between USi$_3$ and Si (containing very little dissolved uranium) at 1315°C and 86 atom % silicon.

The high-temperature region in Fig. 9.3, between 35 and 75 atom % silicon, is uncertain. The region between U and U$_3$Si$_2$, as shown in Fig. 9.3, corresponds to a metastable equilibrium, since it does not show the above-mentioned peritectoid reaction between 4 and 28 atom % silicon at 940°C.

Alloys containing more than 30 atom % silicon are brittle and difficult to polish.

3.2 Crystal Structures of the Uranium Silicides. Zachariasen (1949) has published a detailed analysis of the crystal structures of the uranium silicides. The crystallographic data are summarized in Tables 9.5 and 9.6.

Zachariasen concludes that covalent bonds between silicon atoms are present in all structures except U_3Si. The observed Si-Si distances of about 2.3 A indicate that these are essentially single bonds. This binding between silicon atoms results in Si_2 groups in U_3Si_2, zigzag silicon chains in USi, a graphitelike layer of silicon atoms

Table 9.6 — Interatomic Distances in the Uranium Silicides*

α USi_2	U−12Si = 3.03 A	Si−6U = 3.03 A
		Si−3Si = 2.29 A
β USi_2	U−12Si = 3.01 A	Si−6U = 3.01 A
		Si−3Si = 2.22 A
USi	U−7Si = 2.98 A	Si−7U = 2.98 A
	U−U = 3.62 A	Si−2Si = 2.36 A
U_3Si_2	U_I−4Si = 2.96 A	Si−2U = 2.96 A
	U_I−8U = 3.32 A	Si−6U = 2.92 A
	U_{II}−6Si = 2.92 A	Si−1Si = 2.30 A
	U_{II}−4U = 3.32 A	
U_3Si	U_I−4Si = 3.01 A	Si−4U = 2.92 A
	U_I−8U = 3.04 A	Si−4U = 3.01 A
	U_{II}−2Si = 2.92 A	Si−4U = 3.17 A
	U_{II}−2Si = 3.17 A	
	U_{II}−4U = 3.02 A	
	U_{II}−4U = 3.04 A	

*Zachariasen (1949).

in β USi_2, and a three-dimensional lattice of silicon atoms in α USi_2. With respect to bonds between uranium atoms, Zachariasen considers such bonds to be definitely present in U_3Si and U_3Si_2 and probably absent in USi.

The structure of these compounds has been considered by Zachariasen in the light of Pauling's (1947) ideas on the nature of the metallic bond. The observed interatomic distances in these structures appear to be satisfactorily accounted for in α USi_2, β USi_2, USi, and U_3Si_2 by assuming the normal number of four valence electrons for silicon and 2.3 valence electrons for uranium. The atomic radius of uranium with 2.3 valence electrons and coordination number 12 is 1.636 A instead of 1.516 A found in metal. The larger radius appears in many uranium compounds of metallic nature. The results obtained by this treatment of the uranium silicides structures indicate that in these silicides a low-valence form of uranium is present, quite similar to the situation in UH_3 (cf. Chap. 8, Sec. 2.6a). The U_3Si structure, on the other hand, is satisfactorily accounted for by the presence of uranium with valence 5.78.

REFERENCES

1896   H. Moissan, Compt. rend., 122: 274.
1897   H. Moissan, Bull. soc. chim. Paris, 17: 14.
1908   E. Defacqz, Compt. rend., 147: 1050.
1911   P. Lebeau, Compt. rend., 152: 955, 956.
1911   O. Ruff and A. Heinzelmann, Z. anorg. Chem., 72: 72.
1913   P. Lebeau and A. Damiens, Compt. rend., 156: 1987.
1913   E. K. Rideal, dissertation, University of Bonn.
1913   E. Wedekind and O. Jochem, Ber., 46: 1204.
1914   P. Lebeau and A. Damiens, Bull. soc. chim. France, 15: 367.
1914a   E. Tiede and E. Birnbräuer, Z. anorg. Chem., 87: 165.
1914b   E. Tiede and E. Birnbräuer, Z. anorg. Chem., 87: 166.
1917   P. Lebeau and A. Damiens, Ann. chim., 8: 221.
1918   W. R. Mott, Trans. Electrochem. Soc., 34: 279.
1919   O. Ruff and O. Goecke, Z. angew. Chem., 24: 1461.
1925   O. Heusler, dissertation, University of Frankfurt am Main.
1926a   O. Heusler, Z. anorg. u. allgem. Chem., 154: 353.
1926b   O. Heusler, Z. anorg. u. allgem. Chem., 154: 366.
1929   J. L. Andrieux, Ann. Chim., 12: 423.
1931   N. G. Hägg, Z. physik. Chem., 12: 42.
1934   J. Schmidt, Elektrochem., 40: 171.
1947   L. Pauling, J. Am. Chem. Soc., 69: 542.
1948   J. L. Andrieux, Rev. mét., 45: 49-59; J. four élec., 57 (3): 54.
1948   L. Litz, A. B. Garrett, and F. C. Croxton, J. Am. Chem. Soc., 70: 1718.
1948   R. E. Rundle, N. C. Baenziger, A. S. Wilson, and R. A. McDonald, J. Am. Chem. Soc., 70: 99-105.
1949   J. L. Andrieux and P. Blum, Compt. rend., 229: 210.
1949   F. Bertaut and P. Blum, Compt. rend., 229: 666.
1949   W. H. Zachariasen, Acta Crystallographica, 2: 94-99.

Project Literature

MP Ames 1:   A. I. Snow, Report CT-954, Oct. 2, 1943.
MP Ames 2:   J. H. Carter, Report CT-609, Apr. 24, 1943.
MP Ames 3:   J. H. Carter, Report CT-542, Mar. 27, 1943.
MP Ames 4:   A. I. Snow, Report CT-1102, Nov. 28, 1943.
MP Ames 5:   J. H. Carter, Report CT-490, Feb. 20, 1943.
MP Ames 6:   R. E. Rundle, Report CT-686, May 22, 1943.
MP Ames 7:   R. E. Rundle, A. S. Wilson, and R. A. McDonald, Report CC-1131, Dec. 18, 1943.
MP Ames 8:   R. E. Rundle, Report CT-1270, Mar. 9, 1944.
MP Ames 9:   A. I. Snow, Report CT-816, July 24, 1943.
MP Ames 10: A. H. Daane and A. I. Snow, Report CT-751 A, June 2, 1943.
MP Ames 11: R. P. Baker and J. H. Carter, Report CT-393, Dec. 15, 1942.
MP Ames 12: N. C. Baenziger, Report CT-1515, April, 1944.
MP Ames 13: F. H. Spedding, Report CP-42, Apr. 25, 1942.
MP Ames 14: H. A. Wilhelm and A. H. Daane, Report CC-205, July 16, 1942.
MP Ames 15: A. H. Daane, Report CT-422, Jan. 15, 1942.
MP Ames 16: D. H. Ahmann, Report CT-393, Sec. I. Part B 3, Dec. 15, 1943.
MP Ames 17: F. H. Spedding, Report CC-298, Oct. 16, 1942.
MP Ames 18: T. Powell, Report CC-1778, Aug. 18, 1944.

MP Berkeley 1: L. Brewer, L. A. Bromley, P. W. Gilles, and N. L. Lofgren, Report
CC-3234, Oct. 8, 1945.
MP Berkeley 2: L. Brewer, Report CC-672, May 15, 1943.

MP Chicago 1: R. F. Plott and C. H. Raeth, Report CP-228, Aug. 14, 1942.
MP Chicago 2: R. F. Plott and C. H. Raeth, Report CE-236, Aug. 15, 1942.
MP Chicago 3: T. Davis and M. Burton, Report CC-231, Aug. 15, 1942.

Battelle 1: Battelle Memorial Institute, Report CT-1388, Feb. 10, 1944.

British 1: H. S. Peiser and T. C. Alcock, Report BR-589, Mar. 3, 1945.

MIT 1: P. Gordon and B. Cullity, Report CT-1101, Dec. 4, 1943; M. Cohen, Report
CT-1384, Mar. 1, 1944; B. Cullity, G. Bitsianes, and A. R. Kaufmann, Reports CT-
1696, May 1, 1944; CT-1819, June 1, 1944; CT-1938, July 1, 1944; CT-2106, July 1,
1944; and CT-2145, Aug. 1, 1944.
MIT 2: G. Bitsianes, B. Cullity, R. B. Bostian, and A. R. Kaufmann, Report CT-2699,
Feb. 8, 1945.

UCRL 1: G. E. MacWood and D. Altman, Report RL 4.7.600, Oct. 24, 1944.

Chapter 10

URANIUM COMPOUNDS WITH ELEMENTS OF GROUP Va

1. URANIUM NITRIDES

The reaction of metallic uranium with free nitrogen was discovered by Moissan (1896), who found that the two elements form a yellow nitride at 1000°C. Fifty years earlier Rammelsberg (1842) observed that a brown uranium-nitrogen compound is formed when UCl_4 is heated to red heat in an atmosphere of ammonia; later Uhrlaub (1859) determined the composition of this product as U_3N_4. The same composition was found by Heusler (1926) for the nitride obtained by Moissan's method of direct reaction between uranium and nitrogen. On the basis of these results U_3N_4 was considered until recently as the main uranium nitride with a rational chemical formula.

New methods of preparation and the chemical and physical properties of this compound were described by Kohlshütter (1901), Colani (1903; 1907a), Hardtung (1912), Miner (1922), and Herzer (1927). According to Herzer the nitride is a dark-brown or black powder which decomposes above 1400°C and has a density of 10.09 g/cc.

Heusler (1926) asserted that a lower nitride with the composition U_5N_4 is formed when U_3N_4 is decomposed in vacuum at 1650°C. According to the same author, decomposition at 1900°C leads to a still lower nitride, U_5N_2. Lorenz and Woolcock (1928) denied the existence of these compounds. They observed that the nitrogen pressure over the nitride (at 1280 and 1480°C) changes gradually and without discontinuity as the uranium/nitrogen ratio changes in the range from $UN_{0.1}$ to $UN_{1.0}$. They suggested that products with less than one nitrogen atom per uranium atom are solid solutions of uranium nitride in metallic uranium.

All these observations should be reevaluated in the light of x-ray studies carried out at Ames (MP Ames 1) and Battelle (Battelle 1). These studies showed that the compound U_3N_4 does not exist. The uranium nitride phase with the lowest nitrogen content is the mononitride, UN. Between uranium and UN there are neither intermediate stoichiometric compounds nor extensive solid solubility. The mononitride has no appreciable nitrogen decomposition pressure at the temperatures used in the investigation of Lorenz and Woolcock; therefore the nitrogen pressures measured by these investigators could not have been the equilibrium decomposition pressures of UN_x systems with x less than 1.

Between UN and the next highest uranium-nitrogen compound, the sesquinitride, U_2N_3, the uranium-nitrogen system is heterogeneous. It consists of the two separate phases, UN and U_2N_3. From U_2N_3 up to the dinitride, UN_2, which is the third "stoichiometric" nitride of uranium, the system is homogeneous; in other words, the U_2N_3 crystal structure is transformed gradually with increasing nitrogen content, without discontinuity, into the UN_2 structure.

1.1 Crystal Structure and Phase Relations.   (a) The Range U to UN. The solubility of nitrogen in metallic uranium has not been studied in detail but was reported by Ames investigators as "slight." Because of the stability of UN, the system uranium plus dissolved nitrogen is metastable except at very high temperatures. The measurement of solubility is made difficult by the fact that the rate of formation of the nitride is already high at 600 to 700°C with massive uranium and at 300 to 350°C with the fine uranium powder obtained by the decomposition of the hydride. According to observations at Ames, the solubility of uranium in UN also cannot be large because the lattice spacing of the UN crystals does not change appreciably with change in the ratio of uranium to nitrogen (in preparations made by fusing uranium with UN at high temperature). The small observed variations (e.g., a_0 = 4.899 A instead of a_0 = 4.880 A) are probably due to partial replacement of nitrogen by oxygen or carbon rather than to the uranium content of the UN lattice. The solubility of UN in uranium is probably also small, at least in the solid state. The presence of nitride (or oxide) has been observed to increase the viscosity of uranium melts (Mallinckrodt 1).

The mononitride, UN, is a light-gray powder (MP Ames 2) having a face-centered cubic structure with a lattice constant of 4.880 ± 0.001 A (MP Ames 1,3; Rundle, Baenziger, Wilson, and McDonald, 1948; Battelle 1). Four uranium atoms are contained in a unit cell. The structure is of the rock salt type rather than of the zinc sulfide type; this is indicated by the fact that on powder diagrams the (4 2 0) reflection is markedly more intense than the (3 3 1) reflection (MP Ames 3). The density of UN calculated from x-ray data is 14.32 g/cc. It is entirely isomorphous with the monocarbide, UC (Chap. 9).

If uranium nitride is prepared by the action of nitrogen on uranium at temperatures up to 1300°C, the reaction does not stop or slow down at the composition UN but proceeds to the composition of the U_2N_3 + UN_2 phase which is in equilibrium with the nitrogen atmosphere under the conditions of the experiment. Pure mononitride can be prepared by decomposing the higher nitrides at temperatures above 1300°C, and, if the heating is carried out slowly, nitrogen is evolved smoothly. Traces of oxygen must be avoided, since otherwise UO and UO_2 may be formed. The UN remaining after the decomposition of U_2N_3 is a very stable compound. It does not markedly decompose at 1700°C (MP Ames 4). According to Battelle (Battelle 1) the decomposition pressure of UN at 1200°C is less than 10^{-6} mm Hg. UN has been observed to sinter at 2300°C and to melt in an atomic hydrogen arc at about 2630 ± 50°C (MP Ames 5).

(b) The Range UN to U_2N_3 (Uranium Sesquinitride). Systems containing between 1 and 1½ nitrogen atoms per uranium atom are diphasic systems according to x-ray evidence. The second phase, U_2N_3, is also cubic but is body-centered and has a lattice constant of 10.678 ± 0.005 A (MP Ames 6; Rundle, Baenziger, Wilson, and McDonald, 1948; Battelle 1,2). If only strong maxima in the diffraction pattern are considered, the U_2N_3 lattice appears face-centered with a lattice constant of approximately 5.3 A. Consideration of the weaker maxima leads to a body-centered lattice with a lattice constant twice as large. This structure ($D5_3$) is known for Mn_2O_3 and some rare-earth oxides and is closely related to the fluorite structure.* It allows a smooth transition from the U_2N_3 lattice to the fluorite-type lattice of UN_2 (MP Ames 1).

The positions of the uranium atoms in the elementary cell of U_2N_3, which contains 16 molecules per unit cell, are as follows:

$(0\ 0\ 0;\ \frac{1}{2}\ \frac{1}{2}\ \frac{1}{2})+$

8 UI in (b): $(\frac{1}{4}\ \frac{1}{4}\ \frac{1}{4};\ \frac{1}{4}\ \frac{3}{4}\ \frac{3}{4};\ \supset)$

24 UII in (d): $\pm(x\ 0\ \frac{1}{4};\ \bar{x}\ \frac{1}{2}\ \frac{1}{4})\supset$ x = −0.018

48 N in (e): $\pm(x\ y\ z;\ \frac{1}{2}-x\ y\ \bar{z};\ \bar{x}\ \frac{1}{2}-y\ z;\ x\ \bar{y}\ \frac{1}{2}-z)\supset$

The nitrogen-atom parameters x, y, and z cannot be obtained from the pattern because of the comparative weakness of scattering by nitrogen atoms, but they are presumably similar to those of the oxygens in manganese sesquioxide. In Mn_2O_3 the oxygen parameters are

---

*See Strukturbericht II: 38 (1937).

$x = 0.385$, $y = 0.145$, and $z = 0.380$. The density of U_2N_3 calculated from x-ray data is 11.24 g/cc. The spacing in U_2N_3 and UN remains constant with changing uranium/nitrogen ratio, showing lack of mutual solubility of these two phases.

Table 10.1 — Lattice Constants of U_2N_3 and UN_2

	Lattice constant, A		
Composition	Body-centered cell	Pseudo face-centered cell	Density, g/cc
$UN_{1.435}$	10.678 ± 0.005	5.339 ± 0.003	...
$UN_{1.52}$	10.658 ± 0.005	5.329 ± 0.003	11.24
$UN_{1.75}$	10.580 ± 0.005	5.290 ± 0.003	...
UN_2*	...	5.31 ± 0.01	11.73

*This phase appears to be truly face-centered cubic with the smaller unit cell.

(c) The Range U_2N_3 to UN_2.   The gradual conversion of U_2N_3 to UN_2 without a discontinuous phase transition is possible because the Mn_2O_3 structure of U_2N_3 can be interpreted as a distorted fluorite structure in which UN_2 crystallizes. In the latter compound the metal atoms are in the positions 0 0 0; $0\frac{1}{2}\frac{1}{2}$; $\frac{1}{2}0\frac{1}{2}$; and $\frac{1}{2}\frac{1}{2}0$; and the nitrogen atoms in the positions $\frac{1}{4}\frac{1}{4}\frac{1}{4}$; $\frac{3}{4}\frac{3}{4}\frac{1}{4}$; etc; $\frac{3}{4}\frac{3}{4}\frac{3}{4}$; $\frac{1}{4}\frac{1}{4}\frac{3}{4}$; etc. In reference to the (eight times as large) body-centered unit the 0 0 0 metal atoms are in positions 0 0 0; $\frac{1}{2}$0 0; $\frac{1}{2}\frac{1}{2}\frac{1}{2}$; $0\frac{1}{2}\frac{1}{2}$; $\frac{1}{2}0\frac{1}{2}$; $\frac{1}{2}\frac{1}{2}0$. In the transition to U_2N_3 structure these uranium atoms remain fixed; the other 24 metal atoms are displaced slightly from the fourfold axes, thus requiring a larger unit cell. The gradual change in cell dimensions is illustrated by Table 10.1; this table also gives the smaller pseudo face-centered cell which can be chosen on the basis of the intense reflections alone.

As the nitrogen content increases, the weak diffraction maxima indicating the distortion of the face-centered unit disappear. The distribution of the nitrogen atoms in the intermediate products between U_2N_3 and UN_2 remains unknown, since the whole pattern analyzed is due to the special arrangement of the much heavier uranium atoms.

It will be noted that U_2N_3 ($\rho = 11.24$) is less dense than UN_2 ($\rho = 11.73$) and particularly UN ($\rho = 14.31$). Consequently, U_2N_3 decomposes under high pressures into UN plus UN_2.

While UN is stable in vacuum up to 1700°C and higher, U_2N_3 loses nitrogen in vacuum at temperatures above 700 to 800°C, and UN_2 cannot be prepared at all except under high nitrogen pressures. The equilibrium pressure of nitrogen over the higher nitrides is a smooth function of temperature and of the nitrogen content x in the solid (1.5 < x < 2). According to data compiled at Ames the pressure increases very rapidly with x and has at x = 1.712 a value of 4 mm Hg at 450°C and about 400 mm Hg at 800°C. The slope of the decomposition pressure curve indicates a heat of nitrogenation of 16.0 kcal per mole.

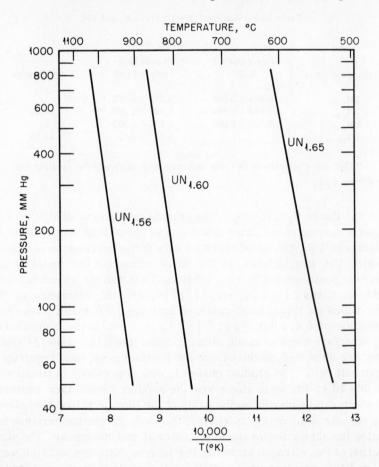

Fig. 10.1 — Nitrogen pressure in equilibrium with three uranium nitrides of different compositions at various temperatures.

The equilibrium is established rapidly and reversibly (MP Ames 7). More extensive measurements were carried out at Battelle (Battelle 1,3), where it was found that the equilibrium was established within 1 hr at 658 to 1048°C but that up to two days was required to establish the equilibrium at 492°C. Here also the rapid change of pressure with composition was noted. For example, at 1048°C the pressure was 50 mm Hg for $UN_{1.52}$ and 825 mm Hg for $UN_{1.56}$; at 492°C it was 50 mm Hg for $UN_{1.65}$ and 825 mm Hg for $UN_{1.68}$. Figure 10.1 shows equilibrium curves, $\log p = f(1/T)$, for three nitrogen compositions.

The above-mentioned value (400 mm Hg at 800°C for $UN_{1.71}$) determined at Ames does not agree well with Fig. 10.1; this difference was ascribed at Battelle to errors of nitrogen determination.

The decomposition of UN_x, which is rapid at x greater than 1.5, slows down perceptibly after the composition U_2N_3 has been reached.

1.2 <u>The Preparation of Uranium Nitrides</u>. The following two methods of preparation of the nitride were mentioned above:

1. Direct reaction of uranium with nitrogen at or below 1000°C (Moissan, 1896).

2. Reaction of UCl_4 with ammonia at red heat (Rammelsberg, 1842). The following methods were described later in the literature:

3. Reaction of U_3O_8 with magnesium in a nitrogen stream (Kohlschütter, 1901).

4. Reaction of $2NaCl \cdot UCl_4$ with ammonia at red heat (Colani, 1907a).

5. Reaction of uranium powder with ammonia at or below 1000°C (Hardtung, 1912).

6. Reaction of uranium metal or its alloys with Mg_3N in vacuum (Miner, 1922).

7. Reaction of UC_2 with nitrogen at 1180°C (Heusler, 1926).

8. Reaction of UH_3 with ammonia at 200°C (MP Ames 1).

9. Reaction of UH_3 with nitrogen at 350°C (MP Ames 1).

According to the last authors, only the reactions of uranium, UH_3, or UCl_4 with nitrogen or ammonia are likely to produce nitride not contaminated by other solids which are difficult to remove. To avoid oxide contamination, oxygen must be carefully excluded from all reactants.

According to the phase relations discussed in Sec. 1.1 (see Fig. 10.1), the preparation of nitride under atmospheric pressure of nitrogen must lead to mixed crystals of U_2N_3 plus UN_2, with an average composition between $UN_{1.5}$ and $UN_{1.8}$, depending on the temperature of the reaction. The same result is obtained when ammonia is used instead of nitrogen. The fact that earlier analyses gave lower nitrogen values (U_3N_4, i.e., $UN_{1.33}$) can be attributed to the presence of oxides or to errors in nitrogen determination by the method of Dumas.

The reaction of uranium metal with nitrogen has recently been studied in some detail. At Ames massive uranium and uranium powder prepared by decomposition of UH_3 were used. As an example of preparations of the first type, uranium turnings cleaned in 8N HNO_3 were converted to $UN_{1.712}$ (a steel-gray powder, $\rho = 11.3$ g/cc) by heating at 450°C at 1 atm of nitrogen (MP Ames 7). As an example of the use of metal powder, a compound $UN_{1.75}$ was obtained by heating uranium powder for three days to 520°C in a stream of pure nitrogen (MP Ames 8). It was observed (MP Ames 9) that a layer of nitride forms around the bright metal core when thick uranium turnings are partly oxidized by ignition in air. It was found elsewhere that the rate of nitrogenation of uranium increases with temperature, particularly rapidly around 800°C, probably in consequence of the $\beta \rightarrow \gamma$ transformation of the metal (MP Chicago 1). According to observations at Battelle the reaction of massive uranium with nitrogen begins to be rapid above 450°C (Battelle 4).

At the National Bureau of Standards the uranium-nitrogen reaction was found to be slight at 400°C, to become marked at 500 to 600°C, and to proceed rapidly above 800°C (Natl. Bur. Standards 1). At or below 750°C the specimen becomes covered by a black adherent scale with a metallic luster. Small bits of this scale pop off upon cooling. When all this scale was removed by grinding, the nitrogen content of the bright core was found to be 0.012 per cent compared to 0.003 per cent before the reaction. At or above 800°C a dark-gray or black nonmetallic powder is formed instead of the scale; analysis gave the composition $UN_{1.6}$ for both scale and powder. It was observed at the National Bureau of Standards that an oxide film affords no protection from nitrogen at 700°C.

The reaction of uranium metal with ammonia, also studied at Ames (MP Ames 1), begins at about 400°C. With turnings a temperature in excess of 700°C is required for rapid conversion. The product $UN_{1.747}$ was obtained by heating uranium in an ammonia atmosphere at 800°C for 24 hr.

The reaction of uranium hydride with nitrogen occurs even at 200°C (MP Ames 10). It also gives a product of approximate composition $UN_{1.75}$. The reaction requires 10 to 12 hr at 250°C and 1 to 2 hr at 350°C.

The reaction of uranium hydride with ammonia occurs, although slowly, even at 100°C (1.5 mg of nitrogen taken up by 5 g of uranium in 1 hr) (MP Ames 1). It becomes rapid above 200°C (316 mg of nitrogen taken up by 5 g of uranium in 1 hr). This is perhaps the easiest way to obtain the nitride. The product is an exceedingly fine powder (bulk density 3.4 g/cc, true density 11.3 g/cc). Contrary to what was thought for a while (MP Ames 11), low-temperature reactions of UH_3 with nitrogen and ammonia do not stop or slow down at the composition UN but proceed to U_2N_3 and beyond. It is doubtful whether the mononitride is formed at all as an intermediate in low-temperature nitrogenation (MP Ames 13). As mentioned before, the way to

prepare the mononitride is to decompose the higher nitrides in vacuum above 1300°C. For example, at Ames, UN was obtained by heating
$UN_{1.67}$ to 1650°C in vacuum in a graphite crucible (MP Ames 2).

The dinitride, UN_2, has not yet been obtained in the pure state. A
mixture of UN plus UN_2 is obtained (as shown by x-ray evidence) when
uranium is combined with nitrogen under 1,800 psi at 600°C (MP
Ames 14). Although the average composition of this product ($UN_{1.75}$)
is the same as that of the nitride obtained under atmospheric pressure, its x-ray diagram proves the absence of U_2N_3. As mentioned in
Sec. 1.1c, disproportionation of U_2N_3 into UN plus UN_2 under high
pressure is to be expected because of the greater density of UN_2 and
particularly of UN, as compared to that of U_2N_3.

1.3 The Physical and Chemical Properties of Uranium Nitrides.
The nitrides have been described as dark-brown, steel-gray, dark-
gray, and black powders. Their x-ray structure was discussed in
Sec. 1.1. The densities calculated from x-ray data are 14.31 g/cc for
UN, 11.24 g/cc for U_2N_3, and 11.73 g/cc for UN_2. The directly determined density of $UN_{1.8}$ (as obtained from UH_3 and ammonia) was 11.3
g/cc. This product had a bulk density of 3.4 g/cc and an electrical
resistivity, without compression, of about 200 ohm-cm (MP Ames 12).

The heats of formation of the nitrides are not well-known. Neumann,
Kroger, and Haebler (1932) found that 68.5 kcal is liberated per gram
atom of absorbed nitrogen in the formation of a nitride containing 3.5
per cent nitrogen. This is a much lower nitrogen content than even
that of UN (5.55 per cent N). It is thus quite uncertain what phase was
actually formed, but we may assume tentatively that it was U_2N_3 and
that the low nitrogen content was due to the presence of unreacted
metal in the product. With this assumption, a value of 256 kcal per
mole can be calculated for the heat of the following reaction:

$$2U + \tfrac{3}{2}N_2 \rightarrow U_2N_3 \qquad - \quad (1)$$

The slopes of the three straight lines in Fig. 10.1 lead to the following values for the integral heats of decomposition of UN_x into UN
and $[(x-1)/2]N_2$:

$$UN_{1.65} \qquad \Delta H = 15.69 \text{ kcal/g atom of U}$$

$$UN_{1.60} \qquad \Delta H = 18.79 \text{ kcal/g atom of U}$$

$$UN_{1.56} \qquad \Delta H = 19.55 \text{ kcal/g atom of U}$$

The change in ΔH is not linear with x, and the extrapolation to $x = 1.5$
is uncertain. A Battelle report (Battelle 3) gives $\Delta H = 21.9$ as the result of this extrapolation, leading to the thermochemical equation

$$2UN + \tfrac{1}{2}N_2 = U_2N_3 + 43.8 \text{ kcal} \qquad (2)$$

If the trend of ΔH vs. x shown by the above figures is correct, the further nitrogenation of U_2N_3 is an endothermal process. Therefore UN_2 can be expected to be more stable at the higher temperatures.

From the above-estimated heat values (256 kcal for U_2N_3 and 44 kcal for the difference $U_2N_3 - 2UN$) 106 kcal per mole is obtained for the heat of formation of UN, a value that explains the stability of this product. We thus have the following preliminary estimates:

$$U + \tfrac{1}{2}N_2 \rightarrow UN + 106 \text{ kcal} \qquad (3)$$

$$2UN + \tfrac{1}{2}N_2 \rightarrow U_2N_3 + 44 \text{ kcal}$$

$$U_2N_3 + \tfrac{1}{2}N_2 \rightarrow 2UN_2 - x \text{ kcal (endothermal)} \qquad (4)$$

Estimates of ΔF and ΔS of nitride formation (MP Chicago 2; MP Berkeley 1) were made on the assumption that U_3N_4 is the nitride formed and are thus obsolete. There seems to be little justification in attempting to correct these estimates to obtain data for UN and U_2N_3 because of the absence of reliable thermochemical information.

1.4 The Chemical Reactions of Uranium Nitrides. The thermal decomposition of the higher uranium nitrides, U_2N_3 and UN_2, was discussed in Sec. 1.3. These nitrides can also be reduced by hydrogen, a reaction that was studied at Ames (MP Ames 8). The formation of ammonia from $UN_{1.75}$ and hydrogen could be noticed even at 100°C. At 500°C, 10 per cent of the nitrogen content was lost after 4 hr. At 800°C, $UN_{1.75}$ was reduced to $UN_{1.52}$ after heating 1 hr in hydrogen; continuation of the treatment for another 18 hr led to a product with the composition $UN_{1.43}$. At the last temperature, however, thermal decomposition may have played an important part in addition to the reduction by hydrogen.

The nitride is easily oxidized. It is not normally pyrophoric, although the finely powdered product obtained from UH_3 has occasionally been described as such (MP Ames 2). It will ignite in air at 150 to 200°C and burn to form UO_2 (or U_3O_8) and nitrogen (but no nitrogen oxides) (MP Ames 12). The formation of ammonia in this process (from reaction of nitride with moisture?) also was noticed (MP Chicago 3). Kohlschütter (1901) had long ago described the decomposition of the nitride by heating with water vapor. The same author has observed the oxidation of the nitride by various compounds such as CuO, $PbCrO_4$, Fe_2O_3, UO_2 (see MP Ames 15), $KClO_3$, and KNO_3.

The nitride is slowly oxidized by concentrated nitric acid. It is not attacked by hot or cold hydrochloric or sulfuric acid or by sodium hydroxide solution (Kohlschütter, 1901; MP Ames 1). However, it reacts with molten alkali, liberating ammonia (Kohlschütter, 1901), and with gaseous hydrogen chloride (MP Ames 16). The reaction with hydrogen chloride gas at 400 to 500°C gave ammonium chloride and converted $UN_{1.75}$ into a hard hygroscopic sintered mass of green and brown color, which contained, by x-ray evidence, UCl_4, U_3O_8, UO_2, and an unknown phase, perhaps the double salt $(NH_4)_2UCl_6$. Analysis showed the presence of 2 per cent ammonia. A similar treatment of UN gave a product with less than 1 per cent ammonia. The nitride reacts readily with hot 85 per cent phosphoric acid to form uranium(IV) phosphate and with concentrated perchloric acid to form uranyl perchlorate (MP Ames 1).

The chemical properties of the mononitride have not been studied in detail but were described as "similar to those of the higher nitrides." The mononitride was found to react with carbon (finely divided graphite) at 2250°C and with uranium dioxide at about 2300°C (MP Ames 15), but the reaction products are unknown.

2. URANIUM PHOSPHIDES

Rammelsberg (1872) supposed that uranium monophosphide was produced by decomposition of $UO_2(H_2PO_2)\cdot H_2O$. At Ames a uranium monophosphide phase was identified by x-ray analysis. It has the sodium chloride structure with $a_0 = 5.589$ A (MP Ames 17).

A triuranium tetraphosphide, U_3P_4, was described by Colani (1907a). This product was obtained in small yield when PH_3 reacted with $2NaCl\cdot UCl_4$. The reaction could be brought about by passing a stream of dry hydrogen over a melt containing aluminum phosphide and excess $2NaCl\cdot UCl_4$. The product is leached with water and hydrochloric acid and washed with water, alcohol, and ether. The residue is a fine black crystalline powder containing some aluminum. It slowly oxidizes in air to yellow uranyl phosphate. It burns in air and is attacked slowly by water, particularly if air is present. Dilute hydrochloric acid has no effect, but boiling concentrated nitric acid, aqua regia, or molten sodium hydroxide decompose the phosphide instantaneously.

Lilliendahl and Driggs (1929) suggested, as a commercial method of preparation of uranium phosphide, heating to 600 to 1000°C of 3 parts of finely powdered metal with 1 part of phosphorus powder. Unreacted phosphorus can be leached out by alcohol and ether or distilled away.

3. URANIUM ARSENIDES AND ANTIMONIDES

Colani (1907b) prepared uranium arsenide and antimonide by methods similar to the one he used for the preparation of phosphide. To obtain the arsenide he passed dry hydrogen charged with arsenic vapor over $2NaCl \cdot UCl_4$. A small quantity of square or hexagonal black lustrous tablets was obtained. Melting of Na_3As with excess arsenic and $2NaCl \cdot UCl_4$ in a stream of dry hydrogen gave a steel-gray microcrystalline arsenide. Ignition of a mixture of arsenic powder, arsenous oxide, uranium trioxide, and aluminum could also be used, although no success was obtained with the aluminothermic method in the case of uranium phosphide preparations. The uranium arsenide, whose composition Colani determined as U_3As_4, is hardly attacked at all by dry air but somewhat more by moist air. It is easily soluble in nitric acid and burns in the flame of a bunsen burner.

The uranium-arsenic system was studied by x-ray analysis at Ames (MP Ames 17). Two compounds were identified: a monoarsenide, UAs (cubic, NaCl structure, a_0 = 5.767 A, ρ = 10.77 g/cc), and U_2As, prepared at Ames by Figard (MP Ames 18). The latter gives an x-ray pattern that is different from those of uranium, UO_2, or UAs.

The uranium antimonide was obtained by Colani (1907a) by ignition of dry $2NaCl \cdot UCl_4$ with an equivalent mixture of antimony and aluminum in a stream of hydrogen. Analysis of the (aluminum-free) product gave 58 per cent antimony and 42 per cent uranium corresponding to an approximate formula U_3Sb_8. Heating this product in a stream of hydrogen caused a gradual loss of antimony, but the composition U_3Sb_4, which Colani expected by analogy with his formulas for phosphide and arsenide, was not reached. The antimonide was attacked by concentrated hydrochloric acid and dissolved by concentrated nitric acid with the formation of an antimony oxide.

REFERENCES

1842   C. Rammelsberg, Pogg. Ann., 55: 323.
1859   G. E. Uhrlaub, dissertation, University of Göttingen, p. 27.
1872   C. Rammelsberg, Sitzber. kgl. preuss. Akad. Wiss., p. 449.
1896   H. Moissan, Compt. rend., 122: 1092.
1901   V. Kohlschütter, Ann., 317: 166.
1903   A. Colani, Compt. rend., 133: 383.
1907a A. Colani, Ann. chim. et phys., 12: 88.
1907b A. Colani, Ann. chim. et phys., 12: 93, 95.
1912   H. Hardtung, dissertation, Hannover Techn. Hochschule.
1922   C. G. Miner, U. S. Patent No. 1631544, transferred to the Anglo-California Trust Co.
1926   O. Heusler, Z. anorg. u. allgem. Chem., 154: 353, 366.
1927   H. Herzer, dissertation, Hannover Techn. Hochschule.
1928   R. Lorenz and J. Woolcock, Z. anorg. u. allgem. Chem., 176: 302.
1929   W. C. Lilliendahl and F. H. Driggs, U. S. Patent No. 1893296, assigned to Westinghouse Electric Corporation, Lamp Division.
1932   B. Neumann, C. Kroger, and H. Haebler, Z. anorg. u. allgem. Chem., 207: 146.
1948   R. E. Rundle, N. C. Baenziger, A. S. Wilson, and R. A. McDonald, J. Am. Chem. Soc., 70: 99-105.

Project Literature

Battelle 1:   Battelle Memorial Institute, Report CC-2700, Feb. 1, 1945.
Battelle 2:   Battelle Memorial Institute, Report CT-1009, Oct. 10, 1943.
Battelle 3:   Battelle Memorial Institute, Report CT-1697, May 1, 1944.
Battelle 4:   Battelle Memorial Institute, Reports CT-688, May 10, 1943, and CT-956, Sept. 10, 1943.

Mallinckrodt 1:   Mallinckrodt Chemical Works, Report A-1072, Feb. 15, 1945.

MP Ames   1:   R. E. Rundle, N. C. Baenziger, A. S. Newton, A. H. Daane, T. A. Butler, I. B. Johns, W. Tucker, and P. Figard, The System Uranium-Nitrogen, in National Nuclear Energy Series, Division VIII, Volume 6.
MP Ames   2:   A. H. Daane and P. Chiotti, Report CT-1775, Aug. 1, 1944.
MP Ames   3:   R. E. Rundle, N. C. Baenziger, A. S. Wilson, and R. A. McDonald, Report CC-2397, Feb. 17, 1945, p. 22.
MP Ames   4:   A. H. Daane and N. Carlson, Report CC-1496, Apr. 10, 1944, p. 14.
MP Ames   5:   A. H. Daane and N. Carlson, Report CC-1500, May 10, 1944, p. 10.
MP Ames   6:   Reports CT-686, May 22, 1943; CN-1495, Mar. 10, 1944; and CC-1514, Mar. 10, 1944.
MP Ames   7:   A. S. Newton, Report CC-1781, Oct. 18, 1944.
MP Ames   8:   A. S. Newton, R. W. Nottorf, and A. H. Daane, Report CC-1524, Apr. 14, 1944.
MP Ames   9:   D. W. Peterson and E. J. Wimmer, Report CC-1779, July 10, 1944.
MP Ames 10:   W. Tucker and P. Figard, Report CC-1500, May 10, 1944.
MP Ames 11:   T. A. Butler, R. Fischer, and A. S. Newton, Report CC-664, May 14, 1943.
MP Ames 12:   I. B. Johns, Report CC-587, Apr. 19, 1943.
MP Ames 13:   P. Figard and W. Tucker, Report CC-1496, May 11, 1944; W. Tucker and P. Figard, Report CC-1500, May 10, 1944.
MP Ames 14:   N. C. Baenziger, Report CC-1984, Dec. 19, 1944.
MP Ames 15:   R. E. Rundle and N. C. Baenziger, Report CC-1524, Mar. 10, 1944.
MP Ames 16:   J. C. Warf and R. P. Ericson, Report CC-1504, Aug. 10, 1944, p. 24.
MP Ames 17:   R. E. Rundle and N. C. Baenziger, Report CC-1778, Aug. 18, 1944.
MP Ames 18:   N. C. Baenziger, Report CC-1781, Oct. 18, 1944.

MP Berkeley 1:   L. Brewer, Report CC-672, May 15, 1943.

MP Chicago 1:   F. Foote and J. P. Howe, Report CT-1269, Jan. 29, 1944.
MP Chicago 2:   J. W. Davis and M. Burton, Report CC-231, Aug. 15, 1944, p. 17.
MP Chicago 3:   G. T. Seaborg, Report CK-591, Apr. 15, 1943.

Natl. Bur. Standards 1:   W. E. Lindlief, Report CT-1101, Dec. 4, 1943; W. E. Lindlief, Report CT-1179, Jan. 1, 1944; V. C. F. Holm and W. E. Lindlief, Report CT-2733, Mar. 19, 1945.

Chapter 11

URANIUM OXIDES, SULFIDES, SELENIDES, AND TELLURIDES

PART A. URANIUM OXIDES AND HYDROXIDES

The three oxides UO_2, U_3O_8, and UO_3 have been known for over 100 years, but a systematic investigation of the uranium-oxygen system was first undertaken by Hüttig, Biltz, and coworkers in 1920 to 1928. Studies at the Manhattan Project and in Britain have provided considerable new information, including the discovery of uranium monoxide, UO.*

The solubility of oxygen in uranium is small, and the same is true of the solubility of uranium in uranium monoxide; in other words, the range U to UO is essentially diphasic. This was also considered to be true of the range UO to UO_2 until Zachariasen found in 1946 a homogeneous phase with the composition $UO_{1.75}$. This may be either a new compound, U_4O_7, or the lower end of a series of solid solutions which previously was known to extend from $UO_{2.00}$ to $UO_{2.30}$. Above $UO_{2.30}$ there follows a diphasic region, the upper limit of which is not definitely established. The existence of an oxide $UO_{2.5}$ (or U_2O_5) was asserted several times in the earlier literature, and at Ames, a homogeneous phase with this composition again was found by x-ray analysis. According to the Ames observers, $UO_{2.5}$ is the lower end of a monophasic region that extends up to and beyond $UO_{2.67}$ (or U_3O_8). On the other hand, Biltz and Müller's tensimetric analysis of the uranium-oxygen system gave no indication of the existence of a compound U_2O_5. According to their observations, the lower limit of existence of a third homogeneous uranium-oxygen phase (after UO and UO_2) is at $UO_{2.62}$ quite close to $UO_{2.67}$ (or U_3O_8). Although it is thus uncertain whether the monophasic range begins at $UO_{2.50}$ or at $UO_{2.62}$,

---

*A systematic study of the uranium-oxygen system also was promised in a note from E. Gleditsch's laboratory at Oslo [Nature, 28: 127 (1940)], but the results of this study, if any, were not available at the time of the writing of this review.

there is no doubt that the upper limit of this homogeneous range extends far beyond $UO_{2.67}$ — in all probability up to UO_3.

The phases UO and UO_2 (more exactly, $UO_{1.75}$ to $UO_{2.30}$) are cubic face-centered. The first phase has the rock salt structure, and the second has the fluorite structure. The phases between $UO_{2.50}$ (or $UO_{2.62}$) and $UO_{3.00}$ show a continuous transition from an orthorhombic to a hexagonal lattice. In addition to the hexagonal form, $UO_3(I)$, uranium trioxide also exists in one amorphous and in at least two, and perhaps four, other (as yet unidentified) crystalline forms.

The affinity for water increases from UO_2 to UO_3. The hydrates $UO_2 \cdot xH_2O$ and $U_3O_8 \cdot xH_2O$ (with x probably equal to 2) are known but have not been very well investigated. Several UO_3 hydrates, on the other hand, have been well identified, both crystallographically and analytically. They include the semihydrate, $2UO_3 \cdot H_2O$ (or $H_2U_2O_7$); four allotropic forms of the monohydrate, $UO_3 \cdot H_2O$ (or H_2UO_4); and two allotropic forms of the dihydrate, $UO_3 \cdot 2H_2O$ (or H_4UO_5).

The oxide UO_2 is strongly basic. The oxide UO_3 is amphoteric; i.e., it can be neutralized either by acids, forming uranyl salts (e.g., UO_2SO_4), or by alkalis, forming uranates (e.g., Na_2UO_4). The oxide U_3O_8 has often been interpreted as urano-uranate, $UO_2 \cdot 2UO_3$, or $U^{+4}(UO_4^{--})_2$. In agreement with this idea, U_3O_8 gives, with acids, mixtures of uranium(IV) and uranyl salts. However, the formula $UO_2 \cdot 2UO_3$ should not be taken as indicating the presence in U_3O_8 of two types of uranium atoms. X-ray analysis shows that all uranium atoms in U_3O_8 occupy equivalent positions; thus they probably carry also the same average charge $(+5\frac{1}{3})$. This fractional charge can be maintained by resonance between two (or more) valence states (e.g., U^{+4} and U^{+6}).

Magnetic investigations (cf. Sec. 2.4g) have been interpreted to show that U_3O_8 contains U^{+5} ions rather than U^{+4} ions, and it was suggested that the formula of this compound should be written $U_2O_5 \cdot UO_3$ rather than $UO_2 \cdot 2UO_3$. However, here again the interpretation must take into account the existence of mesomerism. In other words, U_3O_8 must be described as a mesomeric compound that behaves in the presence of acids as if it contained the ions U^{+4} and U^{+6}, and in a magnetic field as if it contained the ions U^{+5} and U^{+6}.

The nature of oxides with more than three oxygen atoms per uranium atom is not completely understood. Anhydrous oxides with $O/U =$ 3.2 to 3.5 have been obtained by dehydration of $UO_4 \cdot 2H_2O$, or decomposition of ammonium diuranate in a stream of oxygen (cf. Sec. 5.3g) and allegedly also by electrolysis (cf. Sec. 3.2). These oxides decompose to UO_3 and oxygen in contact with water and to UO_2^{++} salts and oxygen in contact with acids. The Brown University group considered $UO_{3.5}$ as an equimolar mixture of UO_3 and anhydrous uranium tetroxide, UO_4; however, it is not clear why $UO_{3.5}$ should not be tentatively considered as a separate oxide, U_2O_7, since all attempts to prepare pure anhydrous UO_4 have so far been unsuccessful (cf. Sec. 3.4a).

Hydrated UO_4 may contain, in the air-dry state, up to 4.5 moles of H_2O. According to most observers, however, the only definite UO_4 hydrate is $UO_4 \cdot 2H_2O$. In contrast to anhydrous $UO_{3.5}$, the hydrate $UO_4 \cdot 2H_2O$ is not decomposed by water or acids; in 4N H_2SO_4 it gives a weak solution capable of reducing $KMnO_4$. Consequently, $UO_4 \cdot 2H_2O$ should perhaps be considered as free "peruranic acid," H_4UO_6 [i.e., $UO(OH)_3(OOH)$], giving, in equilibrium with water, a small amount of H_2O_2.

$$UO(OH)_3(OOH) + H_2O \rightleftarrows H_2O_2 + UO(OH)_4 \qquad (1)$$
$$\text{peruranic acid} \qquad\qquad \text{uranic acid}$$

The interpretation of $UO_4 \cdot 2H_2O$ as free peruranic acid was first suggested many years ago by Fairley (1877a) and Pisarzhevskiĭ (1903) but was opposed by Rosenheim and Daehr (1932), who pointed out that $UO_4 \cdot 2H_2O$ cannot be neutralized by alkalis to form peruranates. They determined the "peroxide" oxygen in $UO_4 \cdot 2H_2O$ by titration with permanganate or iodine according to the method of Riesenfeld and Mau (1911) and found one oxygen atom to be peroxidic. On the strength of this result they suggested the structure

$$\left(\begin{matrix} O \\ O \end{matrix} \right. \!\!\!\!\! > U < \!\!\!\!\! \left. \begin{matrix} O \\ O \end{matrix} \right) \cdot 2H_2O$$

Hüttig and von Schroeder (1922a), on the other hand, formulated the uranium peroxide hydrates as addition compounds of UO_3 and H_2O_2.

$$(UO_3) \cdot xH_2O_2 \cdot yH_2O \quad x = 1, \ y = 1 \text{ for } UO_4 \cdot 2H_2O$$

All these suggestions carry little weight, and a significant description of the constitution of $UO_4 \cdot 2H_2O$ can be based only on a complete x-ray analysis, which has not as yet been carried out.

1. PHASE RELATIONS IN THE URANIUM-OXYGEN SYSTEM

1.1 Range U to UO. Solubility of Oxygen in Uranium. Uranium Monoxide. The solubility of oxygen in uranium is very small, even at temperatures above 2000°C. The best available results, obtained at the National Bureau of Standards (Natl. Bur. Standards 1,2), are plotted in Fig. 11.1. The curve shows a solubility of the order of 0.05 atom % oxygen at the melting point of uranium (1133°C), rising to 0.1 atom % at 1400°C and to 0.4 atom % at 2000°C. The solubility of oxygen (or uranium oxide) in solid uranium could not be determined exactly, but the fact that oxide inclusions appear (after annealing in the γ range), even in metal containing as little as 0.05 atom % oxygen, shows that solubility in the solid phase is even smaller than in the liquid. This is indicated by the broken line in **Fig. 11.1.**

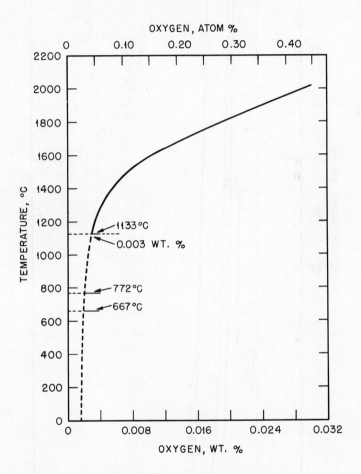

Fig. 11.1—Solubility of oxygen in uranium.

According to the National Bureau of Standards (Natl. Bur. Standards 3), uranium oxides precipitate from oxygen solutions in molten uranium upon cooling; precipitation occurs in either the β or the γ range, depending on the original oxygen concentration.

At Ames (MP Ames 1) uranium oxide samples, UO_x, of varying composition were prepared by heating compressed mixtures of uranium

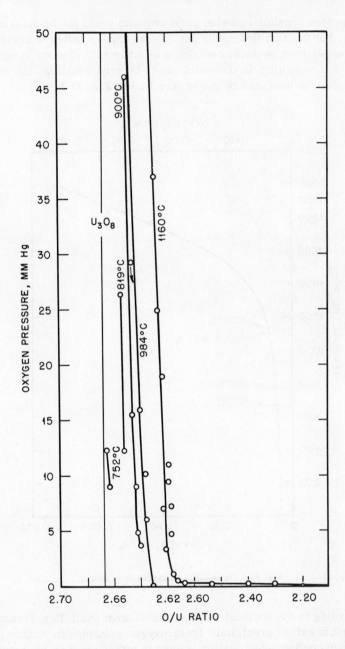

Fig. 11.2—Decomposition isotherms of $UO_{2.665}$ to $UO_{2.200}$ [from W. Biltz and H. Müller, Z. anorg. u. allgem. Chem., 163: 279 (1927)].

metal powder and uranium dioxide in vacuum. In the composition range $x = 0$ to 1, the products gave only uranium and uranium monoxide x-ray diffraction patterns. The lattice spacing of α uranium was not markedly affected by the presence of oxygen; the same was true of the effect of excess uranium on the lattice spacing of uranium monoxide. The first fact is in agreement with the above-mentioned low solubility of oxygen in α uranium; the second fact indicates that the solubility of uranium in uranium monoxide is also very small.

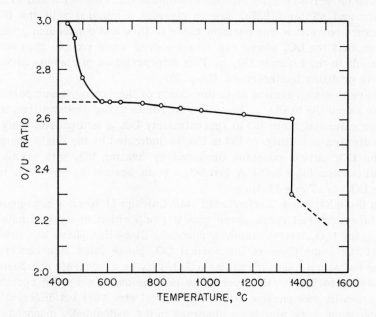

Fig. 11.3—Change in uranium-oxygen ratio with change in temperature under a constant oxygen pressure of 10 mm Hg [from W. Biltz and H. Müller, Z. anorg. u. allgem. Chem., 163: 295 (1927)].

Although pure uranium monoxide, UO, has been observed only as a thin surface layer on uranium metal (except for one accidental observation; see Sec. 4.3), its existence is fully established by x-ray evidence. It possesses a cubic face-centered lattice (of the rock salt type) with the lattice constant $a_0 = 4.91$ A (cf. Table 11.2). This structure is different from that of uranium metal in any of its three allotropic forms but is very similar to the structures of the mononitride, UN, and the monocarbide, UC. The three compounds — monoxide, mononitride, and monocarbide — readily form mixed crystals; in other words, the nonmetallic components—oxygen, nitrogen, and carbon—are capable of replacing each other at random in these compounds.

1.2 <u>Range UO to UO$_{2.3}$</u>. <u>Uranium Dioxide</u>. A range of uranium-oxygen compounds, which do not lose oxygen except at very high temperatures, extends considerably beyond UO$_{2.00}$, probably up to about UO$_{2.15}$. Uranium dioxide can take up, without change in crystal structure, even more oxygen—to a composition of about UO$_{2.3}$, but oxygen taken up in excess of UO$_{2.15}$ is comparatively loosely bound and can be driven out by heating in vacuum (cf. Figs. 11.2 and 11.3).

Biltz and Müller (1927a) saw in cleveite a mineral with the UO$_2$ structure but with a composition close to U$_3$O$_8$ and consequently suggested that the UO$_2$ phase can take up even more oxygen than corresponds to the formula UO$_{2.30}$. This interpretation of cleveite structure is probably incorrect (cf. Chap. 3).

There is disagreement as to the extent of the homogeneous portion of the range UO to UO$_{2.3}$. According to Ames x-ray observations the region extending from UO to approximately UO$_{2.00}$ is diphasic. Only a very limited solubility of UO in UO$_2$ is indicated by the slight change in the UO$_2$ lattice constant produced by heating UO$_2$ with uranium metal powder (a$_0$ = 5.461 A for UO$_2$ + U as against a$_0$ = 5.459 A for pure UO$_2$; cf. Table 11.3).

On the other hand, Zachariasen (MP Chicago 1) found a homogeneous face-centered cubic phase with a composition of approximately UO$_{1.75}$ (or U$_4$O$_7$, tetrauranium heptoxide). Since this phase was present at the same time as the normal UO$_2$ phase (also face-centered cubic but with a smaller spacing), U$_4$O$_7$ may be a separate compound, analogous to Th$_4$S$_7$. However, this conclusion is not quite certain. Occasionally, two phases, similar in crystal structure but different in composition, have also been observed in the undoubtedly monophasic range between UO$_{2.00}$ and UO$_{2.30}$, where their presence may have been due to incomplete equilibration. It is therefore possible (MP Berkeley 1) that the solid-solution range now being discussed extends down to UO$_{1.75}$ and that the simultaneous occurrence of UO$_{1.75}$ and UO$_{2.00}$ also was due to a lack of equilibrium. This hypothesis is supported by x-ray analysis. All oxides between UO$_{1.75}$ and UO$_{2.3}$ crystallize with the cubic face-centered fluorite structure. According to Table 11.3, the lattice constant of UO$_{1.75}$ is 5.477 A, that of stoichiometrically pure UO$_2$ is 5.460 A, and that of UO$_{2.30}$ is 5.430 A. This uniform change of spacing makes the hypothesis of a continuous series of solid solutions plausible. It is noteworthy that the lattice constant decreases with increasing oxygen content. This shrinkage led Rundle and coworkers at Ames to advance the suggestion that the increase in the ratio of oxygen to uranium may be due to deficiency of uranium rather than to presence of extra oxygen. If this hypothesis is correct, the density of the oxides should decrease considerably from $UO_{2.00}$ (or $UO_{1.75}$) to $UO_{2.30}$. On the other hand, Hüttig (1924) suggested that oxygen atoms in excess of the formula UO_2 may be freely movable in the UO_2 crystal lattice; Biltz and Müller (1927a) assumed that they are held in the "interstitial" spaces of the UO_2 lattice. According to these concepts the density of $UO_{2.30}$ should be higher than that of UO_2, since the unit cell of $UO_{2.30}$ must contain, despite its somewhat smaller size, the same number of uranium atoms and 15 per cent more oxygen atoms than the unit cell of $UO_{2.00}$.

The experimental data on the density of uranium oxides between $UO_{2.00}$ and $UO_{3.00}$ are given in Table 11.1. They show, in the first place, that the empirical density of $UO_{2.00}$ agrees well with the value calculated from the x-ray diffraction data on the assumption that each elementary cell contains its full complement of four uranium atoms. In the second place, the experimental densities increase slightly from $UO_{2.00}$ to $UO_{2.30}$. This proves that no substantial fraction of uranium is "missing" either in $UO_{2.00}$ or in the higher oxides up to $UO_{2.30}$ (for the probable effects of a small deficiency of uranium, too small to affect density determinations but enough to account for electrical conductivity, see Sec. 2.2h).

The observed increase in density of the oxides between $UO_{2.00}$ and $UO_{2.30}$ is somewhat less pronounced than would be expected from the x-ray data on the assumption of no "deficit" of uranium or oxygen (cf. column 4 of Table 11.1). The mole volumes calculated by Biltz and Müller (column 6 of Table 11.1) are practically constant from $UO_{2.0}$ to $UO_{2.26}$, while the lattice constants require that these volumes should decrease by about 1.7 per cent. However, the density measurements may not have been accurate enough to reveal the anticipated decrease.

Jolibois (1947) found independently that the low-temperature oxygenation of UO_2 leads in a first step to a bluish-black oxide with $x = 2.33$ and interpreted the latter as a new uranium oxide U_3O_7 (or $UO_3 \cdot 2UO_2$). He noticed, however, that the crystal structure of this new oxide was almost identical with that of UO_2. Similar observations were made by Grønvold and Haraldsen (1948b). They found that at 150°C, in oxygen, UO_2 is oxidized to an oxide $UO_{2.34}$ with a contracted lattice (a = 5.40 A instead of 5.468 A) and partially broadened diffraction lines, indicating some loss of symmetry. This contraction is greater than that found at Ames for the transition from $UO_{2.00}$ to $UO_{2.30}$; nevertheless, it does not seem likely that the oxide observed by Jolibois and by Grønvold and Haraldsen was a new phase. According to above-described measurements, it probably was the oxygen-saturated form of the continuous, monophasic $UO_{1.75}$-$UO_{2.3}$ series.

Table 11.1—Densities of UO_x

	Density, g/cc			Mole vol., cc/mole		Oxygen volume,* cc/g atom
x	Hille-brand, 1893	Biltz and Müller, 1927	Calculated from x-ray data	Hille-brand, 1893	Biltz and Müller, 1927	
1.992	...	10.80	...	...	25.03	6.2
2.000	...	10.82	10.97	...	24.97	6.15
2.005	10.95	... ⎫	11.15	24.69 ⎡	...	...
2.015	10.88	... ⎭		24.85 ⎣	...	...
2.018	...	10.71 ⎫	11.22	... ⎡	25.25	6.2
2.022	10.82	10.82 ⎭		25.01 ⎣	25.01	6.1
2.028	10.86	...	...	24.92	...	...
2.051	10.83	...	...	25.02	...	...
2.078	...	10.72	...	...	25.31	6.1
2.098	...	10.94	...	...	24.89	5.7
2.132	10.89	...	...	25.00	...	...
2.152	...	10.62	...	...	25.58	6.0
2.162	...	11.08	...	...	24.62	5.5
2.193	...	10.87	...	...	25.14	5.7
2.200	...	10.87	...	...	25.15	5.7
2.201	10.98	...	...	24.90	...	...
2.207	...	10.94	...	...	25.00	5.6
2.262	...	10.90 ⎫	11.35	... ⎡	25.17	5.5
2.306	...	10.40 ⎭		... ⎣	26.44	5.9
2.318	10.47	...	...	26.29	...	...
2.333	...	9.754	...	...	28.24	6.65
2.349	...	10.34	...	...	26.67	5.95
2.417	10.15	...	...	27.28	...	...
2.451	...	9.469	...	...	29.30	6.8
2.454	...	9.38 ⎫	8.33†	... ⎡	29.58	6.9
2.537	...	8.773 ⎭		... ⎣	31.77	7.5
2.631	...	8.379	...	...	33.44	7.9
2.643	...	8.369	...	...	33.51	7.8
2.660	...	8.216	...	...	34.18	8.1
2.666	...	8.300	8.39	...	33.83	7.9
2.685	...	7.938	...	...	35.43	8.5
2.920	...	7.074	...	...	40.28	9.45
3.069	...	6.039	...	...	47.57	11.4

*Obtained by subtracting 12.7 cc as the volume of the uranium atom from the mole volume and dividing the remainder by x (Biltz and Müller, 1927).

†U_2O_5, according to Sec. 2.3.

1.3 Range $UO_{2.3}$ to $UO_{2.67}$. The Compound U_3O_8. In this region, too, there is a disagreement between the conclusions reached at Ames from crystallographic measurements and the tensimetric data of Biltz and Müller. The disagreement concerns the composition of the orthorhombic phase, which is present (together with the cubic UO_2 phase) when x in UO_x exceeds 2.30.

According to the Ames x-ray analyses (MP Ames 2) this phase has the composition $UO_{2.50}$. A pure homogeneous phase of this composition was obtained at Ames by heating together equal quantities of UO_2 and U_3O_8. The crystal structure of the new phase was found to be somewhat different from that of U_3O_8 although both are orthorhombic. It may therefore be considered a separate compound, diuranium pentoxide (U_2O_5). Single needle-shaped crystals exhibiting the x-ray pattern of U_2O_5 were obtained accidentally at Ames (accompanied by octahedral $UO_{2.3}$ crystals) during an experiment in which UO_2Cl_2 was decomposed at 900°C. Despite the difference in crystal structure, U_2O_5 can be converted into U_3O_8 by gradual uptake of oxygen without the formation of a new phase.

The existence of the homogeneous phase U_2O_5 also was asserted by several earlier investigators, particularly de Coninck (1901a, 1903), de Coninck and Camo (1901b), de Coninck and Raynaud (1912), Schwarz (1920), and Lydén (1939); but denied by others, e.g., Lebeau (1922a), Jolibois and Bossuet (1922), and, particularly, Biltz and Müller (1927b).

Biltz and Müller measured the pressures of oxygen over uranium oxides of various composition. Between $UO_{2.67}$ (or U_3O_8) and $UO_{2.20}$, decomposition pressures were established rapidly and reversibly and therefore appear to be true equilibrium pressures. Because of the rapid change in oxygen pressure with composition, reliable p vs. x curves can be obtained in this range only if the composition of the solid is determined exactly. This was done by Biltz and Müller, either by direct analysis or by calculation from the amount of oxygen in the gas phase. Earlier investigations of the decomposition pressure of uranium oxides, in which this point was neglected, are of little value, particularly those in which a relatively small amount of solid was used in a large volume of gas, as in the measurements of Schwarz (1920).

Figure 11.2 shows five decomposition isotherms (752 to 1160°C). It reveals a very rapid change of dissociation pressure with composition, particularly between $UO_{2.67}$ (or U_3O_8) and $UO_{2.62}$.

Wagner and Schottky (1931) attempted an interpretation of this decline on the basis of their theory of "ordered mixed phases." They attributed the change in decomposition pressure with decreasing oxygen content to an increased number of "holes" in the oxygen sublattice of U_3O_8. This theory leads to a linear relation between $1/\sqrt{p_{O_2}}$ and the deviation from stoichiometric composition; such a relation is actually found in the 900 and 984°C isotherms (but not in the 1160°C isotherm) in Fig. 11.4, in the region between $UO_{2.67}$ and $UO_{2.69}$.

The undoubtedly monophasic region between $UO_{2.67}$ and $UO_{2.62}$ is followed, according to Biltz and Müller, by a constant-pressure region which, according to the phase rule, must correspond to a diphasic system. This region extends from $UO_{2.61}$ to $UO_{2.30}$; no homogeneous phase with the composition $UO_{2.50}$ exists according to these measurements. The dissociation pressure begins to decline again with decreasing x only in the well-known monophasic region below $UO_{2.30}$.

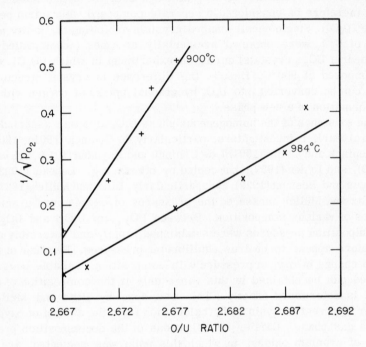

Fig. 11.4—Change of dissociation pressure with change in uranium-oxygen ratio [from C. Wagner and W. Schottky, Z. physik. Chem., (B)11: 200 (1931)].

This decline could be followed at 1160°C down to $UO_{2.20}$, where the pressure was 0.022 mm Hg, the lowest value measured by Biltz and Müller. Further decomposition, leading to the conversion of $UO_{2.20}$ to $UO_{2.00}$, requires temperatures much higher than 1160°C.

The decomposition pressures of the oxides $UO_{2.26}$, $UO_{2.46}$, and $UO_{2.61}$ were found by Biltz and Müller to be identical not only at 1160°C but also at several higher temperatures up to 1240°C. From the observed or extrapolated pressure values, Biltz and Müller constructed a uranium-oxygen isobar corresponding to an oxygen pressure of 10 mm Hg (Fig. 11.3). Under this pressure U_3O_8 is stable between 580°C and somewhere above 650°C. Between 750 and 1360°C, U_3O_8 loses oxygen gradually. The vertical segment of the isobar between $UO_{2.62}$ and $UO_{2.30}$ indicates a diphasic region. The last broken-line portion of the isobar shows the loss of oxygen by $UO_{2.30}$. As mentioned above, it could be followed only down to $UO_{2.20}$.

In pointing out the discrepancy between the tensimetric data of Biltz and Müller and the alleged existence of U_2O_5, Brewer (MP Berkeley 1) suggested that the Ames observation of a homogeneous orthorhombic U_2O_5 phase could perhaps be attributed to the fact (pointed out by Biltz and Müller) that the relatively simple cubic UO_2 diffraction pattern is easily obscured by the more complex orthorhombic pattern of U_3O_8. As to the assertion that single U_2O_5 crystals have been obtained at Ames, Brewer pointed out that no chemical analysis of these crystals has been made. If, however, the Ames observations prove to be correct, then either Biltz and Müller's pressure measurements must be discarded as erroneous, or the lower limit of the monophasic range must be assumed to decline with falling temperature from $UO_{2.62}$ at 1160°C to $UO_{2.50}$ at room temperature, which is unlikely.

The crystal structure attributed to U_2O_5 by the Ames observers is described in Sec. 2.3. It corresponds to a density of 8.33 g/cc. No such low density was observed in the region of $x = 2.5$ by Biltz and Müller (cf. Table 11.1), who found a practically linear decrease of ρ with x from $x = 2.3$ to $x = 2.92$.

Grønvold and Haraldsen (1948b) found that oxidation of UO_2 at 200 to 250°C leads to the formation of a tetragonal phase ("δ phase") which has a narrow range of homogeneity around $UO_{2.40}$. Its lattice constants are a = 5.37 A and c = 5.54 A, and its density (10.00 g/cc) indicates that it has the composition $U_{0.88}O_{2.12}$; i.e., it is derived from UO_2 by partial substitution of uranium atoms by oxygen atoms. This phase (whose composition is close to U_2O_5) is only stable below 270°C. This fact explains why it was not found by Biltz and Müller.

The same observers found that a product containing approximately 2.60 oxygen atoms per uranium atom is obtained by reacting U_3O_8 with $KHCO_3$ (a procedure which Lydén claimed leads to U_2O_5) or by reduction of U_3O_8 with hydrogen at 330°C. The structure of these products is similar to that of U_3O_8; these experiments were taken as indicating the extension of homogeneous U_3O_8 structure down to $UO_{2.56}$. (Biltz and Müller found that the homogeneous U_3O_8 phase extends, at higher temperatures, down to $UO_{2.62}$.)

The crystal structure of U_3O_8 is described in Sec. 2.4a. It is orthorhombic and corresponds to a density of 8.39 g/cc, which is in approximate agreement with the results of direct measurement shown in Table 11.1.

1.4 Range $UO_{2.67}$ to $UO_{3.0}$. Uranium Trioxide. According to ten-simetric data given by Biltz, the U_3O_8 phase can exist from $UO_{2.67}$ down to $UO_{2.62}$. If Ames x-ray observations are correct, its stability may perhaps extend to $UO_{2.50}$. There is no doubt that the same phase can exist also above $UO_{2.67}$. The density curve (Fig. 11.5) shows no

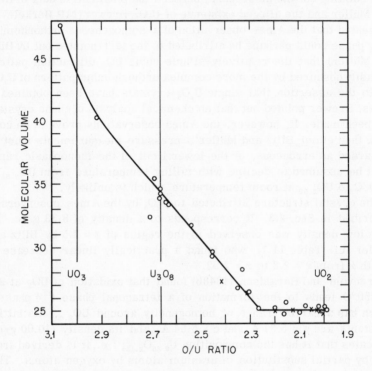

Fig. 11.5—Change in molecular volume as a function of uranium-oxygen ratio [from W. Biltz and H. Müller, Z. anorg. u. allgem. Chem., 163: 293 (1927)].

break at the composition $UO_{2.67}$ but continues to decrease linearly with increasing oxygen content up to $UO_{2.92}$. The rapid density drop between $UO_{2.92}$ and $UO_{3.00}$ may be due to the fact that the density value for UO_3 was obtained with amorphous trioxide.

According to Ames data (MP Ames 3), the x-ray diagrams confirm the gradual transition from U_3O_8 to UO_3. Zachariasen (MP Chicago 2) pointed out that the hexagonal uranium lattice found by him in one crystalline form of UO_3 can be obtained from the orthorhombic uranium lattice of U_3O_8 by a gradual change in the parameters of uranium atoms. This makes a continuous transition between the two oxides both possible and plausible.

For some unknown reason, however, the products of partial thermal decomposition of UO_3, obtained by Fried and Davidson (MP Chicago 3), did not show parameters intermediate between those of UO_3 and U_3O_8 (cf. Sec. 2.5b).

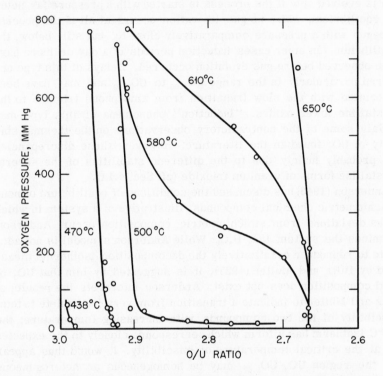

Fig. 11.6—Decomposition pressures of uranium-oxygen system in the range UO_3 to $UO_{2.660}$ [from W. Biltz and H. Müller, Z. anorg. u. allgem. Chem., 163: 266 (1927)].

Boullé and Dominé-Bergès (1948) found that, in the thermal deoxygenation of orange UO_3 to U_3O_8, 0.61 per cent by weight is lost at 520°C and the remaining 1.27 per cent is lost at 610°C. Sudden cooling from temperatures between these two limits indicated the existence of a crystalline phase with the composition $UO_{2.90}$; x-ray investigation indicated that this phase had the same structure as U_3O_8 and probably represents a solid solution.

Biltz and Müller (1927c) have also measured the oxygen pressures obtained by the decomposition of UO_3 to U_3O_8. The results are represented by six isothermals (438 to 650°C) in Fig. 11.6 and by the 10-mm isobar in Fig. 11.3.

In this region, as contrasted to that of oxide compositions below U_3O_8, the oxygen evolution is slow (perhaps in part because of comparatively low temperatures used), and complete equilibrium often cannot be reached even after several days of heating. Sometimes oxygen is evolved only if the process is started with a pressure far below the equilibrium, while no gas liberation occurs at all if the process is begun with a pressure comparatively close to, but still below, the equilibrium. In other cases induction periods of a day or more have been observed before gas evolution occurred. Delays of this type occurred particularly in the range $UO_{3.00}$ to $UO_{2.92}$ and may have been associated with the slow transition from amorphous trioxide to the crystalline lower oxides. "Induction" phenomena of this type may explain some of the contradictory observations on the thermal stability of UO_3 found in the literature. However, these discrepancies are probably mainly due to the different stabilities of the several crystalline forms of uranium trioxide (cf. Sec. 4.1a).

Anderson (1946) has discussed the conditions of equilibrium of nonstoichiometric chemical compounds. Illustrative of a system in which gross deviations from stoichiometric composition occur, Anderson examines the system UO_3-U_3O_8. While Anderson's model is not adequate to reproduce quantitatively the decomposition isotherms measured by Biltz and Müller (1927), it is suggested by him that UO_3 of ideal composition does not exist. Anderson interprets the results of Biltz and Müller to indicate a transition from very incomplete to total miscibility of the two compounds with increasing temperature; the 580°C isotherm has a form which corresponds closely to the expected one at the critical temperature of miscibility. It would thus appear that the region UO_3-$UO_{2.67}$ may be homogeneous or heterogeneous depending on the thermal history of the material.

Because of the slowness of deoxygenation, most pressure values used in the construction of isothermals in Fig. 11.6 were obtained by extrapolation to infinite time, and it may be questioned how closely they approach the true equilibrium pressures. No check by approaching the equilibrium from the oxygenation side could be obtained, since $UO_{2.70}$, for example, took up no oxygen at all in 24 hr under 72 mm Hg pressure at 400°C. (The usual method of determination of uranium as U_3O_8 is based on the assumption that this oxide takes up no additional oxygen when ignited and cooled in air.) However, the conversion of UO_3 to U_3O_8 is not thermodynamically irreversible. Finely dispersed U_3O_8, obtained by igniting uranyl oxalate, was found by Lebeau (1922a) to take up oxygen at 350°C, which changed its color within

12 hr from dark gray to orange-brown. This was confirmed by Biltz and Müller (1927b), who found that the product actually had a composition close to UO_3. Preparations of U_3O_8 ignited at 800°C were, on the other hand, "dead" burnt and took up no oxygen. Fried and Davidson (MP Chicago 3) found that ignited U_3O_8 can be converted to UO_3 by heating to 500°C under 28 atm of oxygen for one or several days. The color changed in this case from black to deep red.

Boullé and Dominé-Bergès (1949) confirmed Lebeau's observation of comparatively rapid reoxidation of U_3O_8, prepared by low-temperature ignition of uranyl oxalate (or by vacuum decomposition of UO_3), and the conversion of this easily oxidizable modification into a practically nonoxidizable microcrystalline modification, with a slightly different x-ray diffraction pattern.

Biltz and Müller saw another illustration of the reversibility of the reaction

$$3UO_3 \rightleftarrows U_3O_8 + \tfrac{1}{2}O_2 \tag{2}$$

in the existence in nature of yellow UO_3 minerals, such as gummite or rutherfordite, which they thought should be interpreted as oxidation products of U_3O_8 minerals (cf. Sec. 4.6a).

Even if the reversibility of the conversion of UO_3 to $U_3O_8 + O_2$ at the temperatures used is to be considered as certain, the question whether the oxygen pressures determined by Biltz and Müller actually were equilibrium decomposition pressures or not remains open. Biltz and Müller were satisfied that they were at least not far off the equilibrium because the heats of decomposition derived from the pressure isotherms agreed well with those determined by direct calorimetric measurements (cf. Sec. 4.1a).

Figure 11.6 shows no region of constant pressure between $UO_{2.67}$ and UO_3 at temperatures above 580°C. This clearly indicates a monophasic system. Below 580°C Brewer (MP Berkeley 1) interpreted the results given by Biltz as indicating the existence of a diphasic region extending from $UO_{2.67}$ to $UO_{2.92}$ and of a monophasic region with rapidly changing pressure between $UO_{2.92}$ and $UO_{3.00}$. He suggested that at the lower temperatures the trioxide used by Biltz and Müller was amorphous, and its composition could be changed continuously only down to $UO_{2.92}$, where a crystalline lower oxide phase began to form. At the higher temperatures (500°C and above) UO_3 might have become crystallized, and this would have made a continuous transition all the way from UO_3 to U_3O_8 possible.

1.5 Range above UO_3. Table 11.1 indicates that some oxygen in excess of the composition $UO_{3.00}$ can be taken up by uranium trioxide, but this uptake leads to a rapid "swelling" of the crystal structure, and the binding becomes very loose. The UO_x range above $x = 3.00$ is not very well-known. Attempts to obtain water-free uranium peroxide, UO_4, have failed. However, Brown University observers (Brown 1) found that when $UO_4 \cdot 2H_2O$ was heated to 130°C for 24 hr and UO_4 was about half decomposed to UO_3, an increase in temperature to 300°C led to the loss of all water without further loss of oxygen, leaving an anhydrous compound with an average composition close to $UO_{3.5}$. This compound is a per-compound different in properties from the peroxide $UO_4 \cdot 2H_2O$; in contact with water or acid it decomposes with evolution of oxygen and formation of hydrated UO_3. An anhydrous uranium per-compound also was obtained at Brown University (Brown 2) by calcination of $(NH_4)_2U_2O_7$ in a rapid stream of oxygen. At 250 to 350°C ammonia and water were evolved. The temperature was then raised to 550°C, and a red powder, which appeared to be stable at high temperatures, was produced. It had the composition $UO_{3.14}$ to $UO_{3.38}$ and liberated oxygen in contact with water.

2. PHYSICAL PROPERTIES OF ANHYDROUS URANIUM OXIDES

2.1 <u>UO</u>. The properties of uranium monoxide are not well-known. It has been described (MP Berkeley 1) as gray and brittle and having a metallic luster. Table 11.2 contains the lattice constants found by different observers.

The lattice constant of UO is not known as exactly as the last figure in Table 11.2 seems to indicate because the sample used may possibly have contained some carbide or nitride.

Differentiation between rock salt and zinc blende structures is not easy when the two components are as different in mass as uranium and oxygen; however, the available evidence favors the rock salt structure (MP Ames 3).

According to Brewer (MP Berkeley 1) UO is more volatile than UO_2. He estimated on the basis of general analogies that its vapor pressure may be of the order of 10^{-5} atm at 2000°K.

2.2 <u>UO_2</u>. (a) <u>Crystal Structure</u>. Table 11.3 gives a summary of the crystal structure analyses of the fluorite-type UO_x phases with x from 1.75 to 2.3. According to Sec. 1.2, $UO_{1.75}$ may be either a separate compound, U_4O_7, or the lower end of a continuous series of solid solutions extending upward to $UO_{2.3}$. If U_4O_7 does exist as a stable compound (in respect to dismutation to UO_2 and UO), its heat of formation must be larger than the sum of the heats of formation of $3UO_2$ and UO, i.e., in excess of 925 ± 30 kcal per mole (132 ± 4 kcal per oxygen atom).

The crystal habitus of UO_2 has been described as follows:

Arfvedson (1824a)	Microscopic regular octahedra
Péligot (1842a)	Metallic lustrous scales
Hillebrand (1893b)	Fine strongly reflecting octahedra
Aloy (1900)	Rhombic tablets, usually rounded on four corners
Hofmann (1915)	Brilliant cubes

Table 11.2 — Lattice Constants of UO

Composition	Lattice constant, A	Density, g/cc	Reference
UO + U	4.91 ± 0.01 4.91 4.93	14.2 ± 0.1 14.2 14.0	MP Ames 3 British 1 Battelle 1
UO + UO_2	4.930 ± 0.001	14.0 ± 0.01	MP Ames 10

Table 11.3 — Crystal Lattice of UO_2

Preparation	Lattice constant, A	Density, g/cc	Reference
?	5.47	...	Goldschmidt, 1923a
?	5.48	...	Van Arkel, 1924
"Very pure sample"	5.4568 ± 0.0005	10.96 ± 0.01	MP Chicago 4
UO + UO_2	5.461 ± 0.001	...	MP Ames 2,4
$UO_{2.00}$	5.4586 ± 0.008	10.97	MP Ames 2,4
$UO_{2.10}$	5.437 ± 0.001	11.18	MP Ames 2,4
$UO_{2.20}$	5.433 ± 0.001	11.27	MP Ames 2,4
$UO_{2.30}$ (with U_2O_5)	5.4297 ± 0.0008	11.36	MP Ames 2,4
UO_2 + 5% U_3O_8	5.464 ± 0.002	12.096	SAM Columbia 1
$UO_{2.00}$	5.460	...	MP Chicago 1
$UO_{1.75}$	5.477	...	MP Chicago 1
$UO_{2.0}$	5.468	10.80	Grønvold and Haraldsen, 1948b
$UO_{2.34}$	5.40	11.05	Grønvold and Haraldsen, 1948b
UO_2	5.4581 ± 0.0005	...	Rundle et al., 1948

The crystals of UO_2 are isomorphous with those of ThO_2. The two oxides form mixed crystals (Hillebrand, 1893b).

(b) <u>Density</u>. In addition to the systematic data by Hillebrand and Biltz and Müller given in Table 11.1, the determinations of UO_2 density given in Table 11.4 can be quoted.

The difference in density between degassed and nondegassed powder is related to the capacity of UO_2 to take up gases (studied by Schmidt, 1928). Density can be increased, e.g., from 10.4 to 10.6 g/cc, by prolonged heating to 130°C (Biltz and Müller, 1927d).

(c) <u>Hardness</u>. The scratch hardness of UO_2 is 3.5 on Mohs' scale. It depends on previous history of the oxide (SAM Columbia 4).

Table 11.4—Density of UO_2

Material	Density, g/cc	Packing density, g/cc	Reference
UO_2 from oxalate	10.15	...	Ebelmen, 1842a
Cf. Table 11.1	10.95−11.0	...	Hillebrand, 1893a
Amorphous product from oxalate + H_2	8.2	...	Raynaud, 1912
Average for $UO_{2.00}$ to $UO_{2.26}$ (cf. Table 11.1)	10.8	...	Biltz and Müller, 1927d
Mallinckrodt powder, not tapped	...	3.76	UCRL 1
Mallinckrodt powder, tapped	...	4.51	UCRL 1
	10.28	4.96	SAM Columbia 2
	10.9−11.1	...	MP Ames 5
Melted, sintered at 2200°C	10	...	MP Ames 5
Sintered at 1400 to 2000°C	8.11	...	MP Ames 6
Mallinckrodt product, 98.5 per cent UO_2; 44 to 53 μ, degassed	10.37	...	SAM Columbia 3
Same, not degassed	9.30	...	SAM Columbia 3
1 to 2 μ, degassed	10.47	...	SAM Columbia 3

Table 11.5—Vapor Pressure of UO_2 Determined by Effusion

Temp., °K	Temp., °C	UO_2 evaporated,* g × 10^8	Evaporation time, min	p, mm Hg × 10^3
1873	1600	45	180	0.071
2023	1750	126	23	1.7
2073	1800	825	60	4.0
2173	1900	875	15	18
2273	2000	2,300	10	72

*Hole 0.058 cm in diameter.

(d) <u>Melting Point</u>. Ruff and Goecke (1911) found the melting point of UO_2 to be 2176°C (under nitrogen), while Friederich and Sittig (1925) gave 2500 to 2600°C.

(e) <u>Vapor Pressure</u>. Vapor pressure measurements of UO_2 were made at Chicago (MP Chicago 5) by the effusion method, using α counting as the analytical method. The results are shown in Table 11.5.

The data in Table 11.5 permit an estimate that the heat of vaporization of UO_2 must be about 137 kcal per mole. Brewer gave, in his

Table 11.6—Specific Heat of UO_2 at Low Temperatures* (U. S. Bur. Mines 1)

Temp., °K	C_p, cal/mole/°C	Temp., °K	C_p, cal/mole/°C	Temp., °K	C_p, cal/mole/°C
15	0.378	31	2.280	170	11.10
16	0.473	32	2.261	180	11.57
17	0.574	35	2.425	190	12.04
18	0.687	40	2.637	200	12.47
19	0.815	50	3.365	210	12.86
20	0.968	60	4.103	220	13.23
21	1.148	70	4.841	230	13.56
22	1.358	80	5.563	240	13.87
23	1.599	90	6.298	250	14.17
24	1.900	100	6.958	260	14.44
25	2.270	110	7.619	270	14.69
26	2.715	120	8.276	280	14.94
27	3.478	130	8.923	290	15.16
28	6.230†	140	9.518	300	15.38
29	8.645†	150	10.07		
30	5.150†	160	10.60		

*Mallinckrodt material, containing 0.7 per cent UO_3.
†This value is uncertain because of the rapid change of heat capacity in this region.

compilation of thermodynamic data (MP Berkeley 1), an estimated value for the vapor pressure of UO_2 (p_{O_2} = 10^{-8} atm at 2000°K) which is much lower than the empirical values in Table 11.5.

Oxides with an average composition $UO_{>2.20}$ have a measurable rate of sublimation at temperatures much lower than those used in the experiments summarized in Table 11.5. According to Biltz and Müller (1927a), brownish sublimates, which were obtained when $UO_{2.30}$, $UO_{2.46}$, and $UO_{2.67}$ were heated to 1300°C, had compositions between $UO_{2.16}$ and $UO_{2.17}$, while the residues had compositions between $UO_{2.16}$ and $UO_{2.21}$.

Although the vapor pressure of the oxide probably increases with increasing oxygen content, the decomposition pressure of oxygen increases even more rapidly, and high oxygen pressure above the oxide interferes with its sublimation. This may explain the reason that (with pumping velocities available to Biltz and Müller) no sublimate was obtained when the composition of the solid was $UO_{>2.6}$. The mechanism of sublimation can perhaps be visualized as follows: In the decomposition of U_3O_8, molecules of a higher and relatively volatile oxide (e.g., UO_3) are formed. In a sufficiently high vacuum these molecules have a chance to evaporate. They decompose, in contact with a cold surface, into oxygen and $UO_{2.15}$ (this is the phase in equilibrium with low-pressure oxygen at moderately high temperatures).

Table 11.7 — Heat Content and Entropy of UO_2 above 298.16°K*

Temp., °K	$H_T - H_{298.16}$, cal/mole	$S_T - S_{298.16}$, cal/mole/°C
400	1,680	4.82
500	3,470	8.82
600	5,340	12.24
700	7,280	15.22
800	9,250	17.84
900	11,250	20.21
1000	13,280	22.35
1100	15,340	24.30
1200	17,420	26.11
1300	19,510	27.79
1400	21,620	29.36
1500	23,750	30.83

*Moore and Kelley (1947).

(f) Specific Heat and Entropy. Regnault (1840) gave 0.619×10^{-3} cal per gram (corresponding to 16.7 cal per mole) as the average specific heat of UO_2 between 0 and 100°C. The first detailed studies of the thermodynamic constants of uranium dioxide were carried out 100 years later at the Pacific Branch of the U. S. Bureau of Mines at Berkeley (U. S. Bur. Mines 1,2; Moore and Kelley, 1947).

The results are represented in Tables 11.6 and 11.7.

The specific-heat curve of UO_2 has a peak between 15 and 50°K (cf. Fig. 11.7) with a maximum value of 9 cal/mole/°C at 28.6°K. The "extra" entropy corresponding to this peak was estimated as 0.87 e.u.; it can be attributed to a magnetic transformation (cf. Sec. 2.2i).

The following equation was derived by Moore and Kelley (1947) to represent the heat content of UO_2 in the range 300 to 1500°K to ±0.1 per cent:

$$H_T = H_{298.16} + 19.20T + 0.81 \times 10^{-3}T^2 + 3.957 \times 10^5 T^{-1} - 7124 \quad (3)$$

By differentiation the heat capacity given in Eq. 4 is obtained.

$$C_p = 19.20 + 1.62 \times 10^{-3}T - 3.957 \times 10^5 T^{-2} \qquad (4)$$

MacWood and Altman (UCRL 2) gave Eq. 5 as an interpretation of the data of Moore and Long (U. S. Bur. Mines 2).

$$C_p = 18.45 + 2.431 \times 10^{-3}T - 2.272 \times 10^5 T^{-2} \qquad (5)$$

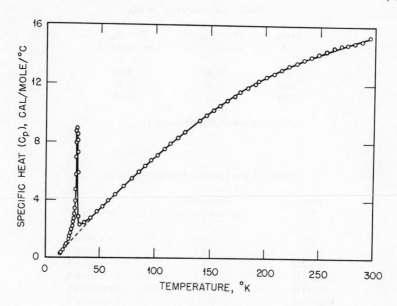

Fig. 11.7—Heat capacity of UO_2.

The entropy at 298.1°K was estimated at the U. S. Bureau of Mines (Pacific Branch) (U. S. Bur. Mines 1) by graphical integration (using, below 15°K, the Debye function with $\theta = 160°K$) as

$$S_{298.1°K} = 18.63 \pm 0.1 \text{ e.u.}$$

Brewer (MP Berkeley 1) calculated the additional entropy values shown in Table 11.8.

(g) Thermal Conductivity. The values shown in Table 11.9 were found at Princeton (Princeton 1,2,3) for UO_2 powder.

(h) Electrical Properties. Similar to all dark-colored oxides, UO_2 is a semiconductor. Its conductivity depends strongly on exact composition, purity, and degree of agglomeration (apparent density). The lowest conductivities were observed by Friederich and Sittig (1925).

At room temperature they found, using a rod made of brown UO_2 sintered in hydrogen at 1100°C, a specific conductivity of only 4×10^{-8} (ohm-cm)$^{-1}$, and a specific conductivity of about 3×10^{-5} (ohm-cm)$^{-1}$ was found with a rod made of "dark-blue" UO_2 (cf. Sec. 2.2j) sintered in nitrogen (also at 1100°C). These results indicate that conductivity

Table 11.8 — Entropy* of UO_2

Temp., °K	$(F_T - H_{298})/T$, cal/mole/°C	$S_T - S_{298}$, cal/mole/°C
500	20.48	8.82
1000	27.67	22.35
1500	33.6	30.83

*After Brewer (MP Berkeley 1).

Table 11.9 — Thermal Conductivity of UO_2

Temp., °C	Conductivity, 10^4 cal/cm sec °C	Reference
50	3.5	Princeton 1
100	3.4	Princeton 1
20 − 225	3.4	Princeton 2
270 − 610	1.9	Princeton 2
18 − 160	3.3*	Princeton 2
Unknown	50 ± 5†	Princeton 3

*Powder pressed to $\rho = 6$.
†Sintered.

decreases with progressive approach to the composition $UO_{2.00}$. It may be questioned whether it was the presence of "excess" oxygen or some other condition (such as a more compact form of the oxide or better electrical contact) that was mainly responsible for the considerably higher conductivities observed by LeBlanc and Sachse (1930), Hartmann (1936), Meyer (1933), and Ames observers (MP Ames 6). The results of these authors are collected in Table 11.10. The values range from 2.4×10^{-4} to as much as 0.1 (ohm-cm)$^{-1}$ at room temperature. Meyer obtained confirmation that uptake of oxygen in excess of the formula $UO_{2.00}$ (by heating a UO_2 sample previously ignited in high vacuum to approximately 500°C in 1 to 100 mm of O_2) increases the conductivity from 3×10^{-4} or 3×10^{-3} (ohm-cm)$^{-1}$ before the treatment to about 1.2×10^{-2} (ohm-cm)$^{-1}$ after it.

Meyer (1933) gave for the electrical resistivity of UO_2

$$R = 0.1804e^{1867/T} \tag{6}$$

and found that the oxygen treatment leaves the factor before the exponential practically unchanged but decreases the "activation energy" by 20 to 30 per cent.

Table 11.10 — Electrical Conductivity of UO_2

Temp., °C	Conductivity, 10^3 (ohm-cm)$^{-1}$	Temp., °C	Conductivity, 10^3 (ohm-cm)$^{-1}$
Hartmann, 1936		MP Ames 6	
−60	2.09	28	46.7;* 61;†
−46	3.16		95;‡ 114§
−25	5.25	110	80.6
−15	7.08	157	118
12	11.7	295	173
20	12.7	478	247
		793	403
		992	1,282
LeBlanc and Sachse, 1930		Meyer, 1933	
20	0.24	22	9.3
50	0.4	40	14.0
100	1.0	60	20.5
150	5.0	80	29.5
200	10	96	37.2
250	15	131	56.8
300	25	155	73.0
350	35	182	94
400	40	223	126
450	55	270	171
500	70	327	235

*Pressed and sintered at 1900 to 2000°C, ρ (apparent) = 8.11 g/cc.
†ρ (apparent) = 7.78 g/cc.
‡ρ (apparent) = 8.11 g/cc.
§ρ (apparent) = 9.17 g/cc.

Hartmann (1936) pointed out that the increase of conductivity with the addition of extra oxygen puts UO_2 into the class of "hole conductors," in which the conductivity is due to the fact that some metal ions are missing from the lattice, their charge being taken over by other metal ions that temporarily assume a higher valence state, e.g., U^{+6}. He confirmed this hypothesis by measurements of the Hall effect in UO_2 (the sign of this effect makes it possible to distinguish hole conductors from "excess-electron" conductors).

The hypothesis that the oxides $UO_{>2.00}$ (up to $UO_{2.30}$) may be deficient in uranium, rather than containing extra oxygen, was discussed in Sec. 1.2 from the point of view of x-ray observations, and it was pointed out that density values do not permit the formulation of $UO_{2.30}$ as $U_{0.85}O_2$. However, it may be that a small deficit of uranium exists (independently of the presence of extra oxygen) and that this deficit is the principal cause of increased electrical conductivity. Hartmann estimated from the Hall effect that his "oxygenated" UO_2 preparations contained about 3×10^{18} "disturbance centers." This means that only about 1 in 10^5 uranium atoms was missing, a fraction that is not detectable by density measurements.

Amrein (1942) studied several oxides whose resistance decreases with temperature in such a way that, above a certain current density, the potential decreases with increasing current ("negative characteristic"). Results of this type were obtained with sintered UO_2 (density 9.2 g/cc) pressed into small rods in vacuum.

Prigent (1949) has studied in some detail the electrical characteristics of UO_2 from the point of view of its utility as a thermistor element. Thermistors are used as detectors of infrared radiation and ultrahigh-frequency waves; they are oxide semiconductors which have large positive temperature coefficients of resistance. The earliest thermistors were made of uranium oxide. Prigent studied the effects of pressure, tension, and temperature on the resistance of crystalline and amorphous UO_2 prepared in a variety of ways. The measurements were difficult to reproduce, and Prigent concluded, in agreement with earlier investigators, that uranium oxides were unsatisfactory as thermistors. There is reason to believe, however, that insufficient attention was paid in this work to the great variability in composition which may readily occur in uranium oxides even under what are thought to be identical experimental conditions.

Wahlin (1932) investigated the ion emission of a UO_2-coated tungsten wire. The ions emitted included U^+ and probably also UO^+. The photoelectric effect of UO_2 was investigated by Pochettino (1932).

(i) Magnetic Properties. The magnetic data on UO_2 are listed in Table 11.11. Figure 11.8 gives the results of Haraldsen and Bakken (1940). The χ-vs.-T curve can be represented by the Curie-Weiss equation $\chi = C/(T - \theta)$, with $\theta = -310°K$ (according to Sucksmith, 1932) and $\theta = -180°K$ (according to Haraldsen and Bakken, 1940).

The value of the constant C found by Sucksmith corresponds to a magnetic moment of $n_B = 4.4$ Bohr magnetons (calculated by Haraldsen and Bakken), while the value given by Haraldsen and Bakken (1940) indicates a moment of n_B = 2.92 magnetons. The latter is very close to the theoretical value for pure spin magnetism of the uranium(IV) ion (n_B = 2.83). It will be noted that Sucksmith used in his calculation of n_B for U^{+4} only the values obtained with $U(SO_4)_2$ and with UCl_4 at low temperatures, which give n_B = 3.2 and 2.8, respectively.

(j) <u>Color</u>. The color of uranium dioxide varies from brown to black. According to Lebeau (1922b) pure crystalline UO_2 is a brown

Table 11.11 — Magnetic Susceptibility of UO_2

Observer	Temp., °C	Specific susceptibility, χ (sp), cgs units $\times 10^6$	Volume susceptibility, χ (vol), cgs units $\times 10^6$	Molar susceptibility,* χ (mol), cgs units $\times 10^3$
Wedekind, 1915	17	7.51	...	...
Meyer, 1899	16	...	2.46	...
Sucksmith, 1932	−183	...	...	5.96
	−75	...	...	4.70
	17	...	...	3.96
	97	...	...	3.51
	157	...	...	3.21
	226	...	...	2.75
	275	...	...	2.43

*Corrected for diamagnetism.

powder, independent of the method of preparation. Biltz and Müller (1927e) obtained, by reduction with hydrogen, brown UO_2 from green U_3O_8, and dark-brown UO_2, with a slight violet tinge, from black U_3O_8. Solid solutions of the type $UO_{2.10}$ to $UO_{2.30}$, obtained by the decomposition in vacuum of U_3O_8, are black. Jolibois (1947) described the oxide U_3O_7 (or $UO_{2.33}$) as bluish black. De Coninck (1902, 1904a) described a "brick-red" modification of UO_2 obtained by decomposition of UO_2Br_2 (cf. Sec. 5.3a), but it is not quite certain that it was not a higher oxide (UO_3?). Friederich and Sittig (1925) obtained a "blue" dioxide by heating U_3O_8 to 1100°C in a stream of nitrogen. From what is known of the decomposition pressure of uranium oxides (Sec. 4.1a) this preparation certainly was not $UO_{2.0}$ but rather $UO_{2.15}$; it might also have contained nitrogen.

2.3 $\underline{U_2O_5}$. (Rundle, Baenziger, Wilson, and McDonald, 1948.) It was mentioned in Sec. 1.3 that Ames x-ray data indicate the existence of the oxide U_2O_5 but that Biltz's tensimetric data, obtained at 1160°C, do not confirm it. If a homogeneous phase U_2O_5 actually does exist at room temperature, it could represent the lower limit of the solid-solution range extending upward to and beyond U_3O_8. The diffraction pattern of U_2O_5 (as observed in a product obtained by heating together equal amounts of U_3O_8 and UO_2) is, according to Ames observers

(MP Ames 7), similar to but not identical with that of U_3O_8. It can be interpreted by means of an orthorhombic unit with the parameters a = 4.135 A, b = 3.956 A, and c = 6.72 A. X-ray examination of a U_2O_5 single crystal (needles obtained accidentally together with octahedral crystals of $UO_{2.30}$ by thermal decomposition of UO_2Cl_2 at 900°C)

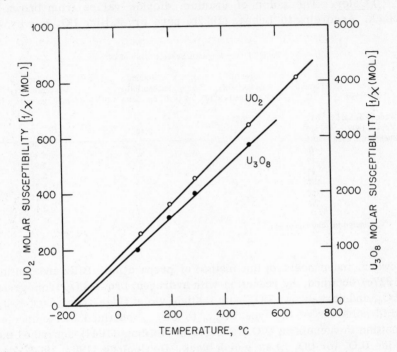

Fig. 11.8 — Magnetic susceptibility of uranium dioxide and triuranium octaoxide [from H. Haraldsen and R. Bakken, Naturwissenschaften, 28: 127 (1940)].

showed that the above dimensions refer to a pseudo unit. In the complete unit the a parameter must be doubled, and the b parameter increased eightfold. The dimensions of the true U_2O_5 unit cell therefore are

$$a = 8.27 \pm 0.02 \text{ A} \qquad b = 31.65 \pm 0.1 \text{ A} \qquad c = 6.72 \pm 0.02 \text{ A}$$

This contains 16 U_2O_5 units per cell. The calculated x-ray density of U_2O_5 is 8.35 g/cc. It was mentioned above that no such low density appears in the density measurements collected in Table 11.1.

A homogeneous phase with composition close to but not identical with U_2O_5 also was observed by Grønvold and Haraldsen (1948b), who obtained it by oxidation of UO_2 with oxygen at 200 to 250°C. This phase (which they called δ phase) is tetragonal, with a = 5.37 A and c = 5.54 A (c : a = 1.03), and has an empirical density of 10.0 g/cc, indicating that its composition is $U_{0.88}O_{1.12}$, i.e., it corresponds to partial substitution of uranium atoms by oxygen atoms and consequent deformation of the cubic UO_2 lattice. According to these observers the homogeneous orthorhombic U_3O_8 phase can be obtained at low temperatures with compositions down to $UO_{2.56}$ (by reduction of U_3O_8 with hydrogen at 330°C).

Table 11.12—Crystal Structure of U_3O_8

Preparation	a_0, A	b_0, A*	c_0, A	ρ, g/cc
$UO_4 \cdot 2H_2O$, heated 40 hr at 700°C	6.7023 ± 0.0005	3.9803 ± 0.0004	4.1385 ± 0.0007	8.39
$UF_6 + H_2O \rightarrow$ ppt, dried 40 hr at 700°C	6.7030 ± 0.0006	3.9812 ± 0.0004	4.1407 ± 0.0006	8.39
$UF_6 + HNO_3 \cdot aq \rightarrow$ ppt, dried 40 hr at 700°C	6.7036 ± 0.0007	3.9815 ± 0.0004	4.1406 ± 0.0005	8.39

*One-third of true value.

It thus seems that, at low temperatures, an additional homogeneous tetragonal phase $U_{0.88}O_{1.12}$ is interpolated between the cubic UO_2 phase (with composition up to $UO_{2.34}$) and the orthorhombic U_3O_8 phase (with composition down to $UO_{2.56}$).

2.4 $\underline{U_3O_8}$. (a) Crystal Structure. The crystal lattice of the oxide U_3O_8 was described by Zachariasen (MP Chicago 6) as containing 2 uranium atoms and $5\frac{1}{3}$ oxygen atoms in an orthorhombic unit, with the dimensions

a = 6.70 ± 0.01 A   b = 3.98 ± 0.01 A   c = 4.14 ± 0.01 A

These values are very close to the dimensions given above for the pseudo unit of U_2O_5. According to Ames data (MP Ames 2,3), the b parameter of U_3O_8 also must be trebled to obtain the correct unit dimensions. Consequently the true unit cell of U_3O_8 contains 6 atoms of uranium and 16 atoms of oxygen and has the dimensions

a = 6.70 ± 0.01 A   b = 11.94 ± 0.03 A   c = 4.14 ± 0.01 A

Table 11.12 shows the results of precision measurements made at Columbia University (SAM Columbia 1). All three measurements agree within ±0.05 per cent. The six uranium atom positions in the

U_3O_8 unit cell are $(0\ 0\ 0)$ $(\frac{1}{2}\ \frac{1}{2}\ 0)$ $(0\ 0\ 0)$ $(0\ \frac{1}{3}\ 0)$ $(0\ -\frac{1}{3}\ 0)$. The probable oxygen positions are

4 O_I atoms in $(0\ 0\ 0)$ $(\frac{1}{2}\ \frac{1}{2}\ 0)$ $(0\ \frac{1}{3}\ \frac{1}{2})$

12 O_{II} atoms in $(0\ 0\ 0)$ $(\frac{1}{2}\ \frac{1}{2}\ 0)$ $(\frac{1}{3}\ 0\ x)$ $(\frac{1}{3}\ \frac{1}{3}\ x)$ $(\frac{1}{3}\ \frac{2}{3}x)$ x = 0.17

The crystal structure of U_3O_8 was investigated independently by Grønvold (1948a). He found himself unable to prepare single U_3O_8 crystals but succeeded in obtaining from one U_3O_8 preparation an

Table 11.13 — Density of U_3O_8

Preparation	ρ, g/cc	Reference
...	7.193	Wedekind and Horst, 1915
...	7.31	Ebelmen, 1842a
From $UO_4 \cdot 2H_2O$	8.30	Biltz and Müller, 1927d (cf. Table 11.1)
Mallinckrodt batch D, particle size 44 to 74 μ, degassed	7.60	SAM Columbia 3
Same, not degassed	6.97	SAM Columbia 3
Same, particle size ½ μ, degassed	8.13	SAM Columbia 3
...	8.34	Grønvold, 1948a

x-ray diffraction diagram indicating the presence of an orthorhombic phase with the dimensions

$$a = 6.703\ A \qquad b = 3.969\ A \qquad c = 4.136\ A$$

Grønvold noted that the density calculated from these cell dimensions does not agree with the empirical density (8.34 g/cc; cf. Table 11.13) if the cell is assumed to contain one molecule of U_3O_8; neither can an agreement be obtained by assuming that the cell contains two molecules of UO_3 with extra interstitial uranium ($U_{1.12}O_3$), or two molecules of UO_3 with oxygen partially replaced by uranium ($U_{1.09}O_{2.91}$), or two similarly modified molecules of UO_2 ($U_{0.82}O_{2.18}$). The only way Grønvold saw to obtain a correct density was to assume the presence, in each cell, of two UO_3 molecules with partially missing oxygen, $UO_{2.67}$, or of two UO_2 molecules with extra interstitial oxygen.

According to Boullé and Dominé-Bergès (1949) the x-ray diffraction pattern of the comparatively easily oxidizable preparations of U_3O_8, obtained at low temperatures (lower than 350°C) from oxalate or trioxide, is somewhat different from that of the (practically non-oxidizable) microcrystalline product obtained at higher temperatures.

(b) Density. The calculated x-ray density of U_3O_8 is 8.39 g/cc. The measured values are somewhat smaller (cf. Table 11.13).

(c) Hardness. According to Columbia University observers (SAM Columbia 4) U_3O_8 has a hardness of 3.5 on Mohs' scale, depending somewhat on the history of the sample.

(d) Thermodynamic Properties. Values that have been determined for specific heat are given in Table 11.14. In a summary report from the University of California Radiation Laboratory (UCRL 2) it is stated that the heat capacity of U_3O_8 is known from 213 to 333°K, but no data or reference is given. This paper gives Eq. 7, said to be

Table 11.14—Specific Heat of U_3O_8

| Temp., °K | C_p, cal/°C | | Reference |
	Per gram	Per mole	
273–373	0.0798	67.2	Donath, 1879
82–195	0.0429	36.2	Russell, 1912
196–250	0.0616	52.0	Russell, 1912
276–314	0.0710	59.8	Russell, 1912
25–100	0.0750	63.3	Natl. Bur. Standards 4

derived from experimental data for U_3O_8 at low temperatures and from interpolated data at the higher temperatures, the interpolation being from the values for UO_2 and UO_3:

$$C_p = 62.6 + 6.6 \times 10^{-3}T - 2.5 \times 10^5 T^{-2} \quad \text{cal/mole/°C} \qquad (7)$$

The entropy of U_3O_8 was first estimated by Kelley (1932) as 72.7 e.u. (at 298°K). Davidson (MP Chicago 7) gave $S_{298} = 66$ e.u. and also estimated values for $S_T - S_{298}$ and $H_T - H_{298}$ for 500, 800, 1100, and 1300°K. Brewer (MP Berkeley 1) interpolated values for 1000 and 1500°K and also gave estimates for the function $(F_T - H_{298})/T$. These estimates are summarized in Table 11.15.

(e) Thermoelectric Power. Bidwell (1914) measured the emf of U_3O_8 against platinum with the cold junction at 165 to 1155°C and the hot junction at 320 to 1265°C. The emf was positive (i.e., the current flowed from platinum to oxide through the hot junction) up to about 700°C (temperature of hot junction) but was negative above this temperature.

(f) Electrical Properties. (1) Dielectric Constant. Keller and Lehmann (1934) found C = 41.77 for U_3O_8 powder preheated 1 hr to 200°C and pressed between two plates (10,000 atm).

(2) Electrical Conductivity. Like UO_2, the oxide U_3O_8 is a semi-conductor. However, according to LeBlanc and Sachse (1930), it differs from UO_2 by the occurrence of polarization, which indicates that the conductance is not purely electronic in nature. This complication, added to the fact that the U_3O_8 preparations used by different observers probably have varied considerably in their oxygen content, gives some explanation of the fact that the conductivities found by different observers differed by as much as a factor of 10^6. As in the case of

Table 11.15 — Estimated Values of Thermodynamic Functions for U_3O_8

Temp., °K	$H_T - H_{298}$, kcal/mole	$S_T - S_{298}$, cal/mole/°C	$(F_T - H_{298})/T$, cal/mole/°C
298	0	0	66*
500	12	28	70
1000	42	72	96
1500	81	105	117

*Equal to S_{298}.

UO_2, the lowest conductivities were observed by Friederich and Sittig (1925), who found that a U_3O_8 rod, sintered in oxygen at 1000°C, had a specific resistance as high as 40×10^6 ohm-cm at room temperature. At 300°C the resistance was ten times smaller. LeBlanc and Sachse (1930) found much better conductivity in a sample obtained by oxidation of UO_2 in air at 500°C. Fischer (1913) observed a still lower specific resistance (only about 2,700 ohm-cm at room temperature) in hydrated U_3O_8, obtained by electrolytic deposition and pressing. Even higher conductivity was observed by Wiegand (1924) in a rod sintered at 1200°C. The results of LeBlanc and Sachse and of Wiegand are shown in Table 11.16.

(g) Magnetic Properties. U_3O_8 is paramagnetic. Early measurements of its magnetic susceptibility gave the values listed in Table 11.17 (uncorrected for diamagnetism).

The results obtained by Haraldsen and Bakken (1940) are shown in Fig. 11.8, which shows that the paramagnetic susceptibility (corrected for diamagnetism) follows the Curie-Weiss law, $\chi = C/(T - \theta)$, with a characteristic temperature $\theta = -170°K$. The molar magnetic moment, calculated from the constant C, is $n_B = 1.39$ Bohr magnetons. This is very close to the theoretical value, $n_B = 1.42$, for $UO_3 \cdot U_2O_5$, assuming spin magnetism of the ground state ^2F of the ion U^{+5} to be the only source of paramagnetism. In other words, U_3O_8 behaves in the magnetic field as if it contained sexivalent and quinquevalent rather than quadrivalent uranium. (For a discussion of this conclusion see the introduction of this chapter.)

(h) <u>Optical Properties</u>. <u>Color</u>. U_3O_8 can be from olive green to black-green or black. However, even black preparations give a green streak on porcelain. The color depends not so much on the exact composition as on the conditions of preparation (Zimmermann, Alibegoff, and Krüss, 1886a). The higher the temperature of ignition,

Table 11.16 — Electric Conductivity of U_3O_8

Temp., °C	Conductivity, 10^4 (ohm-cm)$^{-1}$		Temp., °C	Conductivity, (ohm-cm)$^{-1}$	
	LeBlanc and Sachse, 1930	Wiegand, 1924		LeBlanc and Sachse, 1930	Wiegand, 1924
20	0.001	...	400	0.0007	1.05
50	0.0025	500	450	0.0014	...
100	0.0090	750	500	0.0024	1.88
150	0.030	...	600	...	2.95
200	0.15	2,500	700	...	4.08
250	0.50	...	800	...	5.60
300	1.3	5,000	900	...	7.86
350	3	...	950	...	9.35

Table 11.17 — Paramagnetism of U_3O_8

Observer	Specific susceptibility at 15°C, cgs	Volume susceptibility at 16°C, cgs
Wedekind and Horst, 1915	0.95×10^{-6}	...
Meyer, 1899	...	0.351×10^{-6}

the darker the sample. For example, U_3O_8 prepared by Biltz and Müller (1927a) from $UO_4 \cdot 2H_2O$ below 800°C was moss green, but the oxide prepared at 900 to 1000°C was almost black. The emission spectrum of U_3O_8(?) in the oxyhydrogen flame was observed by Huggins (1870), who found it to be the normal continuous spectrum of hot solids (cf. below). The infrared emission spectrum was observed by Coblentz (1908), who noted weak maxima at 2.8 and 3.4 μ.

The emissivity of "U_3O_8" was first investigated by Burgess and Waltenburg (1914, 1915). (Because of the instability of U_3O_8 at the temperatures used, the results of these experiments relate to oxides of unknown composition but certainly $UO_{<2.67}$.) They found that the emissivity coefficient at 650 mμ was 0.30 for the solid oxide at 1650°C and 0.31 for the liquid(?) at 1700°C. Wiegand (1924) found a total emissivity of 77 to 79 per cent of that of a black body. This ratio remained practically unchanged between 900 and 1400°C. Nichols and Howes (1922) used U_3O_8 as standard in the study of the selective emissivity of other oxides on the assumption that the spectral distribution of the uranium oxide emission parallels closely that of a black body. However, the results obtained by Philipps (1928) indicate that the emissivity of U_3O_8, similar to that of ceria and other oxides, is, in the blue (467 mμ), higher than that of a black body of the same "red brightness."

Table 11.18—Crystallization of UO_3

Substance treated	Temp., °C	p_{O_2}, atm	Time, hr	Phase obtained	Formula of product UO_x	
					x from oxygen uptake	x from weight change by ignition to U_3O_8
UO_3	450−500	28	12	I	...	...
U_3O_8	500−560	28	36	I + II	2.97	3.048
U_3O_8	530−560	30	112	II + III	2.997	2.99
UO_3	530−560	30	112	I + II	...	...
U_3O_8	700−750	70−150	1.5−2	III	2.993	3.01

2.5 UO_3. (a) Crystallization. Formerly, UO_3, usually obtained by ignition of $UO_4 \cdot 2H_2O$, was found to be amorphous (Goldschmidt and Thomassen, 1923b). Microcrystalline UO_3 preparations were first obtained at the Mallinckrodt Chemical Works by ignition of uranyl nitrate. British observers (British 2) found that UO_3 crystallizes on standing; two crystalline forms were identified. Fried and Davidson (MP Chicago 3,8) found that UO_3 can exist in at least three crystalline forms—orange hexagonal UO_3(I), red UO_3(II), and yellow UO_3(III). The last form ("Mallinckrodt oxide") is probably the most stable one.

Because amorphous UO_3 loses oxygen in air above 450°C, the crystallization experiments of Fried and Davidson were conducted under high oxygen pressures (30 to 150 atm). The amorphous trioxide (or U_3O_8) was heated to 500 to 750°C in closed pyrex or quartz tubes.

Table 11.18 shows the nature of the products obtained under different conditions. Phase I was analyzed by Zachariasen (see below). Phase III is that found in Mallinckrodt oxide; it is the most stable of the three. The reason phase I is the first to be formed may be its crystallographic similarity to U_3O_8, which permits its formation from U_3O_8 by gradual uptake of oxygen.

(b) <u>Crystal Structure</u>. Crystalline $UO_3(I)$ obtained by heating amorphous anhydrous UO_3 to 500°C for 8 hr under 20 atm of oxygen pressure has, according to Zachariasen (1948; MP Chicago 2), a hexagonal lattice with the parameters

$$a = 3.963 \pm 0.004 \text{ A} \qquad c = 4.160 \pm 0.008 \text{ A}$$

The space group is $C\bar{3}m$, and the atomic positions are as follows:

$$1 \text{ U in } 1(a)$$

$$1 \text{ O}_I \text{ in } 1(b)$$

$$2 \text{ O}_{II} \text{ in } 2(d), \text{ with } z = 0.17$$

Each uranium atom has two nearest O_I atoms at 2.08 A distance and six O_{II} atoms at 2.39 A distance. The O_I atoms can be considered as "uranyl oxygens" although no UO_2 groups are present in the structure. Instead, endless "uranyl chains," $- O - U - O - U - O - U \ldots$, stretch along the c axis.

Fried and Davidson (MP Chicago 8) reported two x-ray measurements of partly deoxygenated UO_3 which gave somewhat unexpected results (cf. Table 11.19). Contrary to expectation (cf. Sec. 1.4), the lattice constants at $x = 2.82$ and 2.96 were not intermediate between those at $x = 2.67$ and 3.00.

The x-ray diffraction patterns of the forms $UO_3(II)$ and $UO_3(III)$ have not yet been analyzed but are different from those of hexagonal $UO_3(I)$. According to Fried and Davidson (MP Chicago 3,8) three samples of UO_3, prepared at 350 to 400°C by vapor-phase oxidation of U_3O_8 by nitric acid (CEW-TEC 1), gave two different x-ray patterns, both distinct from those of the above-mentioned forms I, II, and III. It is thus possible that UO_3 exists in as many as five crystalline allotropic forms.

(c) <u>Density</u>. The calculated density of $UO_3(I)$ is 8.34 g/cc. Experimental density values are much lower; most of them refer to amorphous oxide, some perhaps to hydrates (e.g., $UO_3 \cdot H_2O$) (cf. Sec. 3.3). For example, Wedekind and Horst (1915) found $\rho = 5.92$ g/cc (16°C); Beck (1928), $\rho = 7.29$ g/cc (15°C); and von Schroeder (1922), $\rho = 7.37$ g/cc (25°C). Biltz and Müller (1927d) gave $\rho = 6.04$ g/cc for $UO_{3.07}$ and $\rho = 7.07$ g/cc for $UO_{2.92}$, while at Columbia (SAM Columbia 2) a value of $\rho = 7.54$ g/cc (25°C) for UO_3 was observed. Jenkins (UCRL 1) gave 3.63 g/cc as the packing density of untapped and 4.26 g/cc as that of slightly tapped UO_3 powder.

(d) <u>Vapor Pressure</u>. As described in Sec. 1.3 a certain volatility of uranium oxide in the two-phase region between $UO_{2.62}$ and $UO_{2.25}$, observed by Biltz and Müller above 1160°C, can be attributed to the disproportionation

$$U_3O_8 \rightarrow UO_2 + 2UO_3 \qquad (8)$$

and evaporation of UO_3. Above $UO_{2.62}$ the partial pressure of oxygen is so much higher than the vapor pressure of UO_3 that the evaporation of

Table 11.19 — Lattice Parameters* of UO_x

x	a_1, A	a_2, A	a_3, A
2.67†	6.70 ± 0.01	3.98 ± 0.01	4.14 ± 0.01
2.82‡	6.90 ± 0.02	3.91 ± 0.02	4.15 ± 0.02
2.96‡	6.90 ± 0.02	3.91 ± 0.02	4.15 ± 0.02
3.00†	6.864 ± 0.004	3.963 ± 0.004	4.160 ± 0.008

*All referred to orthohexagonal axes.
†After Zachariasen; cf. Secs. 2.4a and 2.5b.
‡After Fried and Davidson (MP Chicago 8).

the oxide becomes too slow for observation. Brewer (MP Berkeley 1) estimated from the data of Biltz and Müller that at 1600°K, p_{UO_3} is 10^{-5} atm over $UO_{2.25}$ to $UO_{2.62}$ and 10^{-4} atm over pure UO_3.

(e) Thermodynamic Properties of UO_3. At the U. S. Bureau of Mines at Berkeley (U. S. Bur. Mines 2) the heat capacity of UO_3, probably in the form $UO_3(III)$ (MP Chicago 3), was measured between 15 and 300°K (Table 11.20). The entropy at 298.1°K was evaluated by graphical integration, using the Debye function below 15°K, with $\theta = 140°K$.

$\int C_p \, d(\ln T)$	15 to 300°K	23.378
$\int$ Debye function	0 to 15°K	0.196
		23.57 ± 0.06 e.u.

In the same laboratory Moore and Kelley (1947) determined the heat content at 400 to 900°K, using a sample dehydrated at 600°C for 79 hr (dissociation might have caused an error of approximately 1 per cent). The values lie on a smooth curve (see Table 11.21).

$$H_T - H_{298.16} = 22.09T + 1.27 \times 10^{-3}T^2 + 2.973 \times 10^5 T^{-1}$$

$$- 7696 \, (\pm 0.1\%) \text{ cal per mole} \qquad (9)$$

By differentiation the following may be obtained for C_p:

$$C_p = 22.09 + 2.54 \times 10^{-3}T - 2.973 \times 10^5 T^{-2} \tag{10}$$

Brewer (MP Berkeley 1) derived from these measurements the values of the function $f = (F_T - H_{298})/T$ shown in Table 11.22.

Table 11.20 — Specific Heat of UO_3

Temp., °K	C_p, cal/mole/°C	Temp., °K	C_p, cal/mole/°C
15	0.608	160	14.33
20	1.040	170	14.93
25	1.553	180	15.49
30	2.103	190	16.01
40	3.231	200	16.49
50	4.370	210	16.94
60	5.505	220	17.38
70	6.602	230	17.82
80	7.675	240	18.24
90	8.704	250	18.66
100	9.693	260	19.05
110	10.63	270	19.42
120	11.49	280	19.73
130	12.29	290	20.02
140	13.00	300	20.30
150	13.70		

(f) Thermal Conductivity. U_3O_8 powder, compressed by screws kept tight during the heating, was found by Princeton observers (Princeton 1,2) to have the thermal conductivity given in Table 11.23.

(g) Electrical Properties. LeBlanc and Sachse (1930) found no measurable electrical conductivity of UO_3 up to 300°C. The conductivity which is sometimes observed at room temperature and which disappears at 100 to 150°C can be attributed to uptake of moisture. The conductivity increases strongly at 350 to 490°C, particularly on prolonged heating. This may be due to the loss of oxygen and conversion into the lower, semiconducting oxides. After 30 hr at 400°C a product with an initial composition $UO_{2.9-3.0}$ showed a constant conductivity of 1.4×10^{-5} (ohm-cm)$^{-1}$. Similar values were obtained by Guillery (1932).

The dielectric constant of UO_3 was measured at Tennessee Eastman Corporation (CEW-TEC 2), using an oxide cake washed free of electrolytes. Values obtained ranged from 1.86 (UO_3 dried to orange color at $150\,^{\circ}C$, $\rho = 2.0$ g/cc, 79.19 per cent uranium) to 4.36 (UO_3 "Mallinckrodt" brick red) and to as high as 9.41 to 11.4 (two brown UO_3 samples).

(h) <u>Magnetic Properties</u>. Sucksmith (1932) found UO_3 to be diamagnetic, but other observers found a small temperature-independent

Table 11.21 — Heat Content and Entropy of UO_3 above $298.16\,^{\circ}K$*

Temp., °K	$H_T - H_{298.16}$, cal/mole	$S_T - S_{298.16}$, cal/mole/°C
400	2,090	6.01
500	4,260	10.86
600	6,510	14.96
700	8,820	18.51
800	11,160	21.64
900	13,540	24.44

*Moore and Kelley (1947).

Table 11.22 — The Function $f = (F_T - H_{298})/T$ for UO_3

Temp., °K	f, cal/mole/°C
298	23.57*
500	25.7
1000	34.4

*Equal to S_{298}°.

Table 11.23 — Thermal Conductivity of U_3O_8

Temp., °C	K, cal/cm sec °C
25-150	0.00067
160-340	0.00063
310-600	0.00061

paramagnetism. Wedekind and Horst (1915) gave $+1.08 \times 10^{-4}$ as the specific susceptibility at $16\,^{\circ}C$; Meyer (1899) gave $+0.146 \times 10^{-6}$ as the volume susceptibility at the same temperature. Tilk and Klemm (1939) reported χ (specific) $= +0.26 \times 10^{-6}$ and χ (molar) $= +74 \times 10^{-6}$; corrected for diamagnetism χ (molar) $= +128 \times 10^{-6}$ (average of values obtained with UO_3 samples from nitrate, carbonate, and peroxide). The diamagnetism values used for correction were 20×10^{-6}

for U^{+6} and $11.25 \times 10^{+6}$ for O^{-2}. Haraldsen and Bakken (1940) found χ (molar) = $+157 \times 10^{-6}$ (corrected for diamagnetism).

(i) <u>Color</u>. Amorphous UO_3 powder prepared by ignition of $UO_4 \cdot 2H_2O$ has been described as "sulfur yellow" (Biltz and Müller, 1927) and "bright orange" (Fried and Davidson, MP Chicago 3). Hexagonal crystalline $UO_3(I)$, prepared by Fried and Davidson from the amorphous product, was lighter in color. The phase $UO_3(II)$, obtained from amorphous UO_3 or U_3O_8 by heating to 500°C under 28 atm of oxygen, was deep red. The brick-red product that de Coninck had obtained (cf. Sec. 4.5b) by prolonged heating of orange UO_3 to 600°C in air may have been the same form. The phase $UO_3(III)$, obtained by Fried and Davidson by more prolonged heating under still higher oxygen pressure and also present in Mallinckrodt's microcrystalline product obtained by ignition of uranyl nitrate, is bright yellow. The two new crystalline UO_3 forms, obtained by oxidation of U_3O_8 by N_2O_4 (Sec. 4.6a), were brick red (or orange) and yellow, respectively.

The partially deoxygenated products of UO_3 were described by Fried and Davidson as tan $(UO_{2.96})$, tan-green $(UO_{2.82})$, and green-black $(UO_{2.70})$, thus showing gradual transition to the black U_3O_8 phase.

2.6 <u>UO_4</u>. This oxide is unknown in the anhydrous state. The physical properties of its hydrates are given in Sec. 3.4b. The product $UO_{3.5}$, obtained by Kraus and coworkers (cf. Sec. 1.5) and interpreted as $UO_3 + UO_4$, was described as a red powder.

3. URANIUM OXIDE HYDRATES

3.1 <u>UO_2 Hydrates</u>. Péligot (1842a) observed the formation of a flaky, voluminous, reddish-brown precipitate by the reaction of ammonia with green uranium(IV) salt solutions. According to Zimmermann (1882a) the precipitates formed by the action of alkalis or ammonia on uranium(IV) solutions are light green at first but are rapidly converted in the presence of air into brown U_3O_8 hydrates. Aloy (1901a) found that the best method to obtain alkali-free unoxidized uranium dioxide hydrate is to hydrolyze a dilute air-free uranium(IV) chloride or acetate solution by heating until it becomes colorless. The black precipitate formed is a basic salt, but it can be washed free of anions with boiling water. After drying over H_2SO_4 the hydrate has the composition $UO_2 \cdot 2H_2O$. The uranium(IV) salt solutions can be decomposed also by illumination (Aloy and Rodier, 1922). Rowell and Russell (1925) obtained a slimy black or greenish-black UO_2 hydrate by photochemical decomposition of a uranyl nitrate solution in ether. Black crystalline $UO_2 \cdot 2H_2O$ was obtained by Aloy (1899) by the action of alkali on crystalline uranium(IV) sulfate.

According to Aloy (1899) moist amorphous $UO_2 \cdot 2H_2O$ is oxidized in air, slowly in the cold, but rapidly on heating, to $UO_3 \cdot H_2O$. Crystalline UO_2 hydrate, on the other hand, is stable in air for several days at room temperature. It is converted to green U_3O_8 by heating.

Columbia observers (SAM Columbia 5) found that $UO_2 \cdot xH_2O$ is dehydrated under water above 285°C to brown anhydrous dioxide. In air (SAM Columbia 6) the hydrate lost all water within 0.5 hr at 200°C, suffering at the same time partial oxidation to U_3O_8.

Freshly prepared uranium(IV) hydroxide is readily soluble in acids but loses the property of solubility upon standing (Péligot, 1842a; Rammelsberg, 1843a; Hermann, 1861; Raynaud, 1911). The solutions contain colloidal aggregates or polymeric ions rather than simple uranium(IV) salts.

3.2 U_3O_8 Hydrate. Existence of a U_3O_8 hydrate, first mentioned by Arfvedson (1822a), was definitely established by Ebelmen (1842b). He prepared it by photochemical decomposition of uranyl oxalate. It is a flaky brownish-violet precipitate, easily oxidizable in air. To avoid decomposition it must be dried in vacuum. Aloy (1900) recommended the use of uranyl acetate as starting material instead of the only slightly soluble oxalate. When an aqueous uranyl acetate solution containing ether or alcohol is exposed to light, a violet precipitate is formed which proves less easily oxidizable than the U_3O_8 hydrate obtained by Ebelmen from oxalate. Aloy and Rodier (1920) found that a similar violet precipitate is formed by all uranyl salt solutions (with a concentration of 1 to 5 per cent) when exposed to light in the presence of organic substances that can act as reductants (acetaldehyde, ether, alcohol, glucose). Immediately after precipitation the hydroxide contains anions of the acid, but they can be washed out with boiling water. If uranyl salts of organic acids are used, no addition of extra organic reductant is necessary.

According to Columbia University observers (SAM Columbia 4) U_3O_8 does not react with water even after 11 days at 185°C. However, U_3O_8 prepared by ignition below red heat may contain UO_3 and therefore absorb some water from the air to form $UO_3 \cdot 2H_2O$ (Lebeau, 1922a; Staehling, 1921).

The composition of the violet hydrate, obtained as described above, was given in the literature only as $U_3O_8 \cdot xH_2O$. According to Smith (1879, 1880) and de Coninck and Camo (1901c), electrolysis of uranyl salt solutions leads first to yellow $U_3O_8 \cdot H_2O$ and then to black $U_3O_8 \cdot 2H_2O$. However, Pierlé (1919) asserted that the black deposit has the composition $UO_{3.3} \cdot 2H_2O$ rather than $UO_{2.67} \cdot 2H_2O$ (cf. Sec. 5.3f). He considered the high potential of the electrode coated with this black hydroxide (cf. Sec. 4.1b) as an additional argument against its interpretation as a U_3O_8 hydrate. The formation of a U_3O_8 hydrate deposit on uranium metal by electrolysis of uranyl nitrate solution was again investigated by Francis and Tscheng-Da-Tschang (1935b) and by Francis (1935a).

Aloy (1900) found that amorphous violet $U_3O_8 \cdot xH_2O$ does not become crystalline by freezing or heating under pressure. Drying in vacuum converts it into a solid black mass which can be ground to a black powder. When heated in nitrogen this mass loses water without changing its appearance, but grinding after drying produces a green rather than black powder.

According to Aloy, $U_3O_8 \cdot xH_2O$ is easily oxidized in air to UO_3 hydrate. It dissolves in acids, forming a mixture of uranium(IV) and uranium(VI) salts.

3.3 $\underline{UO_3 \text{ Hydrates}}$. The tendency for hydrate formation increases with the valence of the cation. The UO_3 hydrates are much more stable and therefore better known than the hydrates of UO_2 and U_3O_8.

(a) $\underline{\text{Phase Relations in the } UO_3\text{-}H_2O \text{ System}}$. Very little is known about the UO_3-H_2O system in solution. According to de Forcrand (1913, 1915a), $UO_3 \cdot 2H_2O$ is slightly soluble in water (0.16 g per liter at 27°C). According to Columbia observers, at room temperature $UO_3 \cdot H_2O$ slurries in water have a pH of 4.8 to 5.2, with either the rhombic or the triclinic form as the solid phase.

Several UO_3 hydrates are known in the solid state. The compounds $UO_3 \cdot H_2O$ (or H_2UO_4) and $UO_3 \cdot 2H_2O$ (or H_4UO_5) are well-known from chemical studies. Hüttig and von Schroeder (1922a) deduced from measurements of the hydration isobar the existence of a semihydrate, $2UO_3 \cdot H_2O$ (or $H_2U_2O_7$), and the possible existence of a sesquihydrate, $2UO_3 \cdot 3H_2O$ (or $H_6U_2O_9$).

Hüttig and von Schroeder prepared anhydrous UO_3 by heating the hydrate $UO_4 \cdot 2H_2O$ to 500°C (to decompose all nitrate that might be left from the preparation of the peroxide from uranyl nitrate). The trioxide was completely dehydrated in vacuum at 450°C (no oxygen was lost under these conditions; cf. Sec. 4.1b) and equilibrated with water vapor at room temperature. Hydration occurred with marked evolution of heat and slight increase in volume and resulted at 30°C in formation of the dihydrate $UO_3 \cdot 2H_2O$. This dihydrate was dehydrated by heating under constant H_2O pressure of about 15 mm Hg. Figure 11.9 shows the results. The first half-mole of H_2O was lost gradually between 30 and 100°C; a second half-mole was lost suddenly at 100°C; a third half-mole was lost, again gradually, between 100 and 300°C; and the last half-mole was lost suddenly at 300°C. Following the course of the isobar from right to left, evidence is first found of two stoichiometrically pure compounds, UO_3 and $UO_3 \cdot 0.5H_2O$, with no mixed crystal formation between them. Then follows a region of mixed crystals, $UO_3 \cdot 0.5H_2O + UO_3 \cdot H_2O$. It is not clear from the curve whether between $1H_2O$ and $2H_2O$ there is another continuous region of mixed crystallization, $UO_3 \cdot H_2O + UO_3 \cdot 2H_2O$, or whether no mixed crystals are formed in the region between $UO_3 \cdot H_2O$ and $UO_3 \cdot 1.5H_2O$. If the latter is the correct interpretation, then $UO_3 \cdot 1.5H_2O$ can be considered as another pure hydrate, with a region of mixed crystals extending from $UO_3 \cdot 1.5H_2O$ to $UO_3 \cdot 2H_2O$.

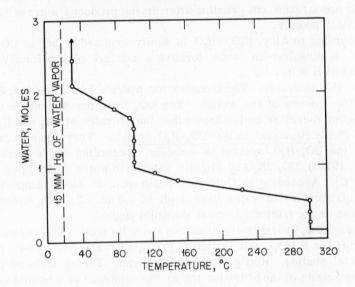

Fig. 11.9—Thermal decomposition of $UO_3 \cdot xH_2O$ at a partial pressure of 15 mm Hg of water vapor [from G. Hüttig and E. von Schroeder, Z. anorg. u. allgem. Chem., 121: 251 (1922)].

(b) **Thermodynamic Properties of UO_3 Hydrates.** Hüttig (1922) used the tensimetric data shown in Fig. 11.9 for the calculation of the heats of hydration by means of Nernst's approximation formula. His results are compared in Table 11.24 with the values obtained by de Forcrand (1913, 1915a) from calorimetric measurements of the heats of neutralization of the oxides UO_3, $UO_3 \cdot H_2O$, and $UO_3 \cdot 2H_2O$ with dilute HNO_3 (cf. Sec. 6.2d).

The heats of neutralization of UO_3 hydrates by alkalis or acids leading to the formation of uranates or uranyl salts will be discussed in Sec. 6.2.

Pierlé (1919) measured the potential of a platinum electrode covered with $UO_3 \cdot H_2O$-gelatin paste. For results see Sec. 4.1b.

(c) <u>Physical Properties of UO$_3$ Hydrates.</u> UO$_3$·0.5H$_2$O. According to the observations made at Columbia University (SAM Columbia 1,7) this hydrate forms monoclinic orange needles 5 to 15 μ thick and 20 to 80 μ long, with a distinct x-ray pattern of their own. Characteristic is their tendency for basal cleavage.

UO$_3$·H$_2$O. The monohydrate is yellow or orange-yellow. Columbia observers (SAM Columbia 1,7) distinguish an amorphous and four allotropic crystalline modifications, all stable at room temperature.

Table 11.24 — Heats of Hydration of UO$_3$

	Reaction	Hüttig, 1922	De Forcrand, 1913
(11)	UO$_3$ + ½H$_2$O (gas) = UO$_3$·½H$_2$O	13.3	...
(12)	UO$_3$ + H$_2$O (gas) = UO$_3$·H$_2$O	23.4	15.5
(13)	UO$_3$ + 1½H$_2$O (gas) = UO$_3$·1½H$_2$O	31.7	...
(14)	UO$_3$ + 2H$_2$O (gas) = UO$_3$·2H$_2$O	39.2	28.5

These four modifications were described as follows:

α: Large six-sided orthorhombic basal tablets.

β: Microscopic orthorhombic prismatic tablets (up to 50 to 100 μ), with slightly larger unit cell than in the α modification.

Zachariasen (MP Chicago 6) gave the following more precise data for an orthorhombic form of UO$_3$·H$_2$O:

$$a = 6.86 \pm 0.03 \text{ A} \quad b = 4.27 \pm 0.03 \text{ A} \quad c = 10.19 \pm 0.06 \text{ A}$$

The unit cell contains four UO$_3$·H$_2$O molecules. The uranium atoms are in positions $(0\ 0\ 0)$ $(\frac{1}{2}\frac{1}{2}0)$ $(\frac{1}{2}0\frac{1}{2})$ $(0\frac{1}{2}\frac{1}{2})$. The calculated density is 6.73 g/cc.

γ: Hexagonal columnar crystals, also of microscopic size with an x-ray pattern similar to that of the α and β form but perhaps belonging to a different system. (Analysis of this form showed <0.9 mole of H$_2$O per mole.)

δ : Triclinic with a complex x-ray diffraction pattern. (This modification may be due to impurities.)

Conditions under which these four forms are obtained are given in Sec. 3.3d.

In earlier investigations UO$_3$·H$_2$O was described as occurring in an amorphous and a rhombic form (Aloy, 1900, 1901b; Lebeau, 1912). Riban (1881, 1882) obtained UO$_3$·H$_2$O in the form of hexagonal prisms by hydrolysis of uranyl acetate; Zehenter (1900) obtained it by the same method in the form of brilliant hexagonal platelets. Ipatieff and Muromtsev (1930) obtained $UO_3 \cdot H_2O$ as transparent prisms with pointed faces by hydrolysis of uranyl nitrate under high hydrogen pressure. Lebeau (1912) converted amorphous $UO_3 \cdot H_2O$ into rhombic lamellae by dissolving it in uranyl nitrate solution (see Sec. 3.3d), evaporating to dryness at 100°C, and extracting with ether.

Table 11.25 — Densities* of $UO_3 \cdot xH_2O$

$UO_3 \cdot xH_2O$	x	Particle size, μ	ρ (degassed), g/cc
$UO_3 \cdot \frac{1}{2}H_2O$	0.47	>105	6.47
$UO_3 \cdot H_2O$ amorphous	1.04	44−74	6.32†
$UO_3 \cdot H_2O$ rhombic‡ (β)	0.82	0−10	6.07
$UO_3 \cdot H_2O$ rhombic (α)		...	6.73§
$UO_3 \cdot H_2O$ triclinic (δ)	1.00	<105	5.74
$UO_3 \cdot H_2O$ triclinic (δ)	1.00	1−2	5.85
$UO_3 \cdot 2H_2O$	2.12	...	4.87

*Kirshenbaum (SAM Columbia 3).
†Malaguti (1843) gave 5.93 g/cc.
‡1 per cent UO_4.
§Zachariasen (MP Chicago 6), by calculation from x-ray data.

$UO_3 \cdot 2H_2O$. According to Columbia University observers (SAM Columbia 1,7) this yellow or greenish-yellow hydrate occurs in two modifications. Form α, obtained only as submicroscopic crystals with a simple diffraction pattern, may be face-centered tetragonal. Modification β, also known so far only in the form of submicroscopic crystals, probably is orthorhombic. Remarkably enough, the diffraction pattern of the β form of the dihydrate is identical with that of the β form of the monohydrate, the unit cell dimensions differing by less than 0.5 per cent. This may mean that the second water molecule is "zeolitic" water.

Mechanical Properties. Table 11.25 gives density values for UO_3 hydrates, mainly according to Kirshenbaum (SAM Columbia 3).

Columbia observers (SAM Columbia 4) found that the scratch hardness of UO_3 hydrates up to $UO_3 \cdot 2H_2O$ is 2.5 on Mohs' scale.

Optical Properties. Color. $UO_3 \cdot 0.5H_2O$ has been described as being orange, $UO_3 \cdot H_2O$ as orange-yellow or yellow, and $UO_3 \cdot 2H_2O$ as pure yellow or greenish yellow.

Fluorescence. Anhydrous UO_3 is nonfluorescent, even at −196°C; the same is true of UO_2, U_3O_8, and $UO_4 \cdot 2H_2O$. On the other hand, all

UO_3 hydrates fluoresce at room temperature and below. The general structure of the fluorescence spectrum is not unlike that of the uranyl ion, but the bands lie further toward the red and are limited to a more narrow region (500 to 600 mμ). The fluorescent light is green rather than greenish yellow, as in the case of uranyl salts. The fluorescence spectra of $UO_3 \cdot 2H_2O$ (α and β), $UO_3 \cdot H_2O$ (α, β, and δ), and $UO_3 \cdot 0.5H_2O$ were examined at Columbia University (Vier, Schultz, and Bigeleisen, 1948; SAM Columbia 8) and found to be similar but easily distinguishable. Each contains three or four bands at intervals of about 810 cm^{-1}. The first two bands are strong; the third and the fourth are increasingly weak. The fluorescence of UO_3 hydrates will be discussed in more detail in the second volume of this treatise in the chapters dealing with fluorescence and photochemistry of uranyl compounds.

(d) Hydration and Dehydration of UO_3. The hydration isobar shown in Fig. 11.9 gives information on the conditions under which the different hydrates can be expected to be formed and to decompose in contact with water vapor under 15 mm Hg partial pressure. Numerous additional data on equilibrium conditions and rates of hydration and dehydration are available in the Manhattan Project literature.

$UO_3 \cdot 2H_2O$. This hydrate is stable at room temperature. According to early Columbia observations (SAM Columbia 4) it is formed from red-orange "active" UO_3 by action of saturated water vapor between 5 and 75°C. Later at the same laboratory (SAM Columbia 7) the α form was obtained by hydration of amorphous UO_3 (prepared by decomposition of $UO_4 \cdot 2H_2O$), and the β form was obtained by hydration of Mallinckrodt's product, which is microcrystalline $UO_3(III)$ (cf. Sec. 2.5a).

At Berkeley (UCRL 4) it was observed that orange or brick-red UO_3 [formed by thermal decomposition of $(NH_4)_2U_2O_7$], when exposed to air, is converted in several days into a yellow-green hydrate, presumably $UO_3 \cdot 2H_2O$. Upon heating, this hydrate is quickly reconverted into the brick-red anhydrous oxide.

De Forcrand (1913, 1915a) found that direct hydration of UO_3 easily leads to products containing as much as $2.5H_2O$ and that dehydration of these products is likely to go beyond the stage of $2H_2O$; therefore the stoichiometric composition $UO_3 \cdot 2H_2O$ can be obtained only by interrupting the dehydration at the right moment.

Instead of hydration by water vapor, $UO_3 \cdot 2H_2O$ can also be obtained by the action of liquid water on UO_3 or $UO_3 \cdot H_2O$. Lebeau (1912) prepared the dihydrate by dissolving $UO_3 \cdot H_2O$ in concentrated aqueous solution of uranyl nitrate (cf. Sec. 3.3c), evaporating over sulfuric acid at room temperature, and removing the uranyl nitrate by ether extraction. The residue was a light-yellow powder with the composition $UO_3 \cdot 2H_2O$. The upper limit of stability of $UO_3 \cdot 2H_2O$ in contact with liquid water is considerably below 100°C, more exactly, at 75°C according to the first Columbia University observations (SAM Columbia 4), or at 60°C according to more recent data (SAM Columbia 7). The fact that the water vapor pressure over $UO_3 \cdot 2H_2O$ is higher than over H_2O at temperatures above approximately 60°C was confirmed by direct measurements at 61 and 77°C.

Table 11.26 — Rate of Dehydration of $UO_3 \cdot 2H_2O$

Preparation	In vacuum 28°C	In water vapor				
		61°C	77°C	87°C	100°C	118°C
$UO_3 \cdot 2H_2O$ from UO_4	...	>518	<95	>12	<2	<1
$UO_3 \cdot 2H_2O$ from Mallinckrodt UO_3	<123	>100	<122; >48	<2	...	...

The figures in Table 11.26 characterize the rate of dehydration of $UO_3 \cdot 2H_2O$ at different temperatures.

Earlier, de Forcrand (1913) had stated that $UO_3 \cdot 2H_2O$ loses H_2O slowly over H_2SO_4 at room temperature, is rapidly dehydrated to $UO_3 \cdot H_2O$ in a stream of dry air at 80°C, and boils at about 135°C. Lebeau (1912) reported that $UO_3 \cdot 2H_2O$ is converted to $UO_3 \cdot H_2O$ by boiling in water. Drenckmann (1861), on the other hand, had asserted that conversion to monohydrate occurs only at 160°C.

$UO_3 \cdot H_2O$. Heating $UO_3 \cdot 2H_2O$ above 60°C must lead, according to Fig. 11.9, first to the formation of mixed crystals of dihydrate and monohydrate (or of dihydrate and sesquihydrate). Conversion to pure monohydrate, $UO_3 \cdot H_2O$, should require heating to about 100°C (under 15 mm Hg H_2O pressure). A still higher temperature should be necessary under water or in contact with saturated water vapor. Formation of mixed crystals was not taken into account at Columbia University (SAM Columbia 4), and the limits of stability of the monohydrate in contact with water were given first (SAM Columbia 7) as 85 to 300°C for the rhombic form and 300 to 310°C for the triclinic form. In the same paper, active red-orange UO_3 was stated to be converted into rhombic $UO_3 \cdot H_2O$ by exposure to saturated water vapor at 85 to 300°C and to triclinic $UO_3 \cdot H_2O$ at temperatures above 300°C; "inactive" orange-yellow UO_3 was stated to require for hydration a temperature of 150 to 300°C and to give triclinic $UO_3 \cdot H_2O$ plates.

In a later investigation at Columbia University (SAM Columbia 7) the yellow orthorhombic α form of $UO_3 \cdot H_2O$ was obtained directly from $UO_4 \cdot H_2O$ by heating in water to 158°C for 117 hr, as well as by heating the β form of $UO_3 \cdot 2H_2O$ in water to 100°C for 100 to 600 hr.

The orange β form of the monohydrate was obtained from the α or β dihydrate by heating to 100°C for 90 hr. The γ form of $UO_3 \cdot H_2O$ was obtained by heating $UO_4 \cdot 2H_2O$ or $UO_3 \cdot 2H_2O$ in water to 185 to 310°C. Finally, the δ form of the monohydrate was obtained by dehydration of α $UO_3 \cdot 2H_2O$ in water at 185 to 310°C.

According to the same investigation, $UO_3 \cdot H_2O$ does not change by heating in water in sealed tubes to temperatures from 60 to 325°C. $UO_3 \cdot H_2O$ crystals do not take up more H_2O after a month in water at 28°C (although $UO_3 \cdot 2H_2O$ is the stable hydrate in water up to 60°C). All four crystallographic forms are very stable at low temperatures; qualitative vapor pressure observations indicate that the hexagonal form is the most stable below 125°C, the rhombic form between 125 and 300°C, and the triclinic form at 300 to 310°C. $UO_3 \cdot H_2O$ loses 85 per cent of its water in 3 hr at 400°C and more than 95 per cent in 1 hr at 450°C.

The following earlier observations can also be quoted: $UO_3 \cdot H_2O$ loses no water in vacuum or dry air at 100°C (Aloy, 1900, 1901b); loses 5 per cent of its water in 30 min at 300°C, 50 per cent at 400°C, and 94 per cent at 500°C (Lebeau, 1912); boils at about 220°C (de Forcrand, 1913); and is transformed into anhydrous UO_3 at 300°C and into U_3O_8 at red heat (Ebelmen, 1842c).

Lebeau (1912) obtained $UO_3 \cdot H_2O$ by hydration of UO_3 in air at 25°C in 24 hr and at 100°C in 1 hr, apparently as an intermediate in the conversion of UO_3 to $UO_3 \cdot 2H_2O$. At the Metallurgical Laboratory (SAM Columbia 9) the growth of hexagonal and platelike crystals (apparently of $UO_3 \cdot H_2O$?) was observed after 30 min of digestion of a suspension of UO_3 in water at 100°C.

$UO_3 \cdot 0.5H_2O$. According to Hüttig and von Schroeder (1922a) (cf. Fig. 11.9) the semihydrate can be obtained under 15 mm Hg H_2O pressure between 160 and 300°C. At Columbia University (SAM Columbia 7) the semihydrate was prepared by heating $UO_4 \cdot 2H_2O$ in water to 310 to 350°C as well as by hydrating inactive orange-yellow UO_3 at 350 to 380°C (SAM Columbia 4). It is stable in sealed tubes under water above 325°C and is slowly hydrated to $UO_3 \cdot H_2O$ below 300°C; hydration is complete in 21 hr at 100°C. According to the same laboratory (SAM Columbia 6) the water content is reduced from 2.8 to 0.5 per cent of the total weight in 4 hr at 450°C in air in an open dish. Heating to 500 to 550°C removes all water. The semihydrate is converted to red UO_3 (containing some U_3O_8) by heating to 450 to 600°C in air and is converted to pure U_3O_8 by heating to 700°C (SAM Columbia 4).

3.4 **UO_4 Hydrates.** (a) <u>Dehydration of $UO_4 \cdot xH_2O$.</u> Uranium peroxide, UO_4, has not been obtained in the pure anhydrous state. When
dehydrated it loses oxygen simultaneously with water until it is converted to UO_3 or lower oxides, depending on temperature and duration
of dehydration. This was first established by Alibegoff (1886a) and
confirmed by experiments of Hüttig and von Schroeder (1922a) and
Rosenheim and Daehr (1929a).

The study of the UO_4-H_2O system is complicated by this simultaneous loss of water and oxygen. Systematic measurements of the dehydration equilibrium as such should be made under oxygen pressures
high enough to prevent deoxygenation as far as possible.

Freshly precipitated air-dried uranium peroxide may contain as
much as four or more molecules of water per atom of uranium. Fair-
ley (1877a) ascribed the composition $UO_4 \cdot 4H_2O$ to air-dried $UO_4 \cdot xH_2O$.
Rosenheim and Daehr (1929a) found that air-dried $UO_4 \cdot xH_2O$ contains
three molecules of water. Hüttig and von Schroeder (1922a) found
about four and one-half molecules of H_2O in peroxide dried in air at
35°C. At the Metallurgical Laboratory (reported in SAM Columbia
10) air-dried uranium peroxide (precipitated from uranyl nitrate at
75°C by 3 per cent H_2O_2) was found to contain $4H_2O$ (see, however, the
discussion of x-ray analysis in Sec. 3.4b).

At Berkeley (MP Berkeley 2), on the other hand, the following analytical results were obtained with air-dried peroxide preparations:

Washed with water, alcohol, and ether	$UO_{3.98} \cdot 2.11H_2O$
Washed with water only	$UO_{3.96} \cdot 2.18H_2O$

Hüttig and von Schroeder (1922a) studied the dehydration of $UO_4 \cdot$
xH_2O under constant H_2O pressure (approximately 15 mm Hg) in the
same way as that of $UO_3 \cdot xH_2O$ (cf. Sec. 3.3a). Figure 11.10 shows dehydration and deoxygenation isobars obtained in this study. They
started with air-dried preparations that contained about four and one-
half molecules of H_2O per atom of uranium. (This phase was interpreted by Hüttig as an addition compound of H_2O_2, H_2O, and UO_3:
$UO_3 \cdot H_2O_2 \cdot 3.5H_2O$.) Curves show that one-half of the peroxide oxygen
is lost at temperatures of about 60°C when the compound still contains three and one-half molecules of water. The resulting phase,
$UO_{3.5} \cdot 3H_2O$, was interpreted by Hüttig as $UO_3 \cdot 0.5H_2O_2 \cdot 2.5H_2O$. Between
54 and 163°C another molecule of water is lost gradually until the
composition $UO_{3.5} \cdot 2H_2O$ is reached (according to Hüttig, $UO_3 \cdot 0.5H_2O_2 \cdot$
$1.5H_2O$). This may correspond to a separate compound or to mixed
crystals of $UO_{3.5} \cdot 3H_2O$ and $UO_3 \cdot 1.5H_2O$. At 163°C the remaining peroxide oxygen is lost together with one and one-half molecules of

water, leaving $UO_3 \cdot 0.5H_2O$ as the solid phase. The last half-molecule of water was found in these experiments to be lost somewhere near 450°C. This agrees with the results of experiments on the dehydration of $UO_3 \cdot 2H_2O$ (described in Sec. 3.3d).

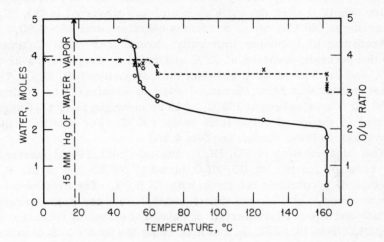

Fig. 11.10—Thermal decomposition of $UO_4 \cdot xH_2O$ at a partial pressure of 15 mm Hg of water vapor. The solid line refers to moles of water, and the dotted line to moles of oxygen [from G. Hüttig and E. von Schroeder, Z. anorg. u. allgem. Chem., 121: 247 (1922)].

Hüttig denied the existence of an oxide UO_4 and interpreted the "UO_4 hydrates" as loose addition compounds of UO_3, H_2O_2, and H_2O. However, this interpretation does not follow necessarily from his observations.

It will be noted that, according to Fig. 11.10, one-half atom of peroxide oxygen is lost before the water content has decreased to $2H_2O$. In other words, a compound $UO_4 \cdot 2H_2O$ (or $UO_3 \cdot H_2O_2 \cdot H_2O$) does not exist, according to Hüttig, under 15 mm Hg of H_2O pressure. This conclusion appears to be in disagreement with the observations of many other investigators who found $UO_4 \cdot 2H_2O$ the most stable form of uranium peroxide.

Fairley (1877a) gave $UO_4 \cdot 2H_2O$ as the composition of uranium peroxide dried at 100°C. This was confirmed by Alibegoff (1886a) and by Rosenheim and Daehr (1929b, 1932).

At Brown University (Brown 3) it was observed that drying $UO_4 \cdot xH_2O$ (precipitated at 75°C) at 50°C in vacuum leads to a rapid loss of water which is in excess of that required by the formula $UO_4 \cdot 2H_2O$;

further dehydration is very slow up to 125°C, where H_2O and O_2 begin to escape simultaneously. At 60 to 70°C, $UO_4 \cdot 2H_2O$ can be dried for 13 hr without appreciable loss of oxygen (Brown 4).

According to Columbia University observations (SAM Columbia 10) the peroxide prepared by Kraus's method, which had the composition $UO_4 \cdot 4H_2O$ after drying in open air, was converted to $UO_4 \cdot 2H_2O$ by drying at 100°C in air or at room temperature over 36N H_2SO_4.

According to Columbia University observations (SAM Columbia 11) the peroxide prepared at 75°C and dried in oxygen for 3 hr at 37°C and for 3 hr at 62°C also has the composition $UO_4 \cdot 2H_2O$. The water content was found unchanged after an additional 3 hr in air at 85°C, or 3 hr in oxygen at 105°C, or 13 hr in oxygen at 75°C. Oxygen started to escape slowly a little below 100°C. (For details of UO_4 conversion to lower oxides, see Sec. 4.6c.)

Heat of Formation of $UO_4 \cdot 2H_2O$. Pisarzhevskiǐ (1900a) measured the heats of reaction of $UO_4 \cdot 2H_2O$ (dried at 100°C) and of $UO_3 \cdot H_2O$ (method of preparation not given) with 2N H_2SO_4. The first reaction was assumed to give uranyl sulfate and H_2O_2, and the second, uranyl sulfate and H_2O. Pisarzhevskiǐ calculated the heat of formation of $UO_4 \cdot 2H_2O$ from $UO_3 \cdot H_2O$, O_2, and H_2O, using the known heat of oxidation of H_2O to H_2O_2.

$$UO_4 \cdot 2H_2O = UO_3 \cdot H_2O + \tfrac{1}{2}O_2 + H_2O + 6.15 \text{ kcal} \tag{15}$$

(b) Physical Properties of UO_4 Hydrates. Density of $UO_4 \cdot 2H_2O$ (Du Pont product, 79 per cent UO_4, 11 per cent H_2O, 10 per cent UO_3) was found at Columbia University (SAM Columbia 3) to be 4.31 g/cc (degassed). According to Berkeley observations (UCRL 1) the bulk density of dried $UO_4 \cdot 2H_2O$ not tapped was 0.72 g/cc and of tapped was 1.00 g/cc.

Crystal Structure. $UO_4 \cdot 2H_2O$ has usually been described as "amorphous." Rosenheim and Daehr (1929a) found that a crystalline peroxide can be prepared by adding an equal volume of 30 per cent H_2O_2 to a 14 per cent solution of ammonium uranyl oxalate at room temperature and heating the gelatinous precipitate with twice its volume of water almost to the boiling point. This leads to a dark-yellow solution from which fine needlelike crystals are formed upon slow cooling. These were found to have a composition $UO_4 \cdot 3H_2O$ and to be converted to $UO_4 \cdot 2H_2O$ by drying in air at 100°C or in vacuum over P_2O_5 at room temperature. However, attempts to repeat the preparation of this "crystalline peroxide" at Berkeley (MP Berkeley 2) gave a compound containing NH_4^+ and oxalate, presumably $UO_4 \cdot 0.5(NH_4)_2C_2O_4 \cdot (1.5$ to $2)$ H_2O. This may also have been Rosenheim and Daehr's product. According to Columbia University (SAM Columbia 7,12) $UO_4 \cdot xH_2O$ precipitated by addition of 1 liter of 3 per cent H_2O_2 to 210 g of uranyl nitrate in 1 liter of H_2O at 75°C was microcrystalline and showed a diffuse x-ray pattern. At the Metallurgical Laboratory (MP Chicago 2,6) an air-dried sample of $UO_4 \cdot xH_2O$ that allegedly contained $4H_2O$ (cf. Sec. 3.4a) was studied by x rays. The lines were diffuse, indicating a crystallite size of 100 to 200 A. The unit cell was found to be orthorhombic and face-centered with the parameters

$$a_1 = 8.74 \pm 0.05 \text{ A} \qquad a_2 = 6.50 \pm 0.03 \text{ A} \qquad a_3 = 4.21 \pm 0.2 \text{ A}$$

The two uranium atoms in the unit are situated in positions (0 0 0) and $(\frac{1}{2}\frac{1}{2}\frac{1}{2})$. Zachariasen pointed out that this structure seems to offer

Table 11.27 — Densities of Uranium Peroxide Hydrates

Composition	ρ, g/cc
$UO_4 \cdot 4H_2O$	5.15
$UO_4 \cdot 3H_2O$	4.90
$UO_4 \cdot 2H_2O$	4.66

insufficient space for four molecules of water. The densities calculated for different possible water contents are given in Table 11.27.

Hardness. Scratch hardness: Mohs' 2.5; if wet and fluffy, Mohs' 2, according to Columbia University observers (SAM Columbia 4).

Paramagnetism. Tilk and Klemm (1939) measured the paramagnetism of $UO_4 \cdot xH_2O$ and found χ (specific) calculated for anhydrous oxide = 0.17×10^{-6}; χ (molar) = 51×10^{-6}; χ (molar) corrected for diamagnetism = 111×10^{-6}. The values used for U^{+6} and O^{-2} in applying the correction for diamagnetism were, respectively, -11.25×10^{-6} and -20×10^{-6}. The paramagnetism is independent of temperature.

Optical Properties. $UO_4 \cdot xH_2O$ is faintly yellow in color. In contrast to $UO_3 \cdot xH_2O$ it does not fluoresce either at room temperature or at the temperature of liquid nitrogen (SAM Columbia 1,8).

3.5 Uranium Oxide Hydrosols. (a) UO_2 Hydrosol. Samsonov (1911) obtained a black precipitate by electrolytic reduction of a solution of 50 g of uranyl chloride in 100 ml of 2N HCl; this precipitate easily formed a hydrosol. The hydrosol was very unstable and coagulated after standing about 24 hr. Its stability could be increased by the addition of uranium tetrachloride. The particles are visible in the ultramicroscope. They carry out a lively Brownian motion and carry a positive charge, as shown by their movement in an electric field and by their coagulation by anions. The hydrosol contains up to 0.2 mole of UO_2 per liter. A similar hydrosol can also be prepared by reduction of UO_2Cl_2 with zinc or copper in dilute hydrochloric acid.

(b) <u>UO_3 Hydrosol</u>. A uranium trioxide hydrosol was first prepared by Graham (1862) by precipitation with KOH from uranyl nitrate (or chloride) solution in the presence of sugar and subsequent dialysis. The orange-yellow hydrosol, completely free of acid or alkali, is very stable. It can be coagulated by salts and repeptized by water.

Mylius and Dietz (1901) obtained colloidal UO_3 from a precipitate obtained by hydrolysis of UO_2Cl_2 with silver oxide. The yellow weakly acid colloid (it contains some hydrochloric acid which cannot be removed by dialysis) is stable at $0°C$ but decomposes at room temperature with precipitation of $UO_3 \cdot 2H_2O$.

Szilard (1907) decomposed a solution of uranyl acetate in water-ether mixture by exposure to light. The violet $U_3O_8 \cdot xH_2O$ precipitate was allowed to oxidize in air to a yellow UO_3 hydrate. The latter was suspended in water, and the suspension was gradually added to hot diluted uranyl nitrate (or thorium nitrate) solution, producing a very stable hydrosol (orange-yellow with uranyl nitrate, greenish yellow with thorium nitrate).

Sen (1928) studied the effect of different sugars on the peptization of UO_3 hydrate. The precipitation of UO_3 hydrate by alkali can be prevented by the presence of sucrose, lactose, glucose, or fructose.

Kargin (1931) found that a UO_3 hydrosol can be obtained simply by precipitating ammonium diuranate from uranyl nitrate solution by ammonia with stirring and peptizing in a large volume of water.

D'yachkovskiĭ (1926, 1931) precipitated alkali uranate from UO_2Cl_2 solution with 0.1N alkali hydroxide. The precipitate was washed with hot water until it began passing through the filter and was then peptized in a large volume of water. Alkali was removed by dialysis. Sol particles were negatively charged. They had a size of about 5×10^{-5} cm and a density of 7.45 g/cc. The sol could be coagulated by KCl and $BaCl_2$, and its charge was reversed by $AlCl_3$. At $15°C$ the viscosity of the sol at 15 per cent was 1.0393 that of water. The hydrosol was stable to boiling but coagulated on freezing. The surface tension of the 1 per cent sol showed a maximum at 30 to $40°C$.

4. OXIDATION AND REDUCTION OF URANIUM AND ITS OXIDES

4.1 <u>Thermodynamics of Formation and Interconversion of Uranium Oxides</u>. (a) <u>Heats of Formation and Interconversion</u>. Direct calorimetric data were obtained by Mixter (1912a) for the combustion of uranium and UO_2 to U_3O_8 and for the formation of sodium uranate,

Na_2UO_4, from uranium, UO_2, U_3O_8, or UO_3 and sodium peroxide (see Sec. 6.2c). The heats of formation of UO_2 and U_3O_8 could therefore be calculated either from combustion with oxygen or from the reaction with Na_2O_2, whereas the heat of formation of UO_3 could be derived only from experiments with Na_2O_2. An independent calculation of the heat of formation of UO_3 was made by Biltz and Fendius (1928). These investigators compared the heat of formation of uranyl chloride solutions from UCl_3 (or UCl_4) with the heat of formation of the same

Table 11.28—Heats of Formation of Uranium Oxides (per Gram Atom of Uranium)

Reaction	$-\Delta H$, kcal/g atom of uranium				
	Mixter, 1912a		Biltz and Fendius, 1928		Brewer,* MP Berkeley 1
	Combustion	Reaction with Na_2O_2	Via UCl_3	Via UCl_4	
(16) $U + O_2 \rightarrow UO_2$	256.6	269.7†	...	...	(257)
(17) $U + \frac{4}{3}O_2 \rightarrow \frac{1}{3}U_3O_8$	281.7	298.5	...	...	285
(18) $U + \frac{3}{2}O_2 \rightarrow UO_3$	...	303.9	294	292 (288)‡	291
(19) $UO_2 + \frac{1}{6}O_2 \rightarrow \frac{1}{3}U_3O_8$	25.1	28.8	...	...	28
(20) $UO_2 + \frac{1}{2}O_2 \rightarrow UO_3$	...	34.2	(27)§	(25)§	34
(21) $\frac{1}{3}U_3O_8 + \frac{1}{6}O_2 \rightarrow UO_3$	...	5.4	(12)§	(10)§	6

*Calculated from Biltz and Müller's tensimetric data; cf. Sec. 4.1a for details of calculation.
†MacWood (UCRL 3) has recalculated Mixter's value, using better data for Na_2O_2 and Na_2O, and gives $\Delta H = -270.4$ kcal per mole.
‡Corrected by MacWood and Altman (UCRL 2).
§By combination with Mixter's combustion values.

solutions from UO_3 and also measured directly the heats of formation of UCl_3 and UCl_4 from the elements. Their results were critically reviewed and corrected at the University of California Radiation Laboratory (UCRL 2). The results of the two calorimetric investigations are shown in Table 11.28.

Biltz and Müller (1927a) made estimates of the heats of reactions 19 and 21 (Table 11.28) from the decomposition pressures of the oxides UO_3 and U_3O_8. Somewhat more detailed calculations were made by Brewer (MP Berkeley 1), using the same experimental data. The tensimetric data of Biltz and Müller are insufficient for exact calculation because of the formation of solid solutions. To obtain exact results the van't Hoff equation would have to be applied not only to the decomposition isochores of U_3O_8 and UO_3 but also to those of all the oxides of intermediate compositions, and the total heat of oxygenation would have to be obtained by integration of the "differential" heats (as derived from this equation) over the ranges $x = 2.0$ to 2.67 and $x = 2.67$ to 3.0, respectively.

No attempts at such exact calculation were made. In the range between U_3O_8 and UO_3, Biltz and Müller applied the van't Hoff equation to "average" pressure values obtained by a rather arbitrary averaging process from the isotherms at 500, 580, and 610°C (cf. Fig. 11.6). They calculated in this way an "average decomposition heat" of 17 kcal per gram atom of oxygen or 5.67 kcal per gram atom of uranium, in satisfactory agreement with Mixter's value in Table 11.28 (5.4 kcal per gram atom of uranium). Brewer (MP Berkeley 1) calculated from the same data a value of 17.5 kcal per gram atom of oxygen.

In the range between UO_2 and U_3O_8 the van't Hoff equation was applied by Biltz and Müller to the diphasic constant-pressure region $UO_{2.30}$ to $UO_{2.62}$ (using the temperature range 1160 to 1240°C). This led to a value $-\Delta H = 39.5$ kcal per gram atom of oxygen for the heat of conversion of $UO_{2.30}$ to $UO_{2.62}$. This figure is close to Mixter's value for the conversion of UO_2 to U_3O_8 in Table 11.28 (25 to 29 kcal per gram atom of uranium or 37.5 to 43.5 kcal per gram atom of oxygen). Brewer (MP Berkeley 1) calculated, also from Biltz's data, $-\Delta H = 40.5$ kcal per gram atom of oxygen.

The "first" differential heat of oxygenation of UO_2 certainly is much greater than 40 kcal per gram atom of oxygen. Brewer estimated, from the one pressure value given by Biltz for $UO_{2.20}$, that the average heat of oxygenation between $UO_{2.00}$ and $UO_{2.25}$ is about 48 kcal per gram atom of oxygen. He further estimated 30 kcal per gram atom of oxygen as the average heat of oxygenation between $UO_{2.62}$ and $UO_{2.67}$. The "last" differential heat of oxygenation of UO_2 to U_3O_8 was estimated by Biltz and Müller to be as low as 15 kcal per gram atom of oxygen.

By combining three estimates (48 kcal per gram atom of oxygen for the range $x = 2.00$ to 2.25, 40.5 kcal per gram atom of oxygen for the range $x = 2.25$ to 2.62, and 30 kcal per gram atom of oxygen for the range $x = 2.62$ to 2.67) with Mixter's value (257 kcal per gram atom of uranium) for the heat of formation of UO_2, Brewer calculated 285 kcal per gram atom of uranium as the heat of formation of U_3O_8. By adding 17.5 kcal per gram atom of oxygen for the heat of oxygenation of U_3O_8 to UO_3, Brewer obtained 291 kcal per gram atom of uranium as the heat of formation of UO_3. These values are given in the last column of Table 11.28.

Avgustinik (1947) obtained a value of 270 kcal per mole for the heat of formation of UO_3 by extrapolation from the heats of formation of the oxides of other elements in the periodic system.

The heat of formation of UO_3 must depend on the allotropic form. According to Brewer the values obtained above probably apply to the hexagonal form $UO_3(I)$; the amorphous form may be slightly less stable, while the crystalline forms $UO_3(II)$ and $UO_3(III)$ may be more stable than $UO_3(I)$ by 1 to 2 kcal per gram atom of uranium.

The fact that the monoxide UO can be formed from uranium and UO_2 indicates that $-\Delta H_{UO}$ is greater than 129 kcal per mole (one-half of the heat of formation of UO_2). On the other hand, since magnesium reduces the higher uranium oxides to metal (and not to UO), $-\Delta H$ must be less than 142 kcal per mole. The most probable value is therefore

$$-\Delta H_{UO} = 135 \pm 5 \text{ kcal per mole}$$

(b) Free Energies and Entropies of Formation and Interconversion of Uranium Oxides. The entropy of the monoxide UO was given by Brewer (MP Berkeley 1) as -20 e.u. at $1000°K$.

UO_2. Brewer (MP Berkeley 3) used Mixter's value of ΔH and an estimate of the entropy of uranium and UO_2 attributed to Latimer,

$$-\Delta S_{UO_2} = 42.1 \text{ e.u.}$$

to calculate the free energy of formation of UO_2. He obtained

$$-\Delta F_{UO_2} (298.15°K) = 244 \text{ kcal per mole}$$

Later Brewer (MP Berkeley 1) gave

$$\Delta S_{UO_2} = 42.5 \text{ at } 298°K, 41.9 \text{ at } 500°K, 40.5 \text{ at } 1000°K, 40.1 \text{ at } 1500°K$$

The above value of the free energy of formation of UO_2 (-244 kcal per mole) permits the estimate that heating to $4200°K$ will be necessary for the equilibrium pressure of oxygen over uranium dioxide to reach 10^{-2} mm Hg.

The entropy of formation of $UO_{2.25}$ (assumed by Brewer to be the upper limit of solid solubility of oxygen in UO_2) was estimated by him as

$$\Delta S = 45.5 \text{ e.u. at } 1000°K$$

U_3O_8. The tensimetric data of Biltz and Müller can be used for the calculation of free energy of formation of this oxide from UO_2 and O_2 (cf. reaction 19). MacWood and Altman (UCRL 2) assumed, as an approximation, the constant decomposition pressures found in the range $UO_{2.30}$ to $UO_{2.62}$ to be valid for the whole range $UO_{2.0}$ to $UO_{3.0}$ and calculated from them the equation

$$-\Delta F^\circ = 74.432 - 2.35T \log T - 4.75 \times 10^{-4}T^2 + 3.096 \times 10^5 T^{-1}$$

$$-28.61T \quad \text{cal per mole of } U_3O_8 \quad (22)$$

leading to

$$-\Delta F^\circ_{298} = 65.07 \text{ kcal per mole of } U_3O_8$$

In deriving Eq. 22, use was made of Eqs. 5 and 7 for the specific heats of UO_2 and U_3O_8, respectively. This gives (for the formation of 1 mole of U_3O_8 from 3 moles of UO_2 and 1 mole of O_2)

$$\Delta C_p = 1.02 + 9.51 \times 10^{-4}T - 6.193 \times 10^5 T^{-2} \quad (23)$$

and hence

$$-\Delta H = 74.432 + 1.02T + 4.75 \times 10^{-4}T^2 + 6.193 \times 10^5 T^{-1} \quad (24)$$

leading to

$$-\Delta H_{298} = 76.75 \text{ kcal per mole of } U_3O_8$$

By combination with the above given value of ΔF°_{298} (65.07 kcal per mole) the following equation is obtained:

$$-\Delta S_{298} = 39.2 \text{ e.u.}$$

The $-\Delta H$ value used (76.8 kcal per mole U_3O_8 or 25.6 kcal per gram atom of uranium) is in satisfactory agreement with the values given in Table 11.28.

Kirshenbaum (SAM Columbia 12) gave for the same reaction

$$-\Delta F^\circ = +76{,}890 - 63.1T + 2.58 \times 10^{-3}T^2 - 0.32 \times 10^{-6}T^3 - 0.62 \times 10^5 T^{-1}$$

$$+ 3.22T \ln T \quad \text{cal per mole of } U_3O_8 \quad (25)$$

leading to the values of ΔF°_T given in Table 11.29. Equation 25 was based on data for the reaction

$$U_3O_8 + 2H_2 \rightarrow 3UO_2 + 2H_2O \text{ (gas)} \quad (26)$$

given by Davis and Burton (MP Chicago 9).

For the formation of U_3O_8 from the elements, MacWood and Altman calculated, using Mixter's value for ΔH_{298} and the above-mentioned specific heat equations,

$$-\Delta H = 850{,}000 - 19.01T + 9.65 \times 10^{-3}T^2 - 3.984 \times 10^5 T^{-1} \quad (27)$$

and obtained by Nernst's approximation

$$-\Delta F^\circ = 850{,}000 + 43.79T \log T - 9.65 \times 10^{-3}T^2 + 1.992 \times 10^5 T^{-1}$$
$$-303.1T \qquad (28)$$

leading to

$$-\Delta F^\circ_{298} = 791.7 \text{ kcal per mole of } U_3O_8$$
$$\Delta S_{298} = 184.2 \text{ e.u.}$$

In Table 11.30 the results of MacWood and Altman are compared with the figures given by Brewer.

The values of the function $(\Delta F_T - \Delta H_{298})/T$ for the formation of U_3O_8 from the elements are also given by Brewer in Table 11.31.

$\underline{UO_3}$. Kirshenbaum (SAM Columbia 1) gave the following equations for the standard free energy of formation of UO_3 from UO_2 or U_3O_8 and oxygen:

$$UO_2 + \tfrac{1}{2}O_2 \rightarrow UO_3$$

$$-\Delta F^\circ = 33.830 - 9.32T + 0.401 \times 10^{-3}T^2 + 1.16 \times 10^5 T^{-1}$$
$$-1.37T \ln T \qquad (29)$$

$$U_3O_8 + \tfrac{1}{2}O_2 \rightarrow 3UO_3$$

$$-\Delta F^\circ = 24{,}600 + 35.1T - 1.38 \times 10^{-3}T^2 + 0.32 \times 10^{-6}T^3$$
$$+ 4.10 \times 10^5 T^{-1} - 7.33T \ln T \qquad (30)$$

leading to values tabulated in Table 11.32.

MacWood and Altman (UCRL 2) used 288 kcal per mole for the $-\Delta H$ of UO_3 formation from the elements at $298°K$, a value that they obtained by substituting a corrected value for the heat of formation of UCl_4 into the calculation of Biltz and Fendius (cf. Table 11.28). They further estimated $S^\circ_{298} = 22$ e.u. (cf. Sec. 2.5e) and, using the specific heat equations for uranium and UO_3, derived the following equations for the heat and free energy of formation of UO_3:

$$\Delta H = -289{,}400 + 6.39T - 5.04 \times 10^{-3}T^2 - 1.29 \times 10^4 T^{-1}$$
$$\text{cal per mole} \qquad (31)$$

$$\Delta F^\circ = -289{,}400 - 14.72T \log T + 5.04 \times 10^{-3}T^2 - 6.5 \times 10^3 T^{-1}$$
$$+ 68.2T \qquad (32)$$

Table 11.29 — Free Energy of Oxidation of UO_2 to U_3O_8

Temp., °K	$-\Delta F_T^\circ$, kcal per mole of U_3O_8
300	63.45
400	59.6
500	55.8
600	52.1
1000	38.2

Table 11.30 — Thermodynamics of Formation of U_3O_8

	MacWood and Altman, UCRL 2	Brewer, MP Berkeley 1
$-\Delta H_{298}$, kcal per mole of U_3O_8	846.6*	855†
$-\Delta F_{298}^\circ$, kcal per mole of U_3O_8	791.7	805
$-\Delta S$, e.u.	184.2	166

*Mixter's value.
†Derived as described in Sec. 4.1a.

Table 11.31 — Entropy of Formation of U_3O_8 at Various Temperatures

Temp., °K	$(\Delta F_T - \Delta H_{298})/T$ for U_3O_8, cal/mole/°C
298	166*
500	167
1000	163
1500	161

*Equal to ΔS_{298}.

Table 11.32 — Free Energy of Oxidation of UO_2 and U_3O_8 to UO_3

Temp., °K	$-\Delta F^\circ$ (for reaction 29), kcal/mole UO_2	$-\Delta F^\circ$ (for reaction 30), kcal/mole U_3O_8
300	29.1	22.6
400	27.2	21.0
500	25.2	19.1
600	23.3	17.2
1000	15.6	8.0

These equations give

$$\Delta H_{298} = -288 \text{ kcal per mole}$$
$$\Delta F_{298}^{\circ} = -269.1 \text{ kcal per mole}$$
$$\Delta S_{298} = -63.6 \text{ e.u.}$$

Brewer (MP Berkeley 1) calculated the values given in Table 11.33 of the function $(\Delta F_T - \Delta H_{298})/T$ for the formation of UO_3 from the elements.

Table 11.33 —Entropy of Formation of UO_3 at Various Temperatures

Temp., °K	$(\Delta F_T - \Delta H_{298})/T$ for UO_3, cal/mole/°C
298	62.0*
500	61.7
1000	60.6

*Equal to S_{298}.

Table 11.34 —Potential of Platinum Electrodes with Uranium Oxides

UO_2 or U_3O_8	$UO_3 \cdot H_2O$*	$U_3O_{10} \cdot 2H_2O$†
−0.773 to −0.787 volt	−0.860 volt	−0.687 volt

*Yellow hydroxide obtained by electrolysis of uranyl acetate.
†Black hydroxide obtained by electrolysis (see Sec. 5.3f).

Pierlé (1919) prepared pastes from gelatin and finely powdered uranium oxides (or hydroxides) and measured the potential of platinum electrodes covered by these pastes and immersed in uranyl nitrate solution [14.3 g of $UO_2(NO_3)_2$ per liter] against a normal calomel electrode. He obtained the values, the meaning of which is not clear, given in Table 11.34.

Free energies of reduction of UO_3 and U_3O_8 by hydrogen are given in Sec. 4.4c.

4.2 Reduction of Uranium Oxides to Metal. The reduction of uranium oxides to metal is treated in Chap. 4. Because of the high stability of UO_2 (Table 11.28) reduction to the metal can be achieved only by very strong reductants, such as the alkaline earth metals.

4.3 Formation of UO by Oxidation of Uranium Metal or Reduction of UO_2. Uranium metal exposed to air at room temperature becomes covered by brownish-black uranium dioxide; green-black U_3O_8 is formed on heating (cf. Chap. 6). However, if the oxygen pressure is very low (10^{-9} atm) and the temperature is above 800°C, a gray film is formed which contains a certain variable proportion of uranium monoxide recognizable by its characteristic x-ray diffraction pattern (MP Ames 8). The amount of UO (relative to that of UO_2) increases upon prolonged heating. After 5 hr of "vacuum" annealing at 800°C, the x-ray pattern of the oxide film showed almost no UO_2 lines (British 1,3).

A mixture of UO_2 and uranium metal powder can be converted into UO.

$$UO_2 + U \rightarrow 2UO \qquad\qquad (33)$$

However, this reaction is so slow, even at 2000°C, that equilibrium could not be reached even in prolonged experiments (MP Ames 9). In the most successful preparation to date of this writing (MP Ames 9,10), heating of a (UO + U) powder mixture to 1900°C gave a product containing 90.6 per cent uranium and 0.24 per cent carbon. Assuming the rest to be oxygen, this preparation must have been approximately one-third UO and two-thirds UO_2.

At Battelle (Battelle 1) the cubic phase with a lattice constant of 4.93 A was observed in the study of uranium-molybdenum alloys and interpreted as UO, formed by oxidation of the metal powder by the residual gas present during the heat-treatment in evacuated quartz tubes.

Remarkably, the UO phase appeared much stronger in samples evacuated by means of a mercury diffusion pump than in those evacuated by a mechanical pump.

The straight thermal decomposition of UO_2 into UO and uranium, which occurs only at very high temperatures, has not yet been studied in detail. According to Greenwood (1908) uranium dioxide first begins to lose oxygen in vacuum at about 1600°C.

4.4 Formation of UO_2 from Uranium or Higher Uranium Oxides. (a) Oxidation of Uranium Metal to UO_2. Oxidation of uranium metal by air at room temperatures leads to the formation of brown dioxide, UO_2; prolonged exposure or heating gives green-black U_3O_8. On massive metal the first oxide layer formed tends to protect the rest from oxidation (MP Ames 5). (For more details of uranium behavior in air, see Chap. 6.)

Since the free energy of formation of UO_2 (−122 kcal per gram atom of oxygen) is much more negative than that of water (−57 kcal per gram atom of oxygen), uranium is thermodynamically capable of reducing water with the formation of UO_2. Massive uranium reacts with cold water only very slowly (Péligot, 1842b; Lely and Hamburger, 1914). Uranium powder reacts slowly at room temperature and more rapidly at 100°C (Moissan, 1893, 1896). Very finely divided uranium (obtained by the decomposition of UH_3) decomposes water rapidly even in the cold.

According to Columbia observers (SAM Columbia 4) UO_2 obtained by treating uranium metal with water is pyrophoric, i.e., it oxidizes spontaneously to U_3O_8 in air.

(b) Thermal Decomposition of U_3O_8 or UO_3 to UO_2 and O_2. According to Sec. 4.1b the standard free energy of decomposition of U_3O_8 at room temperature into UO_2 and O_2 is about 21.5 kcal per gram atom of uranium (21.7 kcal after MacWood and Altman, 21.2 kcal after Kirshenbaum). The free energy of decomposition of UO_3 into UO_2 and O_2 is 29.1 kcal per gram atom of uranium. It can be expected that both U_3O_8 and UO_3 are easily converted to UO_2 by heating in vacuum. However, the above values represent the average or integral free energies of decomposition. The true or differential free energy of decomposition increases far above this average in the solid-solution range below $UO_{2.30}$ (cf. the decomposition isotherms in Fig. 11.6). Experiments by Biltz and Müller (1927a), described in more detail in Sec. 1.3, indicate that because of this the decomposition of U_3O_8 usually stops (in the normally obtainable vacuums and at temperatures up to 1300°C) at an approximate composition of $UO_{2.15}$.

In air the decomposition of U_3O_8 begins at about 900°C, the temperature at which the decomposition pressure of stoichiometrically pure $UO_{2.667}$ reaches the partial pressure of oxygen in the air (Colani, 1907a; Biltz and Müller, 1927a). Since between $UO_{2.667}$ and $UO_{2.61}$ the decomposition pressure drops rapidly with decreasing oxygen content, the decomposition of U_3O_8 stops after only very little oxygen has been lost unless the oxygen pressure over the oxide has been reduced far below that in the open air or the temperature has been raised considerably above 1000°C. According to Fig. 11.2, in the constant-pressure range, $UO_{2.62}$ to $UO_{2.30}$, the decomposition pressure of oxygen is only 0.3 mm Hg at 1160°C. By the time the composition $UO_{2.20}$ has been reached this pressure has dropped to 0.02 mm Hg.

The isotherms reproduced in Fig. 11.2 show how far the decomposition can be expected to progress, according to Biltz and Müller, under various oxygen pressures at temperatures up to 1160°C.

The thermal decomposition of U_3O_8 also was studied by Tennessee Eastman Corporation investigators (CEW-TEC 3). They started with $UO_{3.0}$, $UO_{2.64}$, and $UO_{2.61}$ as material and found that in evacuated sealed tubes no decomposition beyond the formula $UO_{2.61}$ took place at 700°C. The oxygen pressure increase after ignition was 34 mm Hg with UO_3 and 2.4 mm Hg with $UO_{2.64}$. No oxygen was evolved at all with $UO_{2.61}$. At 750 to 830°C after heating UO_3 in high vacuum for 12 hr, the residue still had the composition $UO_{2.55}$ and $UO_{2.62}$. These results confirm the conclusion drawn above from the observations of Biltz and Müller that the thermal decomposition of U_3O_8 at temperatures below 1000°C stops considerably short of the composition $UO_{2.5}$ even in high vacuum.

According to Jolibois and Bossuet (1922) complete decomposition of U_3O_8 to UO_2 and oxygen can be obtained by heating to 2000°C in vacuum.

It has been stated that U_3O_8 can be completely deoxygenated also in a stream of oxygen-free indifferent gas instead of high vacuum. For example, Zimmermann, Alibegoff, and Krüss (1886b) claimed to have obtained pure brown UO_2 by igniting U_3O_8 over the gas flame in a platinum crucible in nitrogen or carbon dioxide, and Friederich and Sittig (1925) described the formation of brown UO_2 by heating UO_3 in a hydrogen stream to about 1000°C and the formation of a dark-blue dioxide (Sec. 2.2j) by heating UO_3 in a stream of nitrogen to 1100°C. The assertion that the latter product also had the exact composition $UO_{2.00}$ appears somewhat doubtful in the light of our present knowledge about the difficulty of complete conversion of U_3O_8 to UO_2.

The deoxygenation of UO_3 will be described in more detail in Sec. 4.5 when dealing with the conversion of this oxide to U_3O_8. Some observations indicate, however, that UO_3 may be converted to UO_2 without the intermediate formation of the U_3O_8 phase (cf. Sec. 4.4c).

(c) Reduction of UO_3 and U_3O_8 to UO_2. Reduction of UO_3 or U_3O_8 by Hydrogen to UO_2. While uranium metal can liberate hydrogen from water to form UO_2, hydrogen is able to reduce the higher uranium oxides, UO_3 and U_3O_8, to UO_2. The free energy of reduction (with formation of H_2O gas), according to the estimates made at Columbia University (SAM Columbia 12), is given by Eqs. 34 and 35.

$$UO_3 + H_2 \rightarrow UO_2 + H_2O \text{ (gas)}$$

$$-\Delta F° = 22,890 + 16.42T - 0.79 \times 10^{-3}T^2 + 0.22 \times 10^{-6}T^3$$
$$-0.69 \times 10^5 T^{-1} - 1.17T \ln T \quad \text{cal per mole} \quad (34)$$

$$U_3O_8 + 2H_2 \rightarrow 3UO_2 + 2H_2O \text{ (gas)}$$

$$-\Delta F° = 36,545 + 77.3T - 3.36 \times 10^{-3}T^2 + 0.762 \times 10^{-6}T^3$$
$$+ 1.56 \times 10^5 T^{-1} + 11.32T \ln T \quad \text{cal per mole} \quad (35)$$

The evaluation of ΔF from these equations gives the values tabulated in Table 11.35.

Related to 1 gram atom of uranium, the free energy of reduction to UO_2 by hydrogen at room temperature is 25.5 kcal for UO_3 and 15.3 kcal for U_3O_8.

The free-energy values given above were based on Davis and Burton's (MP Chicago 9) calculation of the free energy of reaction 35 and were used at Columbia also for the calculation of the free energy

Table 11.35 — Free Energy of Reduction of UO_3 and U_3O_8 to UO_2 by Hydrogen

Reaction	$-\Delta F$				
	300° K	400° K	500° K	600° K	1000° K
(34) Kcal per mole of UO_3	25.5	26.4	27.2	27.9	30.6
(35) Kcal per mole of U_3O_8	45.8	47.5	49.0	50.3	54.0

of formation of U_3O_8 (cf. Eq. 25). The procedures, of course, could also be reversed, and the free energy of formation of U_3O_8 (as estimated by MacWood and Altman; cf. Eq. 28) could be used for the calculation of the free energy of reduction of U_3O_8 by hydrogen.

Preparation of UO_2 by reduction of the higher oxides with hydrogen was first described by Arfvedson (1822b, 1824a), Lecanu (1825), Laugier and Boudet (1825), and later by Sabatier and Senderens (1895). The same method was used by Aloy (1900), de Coninck (1908), Jolibois and Bossuet (1922), Lebeau (1922a), Goldschmidt and Thomassen (1923c), and Biltz and Müller (1927f). Some disagreement exists as to the temperature required for complete reduction to $UO_{2.00}$, particularly in the case of U_3O_8 as starting material. Lebeau (1922a) found that U_3O_8 is rapidly and completely reduced by hydrogen to brown $UO_{2.0}$ at 900 to 1000°C without the formation of intermediary oxides. Biltz and Müller (1927a) also have prepared stoichiometrically pure dioxide by reduction of U_3O_8 by H_2 at 900 to 1000°C. They obtained in this way a brown dioxide from green U_3O_8 and a darker, purplish-brown dioxide from black, strongly ignited U_3O_8 (both products were analyzed as $UO_{2.0}$). Jolibois and Bossuet (1922) asserted that a temperature of 625 to 650°C is sufficient for complete reduction of U_3O_8 by hydrogen if P_2O_5 is used to remove the water vapor.

British observers (British 4) asserted, on the other hand, that the reduction of U_3O_8 by hydrogen at 650 to 680°C stops at $UO_{2.14}$ (or $UO_3 \cdot 6UO_2$), the same composition that is obtained by decomposition of U_3O_8 in vacuum at temperatures up to 1300°C. They suggested that the reduction of U_3O_8 by hydrogen consists of two stages, namely, thermal decomposition of the oxide and reduction of liberated oxygen with hydrogen. Therefore the whole reaction stops when the decomposition pressure of the oxide becomes negligible, which, according to Sec. 4.4b, occurs when the oxygen content is reduced to about $UO_{2.15}$.

It seems unlikely, however, that hydrogen would not react directly with the solid oxide. Even if it proves true that (at 680°C) the reduction of U_3O_8 with hydrogen stops at $UO_{2.15}$, there will be an inclination to attribute this to a high activation energy of the direct reaction of $UO_{2.15}$ with hydrogen, while the thermal decomposition of U_3O_8 stops at the same composition because of the high energy of reaction even at temperatures as high as 1300°C. The statement of the British authors that, according to Newbery and Pring (1916a), a temperature of 2000°C and a hydrogen pressure of 150 atm are required for complete reduction of $UO_{2.15}$ to $UO_{2.00}$ disregards the observations of Lebeau and of Biltz and Müller, who obtained stoichiometrically pure $UO_{2.0}$ by U_3O_8 reduction with hydrogen at 900°C under only 1 atm of hydrogen (cf. above).

Grønvold and Haraldsen (1948b) noted that the reduction of U_3O_8 by hydrogen at low temperature (330°C) leads only to $UO_{\geq 2.56}$, the lowest oxides in which the orthorhombic structure of U_3O_8 is preserved (cf. Sec. 1.3).

In the reduction of U_3O_8 or UO_3 by hydrogen there is apparently no danger of reaction proceeding beyond the composition $UO_{2.0}$. As mentioned in Sec. 4.1a, UO_2 is thermodynamically stable in hydrogen as far as conversion to uranium (and probably also to UO) is concerned. It is less certain whether a reduction to UO_x ($1 < x < 2$) may not be thermodynamically possible (cf. Sec. 4.3), but experiments gave no indications of any reduction of UO_2 by hydrogen even under such extreme conditions as $p = 25$ atm, $t =$ melting point of tungsten (von Wartenberg, Broy, and Reinicke, 1923); $p = 150$ atm, $t = 2500$°C (Newbery and Pring, 1916b); and $p = 100$ atm, $t = 1100$°C (Rideal, 1914). Faehr (1908) observed no reduction of UO_2 in a high-voltage arc in hydrogen. Columbia University investigators (SAM Columbia 4) noted that UO_2 slurries in water were stable in hydrogen up to 250°C.

Reduction of U_3O_8 or UO_3 by Metals and Metal Oxides. According to Gay-Lussac and Thénard (1811) U_3O_8 is reduced to UO_2 by potassium at about 150°C with weak evolution of heat. At the Mallinckrodt Chemical Works (Mallinckrodt 1) a method of reduction of UO_3 by means of magnesium was developed. Magnesium reduces orange UO_3 to brown UO_2 at 500°C; no reduction of UO_2 by magnesium occurs under these conditions. Tammann and Rosenthal (1926) observed the reduction of UO_3 by FeO and at elevated temperatures.

Reduction of UO_3 to UO_2 by Carbon Monoxide. Boullé et al. (1950) found that UO_3 (but not U_3O_8) is rapidly reduced to UO_2 by CO at 350°C.

Reduction of U_3O_8 or UO_3 by Carbon. Uranium dioxide was first prepared by Klaproth (1789) by reducing U_3O_8 with carbon. Because of the black color of the powder and its metallic luster, he took it to be the metal. The error was recognized only fifty years later by Péligot (1842b,c). According to Moissan (1892) reduction of U_3O_8 to uranium by carbon is possible only in an electric furnace at 3000°C.

According to Aloy (1901a) UO_2 is formed in the electric furnace from U_3O_8 (or UO_3) and carbon if the temperature is not high enough for the formation of uranium carbide.

Reduction of U_3O_8 and UO_3 by Methane. At Berkeley (UCRL 3,4) it was found that the conversion of bright-red UO_3 to UO_2, which is slow at 500°C in air, nitrogen, or oxygen, is accelerated and leads rapidly to black UO_2 if CH_4 is introduced. The reaction begins at 400°C. At 450°C the product was a mixture of UO_2 and U_3O_8.

Reduction of U_3O_8 or UO_3 by Sulfur, Hydrogen Sulfide, or Carbon Disulfide. Hermann (1861) obtained UO_2 by igniting U_3O_8 (or sodium uranate) with sulfur and ammonium chloride and washing the product with water, as well as by ignition of UO_3 or U_3O_8 in dry carbon dioxide saturated with carbon disulfide vapor. However, this reaction is now known to be complex and may produce oxysulfides instead of the dioxide. Kohlschütter (1901) found that UO_3 is reduced to UO_2 by hydrogen sulfide in alkaline, neutral, or even weakly acid solution.

Reduction of U_3O_8 with Ammonia or Ammonium Chloride. Smith and Matthews (1895) obtained a reddish-brown uranium oxide containing no nitrogen or chlorine by heating U_3O_8 to white heat for 6 hr with a large excess of ammonium chloride in a porcelain crucible placed in a larger unglazed crucible and surrounded with charcoal.

The reduction of UO_3 or U_3O_8 with ammonia was studied at the Tennessee Eastman Corporation (CEW-TEC 4) and found to represent a convenient method for the conversion of U(VI) to U(IV) compounds.

Almost no reduction of UO_3 by ammonia was found to take place below 250°C (except for the occasional formation of a green surface layer). At 250 to 400°C the reduction appears to proceed only to the composition U_3O_8; even this, however, represents a stronger reduction than could be achieved at these temperatures by straight thermal decomposition. Above 450°C reduction to UO_2 sets in and becomes complete (more complete than with hydrogen as reductant) at 550°C.

Reduction of U_3O_8 with Oxalic Acid. Wertheim (1843) obtained black, easily oxidizable UO_2 by heating U_3O_8 with oxalic acid in the absence of air. De Coninck and Raynaud (1911b) reduced U_3O_8 to UO_2 by using calcium oxalate at red heat.

Reduction by Ethanol Vapor. U_3O_8 can be reduced to UO_2 by ethanol vapor at or above 200°C (referred to in CEW-TEC 3).

4.5 Conversion of UO_2 and UO_3 to U_3O_8. (a) Oxidation of UO_2 to U_3O_8. Oxidation of UO_2 by Oxygen. Thermodynamically, U_3O_8 is stable only in a rather narrow range of temperature, e.g., at p_{O_2} = 10 mm Hg between 580 and about 750°C (cf. Fig. 11.4) or in air (p_{O_2} ≈ 150 mm Hg) between about 650 and 900°C (Colani, 1907a; Biltz and Müller, 1927a). Practically, however, U_3O_8 is obtained in a stoichiometrically more or less pure state by ignition of both lower and higher uranium oxides in air. (The usual method of gravimetric determination of uranium as U_3O_8 is based on this fact.) This practical stability of U_3O_8 is due to the difference between the rates of oxidation of the lower uranium oxides to U_3O_8 and of U_3O_8 to UO_3. The first process occurs easily even at rather low temperatures, but the second process is very slow. According to Sec. 1.4 U_3O_8 is not oxidized to UO_3 in air at low temperatures except perhaps in geological times. At elevated temperatures (of the order of 350°C) U_3O_8 can be oxidized in the laboratory if it is present in an especially active, finely dispersed form. To oxidize ordinary U_3O_8 rapidly to UO_3 it is necessary to use oxygen pressures of the order of 100 psi.

Because of this difference in the rate of oxygen uptake below and above the composition $UO_{2.67}$, ignition of lower oxides in air leads to the formation of practically pure U_3O_8 even if the temperature of heating is one at which UO_3 is the thermodynamically stable oxide (below 650°C). If the decomposition temperature of U_3O_8 (about 900°C in air) has been exceeded in the process of ignition, the missing oxygen is likely to be recovered during cooling (unless the latter is very rapid; cf. Zimmermann, Alibegoff, and Krüss, 1886a), but no oxygen in excess of the formula $UO_{2.67}$ will be taken up.

The temperature to which UO_2 has to be brought in air for oxidation to U_3O_8 depends on the "activity" of the sample, i.e., probably on its degree of dispersion. According to Jolibois and Bossuet (1922) rapid oxidation of ordinary UO_2 begins at about 185°C. However, Jolibois (1947) found that the oxidation at low temperature does not proceed beyond the composition $UO_{2.33}$ (approximately U_3O_7). Heating curves showed a break between 220°C (where the composition U_3O_8 is reached) and 300°C (where the transformation of the cubic UO_2 lattice into an orthorhombic phase, U_2O_5 or U_3O_8, begins).

Grønvold and Haraldsen (1948b) noted that UO_2 preparations reacted with oxygen at temperatures as low as 120°C. They made x-ray studies of the products obtained by oxidation in pure oxygen at temperatures from 100 to 265°C. At 150°C, oxygen uptake led to (at least) the composition $UO_{2.34}$; the product showed displaced and partially broadened x-ray interference fringes indicating lowered symmetry. A specimen with composition $UO_{2.34}$ (assumed to be still cubic in symmetry) gave a = 5.40 A, compared to a = 5.468 A for UO_2 [Table 11.3 shows that a change in the same direction but considerably smaller in extent (from 5.459 to 5.430 A) was found at Ames]. Grønvold and Haraldsen were undecided whether the lattice change from $UO_{2.00}$ to $UO_{2.34}$ is continuous (as we have assumed it to be in Sec. 1.2) or involves the formation of a new phase.

Oxidation at 200 to 250°C leads, according to Grønvold and Haraldsen, to products with x greater than 2.34; a distinct new phase which they call δ phase (in distinction to the cubic "β phase") is formed; it is tetragonal and has a narrow range of homogeneity around $UO_{2.40}$. Its lattice constants are a = 5.37 A and c = 5.54 A (c:a = 1.03); the observed density is 10.00 g/cc, which cannot be explained by addition of oxygen to, or subtraction of uranium from, UO_2 but only by substitution of oxygen for uranium (composition $U_{0.88}O_{2.12}$, giving a theoretical density of 10.04 g/cc). This probably is the compound U_2O_5, postulated by Lydén (1939) and some earlier authors (Sec. 1.3), which Biltz and Müller (1927) were not able to observe tensimetrically; this is understandable since the tetragonal phase seems to exist only at temperatures lower than 270°C.

"Pyrophoric" UO_2, which is oxidized in air at room temperature with incandescence, was prepared by Péligot (1842a) by decomposition of oxalate. According to Columbia University observers (SAM Columbia 4) similarly unstable uranium dioxide is obtained by reaction of uranium metal with boiling water. At the University of California Radiation Laboratory (UCRL 5) it was found that UO_2 obtained by reduction of UO_3 with natural gas also is pyrophoric, and the same observation was made in Britain (British 4) with brown UO_2 obtained by reducing UO_3 by hydrogen at 650°C.

Oxidation of UO_2 to U_3O_8 by Oxidants Other than Oxygen. The free energy of formation of U_3O_8 from UO_2 and oxygen ($\Delta F° = -64$ kcal per mole of U_3O_8 at 300°K and -38 kcal per mole at 1000°K, cf. Sec. 4.1b) is considerably less negative than the free energy of formation of an equivalent quantity of water ($\Delta F° = -109$ kcal for 2 moles of H_2O at 300°K and -92 kcal for 2 moles of H_2O at 1000°K); therefore UO_2 cannot be expected to reduce water. Regnault (1840) thought that steam oxidizes UO_2 to U_3O_8, but Chaudron (1921) was unable to confirm this observation. Columbia University observers (SAM Columbia 4) found no effect of water vapor on UO_2 at 310°C. Sabatier and Senderens (1892, 1895, 1896) studied the oxidation of UO_2 by nitrogen oxides. N_2O did not attack brown UO_2 (prepared by UO_3 reduction with hydrogen at red heat) but reacted slowly with black UO_2 prepared at lower temperatures. NO reacted with brown UO_2 below red heat with evolution of heat and formation of a black product which the authors considered to be U_2O_5. NO_2 reacted with both brown and black UO_2 in the cold with strong evolution of heat.

(b) Decomposition of UO_3 to U_3O_8. Thermal Decomposition. As described in Sec. 4.5a, U_3O_8 can be obtained by thermal decomposition of UO_3. The isotherms (cf. Fig. 11.2) indicate that, in oxygen at 1 atm pressure, decomposition of UO_3 should begin a little below 470°C and the composition U_3O_8 should be reached somewhere above 650°C. With free access to air ($p_{O_2} \approx 150$ mm Hg), complete decomposition to U_3O_8 should occur at about 650°C. In vacuum, decomposition should be complete below 580°C. It was mentioned in Sec. 1.4 that in practice the decomposition of UO_3 may be delayed indefinitely or may begin only after an induction period of one or several days. Sometimes it occurs only when the decomposition temperature has been considerably exceeded (or when the pressure has been made much lower than the equilibrium decomposition pressure).

These complications are due to the existence of at least four (and perhaps as many as six) allotropic modifications of UO_3 of different stabilities. According to Fried and Davidson (MP Chicago 3) the results of Biltz and Müller were obtained at the lower temperatures with amorphous UO_3 formed by thermal decomposition of $UO_4 \cdot xH_2O$. At the higher temperatures they probably dealt with the hexagonal form $UO_3(I)$. De Coninck (1901a, 1904b) noticed that the brick-red form of UO_3 obtained by prolonged heating of amorphous trioxide in air is much more stable than the initial amorphous orange-red oxide. This brick-red form was probably the $UO_3(II)$ of Fried and Davidson. According to the last-named authors the yellow form $UO_3(III)$, which can be obtained by prolonged heating of UO_3 or U_3O_8 under high oxygen pressure and is also formed by ignition of uranyl nitrate (Mallinckrodt oxide), is still more stable. Under 1 atm of oxygen, $UO_3(III)$ remained unchanged even after 1 hr at 700°C. This may be due to delayed decomposition but is more likely to represent true thermodynamic stability. Amorphous UO_3 was found by Fried and Davidson to decompose under 1 atm of oxygen to $UO_{2.70}$ at 650°C in approximate agreement with the results of Biltz and Müller (1927c). Red UO_3 obtained by Fischer (1913) from uranyl salt solutions by electrolysis and dehydration behaved in a similar way. It began to decompose in air at 470°C and was more or less completely reduced to U_3O_8 at 600°C. Hüttig and von Schroeder (1922a) noticed first signs of oxygen liberation from UO_3 in vacuum at 400 to 500°C. Jolibois and Bossuet (1922) observed a rapid dissociation of UO_3 in vacuum at 502°C. Goldschmidt and Thomassen (1923d) noticed U_3O_8 lines in the UO_3 x-ray diffraction pattern after heating to 270°C in air, a result which is difficult to interpret since, according to Zachariasen, UO_3(I) is transformed into U_3O_8 without the formation of a new phase.

Lebeau (1922a) must have been dealing with one of the more stable modifications of the trioxide when he observed that UO_3 does not change at all in air up to 600°C. Above 600°C his UO_3 sample was converted into olive-green or green-black intermediate oxides and only at 800°C into gray-black U_3O_8. Similarly, Tammann and Rosenthal (1926) observed that UO_3 began to decompose in air only at 670°C with the formation of intermediate green oxides and was completely converted to black U_3O_8 first at 865°C.

Von Schroeder (see Biltz and Müller, 1927a) made the peculiar observation that a UO_3 sample that had remained almost unchanged after 12 days at 600°C in air was reduced to $UO_{2.68}$ within 12 hr when a stream of oxygen was passed over it. In this connection Fried and Davidson (MP Chicago 3) observed that brick-red UO_3 (obtained by gas-phase oxidation of U_3O_8 by N_2O_4) that was heated for 0.5 hr to 600 to 650°C under 1 atm of oxygen was found about 87 per cent changed into yellow UO_3(III), the rest being green U_3O_8.

The formation of oxides intermediate between UO_3 and U_3O_8 by heating to 600 to 650°C was observed also by Fried and Davidson (MP Chicago 3), who obtained $UO_{2.96}$ and $UO_{2.82}$ by heating amorphous UO_3 under 1 atm of oxygen for 0.5 to 1 hr to 620 and 650°C, respectively. Taylor (1931) noted that at 580°C $UO_{3.04}$ was converted to $UO_{2.91}$ within 15 min but required 123 hr for conversion to $UO_{2.84}$.

At Columbia University (SAM Columbia 4) it was observed that amorphous anhydrous UO_3 heated to 600°C showed no conversion to U_3O_8; commercial hydrated yellow UO_3 (7 per cent H_2O, 14 per cent UO_4) lost both water and oxygen at 600°C and was converted to a mixture of UO_3 and U_3O_8.

$UO_3 \cdot 0.5H_2O$ lost both oxygen and water above 430°C and was gradually transformed into a mixture of U_3O_8 and UO_3.

The initial rate of deoxygenation of UO_3 (of unspecified origin) was measured by Tennessee Eastman observers (CEW-TEC 3), with the results shown in Table 11.36.

According to Boullé and Dominé-Bergès (1948), orange UO_3, prepared by the dehydration at 380 to 390°C in air of yellow hydrate obtained by electrolysis of $UO_2(NO_3)_2$ (see Sec. 5.3c), is converted at 520°C into an intermediate compound of approximate composition $UO_{2.90}$ and is converted at 610°C into U_3O_8. The formation of the intermediate (probably a solid solution with the same crystal structure as U_3O_8) is revealed by following the loss of oxygen with a thermobalance or by recording heating curves.

To summarize, it seems that the "unstable" forms of UO_3, e.g., amorphous trioxide, begin to decompose in air at 400 to 500°C and can be fully converted into U_3O_8 at 650°C, while the "stable" forms, e.g., $UO_3(III)$, only begin to decompose in air at 600°C and may have to be heated to 800°C for complete conversion to U_3O_8. It remains to be investigated how far the forms $UO_3(II)$ and $UO_3(III)$ have to be de-oxygenated before a transformation into the orthorhombic U_3O_8 phase occurs.

Table 11.36—Initial Rate of Decomposition of UO_3

Temp., °C	$\dfrac{\text{Moles oxygen}}{\text{Moles uranium} \times \text{min}}$
500	0.007
590	0.031
640	0.086
800	0.192

4.6 **Formation of UO_3 and Its Hydrates from Other Uranium Oxides.**
(a) UO_3 Preparation by Oxidation of UO_2 or U_3O_8. UO_3 is the thermo-dynamically stable uranium oxide in air at temperatures up to about 450 to 600°C (depending on the allotropic form). Between there and about 650°C the stable phases have compositions between UO_3 and $UO_{2.67}$, and only above 650°C does U_3O_8 become the stable uranium oxide in air. Nevertheless, oxidation of UO_2 by oxygen usually leads only to U_3O_8.

In Secs. 1.4 and 4.5a we commented on the slowness of oxidation of U_3O_8 to UO_3 by air or oxygen. Only a few cases of direct oxygenation of U_3O_8 to UO_3 have been reported in the literature. Lebeau (1922b) and Biltz and Müller (1927a) both observed that finely dispersed gray-black U_3O_8 obtained by low temperature (350°C) decomposition of uranyl oxalate became orange-brown in air after 12 hr heating to 350°C. Chemical analysis of the product indicated a composition close to UO_3.

Boullé et al. (1949, 1950) studied more closely the oxidation of finely dispersed U_3O_8, obtained either by decomposition of uranyl oxalate at 350°C, by vacuum decomposition of UO_3 at 450°C, or by reduction of UO_3 with CO at 350°C. The reconversion to UO_3 occurs, in air or oxygen, at temperatures between 300 and 500°C and can be followed by means of a thermobalance registering weight increase, as well as by x-ray analysis of samples taken at different stages of the oxidation. The thermobalance shows, for example, that the gain in weight cor-responding to complete conversion of U_3O_8 (prepared from oxalate) to

UO_3 (1.87 per cent) is reached in an oxygen atmosphere in about 12 hr at 350°C. The x-ray diffraction pattern shows gradual disappearance of the U_3O_8 maxima, as this orthorhombic material is converted to amorphous UO_3.

If U_3O_8 is preheated before exposure to oxygen, its lattice changes, beginning at 450°C, and particularly higher than 575°C, when the lines in the diffraction pattern are doubled. This recrystallization into a finely crystallized product leads to increased difficulty of oxidation.

Table 11.37 — Oxidation of U_3O_8 by Oxygen

Initial pressure, psi	Temperature, °C	U(IV) oxidized to U(VI) in 2 hr, %
Effect of Oxygen Pressure at 150°C		
60		23
90		26
120		31
165		38
Effect of Temperature at 120 Psi		
	100	7.5
	150	30
	200	56
	250	76

The solid solution $UO_{2.90}$, described by the same observers (Sec. 1.4), gives a diffraction pattern identical with that of the more easily oxidizable U_3O_8 form; it, too, is converted at 610°C to the practically unoxidizable microcrystalline form of U_3O_8.

At Columbia University (SAM Columbia 13) the oxidation of U_3O_8 by oxygen was studied in water slurries under high oxygen pressure. Stainless-steel bombs fitted with pyrex linings were used. Under these conditions oxidation was found to be comparatively rapid; at 100 to 250°C and oxygen pressures of 45 to 165 psi, up to 75 per cent was oxidized in 0.5 to 2 hr. Table 11.37 shows some typical results.

At the Metallurgical Laboratories (MP Chicago 3,8) U_3O_8 was also converted to UO_3 in the dry state by oxygen under high pressure (30 to 150 atm). Temperatures of 500 to 750°C were used, and $UO_3(I)$, $UO_3(II)$, or $UO_3(III)$ was obtained, depending on the conditions of the experiment (cf. Table 11.18). The most stable form, $UO_3(III)$, was obtained at the highest temperatures (700 to 750°C) and oxygen pressures (70 to 150 atm).

U_3O_8 was converted to UO_3 by vapor-phase oxidation at 350 to 400°C with nitric acid or, better, with nitrogen tetroxide, N_2O_4 (the latter reagent was prepared by bubbling air or oxygen through HNO_3 at 65 to 80°C).

Oxidation of UO_2 to UO_3 by hydrogen peroxide was studied in some detail. The free-energy equation of this process was estimated as follows:

$$UO_2 + H_2O_2 \text{ (liq)} \rightarrow UO_3 + H_2O \text{ (liq)}$$

$$-\Delta F = 58{,}360 - 10.15T + 0.465 \times 10^{-3}T^2 + 0.69 \times 10^5 T^{-1}$$

$$+ 1.10T \ln T \quad \text{cal per mole} \qquad (36)$$

Evaluation of ΔF from this equation gave the results shown in Table 11.38.

(b) Formation of UO_3 Hydrates by Oxidation of UO_2 and U_3O_8. Ebelmen (1842c) asserted that U_3O_8 hydrate (obtained by photochemical decomposition of uranyl oxalate) goes over into $UO_3 \cdot 2H_2O$ when dried in air. Aloy (1900, 1901a) found, however, that this procedure leads to the formation of $UO_3 \cdot H_2O$.

De Coninck (1908) obtained $UO_3 \cdot 2H_2O$ by the action of warm H_2O_2 on UO_2 and $UO_3 \cdot H_2O$ (with only a little dihydrate) by the same reaction at room temperature.

Drenckmann (1861) obtained $UO_3 \cdot 2H_2O$ by heating a mixture of U_3O_8 and $KClO_3$ and then washing with boiling water.

Aloy (1900, 1901b) recommended the preparation of $UO_3 \cdot H_2O$ by oxidation of U_3O_8 hydrates in air. This procedure gives amorphous monohydrate at room temperature and a crystalline form (rhombic platelets and prisms) at 100°C. Pure $UO_2 \cdot 2H_2O$ can also be used, but its oxidation is much slower than that of $U_3O_8 \cdot xH_2O$.

(c) UO_3 Preparation by Decomposition of $UO_4 \cdot 2H_2O$. A widely used method of preparation of UO_3 and of other lower uranium oxides uses uranium peroxide as starting material. Uranium peroxide hydrate $UO_4 \cdot xH_2O$ is easily obtained by precipitation with H_2O_2 from any uranium(IV) or uranyl salt solution.

The deoxygenation temperatures of $UO_4 \cdot xH_2O$ depend not only on oxygen pressure but also on H_2O pressure (since the free energy of deoxygenation changes with the water content). The data on the rate and temperature of decomposition are numerous but not systematic, and the results of various observers are not in complete agreement.

Fairley (1877b), Sieverts and Müller (1928), and Rosenheim and Daehr (1929b) all stated that no loss of oxygen occurs in air at 100°C even after several hours. Hüttig and von Schroeder (1922a) found, on the other hand (cf. Sec. 3.4), that under 15 mm Hg of H_2O pressure one-half atom of oxygen is lost at 54°C and another half atom at 163°C (cf. Fig. 11.9). Alibegoff (1886a) observed that decomposition began at 115°C in an oxygen stream and became strong at 140°C.

Brunck (1895) observed no loss of oxygen in a stream of carbon dioxide up to 150°C. Rosenheim and Daehr (1929a) observed complete decomposition to orange UO_3 at 187°C.

Biltz and Müller (1927a) prepared pure UO_3 by decomposing small quantities of $UO_4 \cdot 2H_2O$ in a stream of oxygen for 0.5 to 1 hr at 350 to 400°C. By heating the same peroxide to 600°C in a nitrogen stream or to 800°C in air, they obtained U_3O_8.

Table 11.38—Free Energy of UO_2 Oxidation by H_2O_2 to UO_3

		\-ΔF, kcal/mole	
	Reaction	300°K	400°K
(28)	$UO_2 + H_2O_2$ (liq) → $UO_3 + H_2O$ (liq)	57.5	57.2
(29)	$UO_2 + H_2O_2$ (aq) → $UO_3 + H_2O$ (liq)	54.3	
(30)	$U_3O_8 + H_2O_2$ (liq) → $3UO_3 + H_2O$ (liq)	51.0	51.0
(31)	$U_3O_8 + H_2O_2$ (aq) → $3UO_3 + H_2O$ (liq)	47.8	

According to Columbia University observations (SAM Columbia 4) $UO_4 \cdot 2H_2O$ is converted to $UO_3 \cdot 2H_2O$ by heating to above 110°C and to red-orange UO_3 by heating to 250°C in oxygen or air.

According to this same laboratory (SAM Columbia 7,11) solid $UO_4 \cdot 2H_2O$ begins to lose oxygen (under 1 atm partial pressure of oxygen) a little below 100°C. The decomposition is complete in 3 hr at 200°C and is 75 per cent complete in 13 hr at 143°C. The temperature dependence of the rate of decomposition indicates a heat of activation of 28 kcal per mole.

4.7 Formation of $UO_{3.5}$ from $UO_4 \cdot 2H_2O$.   An anhydrous uranium "peroxide" with a composition close to $UO_{3.5}$ was obtained at Brown University (Brown 1,3,5) by first decomposing $UO_4 \cdot 2H_2O$ at 125 to 130°C for 24 hr (until UO_4 was about half decomposed to UO_3) and then raising the temperature to 325°C. The product liberates oxygen on contact with water or acids. Similar peroxides were obtained from ammonium uranate (Sec. 1.5).

4.8 Formation of $UO_4 \cdot xH_2O$ by Oxidation of Lower Oxides.   The above-described Columbia University observations on the decomposition of $UO_4 \cdot xH_2O$ in sealed tubes (SAM Columbia 11) indicate that oxidation of UO_3 by oxygen to $UO_4 \cdot xH_2O$ is unlikely to occur even under very high pressure of oxygen. In some experiments at 133°C the final oxygen pressure in the tube was as high as 36 atm, but this did not prevent complete decomposition.

Hydrogen peroxide can oxidize UO_2 or U_3O_8 beyond the UO_3 stage. Oxidation of UO_3 by hydrogen peroxide to $UO_4 \cdot xH_2O$ was first noted by Sieverts and Müller (1928). Solid UO_3 was left standing in neutral solution with 50 ml of 30 per cent H_2O_2 for 12 hr at room temperature; the supernatant liquid was decanted, and the solid was washed first with a little H_2O and then repeatedly with alcohol and ether. It was found to be converted into a light-yellow peroxide.

5. CONVERSION OF OTHER URANIUM COMPOUNDS TO OXIDES

5.1 Conversion of Uranium(III) Compounds to Oxides. In Britain (British 1) x-ray diffraction lines of UO were observed in an oxidation product of uranium trifluoride. This product was formed when a leak occurred in an apparatus in which pure uranium trifluoride was being heated at 1200°C. Previously, the same x-ray pattern was observed in a product obtained by heating UF_3 to 1000°C for seven days, which was shown by analysis to contain 92.8 per cent uranium and 0.1 per cent fluorine; it apparently consisted mainly of UO_2 + UO.

Uranium hydride burns to UO_2 and H_2O in oxygen or air (MP Ames 1). It reacts with liquid water, sometimes explosively, as well as with water vapor forming UO_2 and hydrogen.

5.2 Conversion of Uranium(IV) Compounds to UO_2 and U_3O_8. The conversion of UF_4 to fluorine-free U_3O_8 is discussed in Chap. 12 (cf. MP Ames 11,12; CEW-TEC 5; Brown 2).

When uranium(IV) chlorides or double salts such as Na_2UCl_6 are heated in water vapor, they are hydrolyzed with the formation of UO_2. Air must be excluded to prevent oxidation to U_3O_8. These reactions will be discussed in Chap. 14, which deals with the properties of uranium chlorides.

Ebelmen (1842a) and Hermann (1861) obtained UO_2 by heating uranium(IV) sulfate, $U(SO_4)_2$, in a stream of hydrogen and also by heating the same compound to 200°C with K_2S_5 and washing the product with water.

5.3 Conversion of Uranium(VI) Compounds to Oxides. (a) Conversion of Uranyl Salts to UO_2. Uranyl salts can give UO_2 either by straight thermal decomposition, e.g.,

$$UO_2Br_2 \rightarrow UO_2 + Br_2 \tag{37}$$

or by reduction, e.g.,

$$UO_2Cl_2 + H_2 \rightarrow UO_2 + 2HCl \tag{38}$$

or by conversion to UO_3, combined with thermal decomposition of the latter, e.g.,

$$UO_2Cl_2 + CaO \rightarrow UO_3 + CaCl_2 \qquad (39)$$

$$UO_3 \rightarrow UO_2 + \tfrac{1}{2}O_2 \qquad (40)$$

(See, however, Sec. 4.4c in regard to the difficulties of complete reduction of UO_3 to UO_2.)

De Coninck (1902) obtained brick-red UO_2 by thermal decomposition of uranyl bromide, UO_2Br_2, in air. In contrast to the product obtained from UO_2Cl_2 this material was not reoxidized to U_3O_8 in air. By heating in hydrogen the brick-red product was converted to black UO_2. This description makes it doubtful whether the brick-red oxide might not have been UO_3 rather than UO_2.

De Coninck also prepared UO_2 by reducing uranyl chloride, UO_2Cl_2 (1902), or uranyl sulfate, UO_2SO_4 (1908), with hydrogen, as well as by heating UO_2Cl_2 with magnesium or aluminum powder (1909). Reduction with magnesium was also described by Mallinckrodt investigators (Mallinckrodt 1). With potassium or sodium as reductants, alkali uranates are likely to be formed together with the dioxide (de Coninck, 1909). The same author found that UO_2 (black or brownish red) can also be produced from UO_2Cl_2 by heating with CaO or $Ba(OH)_2$ in the absence of air.

Arfvedson (1822c, 1824a) obtained UO_2 by igniting $2KCl \cdot UO_2Cl_2$ in dry hydrogen and leaching the KCl with water.

Milbauer (1904a) converted uranyl sulfide to U_3O_8 and SO_2 by prolonged ignition in air.

Ipatieff and Muromtsev (1930) reduced uranyl nitrate solution to UO_2 at a temperature as low as 360°C by treatment for five days with hydrogen under a pressure of 50 to 80 kg/cm^2.

When organic uranyl salts are ignited, the anion itself may act as reductant. Thus de Coninck (1908; see Müller, 1915) obtained black pyrophoric UO_2 by dry distillation of uranyl formate, $UO_2(COOH)_2$. De Coninck (1912) also found that UO_2 is formed when a suspension of uranyl formate in methanol is allowed to stand in diffuse light for three months. Berzelius (1824a) prepared UO_2 by igniting uranyl oxalate, $UO_2C_2O_4$, in the absence of air. Péligot (1842a) and Ebelmen (1842a) recommended calcining $UO_2C_2O_4$ in dry hydrogen. Péligot's product was first black, then brown and pyrophoric. Ebelmen's product was copper-red, lustrous, and crystalline and was more stable the higher the temperature of its preparation. Rammelsberg (1843b) found UO_2 obtained from oxalate always to contain carbon. Boullé et al. (1950) obtained UO_2 by heating $UO_2C_2O_4$ in vacuum, or in CO atmosphere, to 275 to 350°C.

(b) Conversion of Uranates to UO_2. Wöhler (1842) treated a solution of $(NH_4)_2U_2O_7$ in hydrochloric acid with excess NH_4Cl and NaCl, evaporated, and melted the residue. After extraction of the melt with water, black UO_2 crystals were left behind [for discussion of whether the product contains N, see Uhrlaub (1859) and Hillebrand (1893a)].

According to Hofmann and Höschele (1915) pure UO_2 in well-formed black lustrous cubes can be obtained by melting 1 part of Na_2UO_4 with 4 parts of anhydrous $MgCl_2$. Parsons (1917) described a technical method of converting $Na_2U_2O_7$ to UO_2 by melting with NaCl and charcoal, which process permits recovery of U_2O_5 contained in the uranate. This reaction finds some application in the extraction of uranium from ores.

(c) Conversion of Uranyl Salts to U_3O_8. All soluble uranyl salts can be converted to oxide by thermal decomposition of ammonium diuranate, obtained either by direct precipitation with ammonia or by treatment of an alkali metal polyuranate with ammonia. Alternatively, $UO_4 \cdot 2H_2O$ can be precipitated by H_2O_2 and decomposed at elevated temperatures.

Uranyl salts of volatile or unstable acids can also be converted to U_3O_8 directly by heating. Moissan (1896) obtained U_3O_8 by igniting uranyl nitrate in a porcelain crucible. McCoy and Ashman (1908) converted pure uranyl nitrate to orange UO_3 by heating below red heat and then to U_3O_8 by raising the temperature to 700°C. Zimmermann, Alibegoff, and Krüss (1886a) prepared pure U_3O_8 by decomposing pure uranyl nitrate or oxalate (also hydroxide or peroxide), reducing the product with hydrogen to UO_2, and reoxidizing to U_3O_8 in pure oxygen. According to Ipatieff and Muromtsev (1930), heating of 10 per cent uranyl nitrate solution to 360°C under 50 to 80 kg/cm^2 initial hydrogen pressure leads in two days to black U_3O_8 (longer treatment gives UO_2, cf. Sec. 5.3a).

Hoffman (1948) described how pure U_3O_8 can be obtained from impure uranyl nitrate by extracting the dry salt with ether (containing 5 vol. % water) and transferring into aqueous phase by successive washings with water. In this way, pure U_3O_8 can be obtained from pitchblende, carnotite, and mixtures containing rare earths.

Péligot (1842a) and, later, de Coninck (1901a, 1903) observed that thermal decomposition of uranyl salts gives, together with green U_3O_8, a black oxide which they considered to be U_2O_5; this was contradicted, among others, by Zimmermann, Alibegoff, and Krüss (1886a). The existence of U_2O_5 is, in general, doubtful (cf. Sec. 1.3).

Boullé and Dominé-Bergès (1948) electrolyzed a 10 per cent solution of $UO_2(NO_3)_2$ by placing it in the anodic compartment of a cell whose cathodic compartment contained pure water. After passing a current of 5 μa (800 volts) through this cell for several hours, yellow, very pure UO_3 hydrate appeared in the cathodic compartment. Heating this product to 380 to 390°C in air gave orange nonhydrated but very hygroscopic trioxide; heating to higher temperatures gave olive or black U_3O_8.

(d) <u>Conversion of Uranates to U_3O_8</u>. A commonly used method of preparation or purification of U_3O_8 is the thermal decomposition of ammonium diuranate, $(NH_4)_2U_2O_7$ (obtained from uranyl nitrate by precipitation with NH_4OH).

$$9(NH_4)_2U_2O_7 \rightarrow 2N_2 + 14NH_3 + 6U_3O_8 + 15H_2O \tag{41}$$

Pierlé (1919) recommended ignition of ammonium diuranate in air as a good method of preparation of U_3O_8. Giolitti and Tavanti (1908) purified U_3O_8 by dissolution in HNO_3, repeated precipitation with NH_4OH, and decomposition of pure ammonium diuranate. At Brown University (Brown 6) ammonium diuranate was washed, dried at 130°C in air, and calcined for 4 hr at 350°C to give U_3O_8. In Britain (British 5) $(NH_4)_2U_2O_7$ was precipitated from uranyl nitrate solution (previously purified from Fe), washed with hot water, dissolved in hydrochloric acid, purified by precipitation of impurities with hydrogen sulfide, and reconverted to ammonium diuranate and then to U_3O_8 by ignition at 700°C in an electric furnace. In another procedure (British 6) UF_6 was converted to $(NH_4)_2U_2O_7$ by first converting it with NaOH to $Na_2U_2O_7$, dissolving the precipitate in HNO_3, and reprecipitating with NH_4OH, after which U_3O_8 was obtained by ignition at 600°C in air with an over-all yield of 99 per cent.

(e) <u>Conversion of Uranyl Salts to UO_3</u>. In the preparation of UO_3 by thermal decomposition of $UO_2(NO_3)_2$, difficulties arise as a result of the necessity of raising the temperature high enough to ensure complete decomposition of the nitrate and of the requirement that the temperature be kept low to prevent decomposition of UO_3 to U_3O_8. Arfvedson (1822a, 1824b, 1825) maintained that the establishment of such conditions is impossible and that nitrate cannot be fully destroyed before some of the oxide is converted to U_3O_8. Péligot (1844), on the other hand, thought he could convert uranyl nitrate to nitrogen-free UO_3 by calcining at 250°C. Mixter (1912b) used a temperature of 400°C; however, Lebeau (1912), Jolibois and Bossuet (1922), and Hüttig and von Schroeder (1922b) found 350°C insufficient and 500°C satisfactory. Lebeau recommended heating in an oxygen stream to avoid formation of U_3O_8.

De Forcrand (1915b) recommended low temperatures (290 to 300°C) since he considered the dark-red oxides obtained at 500 to 600°C as "polymerization products" of UO_3 of unknown degree of polymerization (which may be taken to mean different allotropic modifications of UO_3; cf. Sec. 2.5a).

At Columbia University (SAM Columbia 2) pure UO_3 was prepared by ignition of uranyl nitrate hydrate at 300°C. Long, Jones, and Gordon (U. S. Bur. Mines 1) heated uranyl nitrate for 8 hr at 300°C in a closed furnace in order to maintain an oxidizing atmosphere (NO_2). No impurities capable of reducing $KMnO_4$ were found in the product, and no nitrate could be detected by the $H_2SO_4 + FeSO_4$ test.

The preparation of yellow UO_3 hydrate from $UO_3(NO_3)_2$ and its dehydration to orange UO_3 by electrolysis with pure water in the cathode compartment, according to Boullé and Dominé-Bergès (1948), were described in Sec. 5.3c.

Ebelmen (1842d) converted double ammonium uranyl carbonate to UO_3 by prolonged heating at 300°C. Brunck (1895) recommended that UO_3, obtained by calcination of uranyl nitrate at 300°C and still containing some nitrogen, be purified by conversion to ammonium uranyl carbonate, which is then decomposed by heating to 300°C in air.

De Coninck (1901b, 1908) prepared UO_3 from $UO_2SO_4 \cdot 3H_2O$ by heating to red heat. Boullé et al. (1950) obtained UO_3 by heating $UO_2C_2O_4$ in air to 330°C.

(f) Conversion of Uranyl Salts to UO_3 Hydrates by Hydrolysis or Electrolysis. The hydrates $UO_3 \cdot H_2O$ and $UO_3 \cdot 2H_2O$ (or intermediate solid solution) can be obtained by hydrolysis or electrolysis of uranyl salt solutions. According to the data given in Sec. 3.3a it may be expected that $UO_3 \cdot 2H_2O$ is formed below 60°C and $UO_3 \cdot H_2O$ is formed above 100°C; intermediate solid solutions (or the sesquihydrate, if it exists) should arise in the intermediate range of temperatures. Experimental results are not so clear-cut.

Ebelmen (1842c) and Drenckmann (1861) claimed to have obtained $UO_3 \cdot 2H_2O$ by means of prolonged boiling of ammonium uranyl carbonate solution. Riban (1881, 1882) obtained crystalline $UO_3 \cdot H_2O$ (hexagonal prisms) by heating a 2 per cent uranyl acetate solution to 175°C in a sealed tube for 100 hr in vacuum. Zehenter (1900) found heating to 140 to 150°C for 8 hr to be sufficient. According to this author, refluxing a sodium uranyl acetate solution also gives $UO_3 \cdot H_2O$ slightly contaminated by sodium (rather than a sodium polyuranate). The product consists of lustrous sulfur-yellow hexagonal tablets.

In connection with the results last described it should be remembered that hydrolysis of organic uranyl salts may lead to reduction and thus to the formation of hydrates of UO_2 or U_3O_8 rather than of UO_3 (see Sec. 5.3a).

Hydrolysis of uranyl salts of mineral acids always leads to UO_3 hydrates. The formation of $UO_3 \cdot H_2O$ by evaporation of uranyl nitrate solution and heating was first observed as early as 1843 by Malaguti. Ipatieff and Muromtsev (1930) obtained crystalline $UO_3 \cdot H_2O$ in the form of transparent prisms by hydrogenating under high pressure at 300°C a solution of uranyl nitrate in nitric acid.

Berzelius (1845) found that the monohydrate can also be obtained by the hydrolysis of solid hydrated uranyl nitrate (by heating on a sand bath); de Coninck (1911a) found that careful ignition of $UO_2(NO_3)_2 \cdot 6H_2O$, washing with warm water, and drying gives pure orange $UO_3 \cdot H_2O$ without traces of yellow $UO_3 \cdot 2H_2O$. Hydrolysis explains also the formation of $UO_3 \cdot H_2O$ by reaction of uranyl salt solutions with hydrocyanic acid, according to Fischel (1889).

As mentioned in Sec. 3, Smith (1879, 1880) found that electrolytic decomposition of uranyl salt solutions gives a U_3O_8 hydrate. A yellow $UO_3 \cdot H_2O$ is supposedly produced first but is later converted into a black hydroxide ($U_3O_8 \cdot 2H_2O$?). De Coninck and Camo (1901c) also obtained $UO_3 \cdot H_2O$, together with $UO_3 \cdot 2H_2O$, by electrolysis of uranyl nitrate. Pierlé (1919) obtained the $UO_3 \cdot H_2O$ by electrolysis of uranyl acetate solution at room temperature with a current density of 0.1 amp per 100 cm^2. The "black oxide" obtained by prolonged electrolysis, which was interpreted by Smith as $U_3O_8 \cdot 2H_2O$, has, according to Pierlé, the composition $U_3O_{10} \cdot 2H_2O$.

The electrolytic deposition of U_3O_8 hydrates on uranium electrodes was also studied by Francis and Tscheng-Da-Tschang (1935b) and Francis (1935a).

(g) Conversion of Uranates to UO_3 and $UO_{3.5}$. In order to convert $(NH_4)_2U_2O_7$ to UO_3 without further decomposition to U_3O_8, it is necessary to use low temperatures. Ebelmen (1842d) obtained UO_3 by prolonged heating to 300°C; Brunck (1895) noted that UO_3 so obtained still contains nitrogen and recommended purification by conversion to ammonium uranyl carbonate and its subsequent decomposition (cf. above). Goldschmidt and Thomassen (1923c) heated $(NH_4)_2U_2O_7$ to 230°C for 16 hr and then to 260 to 370°C for 20 hr; the product still contained some water.

Brown University observers (Brown 4) confirmed the fact that UO_3 prepared by low-temperature decomposition of $(NH_4)_2U_2O_7$ contains some ammonia that cannot be removed simply by heating without causing conversion to U_3O_8; preparation from $UO_4 \cdot 2H_2O$ was therefore found more reliable. Similar observations were made at Clinton Laboratory (MP Clinton 1). In a later report (Brown 6) the Brown group reported the preparation of UO_3 from $(NH_4)_2U_2O_7$ by drying in air at 130°C and calcining at 350°C for 4 hr followed by washing with water. The hydrated trioxide obtained in this way is then recalcined for 4 hr at 350°C to obtain anhydrous UO_3. Bright-red UO_3 was obtained by ignition of ammonium diuranate also at the University of California Radiation Laboratory (UCRL 5). In a later report (UCRL 6) from the same laboratory $(NH_4)_2U_2O_7$ was calcined at 500°C.

As mentioned in Sec. 1.5, Brown University observers (Brown 2) found that, if ammonium diuranate is decomposed in a rapid stream of oxygen and the temperature is gradually increased from 250 to 550°C, the evolution of ammonia and water (which occurs at 250 to 350°C) is followed by the formation of a red powder that appears to be stable at high temperatures and has a formula between $UO_{3.14}$ and $UO_{3.38}$, as shown by analysis. This product appears to be a peroxide or rather a "moloxide"; it releases its "excess" oxygen on contact with water or acid. It may be a mixture of UO_3 with $UO_{3.5}$, the anhydrous peroxide obtained at Brown from $UO_4 \cdot 2H_2O$ (cf. Sec. 1.5).

(h) Conversion of Uranyl Salts to $UO_4 \cdot xH_2O$. All soluble uranyl salts are converted to $UO_4 \cdot xH_2O$ by hydrogen peroxide. The same is true of uranium(IV) salts, which are probably first oxidized to uranium(VI) salts and then to uranium peroxide. This important reaction will be dealt with in Part II of this volume.

6. CONVERSION OF OXIDES TO OTHER URANIUM COMPOUNDS

6.1 Conversion of Uranium Oxides to Halides and Oxyhalides (cf. Chaps. 12, 14, 15, and 16). (a) Fluorination. UO_2. Uranium dioxide can be converted to UF_4 by the action of gaseous hydrogen fluoride, fluorinated hydrocarbons ("freons"), and by reaction with $NH_4F \cdot HF$. These reactions are described in detail in Chap. 12.

According to Los Alamos observers (Los Alamos 1) UO_2 does not react with molten NaF or BeF_2 but reacts with other metal fluorides (cf. Chap. 12).

U_3O_8. According to Bolton (1866) U_3O_8 reacts with aqueous hydrofluoric acid to give soluble UO_2F_2 and insoluble UF_4.

UO_3. According to Gore (1869) UO_3 is soluble in pure anhydrous liquid hydrogen fluoride, probably with the formation of UO_2F_2.

(b) Chlorination. UO_2. The oxide UO_2 is insoluble in dilute or concentrated cold or hot hydrochloric acid (Arfvedson, 1822b, 1824a; Péligot, 1842a). Hot fuming HCl converts UO_2 slowly into green UCl_4 (Hofmann and Höschele, 1915). Hydrogen chloride gas does not attack UO_2 even at high temperatures (Péligot, 1842a).

UO_2 can be chlorinated by many inorganic or organic chlorinating agents. Among these are S_2Cl_2 (MP Berkeley 4), $COCl_2$, CCl_4, and C_3Cl_6 (see Chap. 14 for detailed discussion and bibliography).

The conversion of UO_2 to uranyl chloride, UO_2Cl_2, can be achieved by means of dry chlorine at red heat (Péligot, 1842a). According to Brown University observers (Brown 4) brown UO_2 is converted to UO_2Cl_2 also when used as anode in the electrolysis of hydrochloric acid.

In the same laboratory (Brown 2) UO_2 was converted to anhydrous $UOCl_2$ by dissolution in excess molten UCl_4 at 600°C under dry nitrogen.

U_3O_8. This oxide reacts only very slowly with dilute hydrochloric acid, even at elevated temperatures, particularly after calcination (Arfvedson, 1822a, 1824b). Reaction with concentrated hydrochloric acid is more rapid; heating with HCl (specific gravity 1.135) in carbon dioxide atmosphere in a closed tube at 180 to 200°C leads to rapid conversion to a green liquid (Zimmermann, Alibegoff, and Krüss, 1886a).

In the presence of oxidizing agents, e.g., HNO_3, hydrochloric acid fairly readily dissolves U_3O_8 to UO_2Cl_2 (British 7).

Black commercial U_3O_8 was found by Kangro and Jahn (1933) to be attacked by chlorine at 900°C, giving yellow crystals of UO_2Cl_2.

Bourion (1910) converted U_3O_8 to UCl_4 by heating to 230 to 250°C with S_2Cl_2. Rosenheim and Kelmy (1932) converted U_3O_8 to UCl_4 and z UO_2Cl_2 by the action of $COCl_2$ at 500°C. According to Michael and Murphy (1910) heating of U_3O_8 with CCl_4 in closed tubes leads to the formation of dark-red crystals of UCl_5, which upon further heating readily dissociate into UCl_4 and chlorine. Venable and Jackson (1919-1920) gave 360°C as the temperature of chlorination of U_3O_8 by CCl_4.

The chlorination of U_3O_8 is discussed in Chap. 14.

UO_3. Gore (1865) stated that UO_3 changes its color to a lighter yellow in hydrogen chloride gas. It does not dissolve in hydrochloric acid, but the residue is soluble in water. No conversion of UO_3 to UCl_4 could be obtained at Purdue (Purdue 1) by anhydrous HCl up to 400°C. Action of hydrogen chloride in C_2H_5OH on UO_3 apparently gave UO_2Cl_2 but no UCl_4.

According to Brown University investigators (Brown 1,4) gaseous HCl reacts under atmospheric pressure with UO_3 when the latter is hydrated or when moisture is present to form hydrated UO_2Cl_2.

The reaction of UO_3 with aqueous hydrochloric acid will be discussed in Sec. 6.2d.

Chlorination of UO_3 with S_2Cl_2 was carried out at Brown University (Brown 7). The reaction can be written as

$$2UO_3 + 4S_2Cl_2 \rightarrow 2UCl_4 + 5S + 3SO_2$$

At Purdue University (Purdue 1) UO_3 (prepared by decomposition of $UO_4 \cdot 2H_2O$) was converted to UCl_4 by 1 hr treatment with $SOCl_2$ at 350°C.

According to Rauter (1892) UCl_4 and UO_2Cl_2 are obtained, together with SiO_2 and Cl_2, when UO_3 is heated to 370 to 380°C for 8 hr with $SiCl_4$ in a closed tube.

Chlorination of UO_3 by $CO + Cl_2$ was described by Quantin (1888). The same author also carried out the first chlorination of UO_3 with CCl_4 by heating the oxide to red heat in CCl_4 vapor. Camboulives (1910) found that the reaction of UO_3 with CCl_4 vapor occurs at 360°C and gives UCl_4 and UCl_5. Michael and Murphy (1910) also observed the formation of unstable dark-red UCl_5 crystals by heating UO_3 with CCl_4 vapor (with or without Cl_2) in a closed tube.

The liquid-phase chlorination of UO_3 by CCl_4 will be discussed in Chap. 14.

(c) Bromination. UO_2. At the Metallurgical Laboratory (MP Chicago 10) it was found that UO_2 (prepared by reduction of U_3O_8 with hydrogen at 300 or 900°C) could not be converted to UBr_4 by the action of HBr (even at 800°C) or of CH_3Br (at 550°C). Hermann (1861) obtained UBr_4 by the action of bromine vapor on a mixture of UO_2 and carbon.

U_3O_8. Heating of U_3O_8 in Br_2 vapor or in dry HBr gives a mixture of uranyl bromide and lower oxides (Richards and Merigold, 1902).

Further details on the conversion of the uranium oxides to bromides are to be found in Chap. 15.

6.2 Miscellaneous Reactions. (a) Conversion of Uranium Oxides to Sulfides and Oxysulfides. Heating UO_2 or U_3O_8 in carbon disulfide vapor leads, according to Hermann (1861), to the formation of the oxysulfide $UO_2 \cdot 2US_2$. A mixture of carbon dioxide and carbon disulfide showed no reaction with UO_2 but reduced U_3O_8 and UO_3 at a red heat to UO_2 (cf. Sec. 4.4c).

According to observations made at Ames (MP Ames 13), UO_2 gives $UO_2 \cdot 2US_2$ with carbon disulfide at 900°C. The Berkeley observations (MP Berkeley 5) on the conversion of uranium oxides to US_2 and UOS by reaction with hydrogen sulfide and to U_2S_3 by subsequent decomposition of US_2 are reported in Sec. 7.

(b) Conversion of Uranium Oxides to Carbides. Tiede and Birnbräuer (1914) found that UO_2 reacts with carbon in vacuum apparently at temperatures as low as 1600°C forming a uranium carbide. Greenwood (1908) found that the reaction of UO_2 with carbon begins in an electric vacuum furnace at about 1490°C (at this temperature the formation of CO can be deduced from increase in pressure). Heusler (1926) studied quantitatively the equilibrium

$$UO_2 + 4C \rightleftharpoons UC_2 + 2CO \qquad (42)$$

between 1480 and 1800°C. The equilibrium carbon monoxide pressure is 18 mm Hg at 1480°C and reaches 1 atm at 1800°C. The equation

$$\log p_{CO} \text{ (mm Hg)} = -\frac{19,100}{T} + 12.09 \qquad (43)$$

represents the data approximately, but deviations indicate that ΔH must change appreciably in the investigated temperature interval. The average ΔH value is 87.3 kcal per mole.

For more details on these reactions see Chap. 9.

(c) Conversion of Uranium Oxides to Uranates and Peruranates by Reaction with Metal Oxides, Peroxides, and Salts. In contrast with UO_2, which is a basic oxide, UO_3 is amphoteric, giving rise both to the uranyl salt series, UO_2X_2, and to the uranate and polyuranate series, X_2UO_4, $X_2U_2O_7$, etc. U_3O_8 (or $UO_2 + 2UO_3$) can be partly converted to uranates, e.g., by reaction with metal oxides.

$$U_3O_8 + 2BaO \rightarrow 2BaUO_4 + UO_2 \tag{44}$$

BaO reacts vigorously with U_3O_8 at or above 330°C [cf. Balarev (1924)]; the conversion may be complete if an oxidant is present, e.g.,

$$U_3O_8 + 6KClO_4 \rightarrow 3K_2UO_4 + 3Cl_2 + 10\ O_2 \tag{45}$$

According to Fowler and Grant (1890) and Hodgkinson and Lowndes (1888, 1889) U_3O_8 reacts with solid $KClO_4$ at 390°C with liberation of oxygen and chlorine.

With UO_3, $KClO_4$ gives potassium uranate with liberation of chlorine (Brunck, 1895).

Another reaction of the same type is the conversion of UO_2, U_3O_8, and UO_3 to sodium uranate by solid sodium peroxide. This reaction was mentioned in Sec. 4.1a in connection with Mixter's determination of the heats of formation of the various uranium oxides.

According to Mixter (1912a) Na_2UO_4 is formed by heating uranium oxides with Na_2O_2 in the calorimeter bomb. The measured heat effects were*

$$UO_2 + Na_2O_2 \rightarrow Na_2UO_4 + 110.9\ \text{kcal} \tag{46}$$

$$\tfrac{1}{3}U_3O_8 + Na_2O_2 \rightarrow Na_2UO_4 + \tfrac{1}{3}O_2 + 82.1\ \text{kcal} \tag{47}$$

$$UO_3 + Na_2O_2 \rightarrow Na_2UO_4 + \tfrac{1}{2}O_2 + 76.7\ \text{kcal} \tag{48}$$

Mixter finds that no peruranate is formed in the bomb but only normal uranate. However, when the product is extracted with water, it reacts

---

*Mixter "experimental" values are calculated for the neutralization of U_3O_8 and UO_3 by $Na_2O + Na_2O_2$ in appropriate proportions by assuming a heat effect of 19.4 kcal for the decomposition of Na_2O_2,

$$Na_2O_2 \rightarrow Na_2O + \tfrac{1}{2}O_2 - 19.4\ \text{kcal}$$

with excess Na_2O_2 and forms peruranate solution. Peruranates must also be formed when UO_2 or U_3O_8 is dissolved in aqueous Na_2O_2 solution (MP Ames 14).

The solubility of UO_2 in aqueous alkaline peroxide solutions was measured at Brown University (Brown 2). The oxide dissolves readily at 25 to 45°C; warm 10 per cent H_2O_2 + 10 per cent K_2CO_3 is most effective. After shaking 2 g of oxide for 2 min in 100 ml of solvent, the amounts shown in Table 11.39 were dissolved. According to the same report, warm (45°C) 5 per cent Na_2O_2 is the best solvent for U_3O_8.

Table 11.39 — Solubility of UO_2 in Alkaline Peroxides

Reagent	Temp., °C	UO_2 dissolved, %
10% H_2O_2 + 10% $(NH_4)_2C_2O_4$	25	35
10% H_2O_2 + 10% $(NH_4)_2CO_3$	25	69
10% H_2O_2 + 10% K_2CO_3	25	46
10% H_2O_2 + 10% K_2CO_3	35	83
5% Na_2O_2	45	17

The reactions of UO_3 with basic oxides were first studied by Zimmermann (1882b). Tammann and Rosenthal (1926) investigated the reaction of UO_3 with 25 metal oxides. Equivalent quantities of UO_3 and metal oxide powders were mixed and heated to 600°C twice in 10 min; BeO, La_2O_3, CeO_2, and MoO_2 showed no reaction at all. With Li_2CO_3 (or $Li_2O + CO_2$), Ag_2O, CaO, BaO, SrO, MgO, ZnO, CdO, HgO, CuO, PbO, CoO, MnO, NiO, Al_2O_3, Cr_2O_3, Fe_2O_3, and V_2O_3, exothermal reactions took place (beginning sometimes as early as at 125°C, e.g., with SrO, sometimes first at 450°C, as with Al_2O_3). The reaction was from 10 to 90 per cent complete. From the almost exact equivalence of the quantities of UO_3 and XO used up it can be concluded that normal uranate, X_2UO_4, was the only product formed. The fact that a reaction has taken place is usually noticeable from a change in color. The yellow product obtained from CaO and UO_3 becomes green above 505°C and loses oxygen. FeO and SnO reduce UO_3 on heating (cf. Sec. 4.4c).

Hedvall (1915) studied the reaction of UO_3 with CoO at 1100 to 1300°C; a yellow crystallized uranate was obtained in some experiments.

Because of the poor solubility of potassium polyuranate, UO_3 is partly converted into this compound when it is dissolved in potassium chloride solution (see Sec. 6.2d).

According to Hodgkinson and Lowndes (1888, 1889), $UO_3 \cdot 2H_2O$ reacts with potassium chlorate, evolving chlorine and forming potassium diuranate.

The heat of neutralization of $UO_3 \cdot 0.5H_2O$ by NaOH was calculated by Pisarzhevskiĭ (1900b) as 17.8 kcal for the reaction

$$2UO_3 \cdot H_2O + 2NaOH \rightarrow Na_2U_2O_7 + 2H_2O \qquad (49)$$

This value was obtained by subtracting the measurement of the heat of decomposition of $Na_2U_2O_7$ by H_2SO_4 to Na_2SO_4 and UO_2SO_4 (44.5 kcal) from the sum of the heats of neutralization of $2UO_3 \cdot H_2O$ and of 2NaOH by $3H_2SO_4$ (2 × 15.35 kcal, measured by Pisarzhevskiĭ, cf. Sec. 6.2d, and 31.4 kcal, respectively).

(d) <u>Conversion of Uranium Oxides to Uranium(IV) and Uranyl Salts.</u> Conversion of uranium oxides to UO_2Cl_2 and UO_2Br_2 was described in Secs. 6.1b and c.

According to Arfvedson (1822b, 1824a) and Péligot (1842a) UO_2 dissolves in HNO_3 giving a lemon-yellow solution of uranyl nitrate. Very dilute HNO_3 has no effect on UO_2 except causing hydration (de Coninck, 1908). UO_2 is easily soluble in concentrated HNO_3 or aqua regia (Raynaud, 1911, 1912).

UO_2 can be oxidized to uranyl nitrate also by means of $AgNO_3$ (Smith and Shinn, 1894; Isambert, 1875). $AgNO_3$ in ammonia solution is reduced by UO_2 to silver. In neutral solution Ag_2O is precipitated first and then transformed into metallic silver.

Conversion of UO_2 to basic nitrate by the action of NO_2 gas in the cold (accompanied by heat evolution) was noted by Sabatier and Senderens (1896). Whether or not a basic uranyl nitrate is formed, however, is not clear (Katz and Gruen, 1949).

U_3O_8 is dissolved in HNO_3 with the formation of uranyl nitrate and evolution of nitrous oxides. According to British observers (British 4) the conversion of U_3O_8 to uranyl nitrate occurs according to the equation

$$2U_3O_8 + 14HNO_3 \rightarrow 6UO_2(NO_3)_2 + 7H_2O + NO + NO_2 \qquad (50)$$

However, this equation does not take into account the formation of complex ions, which increase the amount of oxide brought into solution by a given quantity of nitric acid (cf. below). Ignited UO_2 is only slightly soluble in dilute H_2SO_4 but dissolves easily in concentrated acids and completely in boiling concentrated H_2SO_4 (Arfvedson, 1822a, 1824a). U_3O_8 is converted completely to uranyl sulfate and $U(SO_4)_2$ by prolonged heating with concentrated H_2SO_4 (Zimmermann, Alibegoff, and Krüss, 1886a). U_3O_8 reduces $AgNO_3$ very slowly to metallic silver (Isambert, 1875). UO_3 is soluble in all mineral acids (Lebeau, 1912), giving the corresponding uranyl salts.

The heat of neutralization of UO_3 by diluted HNO_3, i.e., if complex formation is neglected, in the reaction

$$UO_3 + 2H^+ \text{ (aq)} \rightarrow UO_2^{++} \text{ (aq)} + H_2O \tag{51}$$

was measured by de Forcrand (1913, 1915a). He obtained a value of 19.80 kcal per gram atom of uranium (at 18°C).

The heats of neutralization of $UO_3 \cdot H_2O$ and $UO_3 \cdot 2H_2O$ were found by de Forcrand to be 14.85 kcal and 12.38 kcal per gram atom of uranium, respectively. These values have been used (cf. Sec. 3.3b) for

Table 11.40—Solubility of UO_3 in HCl (aq)

HCl, moles/liter	UO_3, moles/liter	HCl, moles/liter	UO_3, moles/liter
2.785	2.832	0.03140	0.02986
1.013	0.958	0.01082	0.01138
0.2979	0.2718	0.00298	0.00343
0.09909	0.0900	0.00108	0.00150

calculation of the heats of formation of the two hydrates from UO_3 and H_2O (cf. Table 11.24). Pisarzhevskiĭ (1900a) found a value of 15.35 kcal per gram atom of uranium for the heat of neutralization of $UO_3 \cdot H_2O$ by 2N H_2SO_4.

Aloy (1896) gave the following values for the heats of neutralization of $UO_3 \cdot H_2O$ by different 0.5N acids: HCl, 16.8 kcal; HBr, 17.6 kcal; HNO_3, 16.8 kcal; H_2SO_4, 19.0 kcal; CH_3COOH, 15.0 kcal per gram atom of uranium.

UO_3 reacts with molten NH_4NO_3 with the formation of uranyl nitrate and liberation of ammonia and water (Audrieth and Schmidt, 1934).

Quantitative determinations of the amount of UO_3 dissolved by a given amount of hydrochloric acid revealed an "extra" solubility, which must be attributed to complexing. Table 11.40 contains some data obtained at the Rockefeller Institute for Medical Research (Rockefeller 1).

These figures indicate a solubility of approximately 1 mole of UO_3 per mole of HCl while the reaction

$$UO_3 + 2HCl \rightarrow UO_2Cl_2 + H_2O \tag{52}$$

permits the dissolution of only 0.5 mole of base per mole of acid. Electrometric or conductometric titration of UO_3 with HCl showed a sharp end point at 0.5 mole of base per mole of acid, confirming the primary formation of uranyl chloride according to Eq. 52. The "extra" solubility of UO_3 may be due to the complexing with OH^- ions,

$$UO_2^{++} + UO_3 + H_2O \rightleftarrows 2UO_2OH^+ \tag{53}$$

or to a "self-complex ion,"

$$UO_2^{++} + UO_3 \rightleftarrows UO_3 \cdot UO_2^{++} \tag{54}$$

Conductance and transference measurements cannot distinguish between these two mechanisms, but pH measurements appear to favor the second one. The solubility of UO_3 in UO_2^{++} solution, according to

Table 11.41 — Solubility of UO_3 in UO_2^{++} – Salt Solutions at 25°C

Solvent	Equivalents of X^- in 1,000 g of solution	Equivalents of UO_3 in 1,000 g of solution
UO_2Cl_2 (aq)	0.2887	0.5272
	1.453	3.138
	1.780	4.064
	1.927	4.121
	2.299	3.870
	2.392	3.778
	4.004	4.058
	4.415	4.454
	4.550	4.548
	5.747	4.086
$UO_2(NO_3)_2$ (aq)	0.05638	0.1062
	0.1303	0.2290
	0.2878	0.4688
	0.4340	0.6958
	0.7373	1.187
	1.280	1.971
	2.058	2.826
	2.519	3.450
	2.892	2.862
	6.322	1.769
$UO_2(ClO_4)_2$ (aq)	0.09787	0.1627
	0.1764	0.2940
	0.4019	0.6390
	0.7899	1.208
	0.9120	1.368
	1.187	1.717
	1.393	1.974
	1.681	2.246
	2.058	2.568
	2.755	3.206
	3.106	3.028
	4.334	2.088
	5.522	1.401

Eq. 53 or 54, cannot, of course, be restricted to chloride solutions but must occur in all uranyl salt solutions. It was measured at Brown University (Brown 1,4) with UO_2Cl_2, $UO_2(NO_3)_2$, and $UO_2(ClO_4)_2$ solutions as solvents. Table 11.41 shows the results.

UO_3 is soluble also in alkali salt solutions. A possible mechanism for this reaction is

$$2UO_3 + 2KCl + H_2O \rightarrow K_2U_2O_7 \text{ (or other polyuranates)} + 2HCl \quad (55)$$

$$2HCl + UO_3 \rightarrow UO_2Cl_2 + H_2O \quad (56)$$

The potassium polyuranate precipitates while part of the UO_3 goes into solution as uranyl chloride (Brown 4). Table 11.42 shows some solubility data.

Table 11.42—Solubility of UO_3 in KCl Solution

KCl, moles/liter	Temp., °C	UO_3 dissolved, equiv./liter
0.01	Room temp.	0.024
0.025	Room temp.	0.021
1.0	Room temp.	0.054
2.0	Room temp.	0.047
3.0	Room temp.	0.020
0.362	80	0.036
0.375	80	0.025
0.468	80	0.024
0.571	80	0.028

One measurement of UO_3 solubility in KNO_3 solution gave 0.016 mole of UO_3 per liter in 1.0N KNO_3 at room temperature (after shaking for 24 hr).

PART B. URANIUM SULFIDES

7. URANIUM-SULFUR SYSTEM

7.1 Phase Relations in the Uranium-Sulfur System. The disulfide, US_2, discovered by Péligot (1842a) in 1842, is the best-known compound in the uranium-sulfur system. Alibegoff (1886a) described in 1886 the preparation of the sesquisulfide, U_2S_3, by reaction of UBr_3 with hydrogen sulfide and of the monosulfide, US, by reduction of U_2S_3 with hydrogen at red heat, but the correctness of the second statement has been questioned (see Sec. 7.3a). According to more recent investigations the monosulfide can be prepared, for example, by reduction of UOS with carbon at temperatures of the order of 1900°C.

At the Metallurgical Laboratory (MP Chicago 11) samples (prepared by Brewer at Berkeley) with the U/S ratio between 1.2 and 1.6 were studied by x-ray analysis. The sample with $S/U \approx 1.5$ contained a single phase, thus confirming the existence of a sulfide U_2S_3. The sample with $S/U = 1.22$ showed lines of the U_2S_3 phase and those of a cubic phase, US, indicating the absence of other compounds between US and U_2S_3. It is not yet known whether the (orthorhombic) U_2S_3 phase can be converted continuously into the orthorhombic phase $US_2(II)$ or whether a diphasic region separates these two sulfides.

No sulfides derived from uranium(VI) (and thus analogous to the oxides U_3O_8 and UO_3) are known, but an oxysulfide UO_2S has been described (cf. Sec. 7.5b).

7.2 <u>Physical Properties of Uranium Sulfides. Crystal Structure and Density</u>. (a) <u>US</u>. According to Alibegoff (1886b) this is a black amorphous powder, but Berkeley observers described the monosulfide as a grayish metallic-looking solid. The crystal structure of uranium monosulfide was studied at Chicago (MP Chicago 2,11,12). It is cubic and face-centered (NaCl-structure) with four uranium atoms in the elementary cell, and it is isomorphous with CeS and ThS (and other sulfides). The lattice constant is

$$a = 5.473 \pm 0.002 \text{ A}$$

The calculated density is 10.87 g/cc.

According to the same observer, x-ray evidence points to the non-existence of intermediate phases between US and U_2S_3.

(b) $\underline{U_2S_3}$. The sesquisulfide was described by Alibegoff (1886a) as gray-black needle-shaped crystals and by Berkeley observers (MP Berkeley 5) as a black crystalline compound. According to Metallurgical Laboratory observations (MP Chicago 1) the U_2S_3 phase is orthorhombic (isomorphous with Th_2S_3 and closely related to Sb_2S_3 and Bi_2S_3). The unit cell contains eight uranium atoms and has the dimensions

$$a_1 = 10.63 \pm 0.02 \text{ A} \quad a_2 = 10.39 \pm 0.02 \text{ A} \quad a_3 = 3.88 \pm 0.01 \text{ A}$$

(c) $\underline{US_2}$. According to Colani (1903a, 1907a) US_2 forms black or steel-gray tetragonal crystals with metallic luster. According to Metallurgical Laboratory observations, a second orthorhombic form, $US_2(II)$, exists in addition to the tetragonal form $US_2(I)$.

The crystal structure of $US_2(I)$ is, according to Metallurgical Laboratory measurements (MP Chicago 13), tetragonal with the parameters

$$a_1 = 10.25 \pm 0.05 \text{ A} \qquad a_3 = 6.30 \pm 0.03 \text{ A}$$

Assuming that the elementary cell contains 10 uranium atoms, the calculated density is 7.54 g/cc.

The orthorhombic form $US_2(II)$ is isomorphous with ThS_2. The space group is Pmnb. The elementary cell contains four uranium atoms. The positions of the atoms are

$$4 \text{ U atoms in } 4(c) \qquad y_1 = 0.250, \ z_1 = 0.125$$
$$4 \text{ S}_I \text{ atoms in } 4(c) \qquad y_2 = 0.375, \ z_2 = 0.432$$
$$4 \text{ S}_{II} \text{ atoms in } 4(c) \qquad y_3 = 0.465, \ z_3 = 0.820$$

The lattice constants are

$$a_1 = 4.22 \pm 0.02 \text{ A} \qquad a_2 = 7.08 \pm 0.04 \text{ A} \qquad a_3 = 8.45 \pm 0.04 \text{ A}$$
$$\rho = 7.90 \text{ g/cc}$$

Each uranium atom has nine sulfur atoms as neighbors, the mean uranium-sulfur distance being 2.91 A.

Berkeley observers (MP Berkeley 5) suggested that one of the two US_2 forms described by Zachariasen (the tetragonal one) may be the intermediate compound between US_2 and $US_{1.5}$, mentioned in Sec. 7.3c. Because of the analogy between US and ThS, between U_2S_3 and Th_2S_3, and US_2 (orthorhombic) and ThS_2, it may be suggested that the intermediate uranium sulfide is U_4S_7, analogous to Th_4S_7. However, the latter compound is hexagonal (according to Zachariasen), and the "extra" uranium sulfide is tetragonal.

Melting Point. According to Berkeley observers (MP Berkeley 5) the melting point of US is considerably above 2000°C; that of U_2S_3 is 1850 ± 100°C; and that of US_2 is 1850 ± 200°C.

Magnetic Properties. At Berkeley (MP Berkeley 6) the magnetic susceptibility of some uranium sulfide preparations was determined; the results are given in Table 11.43. These results were taken as indicating that the U^{++} ion in US has the configuration $5f^2$.

7.3 Preparation of Uranium Sulfides. (a) US. Alibegoff (1886a) prepared a product, which he considered to be US, by reduction of U_2S_3 with dry hydrogen at red heat. Flatt and Hess (1938) found that reduction of US_2 by hydrogen stops at U_2S_3, and they therefore considered Alibegoff's conclusions as erroneous. At the Metallurgical

Laboratory (MP Chicago 14) no reduction of US_2 by atomic hydrogen in quartz tubes was obtained at 200 and 700°C.

At the Metallurgical Laboratory (MP Chicago 15) US was obtained by reduction of UOS with carbon at 1900°C in high vacuum.

$$UOS + C \rightarrow US + CO \tag{57}$$

It may be expected that the equilibrium of the reaction

$$UO + CS \rightleftharpoons US + CO \tag{58}$$

would lie on the side of uranium sulfide formation.

Table 11.43—Magnetic Susceptibility of Uranium Sulfides

Compound	Molar susceptibility per gram atom of uranium, cgs
US	$4,180 \times 10^{-6}$
U_2S_3	$2,630 \times 10^{-6}$
US_2	$3,050 \times 10^{-6}$

According to Berkeley observers US can be prepared conveniently in the following two ways (MP Berkeley 5,7):

1. The first method consists in interaction of fine uranium powder (obtained by decomposition of uranium hydride) with the calculated amount of hydrogen sulfide. The hydride is decomposed at 300°C, and the residue is reacted with hydrogen sulfide at 400 to 500°C. The circulation of hydrogen sulfide is continued until the whole amount required has reacted. The product is at first not quite uniform but contains higher sulfides and unreacted metal. It can be heated to a higher temperature for homogenization. (US can be heated over 1800°C without decomposition.)

2. The second method is the interaction of US_2 with the calculated amount of uranium hydride. The procedure is the same as above, namely, decomposition of hydride at 300°C, heating to 400 to 600°C for the reduction of US_2, and final heating to high temperature for homogenization.

According to the same observers U_2S_3 can be decomposed to US by dissociation in high vacuum at 1800°C.

(b) $\underline{U_2S_3}$. Alibegoff (1886b) prepared U_2S_3 (grayish-black needles) by heating UBr_3 in a hydrogen sulfide stream for 8 to 10 hr with complete exclusion of air. Flatt and Hess (1938) found that reduction of US_2 by oxygen-free hydrogen at red heat gives a grayish-black pyrophoric powder whose analysis (83.2 per cent uranium) indicates the composition U_2S_3 (82.6 per cent uranium) rather than US (88.4 per cent uranium).

At Berkeley (MP Berkeley 5,8) three methods for the preparation of U_2S_3 were developed. The first two are identical with those described above for US.

1. Reaction of uranium (from UH_3) with calculated amount of hydrogen sulfide.

2. Reaction of uranium (from UH_3) with calculated amount of US_2. Because of instability of U_2S_3 at high temperatures the final homogenization of the product has to be carried out by prolonged heating at temperatures much lower than 1800°C.

3. Thermal decomposition of US_2 (which can be obtained by reduction of uranium oxides with carbon in the presence of hydrogen sulfide; cf. below). In one experiment the decomposition of US_2 to U_2S_3 was complete after 20 min at 1600°C in a vacuum of about 10^{-4} mm Hg. However, since U_2S_3 is itself comparatively easily decomposed in vacuum (it can be converted to US in high vacuum at 1800°C), this method is less suitable than methods 1 and 2 for the preparation of a product with the composition corresponding exactly to the formula U_2S_3. On the other hand, the product obtained by this high-temperature method consists of larger crystals and is much more stable in air than those obtained by the low-temperature methods 1 and 2, the latter preparations having to be made and kept in an inert atmosphere.

(c) $US_{1.5 < x < 2}$. At Berkeley (MP Berkeley 5,8) sulfides with compositions intermediate between $US_{1.5}$ (or U_2S_3) and US_2 were obtained by the three methods described in Sec. 7.3b. Of these, thermal decomposition of US_2 is the easiest method if no exact final composition is desired.

(d) US_2. Péligot (1842d) discovered that uranium burns in boiling sulfur, giving US_2. This is, however, not a convenient method of preparation of this compound. Another method is to react UCl_4 with hydrogen sulfide at red heat (Hermann, 1861).

$$UCl_4 + 2H_2S \rightarrow US_2 + 4HCl \tag{59}$$

Colani (1903a, 1907a) suggested the use of the less volatile $2NaCl \cdot UCl_4$ instead of UCl_4 and the use of hydrogen charged with sulfur vapors instead of hydrogen sulfide (because hydrogen sulfide is difficult to dry). Air must be excluded. The reaction begins at 500°C. The product is washed with air-free water, alcohol, and ether and is dried in vacuum. The disulfide can also be obtained by reaction of $2NaCl \cdot UCl_4$ with the sulfides of Na, Mg, Al, Sb, and, even better, with tin sulfide at high temperatures; but all these methods are not to be recommended. Traces of H_2O make the product impure.

At Ames (MP Ames 16) US_2 was prepared by action of hydrogen sulfide on UH_3 (or rather, on finely divided uranium obtained by dissociation of UH_3). Flatt and Hess (1938) prepared US_2 from U_3O_8 by first converting the oxide to UCl_4. After sublimation UCl_4 was converted to US_2 in the same tube by pure dry hydrogen sulfide. The product was a black mass.

At Berkeley (MP Berkeley 5,7) two methods were used, the above-mentioned Ames method involving UH_3 and the reduction of UO_2 or U_3O_8 with carbon in the presence of hydrogen sulfide. The latter reaction is fairly rapid at 1200 to 1300°C. UOS is the first product formed (cf. Sec. 7.5a), but, when the temperature is raised to its maximum value, UOS is reduced to US_2 by the action of CS, the latter being formed from hydrogen sulfide and carbon.

In order to prevent contamination of US_2 by lower sulfides, which may be caused by thermal decomposition of the disulfide at the temperatures used in the final stage of the preparation process, the temperature must be reduced to less than 1000°C, thus allowing the lower sulfides to pick up sulfur from the circulating hydrogen sulfide.

When pure uranium metal is available, the hydride method seems preferable because of its simplicity. However, in this case too, as in that of U_2S_3, the product obtained by the high-temperature process is more stable than that obtained by the low-temperature hydride reaction.

7.4 Chemical Properties of Uranium Sulfides. (a) US. Alibegoff (1886b) described uranium monosulfide as being hardly attacked by concentrated HCl and dilute HNO_3 and as slightly soluble in HCl + Br_2 solution. Fuming HNO_3 oxidized his material with ignition; aqua regia reacted less violently. However, this material might have been U_2S_3 rather than US (cf. Sec. 7.3a). According to Berkeley observers US is very slowly soluble in dilute acids.

(b) U_2S_3. According to Alibegoff (1886b) and Flatt and Hess (1938) U_2S_3 decomposes in air and is often pyrophoric. In the presence of moisture it liberates hydrogen sulfide. Other U_2S_3 properties, according to Alibegoff, are similar to those described above under US. The same observer found that U_2S_3 can be reduced to US by heating in hydrogen to red heat. According to Berkeley observers (MP Berkeley 5) the decomposition pressure of U_2S_3 above 1800°C is high enough to make conversion to US possible also by heating in vacuum.

(c) US_2. Hermann (1861) observed that US_2 oxidized in moist air very slowly, evolving H_2S and leaving yellowish basic uranium sulfate. According to this author and also to Colani (1903a, 1907a) it is only very slowly attacked by water in the cold but is rapidly decomposed by water vapor. Dilute HCl has little effect; concentrated HCl attacks more rapidly, particularly on heating; HNO_3 acts very rapidly. Picon (1929) found US_2 to dissociate in vacuum at 1300°C, evolving sulfur and forming a lower sulfide.

According to Ames observations (MP Ames 16) US_2, made from uranium turnings via UH_3, reacted with bromine at 300°C, liberating S_2Br_2. After heating to 600°C, fused UBr_4 was obtained.

7.5 <u>Uranium Oxysulfides</u>. Two uranium oxysulfides have been described in the earlier literature, $U_3O_2S_4$ (or $UO_2 \cdot 2US_2$) and UO_2S (or $2UO_3 \cdot US_3$); the existence of the first one has been questioned by Berkeley observers. A third compound, UOS (or $UO_2 \cdot US_2$), was discovered at Berkeley.

(a) <u>Uranium(IV) Oxysulfide</u>. <u>UOS</u>. This compound was first obtained at Berkeley (MP Berkeley 7,9) by the reaction of UO_2 with $(H_2S + H_2)$ at 1000 to 1200°C. The minute amount of water vapor produced by interaction of hydrogen sulfide with the oxide refractory suffices to inhibit the formation of US_2. Later it was found at Berkeley (MP Berkeley 5) that UOS can also be conveniently obtained by heating UO_2 or U_3O_8 with carbon in a stream of hydrogen sulfide. Care must be taken to restrict temperature and duration of reduction so as to prevent formation of US_2.

The UOS is blue-black in color. Its structure was determined at Chicago (MP Chicago 2,16) as tetragonal with two molecules per unit cell (type PbFCl, isomorphous with ThOS), space group $P4/nmm$.

$$2 \text{ U atoms at } (0\ 0\ x_1)\ (\tfrac{1}{2}\ \tfrac{1}{2}\ \overline{x}_1) \quad x_1 = 0.200$$
$$2 \text{ O atoms at } (0\ \tfrac{1}{2}\ 0)\ (\tfrac{1}{2}\ 0\ 0)$$
$$2 \text{ S atoms at } (0\ 0\ x_2)\ (\tfrac{1}{2}\ \tfrac{1}{2}\ \overline{x}_2) \quad x_2 = 0.65$$

The lattice dimensions are

$$a_1 = 3.835 \pm 0.002 \text{ A} \quad a_3 = 6.682 \pm 0.002 \text{ A}$$

The uranium-oxygen distance is 2.34 A, and the uranium-sulfur distance is 2.93 A. The calculated density is 9.60 g/cc.

UOS is soluble in concentrated nitric acid; it reacts with bromine water to give uranate and sulfate (MP Berkeley 5,9).

$UO_2 \cdot 2US_2$ was discovered by Rose (1823) and first analyzed by Hermann (1861). It was described as dark lead-gray to black and was obtained by heating U_3O_8, UO_2, or ammonium uranate in carbon disulfide vapor. At Ames (MP Ames 15) the powder diagram of a product obtained from UO_2 and carbon disulfide was found to be different from those of UO_2 and US_2; the observed phase ($UO_2 \cdot 2US_2$?) was complex, and its structure has not yet been determined.

The compound ignites in chlorine, is decomposed by fuming HNO_3, and is slowly oxidized by chlorine water. It is easily decomposed by concentrated HCl (Hermann, 1861).

(b) Uranyl Sulfide. UO_2S (or $2UO_3 \cdot US_3$). This compound was first obtained by Berzelius (1824b) and then by Hermann (1861) by precipitating UO_2^{++} salts by $(NH_4)_2S$. The precipitate is gradually converted in the presence of air into the complex highly colored products known as "uranium reds" (cf. chapter on uranates in Part II of this volume). Pure crystalline UO_2S was obtained by Milbauer (1904a) from KCNS (12 parts), UO_2 (3 parts), and sulfur (5 parts). The product forms black tetragonal needle-shaped crystals. It is converted to U_3O_8 by ignition in air with liberation of sulfur dioxide. Aqua regia or HNO_3 oxidizes it rapidly. HCl does not attack at room temperature but decomposes the uranyl sulfide on heating.

8. URANIUM SELENIDES AND TELLURIDES

8.1 Selenides. Colani (1903b, 1907b) obtained uranium(IV) selenide, USe_2, by ignition at moderate red heat of $2NaCl \cdot UCl_4$ in hydrogen charged with selenium vapor. It also is formed by the action of SnSe on the sodium uranium(IV) double chloride in a dry hydrogen stream. It forms small, brittle, strongly reflecting, black crystals and is often pyrophoric. Its chemical properties are similar to those of US_2, but it is more readily oxidizable, igniting on treatment with HNO_3. If too little selenium is used and the temperature is raised to 1000°C during preparation, U_2Se_3 is produced.

Uranyl selenide was obtained by Milbauer (1904b) by a method similar to the one he used for the preparation of UO_2S_2, namely, ignition of 1 part of U_3O_8 with 5 parts of KCN and 7 parts of selenium in a double crucible at bright-red heat. The product was washed with water and alcohol and dried at 98°C. It forms small, metallic, lustrous, six-sided prisms which are black with a red tinge. Water decomposes it slowly, giving a red solution. It is not affected by dilute alkalis, is easily soluble in cold HCl (forming UO_2Cl_2), and reacts violently with HNO_3, forming at first selenium which is then oxidized.

8.2 Tellurides. Colani (1903b, 1907c) obtained a uranium telluride by heating $2NaCl \cdot UCl_4$ to 1000°C in a stream of dry hydrogen charged with Te vapor. The product had an approximate composition of UTe.

A second uranium telluride, U_2Te_3, was obtained by the heating to 1000°C of $2NaCl \cdot UCl_4$ with Na_2Te containing excess tellurium. It forms black metallic lustrous lamellae. The expected compound UTe_2 could not be obtained.

Montignie (1947) has described the preparation of a uranium telluride of the approximate composition $UTe_{2.2}$ by fusion of tellurium metal, uranium trioxide, and potassium cyanide at red heat. It is reported to be insoluble in the common solvents, reacts with nitric acid to give uranyl nitrate, and is converted to oxide by ignition in air. Bromine reacts to form the bromides; fusion with potassium carbonate gives K_2TeO_4 and K_2UO_4.

The uranium-selenium and uranium-tellurium systems clearly require further study.

REFERENCES

1789   M. H. Klaproth, Crell Ann., II, p. 396.
1811   Gay-Lussac and Thénard, "Recherches physico-chimiques," Paris, vol. 1, p. 262.
1822a   J. A. Arfvedson, Kgl. Svenska Vetenskapsakad. Handl., p. 413.
1822b   J. A. Arfvedson, Kgl. Svenska Vetenskapsakad. Handl., p. 408.
1822c   J. A. Arfvedson, Kgl. Svenska Vetenskapsakad. Handl., p. 410.
1823   H. Rose, Gilb. Ann., 73: 139.
1824a   J. A. Arfvedson, Pogg. Ann., 1: 250.
1824b   J. A. Arfvedson, Pogg. Ann., 1: 256.
1824a   J. J. Berzelius, Pogg. Ann., 1: 362.
1824b   J. J. Berzelius, Pogg. Ann., 1: 373.
1825   J. A. Arfvedson, Ann. chim. et phys., 29: 160.
1825   Laugier and Boudet, J. pharm. chim., 11: 286.
1825   L. R. Lecanu, J. pharm. chim., 11: 279.
1840   V. Regnault, Ann. chim. et phys., 73: 50; Pogg. Ann., 51: 225.
1842a   J. J. Ebelmen, Ann. chim. et phys., 5: 195; Ann., 43: 290.
1842b   J. J. Ebelmen, Ann., 43: 295.
1842c   J. J. Ebelmen, Ann., 43: 294; Ann. chim. et phys., 5: 198.
1842d   J. J. Ebelmen, Ann. chim. et phys., 5: 207; Ann., 43: 302.
1842a   E. Péligot, Ann. chim. et phys., 5: 24; Ann., 43: 268.
1842b   E. Péligot, Ann. chim. et phys., 5: 5.
1842c   E. Péligot, Ann. chim. et phys., 5: 5; Ann., 43: 255.
1842d   E. Péligot, Ann. chim. et phys., 5: 20.
1842   F. Wöhler, Ann., 41: 345.
1843   F. I. Malaguti, Ann. chim. et phys., 9: 463.
1843a   C. Rammelsberg, Pogg. Ann., 59: 12.
1843b   C. Rammelsberg, Pogg. Ann., 59: 9.
1843   J. Wertheim, J. prakt. Chemie, 29: 211.
1844   E. Péligot, Ann. chim. et phys., 12: 562.
1845   J. J. Berzelius, Jahresbericht über d. Fortschritte der physischen Wissenschaften d. Chemie u. Mineralogie (Berzelius), 24: 118.
1859   G. E. Uhrlaub, dissertation, University of Göttingen.
1861   B. Drenckmann, Z. ges. Naturw., 17: 131.
1861   H. Hermann, dissertation, University of Göttingen, p. 31.
1862   T. Graham, J. Chem. Soc., 15: 254; Ann., 121: 52; Ann. chim. et phys., 65: 183.
1865   G. Gore, Phil. Mag., 29: 546.
1866   H. C. Bolton, dissertation, University of Göttingen, p. 20; Z. Chem., 2: 353; Bull. soc. chim. Paris, 6: 450; Monatsber. d. Kgl. Preuss. Akad. d. Wissenschaften, p. 299.
1869   G. Gore, J. Chem. Soc., 22: 393.
1870   W. Huggins, Proc. Roy. Soc. London, 18: 548; Phil. Mag., 40: 302.

1875   M. Isambert, Compt. rend., 30: 1087.
1877a T. Fairley, J. Chem. Soc., 31: 133.
1877b T. Fairley, J. Chem. Soc., 31: 127.
1879   J. Donath, Ber., 12: 743.
1879   E. F. Smith, Am. Chem. J., 1: 329.
1880   E. F. Smith, Ber., 13: 751.
1881   J. Riban, Compt. rend., 93: 1141.
1882   J. Riban, Bull. soc. chim. Paris, 38: 157.
1882a C. Zimmermann, Ann., 216: 12.
1882b C. Zimmermann, Ann., 213: 290.
1886a G. Alibegoff, Ann., 233: 123.
1886b G. Alibegoff, Ann., 233: 135.
1886a C. Zimmermann, G. Alibegoff, and G. Krüss, Ann., 232: 283.
1886b C. Zimmermann, G. Alibegoff, and G. Krüss, Ann., 232: 290.
1888   W. R. Hodgkinson and F. K. Lowndes, Chem. News, 58: 309.
1888   H. Quantin, Compt. rend., 106: 1075.
1889   V. Fischel, dissertation, University of Bern, p. 10.
1889   W. R. Hodgkinson and F. K. Lowndes, Chem. News, 59: 64.
1890   G. J. Fowler and J. Grant, J. Chem. Soc., 57: 275.
1892   H. Moissan, Compt. rend., 115: 1033.
1892   G. Rauter, Ann., 270: 254.
1892   P. Sabatier and J. B. Senderens, Compt. rend., 114: 1431.
1893a W. F. Hillebrand, Z. anorg. Chem., 3: 243.
1893b W. F. Hillebrand, Z. anorg. Chem., 3: 249; U. S. Geol. Survey Bull., 113: 41.
1893   H. Moissan, Compt. rend., 116: 347.
1894   E. F. Smith and O. L. Shinn, Z. anorg. Chem., 7: 49; J. Am. Chem. Soc., 16: 571.
1895   O. Brunck, Z. anorg. Chem., 10: 246.
1895   P. Sabatier and J. B. Senderens, Compt. rend., 120: 618; Bull. soc. chim. Paris, 13: 870.
1895   E. F. Smith and J. M. Matthews, J. Am. Chem. Soc., 17: 637.
1896   J. Aloy, Compt. rend., 122: 1542.
1896   H. Moissan, Compt. rend., 122: 1088.
1896   P. Sabatier and J. B. Senderens, Ann. chim. et phys., 7: 356, 384, 396.
1899   J. Aloy, Bull. soc. chim. Paris, (3) 21: 613.
1899   St. Meyer, Wied. Ann., 69: 245.
1900   J. Aloy, Bull. soc. chim. Paris, (3) 23: 368.
1900a L. Pisarzhevskiĭ, Z. anorg. Chem., 24: 108.
1900b L. Pisarzhevskiĭ, Z. anorg. Chem., 24: 111.
1900   J. Zehenter, Monatsh., 21: 241.
1901a J. Aloy, Recherches sur l'uranium et ses composés, Thèses Toulouse, No. 21, p. 23.
1901b J. Aloy, Ann. chim. et phys., (7) 24: 418.
1901a W. Oechsner de Coninck, Bull. classe sci. Acad. roy. Belg., p. 226.
1901b W. Oechsner de Coninck and M. Camo, Bull. classe sci. Acad. roy. Belg., p. 321.
1901c W. Oechsner de Coninck and M. Camo, Bull. classe sci. Acad. roy. Belg., p. 332.
1901   V. Kohlschütter, Ann., 314: 333.
1901   F. Mylius and R. Dietz, Ber., 34: 2777.
1902   W. Oechsner de Coninck, Compt. rend., 135: 900; Bull. classe sci. Acad. roy. Belg., p. 1025.
1902   T. W. Richards and B. S. Merigold, Z. anorg. Chem., 31: 235.
1903a A. Colani, Compt. rend., 137: 382.
1903b A. Colani, Compt. rend., 137: 383.
1903   W. Oechsner de Coninck, Ann. chim. et phys., (7) 28: 6.

1903   L. Pisarzhevskiĭ, Z. physik. Chem., 43: 161; J. Russ. Phys. Chem. Soc., 35: 44.
1904a W. Oechsner de Coninck, Bull. classe sci. Acad. roy. Belg., p. 448.
1904b W. Oechsner de Coninck, Bull. classe sci. Acad. roy. Belg., p. 363.
1904a J. Milbauer, Z. anorg. Chem., 42: 448.
1904b J. Milbauer, Z. anorg. Chem., 42: 450.
1907a A. Colani, Ann. chim. et phys., (8) 12: 79.
1907b A. Colani, Ann. chim. et phys., (8) 12: 85.
1907c A. Colani, Ann. chim. et phys., (8) 12: 87.
1907   B. Szilard, J. chim. phys., 5: 493, 641.
1908   W. W. Coblentz, Bull. Bur. Stand., 5: 173.
1908   W. Oechsner de Coninck, Bull. classe sci. Acad. roy. Belg., p. 992.
1908   P. Faehr, dissertation, München Techn. Höchschule, p. 46.
1908   F. Giolitti and G. Tavanti, Gazz. chim. ital., (II) 38: 239.
1908   H. C. Greenwood, J. Chem. Soc., 93: 1492.
1908   H. N. McCoy and G. C. Ashman, Am. J. Sci., 26: 522.
1909   W. Oechsner de Coninck, Bull. classe sci. Acad. roy. Belg., p. 744.
1910   F. Bourion, Ann. chim. et phys., 21: 58.
1910   P. Camboulives, Compt. rend., 150: 177.
1910   A. Michael and A. Murphy, Jr., Am. Chem. J., 44: 384.
1911a W. Oechsner de Coninck, Compt. rend., 152: 1179.
1911b W. Oechsner de Coninck and A. Raynaud, Bull. soc. chim. France, 9: 304.
1911   A. Raynaud, Compt. rend., 153: 1480.
1911   E. H. Riesenfeld and W. Mau, Ber., 44: 3589.
1911   O. Ruff and O. Goecke, Z. angew. Chem., 24: 1459-1461.
1911   A. Samsonov, Z. Chem. u. Ind. Kolloide, 8: 96.
1912   W. Oechsner de Coninck and A. Raynaud, Bull. soc. chim. France, 11: 1037.
1912   P. Lebeau, Compt. rend., 154: 1808; Bull. soc. chim. France, 11: 800.
1912a W. G. Mixter, Am. J. Sci., 34: 155; Z. anorg. Chem., 78: 237.
1912b W. G. Mixter, Z. anorg. Chem., 78: 234.
1912   A. Raynaud, Bull. soc. chim. France, 11: 802.
1912   A. S. Russell, Physik. Z., 13: 61.
1913   A. Fischer, Z. anorg. Chem., 81: 201.
1913   R. de Forcrand, Compt. rend., 156: 1956.
1914   C. C. Bidwell, Phys. Rev., 3: 204.
1914   G. K. Burgess and R. G. Waltenburg, J. Wash. Acad. Sci., 4: 567.
1914   D. Lely, Jr., and L. Hamburger, Z. anorg. Chem., 87: 220.
1914   E. K. Rideal, J. Soc. Chem. Ind. London, 33: 674.
1914   E. Tiede and E. Birnbräuer, Z. anorg. Chem., 87: 165.
1915   G. K. Burgess and R. G. Waltenburg, Bull. Bur. Stand., 11: 591.
1915a R. de Forcrand, Ann. chim., 3: 36.
1915b R. de Forcrand, Ann. chim., 3: 34.
1915   J. A. Hedvall, Z. anorg. Chem., 93: 319.
1915   K. A. Hofmann and K. Höschele, Ber., 48: 21.
1915   A. Müller, Z. anorg. u. allgem. Chem., 93: 269.
1915   E. Wedekind and C. Horst, Ber., 48: 111.
1916a E. Newbery and J. N. Pring, Proc. Roy. Soc. London, (A)92: 276.
1916b E. Newbery and J. N. Pring, Proc. Roy. Soc. London, (A)92: 282.
1917   C. L. Parsons, J. Ind. Eng. Chem., 9: 466.
1919   C. A. Pierlé, J. Phys. Chem., 23: 549.
1919-1920   F. P. Venable and D. H. Jackson, J. Elisha Mitchell Sci. Soc., 35: 88.
1920   J. Aloy and E. Rodier, Bull. soc. chim. France, 26: 101.
1920   R. Schwarz, Helv. Chim. Acta, 3: 345; R. Abbegg, "Handbuch der anorg. Chemie," vol. IV,1, 2d part, p. 918, S. Hirzel, Leipzig, 1921.
1921   G. Chaudron, Ann. chim., 16: 251.

1921   H. Staehling, Compt. rend., 173: 1470.
1922   J. Aloy and E. Rodier, Bull. soc. chim. France, 31: 249.
1922   G. F. Hüttig, Z. angew. Chem., 35: 391.
1922a   G. F. Hüttig and E. von Schroeder, Z. anorg. u. allgem. Chem., 121: 243.
1922b   G. F. Hüttig and E. von Schroeder, Z. anorg. u. allgem. Chem., 121: 250.
1922   P. Jolibois and R. Bossuet, Compt. rend., 174: 386.
1922a   P. Lebeau, Compt. rend., 174: 388.
1922b   P. Lebeau, Compt. rend., 174: 390.
1922   E. L. Nichols and H. L. Howes, Phys. Rev., 19: 300.
1922   E. von Schroeder, dissertation, University of Göttingen.
1923a   V. M. Goldschmidt and L. Thomassen, Videnskapsselskapets-Skrifter, (I) 2: 12.
1923b   V. M. Goldschmidt and L. Thomassen, Videnskapsselskapets-Skrifter, (I) 2: 16.
1923c   V. M. Goldschmidt and L. Thomassen, Videnskapsselskapets-Skrifter, (I) 2: 8.
1923d   V. M. Goldschmidt and L. Thomassen, Videnskapsselskapets-Skrifter, (I) 2: 18.
1923   H. von Wartenberg, J. Broy, and R. Reinicke, Z. Elektrochem., 29: 215.
1924   A. E. van Arkel, Physica, 4: 297.
1924   D. Balarev, Z. anorg. u. allgem. Chem., 136: 217.
1924   G. F. Hüttig, Fortschr. Chem., Physik u. physik. Chem., 18 (1): 12.
1924   E. Wiegand, Z. Physik, 30: 40.
1925   E. Friederich and L. Sittig, Z. anorg. u. allgem. Chem., 145: 127, 138.
1925   S. W. Rowell and A. S. Russell, J. Chem. Soc., 127: 2901.
1926   S. D'yachkovskiĭ, Ukraïn. Khem. Zhur., 2: 340.
1926   O. Heusler, Z. anorg. u. allgem. Chem., 154: 363.
1926   G. Tammann and W. Rosenthal, Z. anorg. u. allgem. Chem., 156: 20.
1927a   W. Biltz and H. Müller, Z. anorg. u. allgem. Chem., 163: 279.
1927b   W. Biltz and H. Müller, Z. anorg. u. allgem. Chem., 163: 263.
1927c   W. Biltz and H. Müller, Z. anorg. u. allgem. Chem., 163: 291.
1927d   W. Biltz and H. Müller, Z. anorg. u. allgem. Chem., 163: 295.
1927e   W. Biltz and H. Müller, Z. anorg. u. allgem. Chem., 163: 261.
1927f   W. Biltz and H. Müller, Z. anorg. u. allgem. Chem., 163: 161.
1928   G. Beck, Z. anorg. u. allgem. Chem., 174: 40.
1928   W. Biltz and C. Fendius, Z. anorg. u. allgem. Chem., 176: 49.
1928   M. L. Philipps, Phys. Rev., 32: 832.
1928   O. Schmidt, Z. physik. Chem., 133: 263, 289.
1928   K. C. Sen, Z. anorg. u. allgem. Chem., 174: 61, 63, 69.
1928   A. Sieverts and E. L. Müller, Z. anorg. u. allgem. Chem., 173: 297.
1929   M. Picon, Compt. rend., 189: 96.
1929a   A. Rosenheim and H. Daehr, Z. anorg. u. allgem. Chem., 181: 181.
1929b   A. Rosenheim and H. Daehr, Z. anorg. u. allgem. Chem., 181: 178.
1930   V. N. Ipatieff and B. Muromtsev, Ber., 63: 164.
1930   M. LeBlanc and H. Sachse, Ber. Verhandl. sächs. Akad. Wiss. Leipzig, 82: 155.
1931   S. D'yachkovskiĭ, Kolloid-Z., 54: 280.
1931   V. A. Kargin, Z. anorg. u. allgem. Chem., 198: 80.
1931   N. W. Taylor, J. Am. Chem. Soc., 53: 4459.
1931   C. Wagner and W. Schottky, Z. physik. Chem., (B)11: 163.
1932   P. Guillery, Ann. phys., 14: 218.
1932   K. K. Kelley, U. S. Bur. Mines Bull. 350, p. 47.
1932   A. Pochettino, Atti. accad. nazl. Lincei, 15: 505.
1932   A. Rosenheim and H. Daehr, Z. anorg. u. allgem. Chem., 208: 81.
1932   A. Rosenheim and M. Kelmy, Z. anorg. u. allgem. Chem., 206: 31.
1932   W. Sucksmith, Phil. Mag., 14: 1122.
1932   H. B. Wahlin, Phys. Rev., 39: 183.
1933   W. Kangro and R. Jahn, Z. anorg. u. allgem. Chem., 210: 335.
1933   W. Meyer, Z. tech. Physik, 14: 126; Z. Physik, 85: 287.

1934   L. F. Audrieth and M. T. Schmidt, Proc. Natl. Acad. Sci. U. S., 20: 223.
1934   F. Keller and W. R. Lehmann, Z. Physik, 88: 682.
1935a M. Francis, Compt. rend., 201: 473.
1935b M. Francis and Tscheng-Da-Tschang, Compt. rend., 200: 1024.
1936   W. Hartmann, Z. Physik, 102: 709.
1938   R. Flatt and W. Hess, Helv. Chim. Acta, 21: 525.
1939   R. Lydén, Finska Kemiststamf. Medd., 48: 124.
1939   W. Tilk and W. Klemm, Z. anorg. u. allgem. Chem., 240: 355.
1940   H. Haraldsen and R. Bakken, Naturwissenschaften, 28: 127.
1942   W. Amrein, Schweiz. Arch. angew. Wiss. u. Tech., 8: 85, 109.
1946   J. S. Anderson, Proc. Roy. Soc. London, A185: 69.
1947   A. I. Avgustinik, J. Applied Chem. U.S.S.R., 20: 327.
1947   P. Jolibois, Compt. rend., 224: 1395.
1947   E. Montignie, Bull. soc. chim. France, 1947: 748-749.
1947   G. E. Moore and K. K. Kelley, J. Am. Chem. Soc., 69: 2105-2107.
1948   A. Boullé and M. Dominé-Bergès, Compt. rend., 227: 1365.
1948a F. Grønvold, Nature, 162: 70.
1948b F. Grønvold and H. Haraldsen, Nature, 162: 69.
1948   J. I. Hoffman, J. Wash. Acad. Sci., 38: 233.
1948   R. E. Rundle, N. C. Baenziger, A. S. Wilson, and R. A. McDonald, J.Am.Chem. Soc., 70: 99-105.
1948   D. T. Vier, M. L. Schultz, and J. Bigeleisen, J. Optical Soc. Am., 38: 811-814.
1948   W. H. Zachariasen, Acta Crystallographica, 1: 265.
1949   A. Boullé and M. Dominé-Bergès, Compt. rend., 228: 72.
1949   J. J. Katz and D. M. Gruen, J. Am. Chem. Soc., 71: 2106.
1949   J. Prigent, J. phys. radium, 10: 58-64.
1950   A. Boullé, R. Jary, and M. Dominé-Bergés, Compt. Rend., 250: 300.

Project Literature

Battelle 1: Battelle Memorial Institute, Report CT-2632, Jan. 1, 1945.

British 1: C. W. Bunn and H. J. Spencer-Palmer, Report BR-502, Sept. 7, 1944.
British 2: Report [B]LRG-42, March, 1945.
British 3: Report [B]LRG-33, June, 1944.
British 4: Report BR-250, July 23, 1943.
British 5: W. N. Haworth and K. F. Chackett, Report B-26, Jan. 5, 1945.
British 6: W. N. Haworth, C. B. Amphlett, and L. F. Thomas, Report B-27, Oct. 10, 1941.
British 7: G. O. Morris, Report B-50, Mar. 20, 1942.

Brown 1: C. A. Kraus, Report A-281, Sept. 7, 1942.
Brown 2: C. A. Kraus, Report A-1096, July 29, 1944.
Brown 3: C. A. Kraus, Report A-328, Oct. 15, 1942.
Brown 4: C. A. Kraus, Report A-360, Oct. 26, 1942.
Brown 5: C. A. Kraus, Report [A]M-7, Apr. 19, 1944.
Brown 6: C. A. Kraus, Report A-2300, Oct. 5, 1944.
Brown 7: C. A. Kraus, Reports [A]M-1060, Nov. 2, 1942, and CC-365, Nov. 17, 1942.

CEW-TEC 1: J. W. Gates, L. J. Andrews, and R. B. Pitt, Report CD-495, Jan.16,1945.
CEW-TEC 2: J. R. van Wazer, Report CD-462, Aug. 28, 1944.
CEW-TEC 3: C. Tanford and R. L. Tichenor, Report CD-3.385.1, July 16, 1945.

CEW-TEC 4: B. M. Pitt, E. L. Wagner, and A. J. Miller, Report CD-2.355.2, Jan. 4, 1946.

CEW-TEC 5: J. W. Gates, G. H. Clewett, L. J. Andrews, and H. A. Young, Report CD-457, Aug. 4, 1944.

Los Alamos 1: M. Kolodony, Report LA-35, Nov. 1, 1943.

Mallinckrodt 1: J. R. Lacher, Report A-1034, Feb. 19, 1944.

MP Ames 1: R. E. Rundle, N. C. Baenziger, and A. S. Wilson, X-ray Study of the Uranium-Oxygen System, in National Nuclear Energy Series, Division VIII, Volume 6; see also J. Am. Chem. Soc., 70: 99-105 (1948).

MP Ames 2: R. E. Rundle and N. C. Baenziger, Report CC-1980, Nov. 23, 1944.

MP Ames 3: R. E. Rundle, N. C. Baenziger, A. S. Wilson, and R. A. McDonald, Report CC-2397, Feb. 17, 1945, p. 35.

MP Ames 4: N. C. Baenziger, Report CC-1781, Aug. 10, 1944.

MP Ames 5: F. H. Spedding, Report CP-42, Apr. 25, 1942.

MP Ames 6: P. Chiotti, Report CT-1985, Jan. 4, 1945.

MP Ames 7: R. E. Rundle and N. C. Baenziger, Report CC-1524, Mar. 10, 1944.

MP Ames 8: A. S. Wilson and R. E. Rundle, Report CN-1495, Apr. 10, 1944.

MP Ames 9: R. Raeuchle and J. C. Warf, Report CC-1496, Apr. 10, 1944.

MP Ames 10: R. E. Rundle, N. C. Baenziger, and P. Chiotti, Report CC-1984, Nov. 10, 1944.

MP Ames 11: A. D. Tevebaugh, R. D. Tevebaugh, W. D. Cline, and J. C. Warf, the Conversion of UF_4 to U_3O_8, in National Nuclear Energy Series, Division VIII, Volume 6.

MP Ames 12: R. D. Tevebaugh, W. D. Cline, J. C. Warf, Report CC-1981, Oct. 10, 1944; W. D. Cline, Report CC-1983, Nov. 10, 1944; J. C. Warf and W. D. Cline, Report CC-2723, June 30, 1945.

MP Ames 13: J. Powell, Report CC-1781, Aug. 10, 1944.

MP Ames 14: L. Warf, Report CC-1194, Dec. 9, 1943.

MP Ames 15: R. E. Rundle and R. McDonald, Report CC-1504, Aug. 10, 1944.

MP Ames 16: J. Powell, Report CC-1778, Aug. 18, 1944.

MP Berkeley 1: L. Brewer, L. A. Bromley, P. W. Gilles, and N. L. Lofgren, The Thermodynamic Properties and Equilibria at High Temperatures of Uranium Halides, Oxides, Nitrides, and Carbides, in National Nuclear Energy Series, Division VIII, Volume 6.

MP Berkeley 2: J. W. Hamaker and C. W. Koch, Note on the Composition of Uranium Peroxides, in National Nuclear Energy Series, Division VIII, Volume 6.

MP Berkeley 3: L. Brewer, Report CC-672, May 15, 1943.

MP Berkeley 4: L. Brewer, Report AECD-2307, July 31, 1948.

MP Berkeley 5: L. Brewer, L. A. Bromley, E. D. Eastman, P. W. Gilles, and N. L. Lofgren, Preparation and Properties of Sulfides and Oxysulfides of Uranium, in National Nuclear Energy Series, Division VIII, Volume 6.

MP Berkeley 6: M. Calvin, Report CK-2411, Oct. 1, 1944.

MP Berkeley 7: L. A. Bromley and N. L. Lofgren, Report CT-1344, Feb. 14, 1944.

MP Berkeley 8: L. Brewer, L. A. Bromley, and N. L. Lofgren, Reports CT-1714, Apr. 14, 1944, and CT-2139, Aug. 15, 1944.

MP Berkeley 9: N. L. Lofgren and L. A. Bromley, Report CK-941, September, 1943.

MP Chicago 1: W. H. Zachariasen, Report N-1973, Apr. 24, 1945.

MP Chicago 2: W. H. Zachariasen, Reports CK-2667, Jan. 15, 1945, and CC-2768, Mar. 12, 1945.

MP Chicago 3: S. Fried and N. R. Davidson, The Ignition of U_3O_8 in Oxygen at High Pressures and the Crystallization of UO_3, in National Nuclear Energy Series, Division VIII, Volume 6.
MP Chicago 4: B. Rosenbaum and G. Pederzani, Report CP-1168, May 25, 1943.
MP Chicago 5: J. Rehn and R. Cefola, Report CK-1240, Jan. 19, 1944.
MP Chicago 6: W. H. Zachariasen, Reports CP-1249, Jan. 22, 1944, and CK-1367, Feb. 25, 1944.
MP Chicago 7: N. R. Davidson, Report N-1617, Sept. 18, 1944.
MP Chicago 8: S. Fried and N. R. Davidson, Reports CN-2689, Mar. 3, 1945, and CN-3053, June 28, 1945.
MP Chicago 9: T. W. Davis and M. Burton, Report CC-231, Aug. 15, 1942.
MP Chicago 10: J. J. Katz and N. R. Davidson, Report CK-1221, Jan. 15, 1944.
MP Chicago 11: W. H. Zachariasen, Report CF-2926, Apr. 15, 1945.
MP Chicago 12: W. H. Zachariasen, Report CP-2160, Sept. 23, 1944; Crystal Structure Studies of Sulfides and Oxysulfides of Uranium, Thorium, and Cerium, in National Nuclear Energy Series, Division VIII, Volume 6.
MP Chicago 13: R. C. Mooney, Report CP-1507, Mar. 27, 1944.
MP Chicago 14: J. Karle, Report CK-1512, Apr. 6, 1944.
MP Chicago 15: G. T. Seaborg and coworkers, Report CS-2793, Apr. 10, 1943.
MP Chicago 16: W. H. Zachariasen, Reports CN-2615, Jan. 9, 1945, and CC-2768, Mar. 12, 1945.

MP Clinton 1: L. S. Goldring, Report N-2294, May 23, 1946.

Natl. Bur. Standards 1: H. E. Cleaves, Holm, and Kimble, Report CT-1696, April, 1944.
Natl. Bur. Standards 2: H. E. Cleaves, M. M. Cron, and J. T. Sterling, Report CT-2618, February, 1945.
Natl. Bur. Standards 3: H. E. Cleaves, Report CT-1819, July 6, 1944.
Natl. Bur. Standards 4: Eggleston, reported by T. M. Snyder, Report CP-36, Apr. 17, 1942.

Princeton 1: T. M. Snyder and R. L. Kamm, Reports CP-76, May 16, 1942, and CP-92, May 23, 1942.
Princeton 2: T. M. Snyder and R. L. Kamm, Report CT-96.
Princeton 3: T. M. Snyder and R. L. Kamm, Report CP-124, June 13, 1942.

Purdue 1: E. T. McBee, Report [A]M-2102, Mar. 21, 1945.

Rockefeller 1: R. J. Best, D. Taub, and L. G. Longsworth, Report A-380, Nov. 24, 1942.

SAM Columbia 1: I. Kirshenbaum, Summary Report SAM Laboratory. Work by A. Grenall, B. Cohen, B. Ostrofsky, and R. Palter.
SAM Columbia 2: H. F. Priest and G. L. Priest, Report A-257, Aug. 10, 1942.
SAM Columbia 3: I. Kirshenbaum, Summary Report SAM Laboratory, Sec. 8. Work by C. F. Hiskey, M. L. Eidinoff, J. Mallan, and J. Nadelhaft, Report A-777, Aug. 14, 1943.
SAM Columbia 4: H. C. Urey and coworkers, Report A-743, June 30, 1943.
SAM Columbia 5: C. F. Hiskey, Report A-123, Aug. 20, 1943.
SAM Columbia 6: R. H. Christ, Report A-146, Sept. 4, 1943.
SAM Columbia 7: D. T. Vier, Report A-1277, May 26, 1944.
SAM Columbia 8: D. T. Vier, M. L. Schultz, and J. Bigeleisen, Report A-2176, Dec. 21, 1944.

SAM Columbia 9:   M. L. Eidinoff, Report CC-1536, Apr. 6, 1944.

SAM Columbia 10:. C. F. Hiskey, Report CC-1198, Jan. 5, 1944; C. F. Hiskey and M. L. Eidinoff, Report CC-1383, Feb. 28, 1944.

SAM Columbia 11: I. Kirshenbaum, Summary Report SAM Laboratory, part 11.

SAM Columbia 12: I. Kirshenbaum, Summary Report SAM Laboratory, part 10.

SAM Columbia 13: D. M. Gillies, Report A-1262, Apr. 7, 1944.

UCRL 1: F. A. Jenkins, Report RL-4.6.47, Mar. 9, 1943.

UCRL 2: G. E. MacWood and D. Altman, Report RL-4.7.600.

UCRL 3: G. E. MacWood, Report RL-4.7.602, Dec. 18, 1945.

UCRL 4: M. J. Polissar and S. B. Kilner, Report RL-4.6.42, Nov. 7, 1942.

UCRL 5: S. Rosenfeld, Report RL-4.6.52, Mar. 22, 1943.

UCRL 6: W. C. Wood, H. Bradley, and H. S. Carroll, Report RL-4.6.287, Aug. 29, 1944.

U. S. Bur. Mines 1: E. A. Long, W. M. Jones, and J. Gordon, Report A-329, Oct. 28, 1942.

U. S. Bur. Mines 2: G. E. Moore and E. A. Long, Report A-502, Dec. 31, 1942.

Part 4

URANIUM HALIDES
AND RELATED COMPOUNDS

Chapter 12

NONVOLATILE FLUORIDES OF URANIUM

1. URANIUM TRIFLUORIDE, UF_3

In contrast to uranium trichloride and tribromide, which are rather easily prepared and have been known for a long time, uranium trifluoride is a compound whose preparation is attended with some difficulty. Its synthesis has only recently been achieved. Early attempts to prepare UF_3 were made by reduction of the tetrafluoride with hydrogen at elevated temperatures, a method that was successfully used earlier for the preparation of UBr_3 and UCl_3. In the first such attempt (Bolton, 1866a,b) the tetrafluoride used was prepared from aqueous solution and therefore probably contained small amounts of water. When this material was treated with dry hydrogen at red heat, hydrogen fluoride was evolved, and a reddish-brown mass was obtained. It was insoluble in water and was scarcely attacked by any acid, with the exception of concentrated nitric acid. No analysis of the product was made, but Bolton surmised that it was a lower fluoride. However, recent work has shown that, in the presence of traces of oxygen or water, reaction of uranium tetrafluoride with hydrogen leads to uranium dioxide rather than to the tribromide. Thus Andrews and Gates (CEW-TEC 1), who treated uranium tetrafluoride with hydrogen at 450 to 550°C, obtained as product only impure UO_2, probably because of the presence of oxygen in the tank hydrogen used. British investigators (British 1) also tried to obtain UF_3 by treating uranium tetrafluoride with hydrogen at 600°C. They, too, observed an evolution of hydrogen fluoride but failed to identify the product as UF_3.

Later workers reinvestigated the reaction of hydrogen with UF_4. The hydrogen was rigorously purified by passage over hot uranium metal (Chap. 6, Sec. 1.1). It was found that if the reaction was carried out in a silica reaction tube, fluorine-containing gases were evolved at temperatures above 600°C. The product, however, was largely UO_2. The following explanation was given for this result: In the presence of traces of moisture, a small amount of tetrafluoride undergoes hydrolysis with the production of UO_2 and hydrogen fluoride; the latter attacks the silica, giving SiF_4 and more water. The cycle is repeated until complete conversion of the UF_4 to UO_2 results. This is particularly likely to occur when the hydrogen flow through the system is slow. When the reaction of UF_4 with hydrogen was carried out in a monel reaction tube, no evolution of hydrogen fluoride was observed at all. Even after treatment for 48 hr with pure hydrogen at 980°C the UF_4 could be recovered substantially unchanged. This result of the Ames investigations is not readily understandable since we will see below that British investigators succeeded in preparing UF_3 by substantially the same method.

A number of unsuccessful attempts were made in various laboratories to prepare UF_3 by halogen exchange reactions with UH_3, UCl_3, or UBr_3. Whereas UBr_3 and UCl_3 are formed easily when UH_3 is treated with hydrogen bromide or chloride, the action of hydrogen fluoride on UH_3 yields only UF_4 (MP Ames 1).

$$UH_3 + 4HF \xrightarrow{400°C} UF_4 + \frac{7}{2}H_2 \tag{1}$$

Andrews and Gates (CEW-TEC 1) treated UCl_3 with anhydrous hydrogen fluoride at 450°C and also obtained UF_4.

$$UCl_3 + 4HF \rightarrow UF_4 + 3HCl + \frac{1}{2}H_2 \tag{2}$$

The Brown University group (Brown 1) treated anhydrous uranium trichloride with liquid anhydrous hydrogen fluoride at 25°C and observed the evolution of hydrogen chloride. No details of the experiment were given, but the statement was made that the product obtained oxidized rapidly in dry air and that presumably it was the trifluoride. However, the easy oxidation of the product does not agree well with the known stability of the trifluoride, and the reaction therefore requires further study.

An earlier observation made by the same group (Brown 2) indicated that a reddish, rapidly oxidizing precipitate is also formed by the addition of aqueous hydrofluoric acid to an aqueous solution of uranium trichloride. This observation was confirmed at Ames and at the University of California Radiation Laboratory (MP Ames 2; UCRL 1). However, only uranium tetrafluoride could be isolated from the product, despite the transitory appearance of a reddish-brown color in the precipitate. Since it appeared reasonable that rapid reaction of UF_3 with water might be responsible for the failure, the experiments were repeated in a nonaqueous solvent (MP Ames 3). A solution of

UBr$_3$ in dimethylformamide was treated with anhydrous hydrogen fluoride. (Previous experiments had shown dimethylformamide to be unaffected by hydrogen fluoride.) However, instead of the expected red precipitate of UF$_3$, only a green tarry paste resulted which must have been formed by reaction with the solvent. Thus uranium trifluoride could not be prepared by any of these halogen exchange reactions.

1.1 <u>Preparation of Uranium Trifluoride</u>. Two successful preparations of uranium trifluoride have thus far been achieved. In the first, by careful attention to the effects of moisture and oxygen and to the purity of the uranium tetrafluoride, British workers succeeded in preparing the trifluoride by reduction of the tetrafluoride with hydrogen (British 2). In the second preparation the Ames group obtained UF$_3$ by the reduction of the tetrafluoride with metallic uranium.

(a) <u>Reduction of UF$_4$ with Hydrogen</u>. (British 2.) The success of this procedure is contingent upon the exclusion of all traces of moisture or oxygen. The hydrogen is rigorously purified. After preliminary drying over calcium chloride, oxygen is removed by palladium-asbestos at 350°C, and final drying is achieved by successive passage through calcium chloride, phosphorus pentoxide, and a trap cooled by liquid nitrogen. Traces of suspended solid impurities are removed from the gas stream by a sintered glass filter.

The reaction tube consists of an outer tube of silica fitted with a liner of stainless steel (18/8/1/1) to prevent the hydrogen fluoride produced in the reaction from attacking the silica tube. In large-scale preparations a stainless-steel tube alone is sufficient. A molybdenum boat can be used for holding the tetrafluoride. All exposed silica surfaces in the cooler parts of the apparatus are protected by a coating of ceresin wax. The entire apparatus is evacuated, thoroughly dried, and degassed before use.

To obtain pure uranium trifluoride it is essential to use tetrafluoride that is free of oxygen-containing impurities. Uranium tetrafluoride prepared from aqueous solution and containing as chief impurities tin, copper, sulfate, and water gives mixtures of UF$_3$ and UO$_2$, since at the high temperatures used both sulfate and water react with UF$_4$. Even uranium tetrafluoride prepared from UO$_2$ by high-temperature hydrofluorination, a material which is ordinarily considered as anhydrous, contains enough moisture to cause contamination of the product with UO$_2$. It has been found that UF$_4$ of sufficient purity can be prepared by sublimation of ordinary UF$_4$ in a very good vacuum at 1000°C, using a molybdenum apparatus. Very pure crystalline material suitable for conversion to UF$_3$ can be obtained in this manner.

The reduction of UF_4 by hydrogen proceeds quite rapidly at 1000°C (above the melting point of UF_4) according to the equation

$$2UF_4 + H_2 \rightleftarrows 2UF_3 + 2HF \tag{3}$$

Below 900°C reduction is negligible. Reduction beyond the trifluoride state is very slow at 1000°C. Prolonged heating at 1000°C is inadvisable as there is danger of thermal disproportionation of the trifluoride (see Sec. 1.1b). The product obtained by reduction of UF_4 with hydrogen is crystalline and has a deep violet-red color. That it is pure UF_3 is shown by chemical analysis and by x-ray crystallography.

Uranium trifluoride has also been prepared, although on a very small scale, by the action of atomic hydrogen on UF_4 in a silica vessel (MP Chicago 1). A product was obtained whose x-ray diffraction pattern indicated the presence of 60 per cent UF_3 and 40 per cent UO_2. The oxygen probably arises from water formed by the reaction of atomic hydrogen with silica.

(b) Reduction of UF_4 by Uranium Metal. (MP Ames 4.) Uranium tetrafluoride can be reduced with finely divided uranium at 1050°C.

$$3UF_4 + U \rightleftarrows 4UF_3 \tag{4}$$

Temperature regulation is important, since reaction 4 is reversible and above 1050°C the equilibrium is markedly displaced to the left. The reaction may be carried out in a nickel vessel by mixing uranium tetrafluoride with the stoichiometric amount of uranium turnings. The mixture is heated to 250°C, and hydrogen is introduced to convert the metal to hydride. The hydride is then decomposed at 400°C, giving an intimate mixture of UF_4 with finely divided uranium metal. This mixture is fused at 1050°C for 2 hr in an argon atmosphere, and a black, dense, cokelike solid is obtained. Analysis indicates that it is substantially pure UF_3 contaminated with about 1 per cent UO_2 and UO_2F_2.

1.2 Physical Properties of Uranium Trifluoride. Uranium trifluoride obtained by the reduction of uranium tetrafluoride with hydrogen is a fused crystalline mass almost black in the massive form. Under the microscope the crystals appear violet-red in color. Since the trifluoride disproportionates (to UF_4 and U) above 1000°C, the melting point could not be determined, but observations indicate that it must lie above 1140°C. The compound appears to be very slightly volatile above 1000°C (MP Ames 4; British 2).

X-ray diffraction studies (MP Chicago 2; MP Ames 5) showed that uranium trifluoride has a crystal structure very different from that of the tetrafluoride; it is isomorphous with the rare earth fluorides LaF$_3$, PrF$_3$, CeF$_3$, and NdF$_3$.

The x-ray powder diagram is characteristic of a hexagonal close-packed arrangement of the metal atoms. The unit cell contains two molecules (space group C6/mmc). Parameters of UF$_3$ and very pure LaF$_3$ determined by Zachariasen (MP Chicago 2) and of CeF$_3$, PrF$_3$, and NdF$_3$ from Oftedal (1929, 1931) are given in Table 12.1. However,

Table 12.1 — X-ray Data on UF$_3$ and Some Isomorphous Compounds

Substance	Lattice dimensions, A*		Density, g/cc	Interatomic distance, A[†]			Ionic[‡] radius of M(III), A
	a$_1$	a$_3$		3F at	2F at	6F at	
LaF$_3$	4.140 ± 0.001	7.336 ± 0.001	...	2.39	2.38	2.71	1.06
UF$_3$	4.138 ± 0.003	7.333 ± 0.004	8.95§	2.39	2.38	2.71	1.06
CeF$_3$	4.107 ± 0.003	7.273 ± 0.005	...	2.37	2.36	2.69	1.04
PrF$_3$	4.077 ± 0.003	7.218 ± 0.007	...	2.35	2.35	2.67	1.03
NdF$_3$	4.054 ± 0.003	7.196 ± 0.005	...	2.34	2.34	2.66	1.01

* Assuming hexagonal structure.

† There are 11 fluorine atoms about each metal atom.

‡ After correction to coordination number 8 and subtraction of the ionic radius for F⁻ (1.33 A).

§ British workers have found a density of 9.18 g/cc by immersion in tetralin.

Oftedal stated that the true unit cell, as determined by Laue photographs, is orthorhombic and is three times larger than the hexagonal unit cell deduced from powder diffraction patterns. In agreement with this view, British workers (British 3) found by goniometric measurements that the crystals of UF$_3$ are probably orthorhombic. The crystals are weakly birefringent with all of the refractive indices close to 1.73. All the crystals seem to exist as triplets, twinned on a submicroscopic scale.

1.3 Chemical Properties of Uranium Trifluoride. (a) Thermal Disproportionation. British workers (British 2) found that uranium trifluoride decomposes above 1000°C (see Eq. 4).

$$4UF_3 \rightleftarrows 3UF_4 + U \tag{5}$$

In vacuum UF$_4$ sublimes and uranium metal powder is left. The decomposition is slight at 1000°C and becomes considerable at 1200°C. The failure of the Ames group to obtain UF$_3$ by the reaction of U and UF$_4$ at 1400°C is in agreement with these results. It has been proposed to utilize the disproportionation of UF$_3$ for the preparation of very pure uranium metal, but the difficulty of finding materials that would resist the action of both metal and fluoride at such high temperatures renders success improbable.

(b) Effect of Air. Uranium trifluoride is only slowly affected by moist air at room temperature (British 2). It is not appreciably hygroscopic, thereby resembling the trichloride in stability. On heating in air, oxidation occurs, presumably to an oxyfluoride. At 900°C uranium trifluoride is quantitatively converted in air to U_3O_8.

(c) Water. Uranium trifluoride is insoluble in water. In cold water it is slowly oxidized to a green gelatinous material; at 100°C attack is much more rapid, and the supernatant solution acquires the yellow color indicative of oxidation to the uranyl state (British 2).

(d) Acids. (British 2; MP Ames 4.) Similarly to the rare earth fluorides, UF_3 is rather inert to acids. Like the rare earth fluorides (and in contradistinction to UF_4), UF_3 is insoluble in ammonium oxalate. Acids with oxidizing power convert UF_3 to uranyl salts and thus dissolve it. Dilute hydrochloric, sulfuric, and nitric acids attack UF_3 only slowly in the cold. Hot nitric acid dissolves UF_3 fairly rapidly with the evolution of nitrogen oxides. Hot dilute sulfuric acid also dissolves UF_3 but more slowly than nitric acid. Hot perchloric acid yields clear solutions of uranyl perchlorate, $UO_2(ClO_4)_2$. Ames investigators thought that the reaction of UF_3 with hydrochloric acid is an oxidation,

$$4UF_3 + 4HCl \rightarrow 3UF_4 + UCl_4 + 2H_2 \qquad (6)$$

but British workers stated that with either dilute or concentrated hydrochloric acid deep-red solutions are obtained, a color indicative of the presence of uranium(III) ions. This observation is in agreement with the known stabilizing effect of hydrochloric acid on uranium(III) solutions. The addition of boric acid to hydrochloric acid or other acids results in the formation of fluoborate complexes which markedly facilitate dissolution.

(e) Oxidizing Agents. (MP Ames 4; British 2.) Cupric chloride–ammonium chloride mixture, which dissolves uranium metal readily, attacks UF_3 only slightly. Ferric nitrate dissolves UF_3. Silver perchlorate solution is reduced rapidly, giving a silver mirror. Chlorine reacts to give UF_3Cl (cf. Chap. 15, Sec. 4) (MP Ames 6). The effect of fluorine has not been investigated; doubtless UF_6 would be the ultimate product although interesting intermediates might be encountered.

(f) Reducing Agents. (MP Ames 7.) Uranium trifluoride has been successfully reduced to metal with calcium. An iodine booster charge was added (0.5 mole of I_2 per mole of UF_3) together with a 30 per cent excess of calcium. Yields of well-formed massive metal as high as 98 per cent have been obtained in this way.

2. URANIUM TETRAFLUORIDE, UF$_4$

Two types of preparative methods are available for the preparation of UF$_4$. The first is based on the precipitation of UF$_4$, which is practically insoluble in water, from aqueous solutions of other uranium(IV) compounds.

$$U^{+4} \text{ (aq)} + 4F^- \text{ (aq)} \rightarrow \underline{UF_4} \tag{7}$$

The second method involves the reaction of various uranium compounds, particularly uranium dioxide, with gaseous fluorinating agents at elevated temperatures; for example,

$$UO_2 + 4HF \xrightarrow{550°C} UF_4 + 2H_2O \tag{8}$$

This method is distinguished by simplicity of procedure and the relative ease with which a pure anhydrous product can be obtained.

In the following section a description of the various methods of preparation of UF$_4$ will be given. Emphasis will be placed on the chemical principles rather than on the details of the actual procedures.

2.1 Early Work. Uranium tetrafluoride was first prepared by the action of aqueous hydrofluoric acid on U$_3$O$_8$ (Hermann, 1861). The reaction is vigorous and yields a yellow solution together with a green insoluble residue. The composition of this residue was correctly determined as UF$_4$ by Bolton (1866a,b).

$$U_3O_8 + 8HF \rightarrow 2UO_2F_2 + \underline{UF_4} + 4H_2O \tag{9}$$

Uranium tetrafluoride prepared in this way is very difficult to filter and wash. Bolton obtained a more easily filterable product by the reduction of a boiling solution of uranyl fluoride, UO$_2$F$_2$, with stannous chloride and periodic addition of hydrofluoric acid. Solutions of ammonium uranate or ammonium uranyl carbonate in hydrofluoric acid also yielded an easily purifiable uranium tetrafluoride when reduced in the same way. Bolton also observed that addition of hydrofluoric acid to a solution of uranium tetrachloride produced a voluminous precipitate of uranium tetrafluoride. Both Hermann and Bolton reported that ignited uranium dioxide reacted very slowly with aqueous hydrofluoric acid; freshly precipitated UO$_2 \cdot x$H$_2$O reacted much more rapidly. This observation finds confirmation in the work of

Braddock and Copenhafer (Mallinckrodt 1) who noted that hot constant-boiling hydrofluoric acid (43.2 per cent HF) gave no uranium tetrafluoride with anhydrous uranium dioxide, even after a long period of time. Conversion of UO_2 to UF_4 is apparently accelerated by dilution of the hydrofluoric acid. Ditte (1880) also studied the action of hydrofluoric acid on U_3O_8 with results which he claimed were not in agreement with those of Bolton. Ditte's results are, however, valueless because of poor analytical technique. A repetition of the experiments by Smithells (1883) confirmed Bolton's original conclusions.

2.2 Preparation of Uranium Tetrafluoride in Aqueous Solutions. Wet methods for the preparation of UF_4 have been developed mostly by British workers. Their various modifications are all based on the above-mentioned early work of Bolton. Essentially, they consist in reducing solutions of uranyl fluoride, chloride, or sulfate to the quadrivalent state and precipitating uranium tetrafluoride by the addition of hydrofluoric acid. Various uranium compounds have been employed as starting materials, and a variety of reducing agents have been suggested.

(a) Stannous Chloride–Hydrofluoric Acid Processes. Grosse (SAM Columbia 1) prepared UF_4 in much the same fashion as did Bolton, by dissolving sodium uranate, Na_2UO_4, in excess hydrofluoric acid and reducing the solution with stannous chloride at the boiling temperature. A yield of 99 per cent was reported by him.

$$Na_2UO_4 + 4HF \rightarrow UO_2F_2 + 2NaF + 2H_2O \tag{10}$$

$$UO_2F_2 + SnCl_2 + 4HF \rightarrow \underline{UF_4} + SnCl_2F_2 + 2H_2O \tag{11}$$

Repetition of this work (British 4,5) showed, however, that the uranium tetrafluoride so obtained was invariably contaminated with sodium fluoride (4 to 5 per cent after drying at 100°C). The amount of sodium fluoride did not correspond to any definite compound of NaF and UF_4. The precipitate was probably a mixture of UF_4 with some double salt of UF_4 and NaF, the exact ratio of the two components being dependent on the relative solubilities of the double salt and of UF_4 in aqueous hydrofluoric acid. Since sodium fluoride is an undesirable contaminant, efforts were made to prepare sodium-free UF_4. It was found that this result can be achieved by using sodium uranate (or the more easily available diuranate, $Na_2U_2O_7$) in hydrochloric or sulfuric acid (rather than hydrofluoric acid) solution. British investigators (British 5,6,7) found, for example, that UF_4 containing less than 0.1 per cent sodium can be obtained by first dissolving $Na_2U_2O_7$ in excess hydrochloric acid.

$$Na_2U_2O_7 + 6HCl \rightarrow 2UO_2Cl_2 + 2NaCl + 3H_2O \tag{12}$$

$$UO_2Cl_2 + SnCl_2 + 4HCl \rightarrow UCl_4 + SnCl_4 + 2H_2O \tag{13}$$

$$UCl_4 + 4HF \rightarrow \underline{UF_4} + 4HCl \tag{14}$$

In order to obtain an easily filterable UF$_4$ precipitate, very efficient stirring and vigorous boiling are required. The hydrogen fluoride must be added slowly and in great excess (100 per cent).

The kinetics of the reduction of uranyl ions by stannous ions Sn(II) have been investigated (British 4); the results are consistent with a slow ionic reaction. The rate increases rapidly with temperature between 17 and 110°C. For rapid reduction, therefore, the solution must be as hot as possible. The presence of excess hydrochloric acid also tends to increase the rate of reduction. Fairly good yields (92 to 97 per cent UF$_4$) can be obtained by this procedure, but the corrosion problems encountered in handling hot solutions containing both hydrochloric and hydrofluoric acids are very troublesome. Ebonite affords a partial solution of the difficulty since it is resistant to solutions of 10 per cent hydrochloric acid and 5 per cent hydrofluoric acid (British 8), but ebonite vessels cannot be heated to a high enough temperature. This process was therefore modified (British 9,10) by using sulfuric acid instead of hydrochloric acid. This permits the employment of lead-lined vessels.

$$Na_2U_2O_7 + 3H_2SO_4 \rightarrow 2UO_2SO_4 + 3H_2O + Na_2SO_4 \tag{15}$$

$$2UO_2SO_4 + 2SnCl_2 + 4H_2SO_4 \rightarrow 2U(SO_4)_2 + 2SnCl_2 \cdot SO_4 + 4H_2O \tag{16}$$

However, a complication is introduced by the fact that sparingly soluble U(SO$_4$)$_2$·4H$_2$O may precipitate and contaminate the tetrafluoride (see Sec. 2.2c).

Sodium diuranate is not the most convenient starting material for the preparation of UF$_4$, and the tin chloride reduction method was modified so that U$_3$O$_8$, which is more readily available, could be used instead of the diuranate. A preliminary investigation of the dissolution of U$_3$O$_8$ by various acids showed that solutions of uranyl chloride or sulfate could be obtained by reaction of U$_3$O$_8$ with the appropriate acid in the presence of an oxidizing agent such as nitric acid (British 11,12).

$$U_3O_8 + 6HCl + 2HNO_3 \rightarrow 3UO_2Cl_2 + 4H_2O + 2NO_2 \tag{17}$$

By using somewhat less than stoichiometric quantities (85 per cent) of nitric acid, it is possible to obtain solutions free of excess nitric

acid. Since, as mentioned above, hydrochloric acid gives rise to corrosion difficulties, here, too, its use, despite some desirable features, was abandoned in favor of sulfuric acid (British 13).

U_3O_8 dissolves very readily in mixtures of sulfuric and nitric acids,
especially at moderately elevated temperatures. If the quantities of
acids used are exactly stoichiometric according to the equation

$$2U_3O_8 + 6H_2SO_4 + 2HNO_3 \rightarrow 6UO_2SO_4 + NO_2 + NO + 7H_2O \qquad (18)$$

a smooth and rapid conversion of U_3O_8 to UO_2SO_4 can be obtained
(British 14). Once a solution of UO_2SO_4 is available, the reduction can
be performed exactly as in the procedures described before (British
15), and UF_4 can be precipitated as described in detail in Sec. 2.2c.

(b) Preparation of UF_4 by Electrolytic Reduction of Uranyl Compounds.   Uranyl sulfate solutions can also be reduced to uranium(IV)
sulfate electrolytically, and UF_4 can then be precipitated by the addition of hydrofluoric acid. This procedure dispenses with the use of
stannous salts and yields a purer product since no contamination with
a chemical reducing agent can occur.

(c) Mechanism of Precipitation of UF_4 from Uranium(IV) Sulfate
Solutions. $U(SO_4)_2 \cdot 4H_2O$ is the solid phase in equilibrium with a ura-
nium(IV) sulfate solution above 70°C, and it is so slightly soluble that
a precipitate forms whenever a uranium(IV) sulfate solution containing
more than 5 g per 100 ml is heated to this temperature. It is found,
however, that $U(SO_4)_2 \cdot 4H_2O$ dissolves in aqueous hydrofluoric acid.
Milky suspensions of $U(SO_4)_2 \cdot 4H_2O$ can also be dissolved by the addition of hydrofluoric acid or even of suspensions of sodium fluoride.
That dissolution is caused by the formation of a sulfate-fluoride complex is confirmed by the observation that more than 2 moles of hydrofluoric acid must be added per mole of uranium(IV) sulfate in solution
before any precipitation of UF_4 occurs (British 16). When a solution
of $U(SO_4)_2$ is treated with less than 2 moles of hydrofluoric acid, it
merely changes color (British 17).

All these observations indicate that the precipitation of UF_4 is preceded by the formation of a soluble complex.

$$U(SO_4)_2 + 2HF \rightarrow \text{soluble complex} \qquad (19)$$

$$\text{Soluble complex} + 2HF \rightarrow \underline{UF_4} \qquad (20)$$

As indicated in Eq. 19, the complex probably contains $U(SO_4)_2$ and
HF in the ratio of 1 to 2. The brutto formula, therefore, must be
$U(SO_4)_2 \cdot 2HF$. By analogy with other quadrivalent uranium compounds,
in which uranium is known to have a coordination number of 6, the

structural formula probably is $H_2[U(SO_4)_2F_2]$. If the complex does have this constitution, it is an acid and it should be possible to prepare its salts. A compound $K_2[U(SO_4)_2F_2]\cdot2H_2O$ has, in fact, been prepared from $U(SO_4)_2$ and KF.

The stability of the complex $H_2[U(SO_4)_2F_2]$ has been studied in some detail. Its solutions give qualitative tests for uranium(IV), e.g., precipitation of $U[Fe(CN)_6]$. Therefore, the complex must be markedly dissociated into UF$_4$ and $U(SO_4)_2$. Including the ionic dissociations of the complex, of HF, and of $U(SO_4)_2$, the following equilibria must therefore exist in the solution containing 2 moles of HF per mole of $U(SO_4)_2$:

$$U(SO_4)_2 + 2HF \rightleftharpoons H_2[U(SO_4)_2F_2] \tag{21}$$

$$U^{+4} + 2SO_4^{--} + 2H^+ + 2F^- \rightleftharpoons 2H^+ + [U(SO_4)_2F_2]^{--} \tag{22}$$

When such a solution is allowed to stand (at any temperature between 20 and 100°C), a slow precipitation of UF$_4$ occurs. Careful study showed that this is not caused by partial oxidation of uranium(IV) to uranium(VI) by air but is probably due to slow decomposition of the complex.

$$2H_2[U(SO_4)_2F_2] \rightarrow UF_4 + U(SO_4)_2 + 2H_2SO_4 \tag{23}$$

At 20°C only UF$_4$ precipitates, probably as $2UF_4\cdot5H_2O$, after about 48 hr. The $U(SO_4)_2$ formed according to Eq. 23 remains in solution since the octahydrate $U(SO_4)_2\cdot8H_2O$, which is the stable solid phase at temperatures up to 70 to 80°C, is highly soluble. At 100°C the monohydrate, $UF_4\cdot H_2O$, is the form precipitated. When the solubility limit of $U(SO_4)_2\cdot4H_2O$ is reached, $UF_4\cdot H_2O$ and $U(SO_4)_2\cdot4H_2O$ precipitate as mixed crystals.

It was observed (CEW-TEC 2) that UF$_4$ suspensions that have been fumed with sulfuric acid for a short time dissolve when the acid is diluted with water. This can be interpreted as additional evidence for the formation of the complex $H_2[U(SO_4)_2F_2]$.

(d) <u>Precipitation and Dehydration of $2UF_4\cdot5H_2O$</u>. (SAM Columbia 2.) UF$_4$, precipitated at room temperature, is an amorphous green mass which is extremely difficult to filter and wash. To obtain a crystalline, easily filterable uranium tetrafluoride, the following procedure has been found useful: The amorphous precipitate is suspended in dilute hydrofluoric acid (0.1 to 3 per cent) for a period ranging from a few hours to a few days. This converts it into a voluminous, sometimes gelatinous mass of long, thin, silky, turquoise-colored needles.

When the conversion is complete, the crystals can easily be filtered and washed. The air-dried crystals have a constant composition and do not lose weight in vacuum at room temperature. Analysis indicates the formula to be $2UF_4 \cdot 5H_2O$. This hydrate is said to be very stable in air, with practically no oxidation to sexivalent uranium even after standing several weeks.

(e) Preparation of Anhydrous Uranium Fluoride from the Hydrates. Moist uranium tetrafluoride can be dried in air at 100°C to give a hydrated uranium tetrafluoride that approximates the monohydrate, $UF_4 \cdot H_2O$, in composition. In air at 400°C the water of hydration is removed, but there is considerable hydrolysis and oxidation (British 18). Drying in a stream of inert gas (nitrogen) at 400°C has been tried but abandoned because of the difficulty in eliminating oxygen, traces of which suffice to cause oxidation. Dehydration of the monohydrate at a temperature of 400°C and a pressure less than 50 mm Hg yields fairly satisfactory results. The period of heating should be as brief as possible, and the surface area of exposed monohydrate should be as small as possible to minimize surface oxidation. A product containing about 0.20 per cent water and an oxygen content equivalent to 3 per cent UO_2F_2 can be obtained in this way.

It has been found that satisfactory dehydration of $UF_4 \cdot H_2O$ can also be achieved by heating in a stream of hydrogen fluoride under atmospheric pressure. Even more satisfactory is dehydration at reduced pressure (50 mm Hg) in the presence of a small amount of hydrogen fluoride (5 mm Hg) at 400°C. Under these conditions practically no oxycompounds are formed; the presence of even small amounts of hydrogen fluoride retards their formation to a very marked degree (British 19).

The uranium tetrafluoride hydrate, $UF_4 \cdot 2.5H_2O$, is reported to lose most of its water readily without side reaction during slow heating at 200°C in vacuum (2 mm Hg pressure) for 24 hr. The last traces, however, are eliminated only at 500 to 550°C (SAM Columbia 2). Occasionally traces of hydrogen fluoride are found in the exit gases during dehydration. It is advisable to employ as good a vacuum as possible and to increase the temperature gradually. According to a more recent study, $UF_4 \cdot 2.5H_2O$ is dehydrated by heating in anhydrous hydrogen fluoride at 450°C (SAM Columbia 3). The anhydrous product retains the voluminous appearance of the hydrate and therefore has a low bulk density (1.5 g/cc). It is rehydrated much more readily than UF_4 prepared by high-temperature hydrofluorination. The large surface area (6 to 8 square meters per gram) of the dehydrated material makes it especially suitable for heterogeneous reactions (i.e., preparation of UF_6, UF_5, U_2F_9, etc).

In Sec. 2.6 it will be shown that it is desirable to have hydrogen fluoride present in the atmosphere in which UF$_4$ is being dried, because of the relative ease with which UF$_4$ can be hydrolyzed.

2.3 <u>Preparation of Uranium Tetrafluoride by Vapor-phase Reactions at Elevated Temperatures</u>. When prepared from aqueous solution, uranium tetrafluoride almost invariably contains small amounts of water and often large amounts of oxycompounds which arise either from oxidation or from hydrolysis during dehydration. Other contaminants, such as tin, zinc, or sulfur-containing compounds (e.g., sulfate), may also be present. The methods now to be described are better adapted to the preparation of an oxygen-free anhydrous product. They consist of the treatment of uranium compounds, usually oxide, with gaseous fluorinating agents, usually anhydrous hydrogen fluoride.

(a) <u>Preparation of UF$_4$ by Hydrofluorination of UO$_2$</u>. The preferred sequence of operations is hydrogen reduction of UO$_3$ followed by treatment of the resulting UO$_2$ with anhydrous hydrogen fluoride at atmospheric pressure.

$$UO_3 + H_2 \rightarrow UO_2 + H_2O \tag{24}$$

$$UO_2 + 4HF \rightarrow UF_4 + 2H_2O \tag{25}$$

The reduction of UO$_3$ is carried out at 600 to 700°C. An experimental determination of ΔH for reaction 25 in the range 600 to 800°C is -42.08 kcal (Los Alamos 1).

Hydrofluorination is a relatively slow reaction, and its rate does not show much temperature dependence above 350°C (MP Chicago 3). The use of a large excess of hydrogen fluoride is recommended.

According to Eqs. 24 and 25 either UO$_2$ or UO$_3$ is used as the initial material for the preparation of UF$_4$. The use of the easily available intermediate oxide U$_3$O$_8$ depends on the completion of the reaction

$$U_3O_8 + 2H_2 \rightarrow 3UO_2 + 2H_2O \tag{26}$$

British investigators (British 20) have claimed that U$_3$O$_8$ cannot be reduced to UO$_{2.0}$ by hydrogen under the conditions usually employed (temperatures of 650 to 680°C and atmospheric pressure). The reduction product obtained under these conditions had the composition UO$_{2.14}$ (or UO$_3 \cdot$6UO$_2$); when treated with hydrogen fluoride, it yielded a mixture of uranium tetrafluoride and oxyfluoride,

$$UO_3 \cdot 6UO_2 + 26HF \rightarrow UO_2F_2 + 6UF_4 + 13H_2O \tag{27}$$

UO$_2$F$_2$ is a very undesirable impurity for some purposes.

(b) Preparation of UF_4 by Reaction of UO_3 with Gaseous Ammonia and Hydrogen Fluoride. A one-step process used for the preparation of UF_4 consists in passing a mixture of ammonia and hydrogen fluoride over uranium trioxide at 500 to 750°C (CEW-TEC 3). The reaction is very rapid and produces UF_4 of high purity. From the point of view of the conversion yield, this ammonia−hydrogen fluoride procedure is distinctly superior to the one-step hydrogen−hydrogen fluoride reaction, which has been advocated by various workers.

(c) Preparation of UF_4 by Reaction of Uranium Oxides with Fluorinated Hydrocarbons (Freons) at High Temperatures. Many metal oxides can be converted to fluorides by reaction with fluorinated hydrocarbons (Ruff and Keim, 1931; Henne, 1938). Booth, Krasny-Ergen, and Heath (1946) studied the behavior of UO_3 in the presence of certain fluorinated hydrocarbons. Their results agree well with the later, more extended investigations of the Brown University group, who have developed satisfactory methods for the preparation of UF_4 by the reaction of fluorinated hydrocarbons with various uranium oxides (Brown 3,4). The purity of the product obtained has not as yet been assessed, but the method appears to have certain advantages. The apparatus required is very simple; glass can be used although graphite is perhaps preferable. The reaction appears to be applicable to all oxides of uranium, and the product, which differs in physical characteristics from that obtained by the hydrogen fluoride procedure, may be more desirable for certain purposes.

The results obtained with a number of fluorinated hydrocarbons and various uranium oxides are given in Table 12.2. These experiments were performed in pyrex and Vycor tubes. Calcium fluoride and polished graphite boats are suitable. Because of its greater availability, graphite ware was used in most cases. The gaseous fluorocarbons were carefully dried by passage over phosphorus pentoxide since traces of moisture caused etching of the glass vessels during the initial stages of the reaction. The uranium oxides were dried by heating at 200 to 300°C either in vacuum or in a stream of dry nitrogen. In all cases a large excess of the fluorocarbons was used. In the case of liquid fluorocarbons, their vapors were delivered into the fluorination apparatus by a stream of dry nitrogen.

The most successful preparation was that which used UO_3 with dichlorotetrafluoroethane, $C_2Cl_2F_4$ (Freon 114), at 700°C. This fluorinated hydrocarbon corrodes silica to a less extent at high temperatures than any of the other materials investigated. UF_4 of good quality may also be prepared by the reaction of difluorodichloromethane (Freon 12) with UO_2, but the product has a lower bulk density than that obtained from Freon 114 and it often adheres tenaciously to the boats. The product obtained from the Freon reactions is almost free of chlorine, and very little volatilization of uranium-chlorine compounds occurs.

Table 12.2—Reactions of Uranium Oxides with Freons*

Compound	Commercial name	Boiling point, °C	Oxide		
			UO$_2$	UO$_3$	U$_3$O$_8$
CCl$_3$F	Freon 11	23.4	Reaction at 350°C; product contained many soluble uranium chlorides	Same as for UO$_2$; graphite boats disintegrated rapidly	...
CCl$_2$F$_2$	Freon 12†	−29.8	Fluorination proceeds fairly well even at 350°C; low bulk density	Rapid reaction at 400°C; much sublimation from boats	Rapid reaction at 400°C; material adheres very firmly to boats
CHCl$_2$F	Freon 21‡	8.9	Undergoes extensive pyrolysis with much deposition of carbon	Same as for UO$_2$	Same as for UO$_2$
C$_2$Cl$_3$F$_3$	Freon 113	48	No reaction below 600°C; incomplete after 1 hr at 600 to 650°C	Same as for UO$_2$; material adheres to boat very tenaciously	...
C$_2$Cl$_2$F$_4$	Freon 114	3.5	Fairly satisfactory reaction at 675 to 700°C; product has low bulk density	Rapid at 650 to 700°C; high bulk density; good conversion	Reaction does not go to completion at 700°C
C$_7$F$_{16}$	Coolant C-716	83	No reaction at 600°C	Small amount of fluorination at 600°C	...

*Brown 4.
†Seems to give more soluble chloride than does Freon 114.
‡Ni tube used since HF was a possible by-product.

It has already been mentioned that polished graphite is the most suitable structural material for fluorination with fluorocarbons. Calcium fluoride is also suitable as a boat material although it seems to absorb UF$_4$ at high temperatures (Brown 5). Metals are attacked appreciably; nickel, copper, monel, platinum, and stainless steels (Nos. 304 and 316) are all corroded, and, in addition, they promote pyrolysis with deposition of carbon. Silica ware is attacked to some extent, least of all by Freon 114 and Freon 12. In order to facilitate the reaction, a rotating kiln type of reactor has been constructed in which the reaction mass is kept in motion, continuously exposing fresh surface. A graphite tube is rotated in a silica tube fitted with a graphite liner to prevent attack on the silica (Brown 6).

The action on UO_3 of trichlorofluoromethane, CCl_3F (Freon 11), in the liquid phase has also been examined; reaction for 7 hr at 120°C gave a 94 per cent yield of a water-soluble product, presumably UO_2F_2 (Brown 3).

(d) Preparation of UF_4 from Uranium Metal or Uranium Hydride by High-temperature Hydrofluorination. Uranium metal can be readily converted to uranium tetrafluoride by two general procedures. The first utilizes the two consecutive reactions

$$U + \tfrac{3}{2}H_2 \rightarrow UH_3 \tag{28}$$

$$UH_3 + 4HF \rightarrow UF_4 + \tfrac{7}{2}H_2 \tag{29}$$

Massive uranium metal is first converted to the hydride at 250°C. To achieve complete conversion of the hydride to tetrafluoride, thorough agitation of the solid is necessary. A temperature of 200°C seems to be sufficient for the second step (MP Ames 1). However, in a static system the reaction of hydrogen fluoride with uranium hydride at 270°C is very incomplete owing to caking. If the hydride is decomposed at 500°C and the resulting finely divided metal is treated with hydrogen fluoride at the same temperature, it is claimed that good conversion to uranium tetrafluoride is obtained (MP Clinton 1).

The second procedure attempts to combine both steps into one. The simultaneous reaction of hydrogen and hydrogen fluoride with uranium metal has been investigated by the Ames group (MP Ames 8). A mixture of approximately 50 mole % hydrogen and 50 mole % hydrogen fluoride will smoothly convert small quantities (50 g) of uranium metal to UF_4 at 250°C. Larger quantities of uranium cannot be treated in this simple way. The reaction between hydride and hydrogen fluoride takes place on the surface of the massive metal with the liberation of enough heat to raise the temperature to a point where the rate of hydride formation is negligible. (This rate falls off rapidly on approaching 430°C. See Chap. 8.) Consequently, in order to allow the reaction to proceed, cooling to 350°C is necessary. When larger quantities of metal are handled, mechanical agitation must be used to remove the uranium tetrafluoride scale as it forms (MP Ames 9). With these precautions the simultaneous reaction of hydrogen and hydrogen fluoride is now regarded as definitely superior to the stepwise reaction.

2.4 Miscellaneous Preparations of UF$_4$. The following reactions in which uranium tetrafluoride appears as a product may be used in special cases.

(a) Reaction of Uranium Trichloride with Hydrogen Fluoride. (CEW-TEC 1.) At 450°C uranium trichloride reacts with anhydrous hydrogen fluoride.

$$UCl_3 + 4HF \rightarrow UF_4 + 3HCl + \tfrac{1}{2}H_2 \qquad (30)$$

(See reaction of UH$_3$ with hydrogen fluoride, Chap. 8, Sec. 2.7a.) Note that tetrafluoride is formed rather than the trifluoride.

(b) Reaction of Uranium Phosphate, UO$_2$(HPO$_4$)·H$_2$O, with Hydrogen Fluoride. (CEW-TEC 4.) At temperatures of 350 to 500°C uranyl fluoride, UO$_2$F$_2$, is the only product of reaction of uranyl phosphate and hydrogen fluoride. Above 500°C a rather unexpected reduction occurs, and at 800°C UF$_4$ is the main product obtained. The reaction is quite rapid.

(c) Reaction of Uranium Oxides with Ammonium Fluoride. (CEW-TEC 3; Mallinckrodt 1,2.) UF$_4$ can be readily prepared by reaction of ammonium fluoride or bifluoride with UO$_3$.

$$3UO_3 + 6NH_4HF_2 \rightarrow 3UF_4 + 9H_2O + N_2 + 4NH_3 \qquad (31)$$

A temperature of 700°C is suitable. The product is free from ammonia. At lower temperatures (450°C) products of the type NH$_4$UF$_5$ are obtained. In carrying out this reaction it is necessary to pass the ammonium fluoride vapor over the oxide; simple fusion is insufficient. The essential similarity between this procedure and that described above (HF + NH$_3$) is evident.

(d) Reaction of Uranium Tetrachloride with Liquid Hydrogen Fluoride. (Brown 1.) The reaction of UCl$_4$ and liquid anhydrous hydrogen fluoride proceeds quantitatively at room temperature to give a deep-green crystalline product. The tetrafluoride obtained is found to retain considerable quantities of hydrogen fluoride which cannot be entirely removed even by prolonged pumping at room temperature. Analysis of one sample indicated a composition close to UF$_4$·HF. If this material is heated (in a nickel tube fitted with a platinum liner) for a short time in vacuum (10^{-3} mm Hg) at 625°C, a product entirely free of hydrogen fluoride can be obtained.

(e) UF$_4$ Formation from UF$_6$. For a discussion of UF$_4$ preparation from UF$_6$ see Chap. 13.

2.5 Physical Properties of Uranium Tetrafluoride. (a) Melting
Point. A melting point of 960 ± 5°C is derived from the plateau on
the heating curve of UF_4 determined by an optical pyrometer (MP
Chicago 4).

(b) Volatility. No reliable vapor pressure data are available. A
single value of 1.9×10^{-4} mm Hg at 760°C, obtained in the course of
other work (MP Chicago 5), appears to be consistent with the general
properties of UF_4. Uranium tetrafluoride is not appreciably volatile
in dry nitrogen at 800 to 880°C. The small amount of volatilization
actually observed can be attributed to the presence of UO_2F_2 (Brown
4). British workers have purified the tetrafluoride by vacuum sublimation (British 2). A high vacuum (10^{-6} mm Hg) and a temperature of
1000°C are required. A silica tube equipped with a molybdenum metal
liner was used.

(c) Crystallography, X-ray Structure, and Density. Uranium tetrafluoride obtained by vacuum sublimation at 1000°C consists of green
needlelike crystals up to several millimeters in length and up to
0.5 mm in diameter. Most of them appear twinned parallel to the
length of the needle. They often show between crossed Nicol prisms
an imperfect extinction that changes from one region to another and
is probably due to distortions. The crystals are triclinic (see Table
12.3). All the crystals are twinned so that monoclinic symmetry is
simulated. Goniometric study of a pseudomonoclinic crystal showing
(3 1 0) and (1 1 1) faces yielded the following axial ratios and angle:

$$a : b : c = 1.282 : 1 : 1.760 \ (\pm 0.001) \qquad \beta = 98°49' \pm 2'$$

X-ray photographs of small single crystals of uranium tetrafluoride show monoclinic symmetry with 12 molecules per unit cell. The
approximate lattice dimensions are

$$a_1 = 12.79 \pm 0.06 \ A \qquad a_2 = 10.72 \pm 0.05 \ A$$

$$a_3 = 8.39 \pm 0.05 \ A \qquad \alpha_2 = 126°10' \pm 40'$$

The space group is $C2/c (C_{2h}^6)$. UF_4 is thus isomorphous with ThF_4,
CeF_4, ZrF_4, and HfF_4 (MP Chicago 6). The x-ray structure of anhydrous UF_4 is the same, whether the material is prepared from
aqueous solution or obtained by high-temperature hydrofluorination.

The density calculated from the x-ray data is 6.70 ± 0.10 g/cc. The
density of uranium tetrafluoride determined by immersion in alcohol
has been reported as 6.95 g/cc (British 21) and 6.43 g/cc (SAM Columbia 4). Values of bulk density varying from 1.50 to 3.5 g/cc have

been observed. As has already been mentioned, the bulk density of UF$_4$ prepared from aqueous solution and dried below 500°C is, in general, considerably lower than that of uranium tetrafluoride prepared by high-temperature reactions.

Table 12.3 — Refractive Indices and Pleochroism of UF$_4$ Crystals*

Crystallographic direction	N	Pleochroic color
α	1.500	Very light green, nearly colorless
β	1.585	Emerald green
γ	1.598	Emerald green

*British 3.

No data appear to be available on the structure of the monohydrate, UF$_4$·H$_2$O. The hydrate 2UF$_4$·5H$_2$O is orthorhombic with eight molecules per unit cell (MP Chicago 6,7).

$$a_1 = 12.75 \pm 0.04 \text{ A} \qquad a_2 = 11.12 \pm 0.04 \text{ A} \qquad a_3 = 7.05 \pm 0.03 \text{ A}$$

The computed density is 4.74 g/cc. The atomic positions are

$$4 \text{ U}_\text{I} \text{ in } (x_1 \ y_1 \ \tfrac{1}{4}) (x_1 + \tfrac{1}{2} \ \tfrac{1}{2} - y_1 \ \tfrac{1}{4}) \qquad x_1 = 0.055, \ y_1 = 0.14$$

$$4 \text{ U}_\text{II} \text{ in } \pm(x_2 \ y_2 \ \tfrac{1}{4}) (x_2 + \tfrac{1}{2} \ \tfrac{1}{2} - y_2 \ \tfrac{1}{4}) \qquad x_2 = -0.025, \ y_2 = -0.139$$

The space group is Pnam.

(d) Thermodynamic Data. The specific heat, enthalpy, and entropy of UF$_4$ from 20 to 350°K have been determined (Natl. Bur. Standards 1; Brickwedde, Hoge, and Scott, 1948). The sample of UF$_4$ used contained 2 per cent UO$_2$F$_2$; if the heat capacities of UF$_4$ and UO$_2$F$_2$ do not differ by more than 10 per cent, an error of only 0.2 per cent could be caused by this impurity. The calorimeter was of the adiabatic type (Southard and Brickwedde, 1933; Bekkedahl and Scott, 1942). The C$_p$ values given in Table 12.4 were obtained from a smooth curve laid through the experimental points (see Fig. 12.1).

The following thermochemical data have been obtained at Columbia (SAM Columbia 5) for the formation of UF$_4$:

$$\text{U (solid)} + 2\text{F}_2 \text{ (gas)} \rightarrow \text{UF}_4 \text{ (solid)} \tag{32a}$$

$$-\Delta H_{298°K} = 446 \pm 2 \text{ kcal per mole}$$

Table 12.4—Specific Heat, Entropy, and Enthalpy of UF_4*

Temp., °K	C_p, joules/mole/°C	$S - S_0$,† joules/mole/°C	$H - H_0$,‡ joules/mole
0	0	0	0
5	0.08	0.03	0.1
10	0.68	0.22	1.7
15	2.24	0.76	8.5
20	5.13	1.7	26.1
22.5	6.34		
25.0	7.93	3.2	58.1
27.5	9.84		
30	11.80	4.9	107.3
32.5	13.98		
35	16.15	7.1	177.2
37.5	18.30		
40	20.49	9.5	268.8
45	24.84		
50	29.22	15.0	517.2
55	33.51		
60	37.72	21.1	852
65	41.84		
70	45.81	27.6	1,270
75	49.55		
80	53.14	34.2	176
85	56.59		
90	59.88	40.8	2,331
95	63.00		
100	65.96	47.4	2,961
105	68.84		
110	71.60	54.0	3,649
115	74.28		
120	76.76	60.5	4,391
125	79.19		
130	81.50	66.8	5,183
135	83.66		
140	85.66	73.0	6,020
145	87.57		
150	89.39	79.0	6,895
155	91.14		
160	92.83	84.9	7,806
165	94.42		
170	95.92	90.6	8,750
175	97.33		
180	98.66	96.2	9,724
185	99.93		
190	101.12	101.6	10,273
195	102.24		
200	103.32	106.8	11,745
205	104.36		
210	105.35	111.9	12,789
215	106.32		
220	107.24	116.9	13,852
225	108.10		
230	108.92	121.7	14,933
235	109.71		

Table 12.4—(Continued)

Temp., °K	C_p, joules/mole/°C	$S - S_0$,[†] joules/mole/°C	$H - H_0$,[‡] joules/mole
240	110.47	126.3	16,030
245	111.20		
250	111.89	130.9	17,142
255	112.56		
260	113.20	135.3	18,267
265	113.81		
270	114.41	139.6	19,405
273.16	114.80	140.9	19,767
275	115.03		
280	115.63	143.8	20,556
285	116.20		
290	116.75	147.8	21,717
295	117.28		
298.16	117.62	151.1	22,674
300	117.81	151.8	22,890
305	118.33		
310	118.84	155.7	24,074
315	119.35		
320	119.85	159.5	25,267
325	120.33		
330	120.81	163	26,470
335	121.28		
340	121.74	166.8	26,683
345	122.20		
350	122.64	170.4	28,905

*The heat capacity data were extrapolated to 0°K by the equation

$$C_p = 3R(D + 4.1 \times 10^{-6}T^3) \qquad \text{joules/mole/°C}$$

where D is the Debye function, with $\theta = 150°K$.

[†] $S - S_0 = \int_0^T (C_p/T)\, dT.$

[‡] $H - H_0 = \int_0^T C_p\, dT.$

Using the UF$_4$ entropy value of 36.13 e.u. (151 joules per degree Kelvin) at 298°K from Table 12.4 together with the entropy value for U and F$_2$, the following is calculated:

$$-\Delta F_{298°K} = 424.6 \text{ kcal per mole} \qquad (32b)$$

The above value of ΔH is obtained by measuring the heat of solution of $U(SO_4)_2$, the heat of formation of UF$_4 \cdot H_2O$ in aqueous solution, and the heat of hydration of UF$_4$, and combining them with the heats of formation of $U(SO_4)_2$, SO_4^{--}, and F^- found in the literature (Table 12.5).

Because of the thermochemical importance of the values, Eqs. 32a and 32b have been redetermined (SAM Carbide and Carbon 1) by a

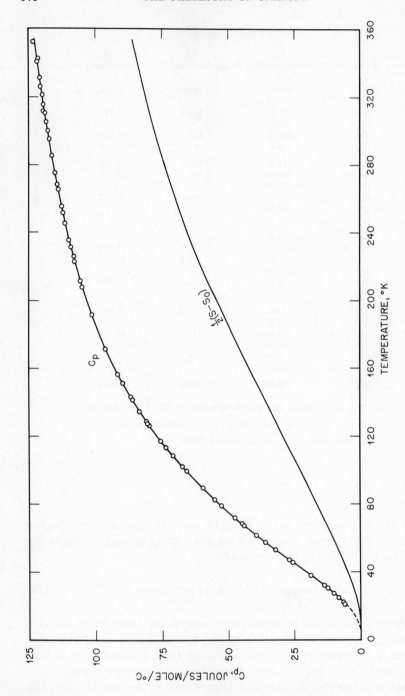

Fig. 12.1 — Molar heat capacity and entropy of UF₄ (Brickwedde, Hoge, and Scott, 1948).

different method. Uranium tetrachloride (solid) was converted to hydrated UF$_4$ (solid) with excess potassium fluoride solution, and the heat of hydration of UF$_4$ (surface area = 6 square meters per gram) was measured in a potassium fluoride solution of the same concentration. A value of -444 ± 2.5 kcal per mole was obtained for the heat

Table 12.5 — Thermodynamic Functions

Equation No.	Reaction	$\Delta H_{298°K}$ for UF$_4$, kcal/mole
(33)	U(SO$_4$)$_2$ (solid) → U(IV) (aq) + 2SO$_4^{--}$ (aq)	-24.1 ± 0.2
(34)	U + 2S + 4O$_2$ → U(SO$_4$)$_2$	-534
(35)	U(IV) (aq) + 4F$^-$ (aq) + H$_2$O → UF$_4$·H$_2$O (solid)	-10.3 ± 0.2
(36)	UF$_4$·H$_2$O (solid) → UF$_4$ (solid) + H$_2$O	$+3.3 \pm 0.3$
(37)	2SO$_4^{--}$ → 2S (solid) + 4O$_2$	$+431.6$
(38)	2F$_2$ → 4F$^-$ (aq)	-312.8

Table 12.6 — Magnetic Susceptibility of UF$_4$ and Some Double Salts

Compound	Curie-Weiss constants*		μ, Bohr magnetons
	C	θ,/°C	
UF$_4$	1.36	147	3.30
KUF$_5$	1.30	122	3.30
K$_2$UF$_6$	1.47	108	3.45
CaUF$_6$	1.31	101	3.25
Na$_3$UF$_7$	1.45	290	3.40

*$\chi = C/(T + \theta)$.

of formation of UF$_4$. We may consider the average of the two determinations, $\Delta H = -443.5$ kcal per mole, as the most probable value. It agrees quite well with that obtained from a study of the high-temperature hydrolysis of UF$_4$ (Los Alamos 1).

(e) Thermal Conductivity of Solid UF$_4$. The thermal conductivity of fused uranium tetrafluoride has been given as 0.0047 ± 4.3 per cent cal/cm sec °C at 60°C (MP Chicago 8).

(f) Magnetic Susceptibility of UF$_4$ and Some Double Uranium Fluorides. Elliott (1949) has determined the magnetic susceptibility of UF$_4$, KUF$_5$, K$_2$UF$_6$, and Na$_3$UF$_7$ in the range 74 to 300°K by the Gouy method. The experimental results are given in Table 12.6.

The calculated moment for a free uranium ion with two 5f electrons (^{3}H$_4$ state) is 3.58 Bohr magnetons. The experimental moments are reasonably close to this value. Although all these compounds except K_2UF_6 obey the Curie-Weiss law over the whole temperature range, the exceptionally large values found for the Weiss temperature make interpretation of the Curie constant difficult.

2.6 <u>Chemical Properties of Uranium Tetrafluoride</u>.   Uranium tetrafluoride is a green crystalline solid. Material prepared by high-temperature reactions usually is darker, denser, and much less hygroscopic than uranium tetrafluoride prepared by Grosse's method (SAM Columbia 2). Chemically, uranium tetrafluoride may in general be considered a stable, rather inert substance. In its physical properties it resembles other quadrivalent fluorides, particularly the isomorphous compounds ZrF_4, HfF_4, and ThF_4. The chemical differences arise principally from the fact that uranium can exist in a number of different valence states whereas the other elements are exclusively quadrivalent in their fluorides.

(a) <u>Water</u>.   The solubility of UF_4 in water at 25°C is approximately 0.10 g, or 0.32 millimole per liter (SAM Columbia 2). Its molar solubility is thus of the same order of magnitude as that of calcium fluoride. British workers reported a solubility of 600 mg per liter (British 22). The most reliable values for the solubility of UF_4 in water probably are those of the SAM Carbide and Carbon Laboratories (SAM Carbide and Carbon 2). They give the solubility at 25°C as 0.004 per cent by weight of UF_4, corresponding to 10^{-4} mole per liter; the solubility at 0°C is 3×10^{-5} mole per liter; and the solubility at 60°C is 4×10^{-4} mole per liter. The solubility increases when hydrofluoric acid is added to water in the amount of about 0.03 g of UF_4 per 100 g of solution at a hydrofluoric acid concentration of 30 wt. %. Whether uranium tetrafluoride reacts with water (apart from hydration) is still doubtful. British workers (British 23) state that uranium tetrafluoride does not lose fluorine when boiled with water for extended periods of time (72 hr) and that no hydrolysis occurs. Other workers (Du Pont 1) state that as much as 85 per cent will decompose and go into solution at room temperature in 24 hr and that considerable reaction occurs at the boiling point in 1 hr. Analysis showed that the uranium in such solutions was no longer in the quadrivalent state. One reason for the discrepancy may be that whereas British workers used 5.0 g per 100 ml the other group used only 0.3 g per 100 ml of tetrafluoride, and the amount of dissolved oxygen in water may have been sufficient to oxidize an appreciable fraction of the smaller sample. In any event it is known that very small amounts of excess hydrofluoric acid repress hydrolysis, as evidenced by the successful precipitation of the tetrafluoride from boiling aqueous solution. No basic uranium fluorides are known.

(b) <u>Water Vapor</u>. Anhydrous UF$_4$ is not appreciably hygroscopic in air at room temperature. In one case an increase in weight of 0.05 per cent in 564 hr was observed (British 24). The previous history of the uranium tetrafluoride is of importance in this connection since material prepared from aqueous solutions, particularly by dehydration of UF$_4 \cdot 2.5H_2O$, is much more hygroscopic than UF$_4$ prepared by high-temperature hydrofluorination.

Table 12.7 — Equilibrium Constants for the
Hydrofluorination of UO$_2$

Temp., °C	1000/T(Kelvin)	K*	ln K†
600	1.145	14.0	2.639
705	1.021	1.47	0.385
800	0.932	0.155	−1.865

*K = $p_{H_2O}^2/p_{HF}^4$.
†ln K = $(42,080/RT) - 21.5$.

(c) <u>Steam</u>. A direct determination of the equilibrium constant of the reaction

$$UO_2 + 4HF \rightleftarrows UF_4 + 2H_2O \tag{39}$$

has been made in the range 600 to 800°C (Los Alamos 1). The results are summarized in Table 12.7. The heat of reaction, which is sensibly constant in this temperature range, is calculated to be $\Delta H = -42.08$ kcal. Domange and Wohlhuter (1949) have also studied this equilibrium in the range 200 to 500°C by passing steam over UF$_4$ at different rates, measuring the gas composition at each rate, and extrapolating to zero rate. From the equilibrium constants so obtained, an average value $\Delta H = -30$ kcal in this temperature range is calculated. This is apparently a fairly good value, since the heat of formation of UF$_4$ derived from these data agrees moderately well with the generally accepted value.

It is obvious that, if uranium tetrafluoride is treated with steam under such conditions that the hydrogen fluoride produced is swept out of the system, the tetrafluoride will be converted to UO$_2$. The equilibrium constants show that hydrolysis of the UF$_4$ will be markedly enhanced with increasing temperatures.

The heats of hydrolysis of a number of bivalent and trivalent fluorides have been determined by Domange (1937). The thermodynamic data available are not sufficiently extensive to make possible a detailed comparison of the free energies of hydrolysis of all the metal fluorides. However, in a general way, all metal fluorides fall into two categories with respect to ease of hydrolysis: (1) Difficultly hydrolyzable fluorides of univalent and bivalent cations, such as NaF, KF, CuF_2, and BeF_2; (2) readily hydrolyzable fluorides of trivalent and quadrivalent cations, such as LaF_3, AlF_3, ThF_4, and UF_4. Extrapolation of the free energy of the hydrolytic reaction of UF_4 to 500°C leads to a value of $\Delta F_{500°C} = +2.8$ kcal. This small value indicates the necessity of maintaining a hydrogen fluoride atmosphere when hydrated UF_4 is being dried at 500°C and suggests that the dehydration should preferably be carried out at lower temperatures.

The dehydration of hydrated UF_4 is further complicated by the extreme sensitivity of the system to traces of oxygen. Should the dehydration process be long-continued and should even traces of oxygen be allowed access to the UF_4, oxidation to uranyl fluoride will occur. Once oxygenation has occurred, no amount of further treatment with hydrogen fluoride will remove the oxygen (British 19; Du Pont 1).

$$UF_4 + HOH \rightleftharpoons UF_3(OH) + HF \tag{40}$$

$$UF_3(OH) + HOH \rightleftharpoons UF_2(OH)_2 + HF \tag{41}$$

$$2UF_2(OH)_2 + O_2 \rightarrow 2UO_2F_2 + 2H_2O \tag{42}$$

The hydrolytic conversion of UF_4 to uranium oxides has been adapted as an analytical method for the determination of fluorine in UF_4 (MP Ames 10). Complete hydrolysis is achieved by the use of superheated steam at 900°C in a platinum apparatus. The steam is first superheated in a quartz heater; the hydrofluoric acid is collected in a platinum condenser and determined by titration.

In an interesting extension of this work it has been found that fluorides (and chlorides) which are normally difficult to hydrolyze can be readily decomposed by steam at 1000°C if mixed with U_3O_8. The equilibrium is shifted to such an extent that substances such as potassium fluoride or calcium fluoride undergo hydrolysis very rapidly under these conditions (MP Ames 11, 12).

(d) Oxidizing Agents. Early workers reported UF_4 to be converted to U_3O_8 by ignition in air and noted that, if free access of air was prevented (heating in a covered crucible), UO_2F_2 would be obtained (Bolton, 1866a,b; Smithells, 1883). More recently UF_4 has been reported to be comparatively stable in air on heating for short periods at 200°C. At higher temperatures decomposition is considerable (British 18; Du Pont 1). The behavior of UF_4 in pure dry oxygen has been investigated more recently, with very interesting results (MP Chicago 9; Brown 4). At 800°C the following reaction has been shown to occur:

$$2UF_4 + O_2 \rightarrow UF_6 + UO_2F_2 \tag{43}$$

By working in metal apparatus the UF$_6$ can be collected. (A detailed discussion of this reaction will be found in Chap. 13, Sec. 1.2.) These observations show that ignition of UF$_4$ in dry air is inadvisable since appreciable amounts of uranium may be lost as the volatile hexafluoride.

Chlorine (Brown 4; MP Ames 13) has very little effect on UF$_4$ at 500 to 675°C; the slight reaction observed can be attributed to oxygen present in the chlorine. A solution of chlorine in carbon tetrachloride has no effect on UF$_4$ at 125 to 130°C (9 hr under pressure). The addition of UCl$_5$ has no effect on this reaction (Brown 7). Fluorine reacts with UF$_4$ above 250°C to give the hexafluoride (MP Chicago 3) (see Chap. 13).

$$UF_4 + F_2 \rightarrow UF_6 \tag{44}$$

Cobaltic fluoride, CoF$_3$, converts UF$_4$ to UF$_6$; antimony pentafluoride, SbF$_5$ (Du Pont 2), on the other hand, forms an equimolar complex with uranium tetrafluoride which is thermally stable up to 200°C in vacuum. When the complex is heated to 650°C, no noticeable quantities of volatile uranium compounds are produced; instead, 15 to 20 per cent water-soluble uranium compounds are found in the residue. The complex decomposes in water to give insoluble uranium tetrafluoride and soluble antimony salts.

Oxidizing acids dissolve UF$_4$ rapidly to give uranyl ion solution. Fuming perchloric acid is one of the best reagents for dissolving the tetrafluoride. Nitric acid attacks UF$_4$ only very slowly; the addition of boric acid (Du Pont 1) enhances the solvent action of nitric acid (as well as of other acids) by removing fluoride ions as they form.

$$2UF_4 + 2H_3BO_3 + 6HNO_3 \rightarrow 2UO_2(NO_3)_2 + 2HBF_4 + N_2O_3 + 5H_2O \tag{45}$$

Other strong oxidizing agents react with UF$_4$ to give solutions of soluble uranyl salts. Thus ceric sulfate solution dissolves UF$_4$; alkali peroxides (e.g., Na$_2$O$_2$) and ammonia—hydrogen peroxide mixtures react vigorously with UF$_4$, yielding soluble peroxyuranates. The reaction with ammonia—hydrogen peroxide is very vigorous and must be carefully regulated. The peroxide is best added to a suspension of

UF_4 in ammonium hydroxide (7N) at such a rate that the solution is maintained just boiling (CEW-TEC 2).

Ferric salts $FeCl_3$ and FeF_3 react with UF_4 with the formation of uranyl solutions; the complexing of fluoride by iron accelerates the reaction. Low pH seems to favor the oxidation and dissolution of UF_4 by FeF_3 (British 22). Uranium tetrafluoride also dissolves readily in boiling aluminum nitrate or aluminum chloride solution, presumably by virtue of the formation of stable aluminum fluoride complex ions, such as AlF_6^{-3}. The uranium is simultaneously oxidized to the sexivalent state (CEW-TEC 5).

The conversion of UF_4 to U_3O_8 by fusion with salts (MP Ames 14) has been studied. The following salts yield easily soluble products, which, however, still contain fluorine: KIO_3, $KBrO_3$, $KClO_3$, and mixtures of $KClO_3$ and $KHSO_4$. The most satisfactory reagent for this conversion is KNO_3. The following salts yield easily soluble products from which the fluorine is volatilized away: $(NH_4)NO_3$, $(NH_4)_2C_2O_4 \cdot H_2O$, KNO_3, $(NH_4)_2S_2O_8$, and NH_4Br. Ammonium oxalate is stated to be the most efficient reagent in this group. Boric acid fusion also finds frequent use in converting UF_4 to U_3O_8. Uranium tetrafluoride undergoes reaction with uranium hexafluoride, UF_6 (see Sec. 3 for a discussion of the preparation of U_2F_9, UF_5, and U_4F_{17}).

(e) Reactions of UF_4 in Which Uranium Remains Quadrivalent. UF_4 is scarcely attacked by cold concentrated hydrochloric, sulfuric, phosphoric, or perchloric acid. In hydrochloric acid solutions the solubility becomes very large with increasing HCl concentration. Thus 0.18 per cent UF_4 may be dissolved in 1N HCl, 1.1 per cent in 6N HCl, and 3.6 per cent in 12N HCl. The addition of boric acid, as has been mentioned above, greatly accelerates the attack of these acids, yielding corresponding solutions of uranium(IV). Fuming with sulfuric acid leads to solution on subsequent dilution with water (CEW-TEC 2).

A mixture of dilute sulfuric acid and silica dissolves UF_4 quite readily with the formation of $U(SO_4)_2$ solution and fluosilicic acid. Hot phosphoric acid dissolves UF_4 to give uranium(IV) phosphate. UF_4 is also moderately soluble in ammonium oxalate (0.4 per cent) solution. On heating UF_4 with sodium hydroxide solution, metathesis to a hydrated oxide, $UO_2 \cdot xH_2O$, occurs (Bolton, 1866a,b).

Uranium tetrafluoride reacts with anhydrous aluminum chloride at moderately elevated temperatures to form uranium tetrachloride (MP Chicago 10).

$$3UF_4 + 4AlCl_3 \xrightarrow{250-500°C} 3UCl_4 + 4AlF_3 \qquad (46)$$

The double salt NaAlCl$_4$ can also be used for this purpose (CEW-TEC 6). This is a satisfactory laboratory method for preparing uranium tetrachloride. The metal chlorides KCl, SiCl$_4$, MnCl$_2$, and ZnCl$_2$, as may be anticipated from thermochemical considerations, fail to react.

(f) <u>Reducing Agents</u>. Pure hydrogen reduces UF$_4$ at 1000°C to UF$_3$ (see Sec. 1). Atomic hydrogen (MP Chicago 11) at 700°C gives a mixture of UF$_3$ and UO$_2$ owing to the presence of traces of moisture from the attack of atomic hydrogen on the quartz reaction vessel.

Metallic uranium reduces UF$_4$ to UF$_3$ (see Sec. 1.1b). Alkali metals reduce UF$_4$ to uranium metal as do the alkaline earth metals, particularly calcium and magnesium (see Chap. 4).

Table 12.8 — Phase Compositions in the NaF-UF$_4$ System*

UF$_4$,[†] mole%	Phases present	
	Major	Minor
67	NaUF$_5$ + UF$_4$	
50	NaUF$_5$	
40	α Na$_2$UF$_6$	
36	α Na$_2$UF$_6$	
33 (a)	γ Na$_2$UF$_6$	
(b)	γ Na$_2$UF$_6$	β_2 Na$_2$UF$_6$ + α Na$_2$UF$_6$
(c)	β_2 Na$_2$UF$_6$ + α Na$_2$UF$_6$ + γ Na$_2$UF$_6$	
31	γ Na$_2$UF$_6$	
29	Na$_3$UF$_7$	
27	Na$_3$UF$_7$	
25	Na$_3$UF$_7$	
20	Na$_3$UF$_7$	NaF

*Zachariasen (1948a).
[†]Symbols (a), (b), and (c) are used to indicate different rates of cooling, (a) indicating the lowest rate.

(g) <u>Complex Compounds of UF$_4$</u>. Uranium tetrafluoride forms a series of double salts with metal fluorides. Compounds of this type were first encountered at an early stage in the development of uranium chemistry. Thus in 1866 Bolton observed the formation of a green water-insoluble compound when a uranyl fluoride solution containing potassium fluoride and formic acid was exposed to strong sunlight (Bolton, 1866a,b). Photochemical reduction of solutions of uranyl ion containing sodium or potassium fluoride can also be effected (in the presence of sunlight) with alcohol, ether, or glucose (Aloy and Rodier, 1922). In each case a green substance, rather closely resembling UF$_4$ in appearance, is produced. The compound melts in air with evolution of hydrogen fluoride; after long-continued heating, a residue of potassium or sodium uranate remains. The complex salts react only slowly with dry hydrogen, are practically insoluble in water or dilute acids, dissolve slowly in concentrated

Table 12.9 — Phase Compositions in the $KF-UF_4$ System*

UF_4,[†] mole %	Phases present		
	Major	Minor	Trace
89	$KU_6F_{25} + UF_4$		
86	KU_6F_{25}	$KU_2F_9 + UF_4$	
83	KU_6F_{25}	KU_2F_9	UF_4
80	$KU_2F_9 + KU_6F_{25}$		$UF_4 + KU_3F_{13}$
75	$KU_2F_9 + KU_6F_{25}$		$KU_3F_{13} + UF_4$
67	KU_2F_9		$KU_6F_{25} + KU_3F_{13} + KUF_5$
60	$KU_2F_9 + UF_5$		
50	KUF_5		
45	KUF_5	$\beta_1 K_2UF_6$	
40	$KUF_5 + \beta_1 K_2UF_6$		
36	$\beta_1 K_2UF_6$	αK_2UF_6	
33 (a)	$\beta_1 K_2UF_6$		
(b)	$\beta_1 K_2UF_6$	αK_2UF_6	
(c)	$\beta_1' K_2UF_6$[‡]		
(d)	$\beta_2 K_2UF_6$		
29	$\beta_1 K_2UF_6 + \alpha K_3UF_7$		
25 (a)	$\alpha' K_3UF_7$		
(b)	αK_3UF_7		
22	αK_3UF_7		
20	αK_3UF_7		KF
17	αK_3UF_7	KF	
14	αK_3UF_7	KF	

* Zachariasen (1948a).

† Symbols (a), (b), (c), and (d) are used to indicate different rates of cooling, (a) indicating the lowest rate.

‡$\beta_1' A_2XF_6$ is a disordered form of $\beta_1 A_2XF_6$, involving isomorphous replacement between A atoms and X atoms.

hydrochloric acid, and dissolve in hot concentrated sulfuric acid with evolution of hydrogen fluoride. Early workers formulated these compounds as $NaUF_5$ and KUF_5.

The $NaF-UF_4$ system has more recently received closer study (Brown 8). Methods of thermal analysis were employed, and it was reported that only one compound, $NaUF_5$ (m.p. 714 ± 10°C), existed. Two eutectic mixtures were found; one was $NaF-NaUF_5$ (m.p. 600 ± 10°C) at 26 mole % UF_4, the other $NaUF_5-HF_4$ (m.p. 650 ± 10°C) at 67 mole % UF_4.

The NaF-UF$_4$ and KF-UF$_4$ systems have been reexamined by x-ray crystallographic techniques, and it is certain from this study that these and similar systems are much more complex than had previously been assumed (MP Chicago 12,13; Zachariasen 1948a). The

Table 12.10 — Crystal Structure Data of the KF-UF$_4$ and NaF-UF$_4$ Systems*

Compound	Symmetry	Unit-cell dimensions, A	n†	ρ, g/cc
KU$_6$F$_{25}$	Hexagonal; C6/mmc	$a_1 = 8.18 \pm 0.01$ $a_3 = 16.42 \pm 0.02$	2	6.73
KU$_3$F$_{13}$	Orthorhombic; Pmcm	$a_1 = 8.03 \pm 0.03$ $a_2 = 7.25 \pm 0.03$ $a_3 = 8.53 \pm 0.04$	2	6.64
KU$_2$F$_9$	Orthorhombic; Pnam	$a_1 = 8.68 \pm 0.01$ $a_2 = 7.02 \pm 0.01$ $a_3 = 11.44 \pm 0.04$	4	6.49
NaUF$_5$	Rhombohedral; R$\bar{3}$	$a = 9.08 \pm 0.01$ $\alpha = 107° 56'$	6	5.81
KUF$_5$	Rhombohedral; R$\bar{3}$	$a = 9.387 \pm 0.002$ $\alpha = 107° 15' \pm 2'$	6	5.38
α Na$_2$UF$_6$	Cubic; fluorite	$a = 5.565 \pm 0.004$	⅓	5.08
α K$_2$UF$_6$	Cubic; fluorite	$a = 5.934 \pm 0.001$	⅓	4.53
β_1 K$_2$UF$_6$	Hexagonal; C6$\bar{2}$m	$a_1 = 6.54 \pm 0.01$ $a_3 = 3.76 \pm 0.01$	1	5.10
β_2 K$_2$UF$_6$	Hexagonal; C32	$a_1 = 6.53 \pm 0.02$ $a_3 = 4.04 \pm 0.01$	1	4.77
β_2 Na$_2$UF$_6$	Hexagonal; C32	$a_1 = 5.94 \pm 0.01$ $a_3 = 3.74 \pm 0.01$	1	5.74
γ Na$_2$UF$_6$	Orthorhombic; Immm	$a_1 = 5.56 \pm 0.02$ $a_2 = 4.01 \pm 0.01$ $a_3 = 11.64 \pm 0.04$	2	5.06
α K$_3$UF$_7$	Cubic	$a = 9.21 \pm 0.01$	4	4.12
α' K$_3$UF$_7$	Tetragonal; I4/amd	$a_1 = 9.20 \pm 0.02$ $a_3 = 18.40 \pm 0.06$	8	4.13
Na$_3$UF$_7$	Tetragonal; I4/mmm	$a_1 = 5.448 \pm 0.07$ $a_3 = 10.896 \pm 0.014$	2	4.49

*Zachariasen (1948a).
†n equals the number of molecules per unit cell.

phases which have been identified in these systems are indicated in Tables 12.8 and 12.9. They were prepared by melting the components together rapidly in a platinum vessel and allowing them to cool. In some cases the melts were quenched. The phases were identified by x-ray methods, and it proved possible to deduce the chemical formulas from the x-ray data. The crystal structure data are summarized in Table 12.10.

It may be useful to indicate briefly some salient features of the crystal structures of some of these compounds (Zachariasen, 1948c).

$NaUF_5$. This phase is rhombohedral. In the unit cell one set of six sodium atoms, one set of six uranium atoms, and five sets of fluorine atoms are in positions $\pm(x\ y\ z)\ (y\ z\ x)\ (z\ x\ y)$ with parameters

	x	y	z
Sodium	$6/13$	$2/13$	$5/13$
Uranium	$3/13$	$1/13$	$9/13$

The arrangement of the sodium and uranium atoms is that of a face-centered pseudocubic lattice in which every thirteenth site is vacant. The pseudo cell is a rhombohedron with a volume $4/13$ that of the true cell and dimensions

$$a = 5.7\ A \qquad \alpha = 90°51'$$

The structure of $NaUF_5$ is closely related to fluorite.

$\alpha\ Na_2UF_6$. The cubic alpha phases are formed by rapid cooling of melts containing slightly more than 33 mole % UF_4. The homogeneity range of the alpha phase extends to 40 mole % UF_4. The chemical composition of this phase can be expressed as $M_{4-x}U_xF_{4+3x}$, where x can have values between 1.33 (corresponding to Na_2UF_6) and 1.60. The phase is cubic face-centered. In the case of the sodium compound Na_2UF_6 the lattice constant varies with the value of x as follows:

x	a, A	ρ, g/cc
1.33	5.565 ± 0.005	5.08
1.45	5.578 ± 0.005	5.35
1.60	5.601 ± 0.001	5.66

This is a disordered structure of the fluorite type. In each unit cell there are 4 Na + U atoms distributed at random over the sites (0 0 0) $(\frac{1}{2}\frac{1}{2}0)\ (\frac{1}{2}0\frac{1}{2})\ (0\frac{1}{2}\frac{1}{2})$. There are also 8.0 to 8.8 fluorine atoms per unit cell. (The unit cell for x = 1.60 must contain more than 8 fluorine atoms rather than less than 4 metal atoms; this conclusion is a direct consequence of the expansion in lattice period as the fluorine content increases.) Of these, 8 fluorine atoms are at $\pm(\frac{1}{4}\frac{1}{4}\frac{1}{4})$ $(\frac{1}{4}\frac{3}{4}\frac{3}{4})\ (\frac{3}{4}\frac{1}{4}\frac{3}{4})\ (\frac{3}{4}\frac{3}{4}\frac{1}{4})$, i.e., they occupy the anion sites of the fluorite structure. The excess fluorine atoms are distributed at random over the sites $(0\ 0\ \frac{1}{2})$ $(0\ \frac{1}{2}\ 0)$ $(\frac{1}{2}\ 0\ 0)$ $(\frac{1}{2}\ \frac{1}{2}\ \frac{1}{2})$. This structure becomes unstable relative to NaUF$_5$ if more than 20 per cent of these sites are occupied.

The isomorphous compound α K$_2$UF$_6$ is prepared by precipitation from aqueous solution containing a large excess of potassium.

$\underline{\beta\ K_2UF_6}$. The β phases of K$_2UF_6$ are formed as pure phases from melts of 33 mole % UF$_4$. By slow cooling, the β_1 phases are obtained. The disordered β_1' phase of K$_2$UF$_6$ is formed by rapid cooling, the β_2 phase by quenching. Rapidly cooled or quenched melts of Na$_2$UF$_6$ yield a mixture of α, β_2, and γ phases of Na$_2$UF$_6$.

β_1 K$_2$UF$_6$ has hexagonal symmetry, and the atomic positions are

> 1 U in (0 0 0)
> 2 K in $\pm(\frac{1}{3}\ \frac{2}{3}\ \frac{1}{2})$
> 3 F$_I$ in $(x_1\ 0\ \frac{1}{2})$ $(0\ x_1\ \frac{1}{2})$ $(\overline{x}_1\ \overline{x}_1\ \frac{1}{2})$
> 3 F$_{II}$ in $(x_2\ 0\ 0)$ $(0\ x_2\ 0)$ $(\overline{x}_2\ \overline{x}_2\ 0)$

with $x_1 = 0.220$ and $x_2 = 0.640$.

In the completely disordered β_1' structure, the uranium and potassium atoms are distributed over the sites (0 0 0) and $\pm(\frac{1}{3}\ \frac{2}{3}\ \frac{1}{2})$ at random. Samples of varying degrees of disorder have been prepared.

The β_2-phase structure is also hexagonal, but it is of a different, but closely related, structure type of lower symmetry.

> 1 U in (0 0 0)
> 2 K or 2 Na in $\pm(\frac{1}{3}\ \frac{2}{3}\ u)$
> 3 F$_I$ in $(x_1\ 0\ \frac{1}{2})$ $(0\ x_1\ \frac{1}{2})$ $(\overline{x}_1\ \overline{x}_1\ \frac{1}{2})$
> 3 F$_{II}$ in $(x_2\ 0\ 0)$ $(0\ x_2\ 0)$ $(\overline{x}_2\ \overline{x}_2\ 0)$

For the potassium compound β_2 K$_2$UF$_6$, $u = 0.450$, $x_1 = 0.190$, and $x_2 = 0.640$; for β_2 Na$_2$UF$_6$, $u = 0.385$, $x_1 = 0.250$, and $x_2 = 0.600$.

The crystal radius of sodium is such as to prevent it from having a coordination number of nine with respect to fluorine, which potassium exhibits in β_1 K$_2$UF$_6$. The structure therefore changes from β_1 to β_2 when potassium is replaced by sodium.

$\underline{\gamma\ Na_2UF_6}$. The chemical composition of this phase is sharply defined, and the compound is obtained in pure form when melts of this composition are cooled slowly without quenching. The phase exhibits orthorhombic symmetry. The atomic positions are

> (0 0 0) $(\frac{1}{2}\ \frac{1}{2}\ \frac{1}{2})$ + 2 U in (0 0 0)
> 4 Na in $\pm(0\ 0\ u)$ with $u = \frac{1}{3}$
> 4 F in $\pm(v\ \frac{1}{2}\ 0)$ with $v = \frac{1}{4}$
> 8 F in $\pm(x\ 0\ z)$ $(\overline{x}\ 0\ z)$ with $x = \frac{1}{4}$ and $z = \frac{1}{6}$

Each uranium and sodium atom is bonded to eight fluorine atoms, four atoms at a distance Na-F = U-F = 2.38 A and four atoms at a distance Na-F = U-F = 2.44 A.

Na_3UF_7. This phase is obtained as the only phase from melts containing 25 to 29 mole % UF_4. The thin green crystals are optically uniaxial. The chemical formula for this phase can be expressed as $(Na_{6-x} + U_x)U_2F_{14+3x}$ with x in the range $0 < x < 0.30$. For $x = 0$, the composition is Na_3UF_7; where x exceeds 0.3, the Na_3UF_7 phase becomes unstable relative to the Na_2UF_6 phases. The phase is tetragonal, and the lattice periods vary with x. For the extreme ends of the range the parameters are

x	a_1, A	a_3, A	ρ, g/cc
0	5.448 ± 0.007	10.896 ± 0.014	4.49
0.29	5.503 ± 0.005	11.006 ± 0.100	4.85

Each metal atom is linked on the average to 7.0 to 7.4 fluorine atoms with Na-F = U-F = 2.36 to 2.38 A. The unit cell and structure of Na_3UF_7 correspond to two fluorite unit cells placed one on top of the other, with uranium atoms occupying one-fourth and sodium atoms occupying three-fourths of the metal sites. Two of the 16 fluorine atoms are removed.

By precipitation from aqueous solution, a series of compounds of the type MUF_6 (M = Ba, Pb, Sr) have been prepared. All these compounds have the LaF_3 type of structure with the following lattice dimensions (MP Chicago 14):

	a_1, A	a_3, A
$BaUF_6$	4.265 ± 0.005	7.456 ± 0.015
$PbUF_6$	4.175 ± 0.005	7.337 ± 0.015
$SrUF_6$	4.103 ± 0.005	7.290 ± 0.015

Because of their low melting points (as compared to UF_4), the $NaF-UF_4$ and $KF-UF_4$ systems have received much attention as possible electrolytes for the preparation of uranium metal by high-temperature electrochemical methods (see Chap. 4).

3. URANIUM PENTAFLUORIDE, UF_5, AND THE INTERMEDIATE FLUORIDES U_2F_9 AND U_4F_{17}

Uranium pentafluoride was first described by Ruff and Heinzelmann (1911), who prepared it by the reaction of anhydrous hydrogen fluoride with uranium pentachloride. For a long time this was the only known compound with a composition between UF_4 and UF_6. However, at least two intermediate compounds exist between UF_4 and UF_5; they have been assigned the formulas U_2F_9 and U_4F_{17}. The discovery of these compounds arose from the study of the reaction between uranium tetrafluoride and uranium hexafluoride which proved unexpectedly complex.

3.1 <u>Historical.</u> In 1941 Grosse (SAM Columbia 6,7) proposed the following reaction for the preparation of uranium pentafluoride:

$$UF_4 + UF_6 \xrightarrow{125°C} 2UF_5 \tag{47}$$

An attempted preparation of UF_5 by this procedure at Chicago yielded a black product instead of the expected white or gray one (MP Chicago 15). Chemical analysis indicated that the black substance possessed the approximate formula UF_4, but x-ray analysis (MP Chicago 1) showed it to be cubic ($a_0 = 8.455$ A) whereas UF_4 is monoclinic and UF_5 tetragonal. Volume considerations, based on the ionic radius of F^- in other uranium compounds, indicated that the atomic ratio of F to U in the black substance was 4 ± 0.3. It was therefore suggested that this new substance was a crystal modification of UF_4, and it received the designation "black UF_4." Some support was given to this view by the observation that black UF_4 was converted to ordinary UF_4 on standing or by moderate heating after inoculation with a crystal of monoclinic UF_4 (MP Chicago 16).

Once attention was directed to the "black fluoride," it was soon realized that a substance of this character is formed under a wide variety of circumstances. Thus when a mixture of UF_6 and hydrogen was passed through a hot tube (400 to 600°C), a small amount of a jet-black solid was obtained with a composition approximating UF_4. In this case, too, it was concluded that a new crystal modification of UF_4 had been found (MP Ames 15).

A similar substance was encountered by British workers in the course of corrosion studies. Mild-steel samples exposed to UF_6 vapor formed a black scale with the composition $FeF_2 \cdot 2UF_5$ (or $FeF_3 \cdot U_2F_9$) (British 25). These workers also studied what was referred to as a "UF_4-UF_6 complex." This complex was prepared by treating UF_4 with liquid UF_6 at 150 to 200°C; it was a brown powder which appeared to have the composition $UF_{4.4}$ (British 26).

A study of the UF_4 (solid) + UF_6 (gas) reaction led to the discovery that under appropriate conditions the reaction of UF_4 with UF_6 gas can give rise to α or β forms of UF_5, U_2F_9, or U_4F_{17}. The compositions of the compounds U_2F_9 and U_4F_{17} were established by chemical analysis. For reference the calculated values of total uranium, uranium(IV), and fluorine for various uranium fluorides have been collected in Table 12.11.

The total uranium and fluorine contents are relatively insensitive to changes in composition between UF_4 and UF_5. The content of quadrivalent uranium is therefore a much more satisfactory index of composition (Table 12.11). The results of a number of analyses of black fluoride are given in Table 12.12 (SAM Carbide and Carbon 3).

Table 12.11—Composition of Various Fluorides

Compound	Total U, %	U(IV), %	F, %	F/U	$\frac{U(IV)}{U} \times 100$
UF_4	75.8	75.8	24.2	4.00	100.0
U_4F_{17}	74.7	65.4	25.3	4.25	87.5
U_2F_9	73.6	55.2	26.4	4.50	75.0
UF_5	71.5	35.8	28.5	5.00	50.0
UF_6	67.7	0.0	32.3	6.00	0.0

Table 12.12—Composition of "Black Fluoride," U_2F_9

Sample	U, %	U(IV), %	F, %	F/U*
I	72.8	54.0	25.6	4.41
II	72.8	53.3	27.1	4.67
III	72.1	52.0	26.4	4.59

*Excellent agreement was obtained by comparing the chemical ratio, F/U, with the ratio calculated by weight loss (by escape of UF_6) on heating at 340°C for 24 hr at 10^{-5} mm Hg pressure. The weight loss was 4.39 per cent.

Analyses of a number of samples of the substance designated as U_4F_{17} (first detected by x-ray methods) showed that this composition agrees best with the analytical data (Table 12.13) (SAM Carbide and Carbon 4).

That samples of U_2F_9 prepared in different ways have an identical lattice constant as indicated by precision x-ray studies is strong evidence for the existence of a true compound of this composition. The existence of U_4F_{17} as a pure compound is not quite as definite, but there is little reason to doubt it.

3.2 Preparation. The UF_4-UF_6 Reaction. The reaction of gaseous UF_6 with UF_4 proceeds in stepwise fashion and can yield U_4F_{17}, U_2F_9,

α UF$_5$, or β UF$_5$, depending on the temperature and the partial pressure of the UF$_6$ in the system. Equilibrium conditions for the formation of the various products are summarized in Table 12.14 and Fig. 12.2.

The rate at which equilibrium is attained (and consequently the purity of the product under any given set of conditions) is primarily dependent on the nature of the UF$_4$; UF$_4$ prepared from aqueous solu-

Table 12.13 — Composition of U$_4$F$_{17}$ (UF$_{4.25}$)

Element	From weight gain*	From chemical analysis	Calculated composition		
			UF$_{4.20}$	UF$_{4.25}$	UF$_{4.30}$
Total U, %	74.8	74.8	74.9	74.7	74.5
U(IV), %	65.7	64.1 ± 1.1†	67.4	65.4	63.4
F, %	25.2	25.3 ± 0.2‡	25.1	25.3	25.5

*From gain in weight when solid UF$_4$ is treated with gaseous UF$_6$ at 17 mm Hg pressure (300°C).
†Mean of six determinations.
‡Mean of two determinations.

Table 12.14 — Conditions for the Preparation of UF$_5$ and the Intermediate Fluorides by Reaction of Solid UF$_4$ with Gaseous UF$_6$

p_{UF_6}, mm Hg*	100°C	200°C	320°C
17.7	...†	U$_2$F$_9$	U$_4$F$_{17}$
120 – 140	β UF$_5$	α UF$_5$	U$_2$F$_9$

*The two pressures given correspond to the vapor pressure of solid UF$_6$ at 0°C and room temperature, respectively.
†Presumably β UF$_5$ would be formed under these conditions.

tion and then dehydrated has a surface area of 6 to 8 square meters per gram as compared to 0.25 square meter per gram for material prepared by high-temperature hydrofluorination. Large-surface-area material is to be preferred for rapid attainment of equilibrium. The rate is also strongly affected by sintering, which is observed at temperatures above 200°C when either UF$_5$ or U$_2$F$_9$ are the stable phases.

The reaction of liquid UF$_6$ with UF$_4$ has not been studied in detail. The product at 100 to 125°C is β UF$_5$, provided large-surface-area UF$_4$ is used. With small-surface-area UF$_4$ or when insufficient time for the attainment of equilibrium is permitted, U$_2$F$_9$ is the product.

This difference in the reactivity of the UF_4 appears to be responsible for the original discovery of the black fluoride. With large-surface-area UF_4 and reaction periods up to one week the reaction of UF_4 with

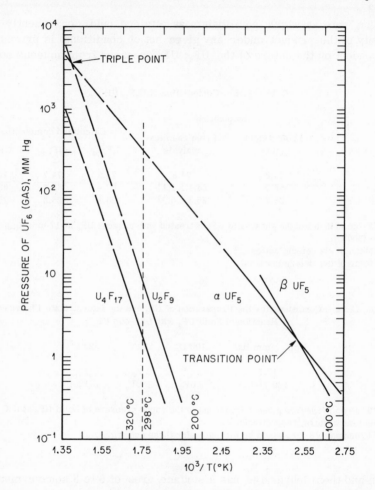

Fig. 12.2—Disproportion pressures of U_4F_{17}, U_2F_9, $\alpha\,UF_5$, and $\beta\,UF_5$.

liquid UF_6 at 125°C affords a good method for the preparation of $\beta\,UF_5$. In general, however, the reactions with gaseous UF_6 are to be preferred, at least pending more detailed information on the liquid-phase reactions.

The original method of Ruff and Heinzelmann (1911) for the preparation of UF_5 has been reinvestigated by the Brown University group

(Brown 4,7,9). Uranium pentachloride is treated with excess liquid hydrogen fluoride for 12 hr at room temperature in either a nickel or a platinum vessel, after which the excess hydrogen fluoride and hydrogen chloride are removed in a stream of nitrogen. The UF_5 produced is the β form. Uranium hexachloride may also be used. In this case the reaction proceeds as follows:

$$2UCl_6 + 10HF \rightarrow 2UF_5 + 10HCl + Cl_2 \qquad (48)$$

The UCl_6 must be very carefully purified from organic matter (by sublimation), and the hydrogen fluoride must be free of all traces of moisture (by pretreatment with UCl_5) to obtain UF_5 free of UF_4. The action of gaseous hydrogen fluoride at 300°C on UCl_5 is another effective method for preparing α UF_5.

Uranium pentafluoride may also be prepared by the reaction of uranium tetrafluoride with fluorine.

$$UF_4 + \tfrac{1}{2}F_2 \xrightarrow{\;150-250°C\;} UF_5 \qquad (49)$$

At higher temperatures (approximately 370°C) UF_6 is the principal product, but at temperatures of 150 to 250°C, α UF_5 of very high purity can be obtained, especially if large-surface UF_4 is employed. The conversion of UF_4 to UF_5 has been accomplished at temperatures below 100°C by using fluorine under pressure (100 lb/in.²) (SAM Columbia 7). This reaction has not been thoroughly studied; occasionally it, too, gives dark-colored products. The reaction can then be completed by mild treatment in a fluorine stream.

3.3 Disproportionation of UF_5 and the Intermediate Fluorides. (SAM Carbide and Carbon 5.) The disproportionation pressures of gaseous UF_6 over α and β UF_5, U_2F_9, and U_4F_{17} have been measured at SAM Carbide and Carbon. The results are summarized in Table 12.15 and Fig. 12.2.

3.4 Thermochemistry of the Intermediate Fluorides. (SAM Carbide and Carbon 5.) From the data in Tables 12.15 and 12.16 the thermodynamic functions of the intermediate fluorides may be computed (Table 12.17).

The heat capacity of α UF_5 has been measured at the National Bureau of Standards (Natl. Bur. Standards 2). The sample contained only 82.82 per cent UF_5, and consequently results are not very precise. The value of $S_{298°K}$ obtained from the heat capacity measurements (45 ± 3 e.u.) falls between the values calculated above for α and β UF_5. The heat capacity measurements are summarized in Table 12.18.

3.5 Physical Properties and Crystal Structures of the Intermediate Fluorides.

Pure UF_5, either α or β, is probably colorless since samples have been prepared which are almost white. However, ma-

Table 12.15 — Equilibrium Decomposition Pressures and Thermodynamics of Disproportionation Reactions

Reaction	Equilibrium vapor pressures*		Temp. range, °C	Thermochemical constants		
	A	B		$\Delta H_{298°K}$, kcal/mole UF_6 (gas)	$\Delta S_{298°K}$, e.u.	$\Delta F_{298°K}$, kcal/mole UF_6 (gas)
$3\,\alpha\ UF_5 \rightleftharpoons U_2F_9 + UF_6$ (gas)	2,942	7.634	100–200	13.46 ± 0.2	21.7 ± 0.3	6.99 ± 0.4
$3\,\beta\ UF_5 \rightleftharpoons U_2F_9 + UF_6$ (gas)	4,166	10.71	100–152	19.06 ± 0.5	35.8 ± 0.7	8.40 ± 0.8
$\frac{7}{2}U_2F_9 \rightleftharpoons \frac{3}{2}U_4F_{17} + UF_6$ (gas)	7,315	13.68	225–320	33.47 ± 1	49.4 ± 1.4	18.8 ± 1.7
$2U_4F_{17} \rightleftharpoons 7UF_4 + UF_6$ (gas)	7,143	12.75	270–350	32.68 ± 0.4	45.2 ± 0.6	19.2 ± 0.7

* $\log p_{mm} = -\dfrac{A}{T} + B.$

Table 12.16 — Thermochemical Constants of Uranium and Certain Fluorides

Substance	$\Delta H^{\circ}_{298°K}$, kcal/mole	$S_{298°K}$, e.u.	$\Delta F^{\circ}_{298°K}$, kcal/mole
UF_6 (gas)	−504 ± 3	90.3 ± 1	483.9 ± 3
UF_4	−443.5 ± 2	36.1 ± 0.1	421.7 ± 2
U		12.0 ± 0.03	
F_2 (gas)*		48.58 ± 0.1	

*From K. K. Kelley, U. S. Bur. Mines Bull. 434 (1941).

Table 12.17 — Thermochemical Constants of the Intermediate Fluorides

Substance	$\Delta H^{\circ}_{298°K}$, kcal/mole	$S_{298°K}$, e.u.	$\Delta F^{\circ}_{298°K}$, kcal/mole
$\alpha\ UF_5$	−483.7 ± 1.3	48.0 ± 0.4	−458.2 ± 1.4
$\beta\ UF_5$	−485.2 ± 1.4	43.3 ± 0.5	−458.7 ± 1.5
U_2F_9	−933.8 ± 3	75.3 ± 0.6	−884.0 ± 3
U_4F_{17}	−1820.5 ± 4	149.0 ± 0.6	−1727.5 ± 4

terial within analytical limits for UF_5 has been prepared which is green [$UF_4 + UF_6$ (gas)], pale gray ($UF_4 + F_2$), or even brown (HF + UCl_5). All these variously colored samples give x-ray patterns showing only UF_5. Both U_2F_9 and U_4F_{17} are black. As in the case of

U_3O_8, the black color of these compounds is possibly due to resonance in the crystal between uranium ions in different valence states.

(a) <u>Structure of α UF$_5$</u>. (MP Chicago 17; Zachariasen, 1949.) α UF$_5$ exhibits tetragonal symmetry with unit cell dimensions

$$a_1 = 6.512 \pm 0.001 \text{ A} \qquad a_3 = 4.463 \pm 0.001 \text{ A}$$

There are two molecules per unit cell, and the calculated density is 5.81 g/cc. On the basis of spatial considerations, Zachariasen has

Table 12.18 — Heat Capacity and Entropy of α UF$_5$*

Temp., °K	C_p, cal/mole/°C	$H-H_0$, cal/mole	$S-S_0$, cal/mole/°C	$(H-H_0) - T(S-S_0)$, cal/mole
0	0	0	0	0
25	4.23	39	2.23	−17
50	10.24	219	6.99	−130
75	15.43	544	12.18	−370
100	19.11	978	17.16	−737
125	21.79	1490	21.72	−1224
150	23.97	2063	25.89	−1821
175	25.77	2685	29.72	−2516
200	27.30	3349	33.27	−3305
225	28.61	4049	36.56	−4178
250	29.71	4778	39.63	−5131
275	30.74	5534	42.52	−6158
300	31.69	6314	45.23	−7255
310	32.04	6633	46.28	−7713

*Crystallographic modification of this sample is unknown.

deduced a structure belonging to the space group I4/m (C_{4h}^5), with the following atomic positions:

2 U atoms in (0 0 0) ($\frac{1}{2}$ $\frac{1}{2}$ $\frac{1}{2}$)
2 F$_I$ atoms in (0 0 $\frac{1}{2}$) ($\frac{1}{2}$ $\frac{1}{2}$ 0)
8 F$_{II}$ atoms in $\pm$(u v 0) (v $\bar{u}$ 0) (u+$\frac{1}{2}$ v+$\frac{1}{2}$ $\frac{1}{2}$) (v+$\frac{1}{2}$ $\frac{1}{2}$−u $\frac{1}{2}$)

with u = 0.315 and v = 0.113.

Each uranium atom is surrounded by an octahedral arrangement of six fluorine atoms, with interatomic distances U-2F$_I$ = 2.23 A and U-4F$_{II}$ = 2.18 A. The octahedra are thus elongated along the a_3 axis; each F$_I$ atom is joined to two consecutive octahedra, and thus there are long strings of octahedra along the a_3 axis. From the uranium-fluorine distances, assuming essentially ionic bonding, the ionic crystal radius of U^{+5} is 0.87 A, slightly smaller than the value 0.89 A found for U^{+4}.

Since all the uranium atoms are equivalent in this structure, the compound cannot be regarded as a double fluoride of quadrivalent and sexivalent uranium, i.e., $UF_4 \cdot UF_6$. This conclusion is reinforced by the fact that uranium pentafluoride is colorless, showing no resonance such as might be anticipated in a fluoride of mixed oxidation state.

(b) Structure of β UF_5. (Zachariasen, 1949.) β UF_5 is also tetragonal with unit cell dimensions

$$a_1 = 11.450 \pm 0.002 \text{ A} \qquad a_3 = 5.198 \pm 0.001 \text{ A}$$

There are eight molecules per unit cell, and the calculated density is 6.45 g/cc, approximately 10 per cent higher than for the α form. Here, also, intensity measurements are not sufficient fully to index the structure. However, on the basis of arguments involving ionic volumes, Zachariasen has suggested space group $I\bar{4}2d(D_{2d}^{12})$ for the structure of β UF_5. In the proposed structure, each uranium atom is bonded to seven fluorine atoms at the following distances:

$$U\text{-}1F_I = 2.18 \text{ A} \quad U\text{-}2F_{II} = 2.23 \text{ A} \quad U\text{-}2F_{II} = 2.29 \text{ A} \quad U\text{-}2F_{III} = 2.18 \text{ A}$$

The mean distance is U-7F = 2.23 A. Of the seven corners of the fluorine polyhedron about each uranium atom, four are shared with adjacent polyhedra.

As in the case of α UF_5, β UF_5 cannot be considered to be $UF_4 \cdot UF_6$, since here, also, all the uranium atoms are equivalent.

(c) Interconversion of α and β UF_5. From the equations for equilibrium pressure of α and β UF_5, a transition temperature of 125°C (under 1.76 mm UF_6 pressure) is calculated, with α UF_5 the stable form above 125°C. To effect complete conversion of the β form, 12 hr at 185°C may be required since at lower temperatures the rate is very slow. The reverse transformation, α into β, has not as yet been achieved.

(d) Structure of U_2F_9. (MP Chicago 18; Zachariasen, 1948b.) This crystal has a cubic structure with $a_0 = 8.4538_4 \pm 0.001_9$ A. The unit cell contains eight molecules with a calculated density of 7.06 g/cc. A structure with the symmetry $I\bar{4}3m$ has been suggested with atom positions (using the notation of the "International Tables for the Determination of Crystal Structure," Chemical Catalog Company, New York, 1935):

$$8 \text{ U} \quad \text{in positions} \quad 8(c) \quad x \approx 0.187 \pm 0.004$$
$$12 \text{ F}_I \quad \text{in positions} \quad 12(e) \quad x \approx 0.225$$
$$24 \text{ F}_{II} \quad \text{in positions} \quad 12(g) \quad x \approx 0.20, z \approx 0.46$$

In this structure each uranium is bound to nine fluorine atoms, and each fluorine atom is linked to two uranium atoms. The three nearest F_I atoms about each uranium atom are at a distance $U\text{-}F_I = 2.26$ A; three of the six F_{II} atoms are at a distance $U\text{-}F_{II} = 2.31$ A, and three others at 2.34 A. By trivial changes in the fluorine parameters the three distances can be made equal. Since each fluorine atom is bound to two uranium atoms, U_2F_9 should be much less volatile than UF_6 or UF_5, in which all or some of the fluorine atoms are bound directly to only one uranium atom. The volatility of U_2F_9 is comparable to that of UF_4.

The uranium atoms are all structurally equivalent. The formulas $UF_4 \cdot UF_5$ or $3UF_4 \cdot UF_6$, which indicate the presence of two kinds of uranium atoms, are therefore inadmissible. The substance may be regarded as a mixed crystal with rational proportions in which resonance renders all the uranium atoms equivalent,

$$(U_{0.5}^{+4} + U_{0.5}^{+5})\ F_{4.5} \quad \text{or} \quad (U_{0.75}^{+4} + U_{0.25}^{+6})\ F_{4.5}$$

(e) Structure of U_4F_{17}. (SAM Carbide and Carbon 4.) The crystal structure of this compound has not yet been worked out. The diffraction pattern resembles that of UF_4 in a general way but differs sufficiently in detail to be designated as "distorted UF_4" and to be considered as belonging to a definite compound. The density of U_4F_{17} (by liquid displacement) is 6.94 g/cc.

(f) Melting Point and Volatility of Uranium Pentafluoride. The melting point of UF_5, although not yet determined with precision, is estimated to lie below 400°C. There is also some evidence for the vaporization of UF_5 without chemical change, but the data are fragmentary, and no definite value for the vapor pressure of UF_5 can be given.

3.6 Chemical Properties. Apart from thermal disproportionation, which has already been discussed above, little can be said about the chemical reactions of the intermediate fluorides. All these compounds undergo hydrolysis. UF_5 resembles UF_6 in its sensitivity to traces of moisture. In moist air α UF_5 appears to be more susceptible to hydrolysis than β UF_5, a behavior which is independent of particle size. With water, uranyl fluoride and UF_4 are formed.

$$2UF_5 + 2H_2O \rightarrow \underline{UF_4} + UO_2^{++} + 2F^- + 4HF \qquad (50)$$

U_2F_9 appears to be much more stable to hydrolysis than UF_5. In water the black color of the compound may persist for 1 hr or more; in air a sample may remain black for two weeks.

U_4F_{17} reacts with water, but no rate comparison with the other intermediate fluorides is available. Presumably it would be less reactive than U_2F_9. The reaction with water proceeds as follows:

$$2U_4F_{17} + 2H_2O \rightarrow UO_2^{++} + 7UF_4 + 4HF + 2F^- \qquad (51)$$

REFERENCES

1861   H. Hermann, dissertation, University of Göttingen, S. 32 (Gmelin, p. 120).
1866a   H. C. Bolton, Z. Chem., (2) 2: 353.
1866b   H. C. Bolton, Bull. soc. chim. Paris, (2) 6: 450.
1880   A. Ditte, Compt. rend., 91: 116.
1883   A. Smithells, J. Chem. Soc., 43: 125-135.
1911   O. Ruff and H. Heinzelmann, Z. anorg. Chem., 72: 64, 71.
1922   J. Aloy and E. Rodier, Bull. soc. chim. France, (4) 31: 247.
1929   I. Oftedal, Z. physik. Chem., 5: 272.
1931   I. Oftedal, Z. physik. Chem., 13: 190.
1931   O. Ruff and R. Keim, Z. anorg. u. allgem. Chem., 201: 245.
1933   J. C. Southard and F. G. Brickwedde, J. Am. Chem. Soc., 55: 4378.
1937   L. Domange, Ann., 7: 225-297.
1938   A. L. Henne, U. S. Patent No. 2136741.
1942   N. Bekkedahl and R. B. Scott, J. Research Natl. Bur. Standards, 29: 87.
1946   H. S. Booth, W. Krasny-Ergen, and R. Heath, J. Am. Chem. Soc., 68: 1969-1970.
1948   F. G. Brickwedde, H. J. Hoge, and R. B. Scott, J. Chem. Phys., 16: 429.
1948a   W. H. Zachariasen, J. Am. Chem. Soc., 70: 2147.
1948b   W. H. Zachariasen, J. Chem. Phys., 16: 425.
1948c   W. H. Zachariasen, Reports AECD-1798, AECD-2089, AECD-2162, and AECD-2163.
1949   L. Domange and M. Wohlhuter, Compt. rend., 228: 1591-1592.
1949   N. Elliott, Phys. Rev., 76: 431.
1949   W. H. Zachariasen, Acta Crystallographica, 2: 296.

Project Literature

British 1:   J. Ferguson, Report B-55, March, 1942, p. 5.
British 2:   H. J. Spencer-Palmer, Report BR-422, date received, June 10, 1944.
British 3:   Report [B]LRG-33, June, 1944.
British 4:   B. G. Harvey, Report B-52, Mar. 12, 1942, p. 3.
British 5:   G. O. Morris, Report B-66, Feb. 17, 1942, p. 2.
British 6:   N. McLeish, Report B-54, Mar. 12, 1942.
British 7:   A. M. Roberts, Report B-60, Feb. 6, 1942.
British 8:   A. M. Roberts, Report B-65, Feb. 6, 1942.
British 9:   C. G. Harris, A. M. Roberts, and E. J. Jefferies, Report B-59, Apr. 8, 1942.
British 10:   A. M. Roberts, Report B-61, Feb. 6, 1942.
British 11:   G. O. Morris, Report B-50, Mar. 20, 1942.

British 12: G. O. Morris, Report B-51, Feb. 25, 1942.
British 13: A. M. Roberts and B. G. Harvey, Report B-62, Mar. 4, 1942.
British 14: G. O. Morris and B. G. Harvey, Report B-69, May 4, 1942.
British 15: A. M. Roberts and G. O. Morris, Report B-63, Mar. 20, 1942.
British 16: Report BR-446, April, 1944.
British 17: Report BR-454, date received, July 6, 1944.
British 18: A. M. Roberts, Report B-67, Feb. 16, 1942.
British 19: T. Wathen, Report BR-483, Aug. 22, 1944.
British 20: G. O. Morris, Report BR-250, July 23, 1943.
British 21: A. M. Roberts and B. G. Harvey, Report B-58, Mar. 30, 1940.
British 22: Report [B]LRG-27, December, 1943.
British 23: A. M. Roberts, Report B-64, Feb. 6, 1942.
British 24: A. M. Roberts, Report B-68, Feb. 16, 1942.
British 25: J. Parks, Report BR-480, Aug. 8, 1944.
British 26: Report [B]LRG-34, July, 1944.

Brown 1: C. A. Kraus, Report CC-1772, Apr. 28, 1944, p. 4.
Brown 2: C. A. Kraus, Report CC-342, Nov. 15, 1942, Part B, p. 8.
Brown 3: C. A. Kraus, Report A-1090, Dec. 1, 1943, Part II.
Brown 4: C. A. Kraus, Report CC-1717, July 29, 1944.
Brown 5: C. A. Kraus, Report [A]M-11, June 15, 1944.
Brown 6: C. A. Kraus, Report A-2309, Mar. 24, 1945.
Brown 7: C. A. Kraus, Report CC-1519, Feb. 22, 1944.
Brown 8: C. A. Kraus, Report [A]M-251, July 13, 1943.
Brown 9: C. A. Kraus, Report A-1090, Dec. 1, 1943, Part I.

CEW-TEC 1: L. J. Andrews and J. W. Gates, Report CD-477, Nov. 9, 1944.
CEW-TEC 2: J. W. Gates, L. J. Andrews, and W. B. Schaap, Report CD-4005, Feb. 22, 1945.
CEW-TEC 3: B. M. Pitt, E. L. Wagner, and A. J. Miller, Report CD-2.355.3, Jan. 3, 1946.
CEW-TEC 4: J. W. Gates, L. J. Andrews, and B. M. Pitt, Report CD-4001, Jan. 30, 1945.
CEW-TEC 5: E. J. Lord, L. J. Andrews, and J. W. Gates, Report CD-1.365.4, Aug. 22, 1945.
CEW-TEC 6: V. P. Calkins, Reports CD-0.385.2, Oct. 6, 1945, and CD-0.350.4, Sept. 25, 1945.

Du Pont 1: J. W. Dobratz, R. C. Downing, C. L. Richardson, and L. Spiegler, Report N-34, Apr. 9, 1943.
Du Pont 2: F. B. Downing, A. F. Benning, and A. L. Finch, Report A-728, June 17, 1943.

Los Alamos 1: I. B. Johns and K. A. Walsh, Report LA-381, Aug. 30, 1945.

Mallinckrodt 1: T. R. Braddock and D. Copenhafer, Report N-17, Jan. 30, 1943, p. 6.
Mallinckrodt 2: T. R. Braddock and D. Copenhafer, Report N-24, Jan. 23, 1943, p. 9.

MP Ames 1: A. S. Newton, J. C. Warf, O. Johnson, and R. Nottorf, Report CC-1201, Jan. 1, 1944.
MP Ames 2: H. D. Brown, Report CC-1525, Mar. 14, 1944.
MP Ames 3: H. Lipkind, Report CC-1525, Mar. 14, 1944.
MP Ames 4: P. Chiotti and R. Raeuchle, Report CC-1525, Mar. 14, 1944.
MP Ames 5: N. C. Baenziger and R. E. Rundle, Report CC-1525, Mar. 14, 1944.

MP Ames 6: N. C. Baenziger and R. E. Rundle, Report CN-1495, Mar. 10, 1944, pp. 26-27.

MP Ames 7: W. Keller and W. Lyon, Report CC-1525, Mar. 14, 1944.

MP Ames 8: Report CC-1059, Sept. 9, 1943, p. 23.

MP Ames 9: A. D. Tevebaugh, K. Walsh, J. Illif, and I. B. Johns, Reports CN-1199, Dec. 10, 1943, p. 7, and CC-1063, Nov. 16, 1943, p. 8.

MP Ames 10: W. D. Cline, R. Tevebaugh, and J. C. Warf, Report CC-1981, Oct. 10, 1944.

MP Ames 11: J. C. Warf and W. D. Cline, Report CC-2723, June 30, 1945.

MP Ames 12: J. G. Feibig and J. C. Warf, Report CC-2939, June 29, 1945.

MP Ames 13: A. Daane, Report CC-298, Oct. 16, 1942.

MP Ames 14: A. D. Tevebaugh, Report CC-682, May 15, 1943.

MP Ames 15: A. D. Tevebaugh, K. Walsh, and I. B. Johns, Report CC-1245, Jan. 8, 1945.

MP Chicago 1: J. Karle, Report CK-1367, Feb. 25, 1944.

MP Chicago 2: W. H. Zachariasen, Report CN-2526, Jan. 2, 1945.

MP Chicago 3: H. S. Brown, Report CC-408, Jan. 8, 1943, p. 5.

MP Chicago 4: E. Creutz, Report CE-301, Oct. 15, 1942.

MP Chicago 5: N. R. Davidson, Report CK-932, Sept. 11, 1943, p. 5.

MP Chicago 6: W. H. Zachariasen, Report CC-2768, Mar. 12, 1945.

MP Chicago 7: W. H. Zachariasen, Report CP-1954, July 29, 1944, p. 22.

MP Chicago 8: R. F. Plott and C. H. Raeth, Report CP-228, Aug. 14, 1942.

MP Chicago 9: S. Fried and N. R. Davidson, Report N-1722, Nov. 14, 1944.

MP Chicago 10: S. Fried, Report N-2128, Oct. 12, 1945.

MP Chicago 11: J. Karle, Report CK-1512, Apr. 6, 1944.

MP Chicago 12: W. H. Zachariasen, Report CC-3401, Jan. 10, 1946 in National Nuclear Energy Series, Division VIII, Volume 6.

MP Chicago 13: W. H. Zachariasen, Report CC-3426, Feb. 9, 1946 in National Nuclear Energy Series, Division VIII, Volume 6.

MP Chicago 14: W. H. Zachariasen, Report CF-2796, Mar. 15, 1945, p. 12.

MP Chicago 15: R. Livingston and W. Burns, Report CN-982, Oct. 9, 1943.

MP Chicago 16: R. Livingston, Report N-781, Mar. 10, 1944.

MP Chicago 17: W. H. Zachariasen, in National Nuclear Energy Series, Division VIII, Volume 6.

MP Chicago 18: W. H. Zachariasen, Report CC-2753, Feb. 28, 1945.

MP Clinton 1: M. Lindner, Report CN-1025, Nov. 8, 1943, p. 15.

Natl. Bur. Standards 1: R. B. Scott, H. J. Hoge, and F. G. Brickwedde, Report A-201, July 14, 1942.

Natl. Bur. Standards 2: F. G. Brickwedde, personal communication to W. F. Libby; these data appear in SAM Carbide and Carbon document DA-R-53, Sec. VI, 1.23.9 (no date).

SAM Carbide and Carbon 1: H. Zeldes, reported by F. T. Miles in Division D, Monthly Report M-2589 (DE), Jan. 5, 1946.

SAM Carbide and Carbon 2: R. Kunin, to be published.

SAM Carbide and Carbon 3: S. Weller, A. Grenall, and R. Kunin, Report A-3226, Mar. 19, 1945.

SAM Carbide and Carbon 4: A. Grenall, P. Agron, R. Kunin, and S. Weller, Report A-3271, June 27, 1945.

SAM Carbide and Carbon 5: P. Agron, Thermodynamics of Intermediate Uranium Fluorides from Measurements of the Disproportionation Pressures, in National Nuclear Energy Series, Division VIII, Volume 6.

SAM Columbia 1: A. V. Grosse, S.F. 314, June, 1941 (from B-54, p. 1).

SAM Columbia 2: A. V. Grosse, Report A-99, March, 1941.

SAM Columbia 3: G. T. Seaborg, Report CK-1514, Mar. 6, 1944.

SAM Columbia 4: H. F. Priest and G. R. Priest, Report A-257, Aug. 10, 1942.

SAM Columbia 5: W. F. Libby, Report A-1228x, Aug. 29, 1944; H. C. Urey, Report A-330, Oct. 20, 1942.

SAM Columbia 6: A. V. Grosse, Report A-83, June 25, 1941, p. 11.

SAM Columbia 7: H. F. Priest, Report A-719, June 3, 1943.

REFERENCES

Chapter 13

URANIUM HEXAFLUORIDE

Uranium hexafluoride was discovered by Ruff and Heinzelmann (1909). Earlier, Moissan (1900) had observed that fluorine reacted vigorously with uranium to give a volatile uranium compound; the quantity obtained, however, was too small for study. Ruff, who had previously studied the volatile heavy-metal fluorides MoF_6 and WF_6 (Ruff, 1909), suspected the existence of an analogous compound of uranium and succeeded in preparing it by the reaction of fluorine with uranium metal, uranium carbide, or uranium pentafluoride. He determined some of the physical and chemical properties of UF_6 (Ruff, 1911), but recent work has necessitated revision of some of his early results.

For thirty years after Ruff's investigation uranium hexafluoride received little attention, and the few recorded preparations were all made according to his method. Grosse (1932) prepared a small quantity for Aston's mass-spectrographic study of the uranium isotopes (Aston, 1931), and small amounts were also made for various magnetic and electron-diffraction studies. The discovery of uranium fission in 1939 naturally directed new attention to UF_6 as the only known stable gaseous compound of uranium. Much progress has since been made in the search for new preparative methods suitable for laboratory and industry, in the development of methods of handling uranium hexafluoride, and in the determination of the physical properties of this compound. The chemistry of uranium hexafluoride has not been investigated with the same vigor, and much of chemical interest still remains to be done.

1. PREPARATION OF URANIUM HEXAFLUORIDE

There are two general methods of preparing uranium hexafluoride; one necessitates, either directly or indirectly, the use of elemental fluorine, and the other depends on disproportionation reactions. Preparations of the first class are based on the observation that all compounds of uranium yield UF_6 when treated with fluorine under appropriate conditions. Among the methods of this type, the most useful in both the laboratory and on an industrial scale are those requiring the least amount of fluorine. Preparations of the second class have as yet received little study but do offer considerable promise for the future.

1.1 Fluorination of Uranium Compounds. Ruff and Heinzelmann (1911) first prepared uranium hexafluoride by the action of fluorine on uranium metal or carbide.

$$U + 3F_2 \rightarrow UF_6 \quad \Delta H = -505 \pm 3 \text{ kcal per mole} \tag{1}$$

(see Sec. 2.2a)

$$UC_2 + 7F_2 \rightarrow UF_6 + 2CF_4 \tag{2}$$

Reaction 1 proceeds very vigorously with evolution of considerable heat. Heating is usually required to start the reaction unless the metal is finely divided, but once the reaction has been initiated it progresses as rapidly as fluorine is admitted to the system. Ruff claimed that gaseous chlorine was required as a catalyst for this reaction and introduced chlorine or calcium chloride into the reactor. This procedure was followed by several later investigators (e.g., Grosse, Henkel, and Klemm). Later work, however, has proved that chlorine or other catalysts are not necessary (MP Chicago 1; British 1) and that the presence of chlorine leads to an impure yellow product instead of pure white uranium hexafluoride. Reaction 2 is less vigorous than reaction 1 but requires much more fluorine. A temperature of 350°C is suitable for reaction 2.

Ruff and Heinzelmann also used the reaction of fluorine with uranium pentachloride at $-40°C$ to prepare uranium hexafluoride. They formulated this reaction as follows:

$$2UCl_5 + 5F_2 \rightarrow UF_4 + UF_6 + 5Cl_2 \tag{3}$$

However, the exact composition of the reaction products has not been determined. The possible formation of explosive chlorine fluorides is a hazard. Intense cooling is required, and the UF_6 is separated by distillation after the completion of the reaction. Other uranium halides can be converted to UF_6 in a similar way. On the other hand, Ruff and Heinzelmann were unable to convert a suspension of UCl_5 in liquid chlorine to UF_6 with free fluorine. They concluded that either the rate of the reaction is very slow at the temperatures of liquid chlorine ($-33.7°C$) or that fluorine must be practically insoluble in liquid chlorine.

The fluorination method of uranium hexafluoride preparation was substantially improved by Abelson at the National Bureau of Standards (Natl. Bur. Standards 1), who introduced the use of uranium tetrafluoride instead of uranium metal.

$$UF_4 + F_2 \rightarrow UF_6 \tag{4}$$

Abelson recommended a temperature of $275°C$ for this reaction and used fused sodium chloride as a catalyst. British workers, however, found that sodium chloride or any other catalyst is unnecessary and even detrimental to successful fluorination (British 1,2). Pending further study, it may be suggested that the uranium tetrafluoride used in reaction 4 should be as pure as possible.

Uranium tetrafluoride is readily fluorinated at relatively low temperatures. At room temperature the reaction is slow, but at $250°C$ it is very rapid. Below $250°C$ the reaction appears to occur in two steps:

$$2UF_4 + F_2 \rightarrow 2UF_5 \tag{5}$$

$$2UF_5 + F_2 \rightarrow 2UF_6 \tag{6}$$

Reaction 5 proceeds much more rapidly than reaction 6, so that little or no UF_6 appears until all the UF_4 has been converted to UF_5. Reaction 5 is efficient in utilization of fluorine, and reaction 6 has poor efficiency resulting in large amounts of unused fluorine. Increasing the temperature is of little practical value; the rate of reaction 6 appears rather insensitive to temperature variations between 250 and $650°C$. The reason for this peculiar behavior is unknown. One method of increasing the amount of fluorine consumed is to pass the gas from the reactor where reaction 6 takes place through a second reactor where reaction 5 is carried out. The latter will absorb practically all the residual fluorine. The direction of flow can then be reversed. However, doubts have been expressed as to the practicability of this procedure, since experiments show that reaction 5 will be finished in the second reactor before reaction 6 is complete in the first reactor (British 1).

It appears that all uranium compounds when heated with fluorine to a sufficiently high temperature will give UF_6 (CEW-TEC 1). Uranyl phosphate, for example, readily undergoes conversion to UF_6 and PF_3 or PF_5 at $400°C$. Oxides of uranium are said to require heating with fluorine to a temperature of 400 to $500°C$ and to be, in general, rather difficult to fluorinate. However, triuranium octaoxide (U_3O_8) can be converted to UF_6 at 300°C if carbon is added to the mixture (MP Chicago 1; Johns Hopkins 1).

$$U_3O_8 + 4C + 9F_2 \rightarrow 3UF_6 + 4CO_2 \qquad (7)$$

According to some reports, pure U_3O_8 reacts with fluorine at 360 to 370°C (SAM Columbia 1).

$$U_3O_8 + 17F_2 \rightarrow 3UF_6 + 8OF_2 \qquad (8)$$

Another method of converting U_3O_8 to UF_6 involves preliminary hydrofluorination (MP Chicago 1).

$$U_3O_8 + 8HF \xrightarrow{500°C} 2UO_2F_2 + UF_4 + 4H_2O \qquad (9)$$

$$UF_4 + 2UO_2F_2 + 5F_2 \rightarrow 3UF_6 + 2O_2 \qquad (10)$$

Figure 13.1 summarizes conditions that assure formation of UF_6 from a variety of uranium compounds.

Fluorination reactions can be carried out in monel vessels; copper, aluminum, and magnesium can also be employed. Platinum cannot be used since it is rapidly attacked by fluorine above 450°C. Calcium fluoride reaction tubes and boats are very useful but are not readily available. Standard flare fittings, such as are used in the refrigeration industry, are convenient for assembling copper apparatus. At low temperatures, fluorine can be handled without serious attack for considerable periods in thoroughly dried and outgassed pyrex or quartz vessels. Soft glass is reputed to be superior to pyrex in its resistance to fluorine. Further manipulative details will be found in Sec. 1.3.

Fluorination of uranium compounds to UF_6 can also be successfully achieved with agents other than elemental fluorine (CEW-TEC 2; SAM Columbia 2; Yale 1). These strong fluorinating agents, however, require fluorine for their preparation and consequently do not dispense altogether with the use of free fluorine. Their use for the synthesis of UF_6 sometimes has additional advantages besides avoiding free fluorine. Thus uranium tetrafluoride can be converted to UF_6 by reaction with cobaltic fluoride.

$$UF_4 + 2CoF_3 \xrightarrow{250-275°C} UF_6 + 2CoF_2 \qquad (11)$$

The reaction mixture is prepared by grinding the tetrafluoride with a fivefold excess of cobaltic fluoride in a dry box. The reaction can be carried out in an all-glass apparatus with the reaction mixture placed in a platinum boat. A temperature higher than 275°C is to be avoided as CoF_3 may undergo thermal decomposition. Cobaltic fluoride can be prepared in a variety of ways (CEW-TEC 3).

$$CoCl_2 \cdot 6H_2O \xrightarrow{400°C} CoCl_2 + 6H_2O \qquad (12)$$

$$CoCl_2 + 2HF \xrightarrow{250°C} CoF_2 + 2HCl \qquad (13)$$

$$2CoF_2 + F_2 \xrightarrow{250°C} 2CoF_3 \qquad (14)$$

$$2CoCl_2 + 5F_2 \xrightarrow{400°C} 2CoF_3 + 4ClF \qquad (15)$$

Silver difluoride or palladium trifluoride would probably be as satisfactory as CoF_3. Antimony pentafluoride, on the other hand, a compound that can be prepared without the use of elemental fluorine, does not convert UF_4 to UF_6 (Du Pont 1).

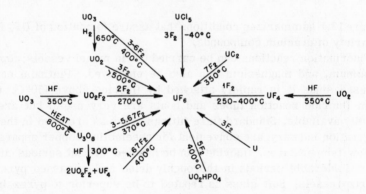

Fig. 13.1—Conversion of uranium compounds to UF_6 (showing number of moles of F_2 required for the formation of 1 mole of UF_6).

1.2 Reactions Not Involving the Use of Elemental Fluorine. Two reactions have been discovered which produce UF_6 without requiring free fluorine. The first is the thermal disproportionation of UF_5.

$$3UF_5 \xrightarrow{175°C} U_2F_9 + UF_6 \qquad (16)$$

$$2U_2F_9 \xrightarrow{>200°C(?)} 3UF_4 + UF_6 \qquad (17)$$

Uranium pentafluoride for reaction 16 can be prepared from the readily available pentachloride.

$$UCl_5 + 5HF \rightarrow UF_5 + 5HCl \qquad (18)$$

Ruff and Heinzelmann (1911), who first used this procedure, experienced considerable difficulty in separating the UF_6 from hydrogen fluoride formed from the compound $UF_5 \cdot HF$, which they used. Subsequent work has indicated that UF_6 can be separated from hydrogen fluoride by fractional distillation or, even more simply, by decantation, since UF_6 is only sparingly soluble in liquid anhydrous hydrogen fluoride at low temperatures (SAM Columbia 3; SAM Carbide and Carbon 1). The disadvantage of this process is that only one-third of the uranium is utilized and the remainder must be reworked.

The second method is based on an interesting reaction that was discovered at Chicago (MP Chicago 2). When uranium tetrafluoride is treated with dry oxygen at an elevated temperature (approximately 800°C), the following reaction occurs:*

$$2UF_4 + O_2 \rightarrow UF_6 + UO_2F_2 \tag{19}$$

At Chicago reaction 19 was carried out in a nickel tube. (Calcium fluoride, being more inert, may be even better.) At 800°C a yield of 20 to 40 per cent of the calculated amount of UF_6 was obtained. The comparatively low yield is probably due to the reaction of UF_6 with nickel. Reaction 19 provides a convenient method for laboratory preparation of UF_6 when no fluoride is available.

Attempts were made by Ergen (SAM Carbide and Carbon 2) to prepare UF_6 by reactions avoiding free fluorine. Mercuric fluoride was examined as a fluorinating agent. UF_4 and UCl_5 were made to react with HgF_2 or with Hg_2F_2 and chlorine, and in both cases UF_6 was obtained in small quantity. [The mercuric fluoride, HgF_2, was prepared by reaction of mercurous fluoride (Henne, 1938) with chlorine according to the method of Ruff and Bahlau (1919).] Uranium pentachloride reacts with HgF_2 at "moderate" temperatures to give a white sublimate, presumably UF_6. An intimate mixture of mercurous fluoride and UF_4 when treated with dry chlorine at 400°C in a glass apparatus yielded a small quantity of UF_6. The low yield was attributed to side reactions with the glass. An attempt was also made to obtain UF_6 by passing a mixture of chlorine and dichlorodifluoromethane over a mixture of UO_3 and mercurous fluoride, but the UO_3 was apparently unattacked.

It is clear from these observations that the preparation of UF_6 without the use of free fluorine is quite feasible.

---

*A similar experiment in a Vycor tube had been carried out earlier by workers at Brown University. They observed the formation of a heavy brown deposit and noticed the weight loss of the residue. The possibility that F_2 or UF_6 might be formed was suggested, but no further work appears to have been done at that time (Brown 1).

1.3   Purification and Handling of Uranium Hexafluoride. Since UF_6 is a very reactive substance, handling problems are encountered in both laboratory and large-scale use. Usually metal apparatus is employed, but glass may be used successfully if certain precautions are taken. UF_6 attacks glass quite rapidly unless it is absolutely free from water or hydrogen fluoride. Phosphorus pentoxide has been used to dry UF_6, but this is a troublesome procedure (Grosse, 1932). The attack on glass is probably due to the following cycle:

$$UF_6 + 2H_2O \rightarrow UO_2F_2 + 4HF \tag{20}$$

$$SiO_2 + 4HF \rightarrow SiF_4 + 2H_2O \tag{21}$$

The water consumed in reaction 20 is regenerated in reaction 21, and therefore traces of moisture are capable of destroying large amounts of UF_6. The use of "getters" to stop this cycle was introduced by Grosse (SAM Columbia 3). Sodium or potassium fluoride will combine with all the components of reaction cycle 20 and 21 (giving $KF \cdot HF$, $KF \cdot 2H_2O$, K_2SiF_6, etc.) and thereby stop the cycle. Pure UF_6 does not form a stable addition compound with KF, but UF_6, HF, and KF give stable triple compounds. However, the amount of UF_6 lost in this way corresponds only to the stoichiometric quantity of hydrogen fluoride present, which will in most cases be small. Therefore in order to keep UF_6 in glass it is only necessary to have present some dry powdered potassium or sodium fluoride amounting to about 5 per cent by weight of the UF_6. Since potassium fluoride is very hygroscopic, it should be fused before use; sodium fluoride is not deliquescent and may therefore be preferable to potassium fluoride (British 3). It has been reported that pure UF_6 can be kept indefinitely without decomposition, even in the absence of a "getter," in a glass container that has been vacuum-baked. The UF_6 must first be purified by multiple vacuum distillation through a train of U tubes containing potassium fluoride (SAM Columbia 4). Amphlett, Mullinger, and Thomas (1948) described in some detail laboratory procedures for the purification and manipulation of uranium hexafluoride.

Care must be taken in condensing UF_6 in a glass apparatus. Traps cooled with liquid air are inadvisable since they often crack when warmed. To prevent such breakage, traps should first be cooled with dry ice so that the material deposits as a fluffy snow. If liquid air is to be used, a trap with an internal cold finger is recommended (SAM Columbia 5).

The following sequence of operations has been found to give clean glass resistant to the action of pure UF_6 (British 4): (1) washing with xylene or trichloroethylene, (2) flaming, (3) pretreating for a few hours with UF_6 vapors at room temperature. The most important step in this sequence is pretreatment with UF_6 vapor, which appears to be much more effective than flaming. Since all-glass apparatus is necessary for complete baking out, stopcocks and joints should be avoided by the use of magnetic breaker seals (referred to by British authors as "Briscoe seals"). If glass stopcocks and ground-glass joints are used, Apiezon grease or wax, or preferably fluorinated hydrocarbons, may be used for lubrication, but too much of the lubricant must not be exposed to the UF_6.

Satisfactory materials for metal equipment are copper, nickel, and aluminum. With copper tubing, ordinary refrigeration flare fittings are used for making connections; if properly made, such joints are vacuum-tight. Pressure gauges should be of the Bourdon type, with tubes made of a UF_6-resistant metal.

Hydrogen fluoride has been found by British workers to be a common impurity in UF_6 (British 5). They recommend much larger quantities of potassium fluoride for the removal of hydrogen fluoride, considering about 50 per cent of the weight of UF_6 necessary to remove 2 mole % hydrogen fluoride. These observers found that a constant evaporation mixture of UF_6 and hydrogen fluoride exists (vapor pressure of 17.91 mm Hg at 0°C) and that the hydrogen fluoride content of an impure UF_6 sample can be reduced to the composition of this mixture (about 2 mole % hydrogen fluoride) by pumping at −80°C. This indicates that potassium fluoride purification should not be applied directly to mixtures with a hydrogen fluoride content greater than 2 mole % until the mixtures have been treated by pumping at −80°C.

The presence and amount of hydrogen fluoride in UF_6 can be determined by either vapor pressure or freezing point measurements. Since hydrogen fluoride is only slightly soluble in UF_6, the vapor pressures are additive. The presence of 1 mole % hydrogen fluoride at 0°C adds 12.0 mm Hg to the vapor pressure (SAM Columbia 6). To detect very small amounts of hydrogen fluoride the freezing point depression method is more accurate. The experimentally determined cryoscopic constant of UF_6 is 0.065°C per 0.01 wt. % hydrogen fluoride. The theoretical value, computed from Raoult's law, is 0.0839°C (SAM Columbia 7).

British workers reexamined various methods proposed for the determination of hydrogen fluoride in UF_6. They concluded that the most satisfactory method is one in which the uranium hexafluoride–hydrogen fluoride mixture is refluxed under such conditions that the hydrogen fluoride remains in the vapor phase. The equilibrium volume can be measured to within 0.01 mole % hydrogen fluoride (British 6).

2. PHYSICAL PROPERTIES OF URANIUM HEXAFLUORIDE

At room temperature uranium hexafluoride is a colorless volatile solid which forms transparent crystals of high refractive index. The crystals sublime without melting under atmospheric pressure; at higher pressures they melt to form a clear, colorless, mobile liquid of high density. Despite the high molecular weight of UF_6, the properties of its vapor closely approximate those of an ideal gas.

The physical properties of UF_6 in its various states will be considered under the following headings:

Phase Relations: Melting point, vapor pressure of solid and liquid, boiling point, triple point, and critical constants.

Thermal Properties: Specific heat, entropy and enthalpy, velocity of sound, heats of sublimation, evaporation and fusion, and thermal conductivity.

Mechanical Properties: Density, kinetic properties of the vapor, viscosity, and surface tension.

Optical Properties: Refractive index, absorption spectrum, and Raman spectrum.

Electrical and Magnetic Properties: Dielectric constant, magnetic susceptibility, and ionization potential.

Structure of the Molecule: Derived from electron diffraction, x-ray crystallography, and dipole-moment measurement.

The graphs included in this chapter are taken from the summary report of Kirshenbaum (SAM Columbia 8), which was extensively used by the writers.

2.1 Phase Relations. (a) Melting Point. The earliest value of the melting point (69.2 to 69.5°C), given by Ruff and Heinzelmann (1911) and quoted by Grosse (SAM Columbia 3), has since been shown to be considerably in error. Brickwedde, Hoge, and Scott (1948; Natl. Bur. Standards 2) found a value of 64.052°C (337.212°K) for material containing 2×10^{-2} mole % impurities, and British workers gave 64.5 ± 0.3°C and 64.8 ± 0.4°C for two samples containing less than 1.5×10^{-2} mole % impurities (British 7). From the knowledge of the specific volumes of the liquid and of the solid, V_L and V_S, at the triple point (see Secs. 2.3a and b), the change in melting point with pressure can be calculated.

$$\frac{dT}{dp} = \frac{T(V_L - V_S)}{\Delta H \text{ (fusion)}} = 0.00505°C/atm \qquad (22)$$

(b) Vapor Pressure of Solid Uranium Hexafluoride. The early vapor pressure values of Ruff and Heinzelmann (1911) were rather erratic, probably owing to contamination by hydrogen fluoride or silicon tetrafluoride. Several other vapor pressure determinations have been made by British workers and by workers at Columbia University. On the whole, taking into account the difficulty of the meas-

Table 13.1 — Vapor Pressures of Solid Uranium Hexafluoride

Temp., °C	Pressure, mm Hg					Temp., °C	Pressure, mm Hg			
	a	b	c	d	e		b	c	d	e
−14.2	6.85		...	...	...	30.4	158			...
−12.4	7.80		...	...	...	31.2		...	178.4	...
−1.5	17.80		...	...	...	32.7		...	199.0	...
0.0	...		...	...	16.9	34.9		...	224.3	...
0.1	20.10		...	...	...	35.3	216.8	...	...	
1.7	23.30		...	...	...	35.5	...	...	...	216.8
3.6	26.60		...	...	...	36.8	...	...	259.5	...
5.5	30.40		...	...	...	38.3	...	...	282.8	...
6.8	32.90		...	...	...	39.7	...	...	310.5	...
8.5	37.00		...	...	...	40.0	...	...	...	295.4
11.3	44.00		...	...	...	40.5	...	...	316.5	...
12.5		45.3	...	...	...	41.7	...	...	348.9	...
14.3		52.0	...	...	...	44.7	...	...	...	...
14.5		52.5	...	...	...	45.0	...	...	...	395.8
14.7		53.6	...	...	...	45.2	...	...	418.8	...
15.0		54.4	...	...	...	47.2	...	...	484.1	...
17.6		66.1	...	...	...	49.6	...	...	549.1	...
20.4		80.8	...	...	...	50.0	...	...	...	522.1
21.0		...	83	...	...	50.2	521.0	...	559.2	...
21.6		...	...	...	88.3	51.6	...	...	605.0	...
22.7		...	93	...	...	51.8	...	...	612.5	...
24.9		...	108	...	...	53.9	...	...	685.5	...
25.0		...	...	119.5	...	55.0	...	...	...	697.2
25.1		113.5	...	...	...	55.7	...	...	751.2	...
26.4		...	123	...	...	60.0	...	...		910.0
28.2		...	138	...	...	63.1	...	...	...	1,072
29.0		...	...	153.3	...					
29.5		149.5	...		...					

[a]British 11.
[b]British 8; Amphlett, Mullinger, and Thomas (1948).
[c]British 12.
[d]British 13.
[e]SAM Columbia 9; Weinstock and Crist (1948).

urements, the results can be said to agree satisfactorily. It is difficult to prepare UF_6 completely free of hydrogen fluoride, and this tends to give too high values for the vapor pressure. There is also the difficulty of assuring complete thermal equilibrium between the thermostat and a loose mass of solid UF_6. Finally, the corrosion of construction materials poses a serious problem. American workers have generally used copper apparatus, whereas the British seem to have preferred glass.

At Columbia University a brass sylphon bellows was used to measure pressure by a null method (SAM Columbia 9; Weinstock and Crist, 1948). Pressures above atmospheric were read on a multiple mercury manometer, using dibutyl phthalate as piston liquid. The error in reading the manometer was about 2 mm Hg in the higher pressure range and less than 0.5 mm Hg in the lower range.

Table 13.1 contains a summary of all the directly measured vapor pressures of solid UF_6.

The following linear relations between log p and $1/T$ have been suggested for the representation of the vapor pressure curve over a limited temperature range:

$$\log p_{mm\ Hg} = 10.74 - \frac{2592}{T} \quad (12.5 - 50.2\,°C) \tag{23}$$

(British 8; Amphlett, Mullinger, and Thomas, 1948)

$$\log p_{mm\ Hg} = 21.87103 - \frac{3123.479}{T} - 3.77962 \log T \quad (0 - 63.1\,°C) \tag{24}$$

(Weinstock and Crist, 1948)

$$\log p_{mm\ Hg} = 11.19 - \frac{2714}{T} \quad (\text{low temperatures}) \tag{25}$$

(British 9)

Two more elaborate relations, intended to represent precisely the vapor pressure curve in a wide temperature region, have been derived. The equation

$$\log p_{mm\ Hg} = -\frac{2751}{T} - 75.0\ e^{-2560/T} - 1.01 \log T + 13.797 \tag{26}$$

represents all British measurements in the range 0 to 65°C with a precision of ±0.5 per cent (British 10); Columbia measurements agree within 1.5 per cent, but all American points lie a little above the curve defined by Eq. 26.

At Columbia (SAM Columbia 8) another equation based on all available data was derived. All experimental points were plotted on a large scale, and a curve was fitted to these points by a series of analytical and graphical approximations, the higher ones involving the application of heat capacity and enthalpy equations for the solid. The resulting equation is said to represent the experimental results to ±0.5 per cent and perhaps even better (see Fig. 13.2).

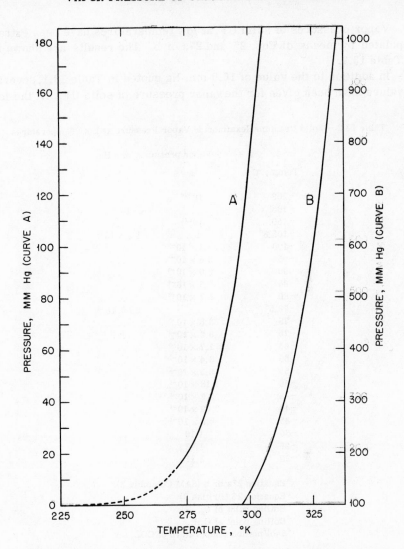

Fig. 13.2—Vapor pressure of UF_6 (solid) (see Eq. 27a or b).

$$\ln p_{mm\,Hg} = -132.8690 - \frac{344,490}{T^2} + \frac{707.31}{T} + 26.436 \ln T$$
$$- 0.038674T \quad (27a)$$

or, in Briggsian logarithms,

$$\log p_{mm\,Hg} = -57.7043 - \frac{149,610}{T} + \frac{307.18}{T} + 26.436 \log T$$
$$- 0.016796T \quad (27b)$$

Vapor pressures of solid UF_6 at low temperatures have been extrapolated by means of Eqs. 25 and 27a or b. The results are shown in Table 13.2.

In addition to the value of 16.9 mm Hg quoted in Table 13.1, several values have been given for the vapor pressure of solid UF_6. At the ice

Table 13.2 — Solid Uranium Hexafluoride Vapor Pressure at Low Temperatures

| Temp., °C | Vapor pressure, mm Hg | |
	a	b
−200	10^{-26}	...
−183[c]	...	10^{-19}
−150	10^{-11}	...
−103.8[d]	...	1.3×10^{-5}
−100	4.3×10^{-5}	...
−95	8.6×10^{-5}	...
−90	3.0×10^{-4}	...
−85	7.3×10^{-4}	...
−80	1.7×10^{-3}	...
−78.5[e]	...	1.7×10^{-3}
−75	3.8×10^{-3}	...
−70	8.2×10^{-3}	...
−65	1.7×10^{-3}	...
−60	3.4×10^{-2}	...
−55	6.5×10^{-2}	...
−50	1.18×10^{-1}	...
−45	2.2×10^{-1}	...
−40	3.69×10^{-1}	...
−35	6.9×10^{-1}	...
−30	1.16	...
−25	1.92	...
−20	3.11	...

[a] Equation 27a or b (SAM Columbia 8).
[b] Equation 25 (British 9).
[c] Boiling point of O_2.
[d] Boiling point of C_2H_4.
[e] Sublimation temperature of CO_2.

point (0°C) Columbia workers found a value of 18.0 mm Hg (SAM Columbia 6); however, British workers gave as a result of a very careful measurement $p_{mm\ Hg}$ (0°C) = 17.54 ± 0.02 (British 5). The latter value probably is more reliable, since it has been found by British workers that a constant evaporation mixture of UF_6 and HF exists, with a vapor pressure at 0°C of 17.91 ± 0.02 mm Hg.

(c) Vapor Pressure of Liquid Uranium Hexafluoride. Values for the liquid-vapor equilibrium of UF_6 are more accurate than those for the solid-vapor equilibrium, perhaps because of the greater ease with which thermal equilibrium can be achieved. Columbia workers, using the same apparatus as for the determination of the solid-vapor equilibrium, found the values listed in Table 13.3 (SAM Columbia 9). This table also includes results of the British observers (British 5,10).

Table 13.3 — Vapor Pressure of Liquid Uranium Hexafluoride

Temp., °C	Pressure, mm Hg			Difference between calculated and observed values
	a	b	c	
65.0	1,175	1,169	1,167	−2
67.9	...	1,273	1,274	+1
70.0	1,385	...	...	...
70.1	...	1,360	1,361	+1
70.2	...	1,366, 1,370, 1,376	1,365	−1, −5, −11
75.0	1,620	...	...	...
75.2	...	1,582	1,577	−5
75.3	...	1,568	1,582	+14
80.0	1,870	...	...	...
80.3	...	1,830, 1,838	1,820	−10, −18
85.0	2,165	...	...	...
85.4	...	2,087	2,089	+2
90.0	2,455	...	...	...
95.0	2,765	...	...	...

[a]Experimental values (British 5,10).
[b]Experimental values (SAM Columbia 9; Weinstock and Crist, 1948).
[c]Calculated from Eq. 30.

Equations 28 and 29 were given by British observers, and Eq. 30 was derived by Weinstock and Crist (1948).

$$\log p_{mm\ Hg} = \frac{2170}{T} + 56.0 \log T - 0.0394T - 131.64 \qquad (28)$$

$$\log p_{mm\ Hg} = 7.73 - \frac{1575}{T} \qquad (29)$$

$$\log p_{mm\ Hg} = 18.60033 - \frac{2065.679}{T} - 3.72662 \log T \qquad (30)$$

Equation 30 was found to reproduce the results to better than ±1 per cent.

A precision formula for the liquid-gas equilibrium was obtained by Kirshenbaum by the same method as that given in Eqs. 27a and b for the solid-gas equilibrium, i.e., by fitting a curve to a large-scale graph of the available data by successive approximations (see Fig. 13.3).

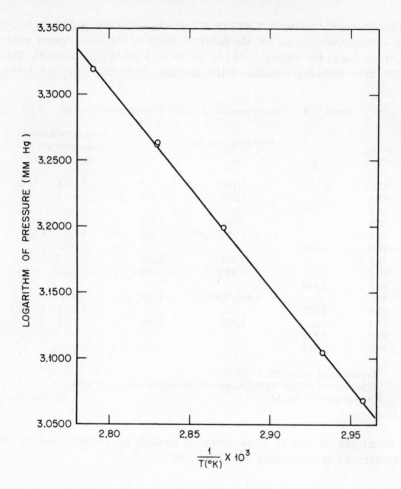

Fig. 13.3—Vapor pressure of UF₆ (liq) (see Eq. 31a or b).

$$\ln p_{mm\ Hg} = -24.6602 - \frac{248,608}{T^2} - \frac{1245.2}{T} + 7.2876 \ln T$$
$$- 0.014371T \quad (31a)$$

or, in Briggsian logarithms,

$$\log p_{mm\ Hg} = -107,098 - \frac{107,969}{T^2} - \frac{540.8}{T} + 7.2876 \log T$$
$$- 0.006241T \quad (31b)$$

This equation reproduces the experimental results in the range 65 to 90°C to better than ±0.4 per cent.

(d) <u>Boiling Point</u>. The temperature at which the sublimation pressure of UF$_6$ reaches 760 mm Hg was reported by Ruff and Heinzelmann (1911) as 56°C. More recently Columbia observers found 56.2°C (SAM Columbia 3), and the British investigators found 55.3 and 56.7°C, with the higher value considered more nearly correct (British 8; Amphlett, Mullinger, and Thomas, 1948). The value 56.5°C was found by Weinstock and Crist (1948). Thus 56.5°C appears to be a reasonable mean value for the "boiling point" of UF$_6$. The triple point of UF$_6$ was found by Brickwedde, Hoge, and Scott (1948) to be

$$t_{tr} = 64.052°C$$

$$p_{tr} = 1,134 \text{ mm Hg}$$

Weinstock and Crist (1948) report the vapor pressure at the triple point to be 1,133 ± 7 mm Hg.

(e) <u>Critical Temperature, Pressure, and Density</u>. British workers reported a value of 245 ± 5°C for the critical temperature of UF$_6$ (British 14). Columbia workers found for this constant a value of 232°C (SAM Columbia 8), and the National Bureau of Standards (Natl. Bur. Standards 3) estimated from the graph of densities of the liquid vs. temperature that the critical temperature must be about 217°C. Another approximate value of the critical temperature (243°C) was calculated from the surface tension, γ, by means of the relation

$$\gamma = \frac{2.12 (t_c - 6 - t)}{(MV_L)^{2/3}} \tag{32}$$

$\gamma = 15.6$ dynes at 80°C, d = 3.57 g/cc at 80°C.

Finally, from boiling point data and the relation

$$\frac{T_b}{T_c} = 0.63$$

the following is obtained:

$$T_c = 249°C$$

British observers estimated the critical pressure as 63 atm (British 15); the Columbia group as about 50 atm (SAM Columbia 10); and

the National Bureau of Standards (Natl. Bur. Standards 3) as 44 atm.
The critical density has been estimated at 1.9 g/cc (British 15) and
1.39 g/cc (Natl. Bur. Standards 3). The latter value is probably
more reliable. Obviously, further work on the critical constants is
required.

2.2 Thermal Properties. (a) Heat and Free Energy of Formation
of Uranium Hexafluoride. The heat of hydrolysis of UF_6 in water was
found calorimetrically to be -50.5 kcal per mole at $298.16°K$ (SAM
Columbia 11).

$$UF_6 \text{ (crystal)} + 2H_2O \text{ (liq)} \rightarrow UO_2^{++} + 2F^- + 4HF \text{ (aq)} \qquad (33)$$

$$\Delta H = -50.5 \pm 1.7 \text{ kcal per mole}$$

Using the value

$$\Delta H \text{ (subl)} = -11.2 \pm 1 \text{ kcal per mole}$$

for the heat of sublimation of UF_6 (SAM Columbia 9) and combining
these results with the following data from the literature,

$$UO_2^{++} + 3H_2 \rightarrow 2H^+ + 2H_2O + U \qquad \Delta H = 106.3 \text{ kcal}$$

$$2H^+ + 2F^- \rightarrow 2HF \text{ (aq)} \qquad \Delta H = 4.0 \text{ kcal}$$

$$6HF \text{ (aq)} \rightarrow 3H_2 \text{ (gas)} + 3F_2 \text{ (gas)} \qquad \Delta H = 456 \text{ kcal}$$

the heat of formation of UF_6 given in Eq. 34 is found.

$$U \text{ (solid)} + 3F_2 \text{ (gas)} \rightarrow UF_6 \text{ (gas)} \qquad \Delta H = -505 \pm 3 \text{ kcal per mole} \qquad (34)$$

Using the entropy values $S_{298°K} = 90.3 \pm 0.1$ e.u. for UF_6 and $S_{298°K} =$
12.0 e.u. for uranium metal, the following is calculated for the free
energy of formation of UF_6:

$$\Delta F_{298°K} = -485 \text{ kcal per mole}$$

(b) Heat Capacity, Entropy, and Enthalpy of Uranium Hexafluoride.
The heat capacities of solid and liquid UF_6 in equilibrium with the
vapor have been very accurately determined at the National Bureau
of Standards (Natl. Bur. Standards 2; Brickwedde, Hoge, and Scott,
1948). Entropies and enthalpies calculated from these data are given
in Table 13.4. The values are probably accurate to within ±1 per
cent.

The following empirical equations represent the results given in
Table 13.4 (SAM Columbia 8):

Solid uranium hexafluoride:

$$H - H_0 = 9865.0 - 20.082T + 0.080790T^2 - 1,047,920T^{-1} \qquad (35)$$

(see Fig. 13.4)

(Precision: ± 0.01 per cent from triple point to $265°K$; ± 1 per cent from 265 to $225°K$)

$$S - S_0 = 126.59 - 20.082 \ln T + 0.16162T - 523,960T^{-2} \qquad (36)$$

(see Fig. 13.5)

(Precision: ± 0.01 per cent from triple point to $273.16°K$)

Liquid uranium hexafluoride:

$$H - H_0 = 5986.6 + 17.954T + 0.032514T^2 - 666,990T^{-1} \qquad (37)$$

(Precision: ± 0.01 per cent)

$$S - S_0 = -50.33 + 17.954 \ln T + 0.065028T - 333,490T^{-2} \qquad (38)$$

(Precision: ± 0.01 per cent)

The errors given are the maximum deviation of the experimental values from those calculated from the equations.

British workers have determined a set of values for the specific heat of solid uranium hexafluoride (British 16,17), but these values probably are less precise than those given in Table 13.5 for liquid uranium hexafluoride. They also presented some values for the specific heat of liquid uranium hexafluoride (British 17) which are somewhat higher than the American values (e.g., $C_p = 50$ cal/mole/$°C$ at $353°K$) and which can be represented by the formula

$$C_p = 47.3 + 0.347(t - 70) \text{ cal/mole/}°C \qquad (39)$$

where t is in degrees centigrade. Additional British data for liquid uranium hexafluoride, quoted by Kirshenbaum (SAM Columbia 12), are given in Table 13.5.

The specific heat and entropy of gaseous uranium hexafluoride have been calculated from infrared data (see Sec. 2.4d), assuming a regular octahedral structure with the six uranium-fluorine distances as 2.0 A (Bigeleisen, Mayer, Stevenson, and Turkevich, 1948). The specific heats so calculated are given in Table 13.6, and the entropies in Table 13.7. An empirical equation has been derived from the ther-

Table 13.4—Heat Capacity, Entropy, and Enthalpy of Uranium Hexafluoride

Temp., °K	C_p, cal/mole/°C	$H - H_0$, cal/mole	$S - S_0$, cal/mole/°C	$T(S - S_0)$, cal/mole
		Solid		
0	0	0	0	0
20	4.06	28.0	1.96	39
25	5.53	52.0	3.04	76
30	6.92	83.2	4.16	125
35	8.27	121.1	5.33	187
40	9.64	165.9	6.53	261
45	11.00	217.5	7.72	347
50	12.34	275.9	8.96	448
55	13.66	340.9	10.21	562
60	14.90	412.4	11.45	687
65	16.07	489.8	12.69	825
70	17.16	572.8	13.91	974
75	18.22	661.4	15.13	1,135
80	19.20	754.9	16.35	1,308
85	20.14	853.9	17.55	1,492
90	20.99	956.2	18.72	1,685
95	21.76	1,063	19.86	1,887
100	22.49	1,174	20.99	2,099
105	23.17	1,288	22.11	2,322
110	23.82	1,405	23.21	2,553
115	24.46	1,526	24.29	2,793
120	25.08	1,650	25.34	3,041
125	25.66	1,777	26.37	3,296
130	26.23	1,907	27.39	3,561
135	26.77	2,039	28.40	3,834
140	27.30	2,174	29.38	4,113
145	27.80	2,312	30.33	4,398
150	28.30	2,452	31.29	4,694
155	28.78	2,595	32.22	4,994
160	29.24	2,740	33.16	5,306
165	29.71	2,887	34.04	5,617
170	30.15	3,037	34.95	5,942
175	30.60	3,189	35.83	6,270
180	31.03	3,343	36.69	6,604
185	31.42	3,499	37.55	6,947
190	31.87	3,657	38.39	7,294
195	32.27	3,818	39.23	7,650
200	32.65	3,980	40.04	8,008
205	33.05	4,144	40.85	8,374
210	33.46	4,310	41.67	8,751
215	33.78	4,479	42.45	9,127
220	34.22	4,648	43.24	9,513
225	34.60	4,820	44.01	9,902
230	34.96	4,994	44.77	10,296
235	35.34	5,170	45.51	

Table 13.4 — (Continued)

Temp., °K	C_p, cal/mole/°C	$H-H_0$, cal/mole	$S-S_0$, cal/mole/°C	$T(S-S_0)$, cal/mole
		Solid		
240	35.70	5,348	46.28	11,107
245	36.07	5,527	47.02	11,520
250	36.43	5,708	47.74	11,935
255	36.78	5,892	48.48	12,362
260	37.14	6,067	49.20	12,792
265	37.50	6,263	49.89	13,221
270	37.86	6,451	50.61	13,665
273.16	38.08	6,571	51.05	13,942
275	38.22	6,640	51.30	14,108
280	38.56	6,833	51.99	14,557
285	38.90	7,027	52.69	15,017
290	39.24	7,223	53.36	15,474
295	39.61	7,420	54.02	15,936
298.16	39.86	7,545	54.45	16,235
300	40.00	7,619	54.69	16,407
305	40.42	7,820	55.36	16,885
310	40.87	8,023	56.03	17,369
315	41.33	8,229	56.68	17,854
320	41.80	8,437	57.35	18,352
325	42.27	8,648	57.99	18,847
330	42.77	8,860	58.64	19,351
335	43.27	9,076	59.28	19,859
337.212	43.49	9,172	59.57	20,088
		Liquid		
337.212	45.59	13,760	73.17	24,674
340	45.68	13,888	73.55	25,007
345	45.84	14,118	74.22	25,606
350	46.02	14,348	74.89	26,212
355	46.20	14,579	75.54	26,817
360	46.33	14,811	76.18	27,425
365	46.48	15,044	76.83	28,043
370	46.62	15,278	77.45	28,657

modynamic functions in Eqs. 35 to 38 and the vapor pressures of Eqs. 27a or b and 31a or b:

$$C_p \text{ [for } UF_6 \text{ (gas)]} = 32.43 + 0.007936T$$
$$- 320,680T^{-2} \quad \text{cal/mole/°C} \quad (40)$$

(see Fig. 13.6)

The C_p values derived from this equation agree well with the spectroscopic results. British calorimetric determinations (British 7,16, 17) appear to be considerably higher; however, sufficient details are

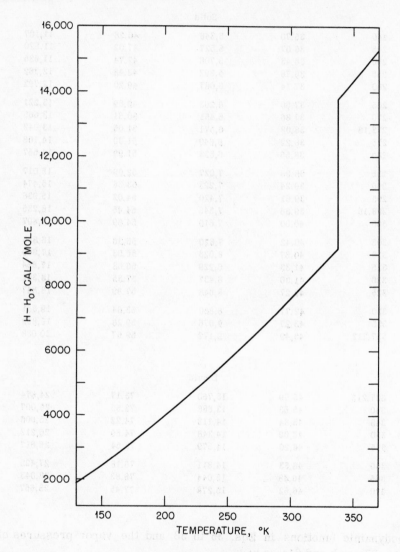

Fig. 13.4 — Enthalpy of UF$_6$ (solid) (see Eq. 35).

not available to resolve this discrepancy, and either the spectroscopic values or those derived from Eq. 40 must be presumed to be more reliable.

The heat content and the entropy of uranium hexafluoride gas have also been derived both by combination of the thermodynamic functions

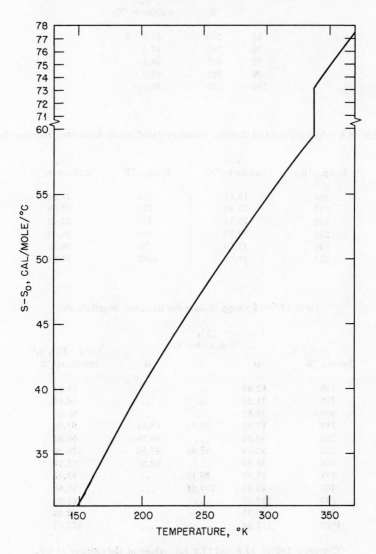

Fig. 13.5 — Entropy of UF$_6$ (solid) (see Eq. 36).

of the condensate with the vapor pressure equations (SAM Columbia 8) and by theoretical considerations from the spectroscopic frequencies. The first method leads to Eqs. 41 and 42.

Table 13.5 — Specific Heat of Liquid Uranium Hexafluoride

Temperature		C_p,
°C	°K	cal/mole/°C
65	338	47.0 ± 1
70	343	47.4
80	353	48.1
90	363	48.8
100	373	49.5

Table 13.6 — Specific Heat of Gaseous Uranium Hexafluoride from Spectroscopic Data

Temp., °K	C_p, cal/mole/°C	Temp., °K	C_p, cal/mole/°C
100	18.93	348	32.38
150	23.45	373	32.93
200	26.74	400	33.46
273	30.13	500	34.80
298	31.00	750	36.35
323	31.75	1000	36.94

Table 13.7 — Entropy of Gaseous Uranium Hexafluoride[a]

Temp., °K	$S°$, cal/mole/°C			$-(F_0° - E_0°)/T$,[b] cal/mole/°C
	b	c	d	
100	62.99	...	...	50.91
150	71.58	...	...	56.41
200	78.81	...	...	61.14
273	87.65	82.27	88.21	67.08
298	90.34	...	90.76	68.92
323	92.89	93.59	92.96	70.69
348	95.22	...	94.97	72.33
373	97.47	98.46	...	73.93
400	99.86	100.46	...	75.65
500	107.45	...	...	81.25
750	122.37	...	...	92.99
1000	132.58	...	...	101.37

[a]Compare Tables 13.8 and 13.9 for values of the entropy of UF_6 (gas) based on the heat of sublimation of UF_6. The values so derived by Masi (1949) agree very well with the spectroscopic values given here in column b.

[b]Bigeleisen, Mayer, Stevenson, and Turkevich (1948).

[c]Values determined from Eq. 40.

[d]Calculated from vapor pressure measurements of Weinstock and Crist (1948).

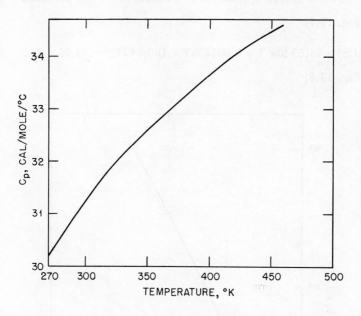

Fig. 13.6—Heat capacity of UF_6 (gas) at constant pressure (see Eq. 40).

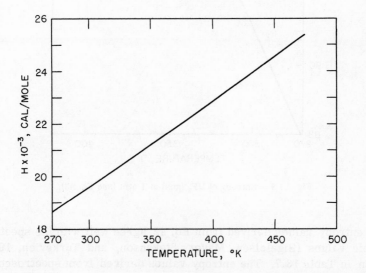

Fig. 13.7—Enthalpy of UF_6 (gas) (see Eq. 41).

$$H = 8460 + 32.43T + 0.003968T^2 + 320{,}680T^{-2} \quad \text{cal per mole} \qquad (41)$$

(see Fig. 13.7)

$$S \text{ (1 atm)} = 74.69 \log T + 0.007935T + 160{,}340T^{-2} - 98.05 \quad \text{e.u.} \qquad (42)$$

(see Fig. 13.8)

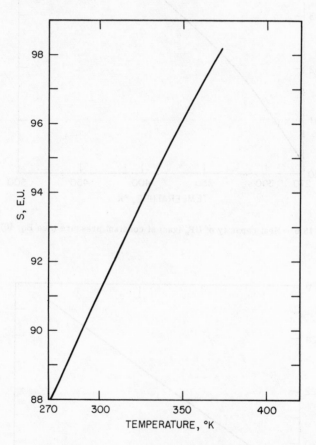

Fig. 13.8—Entropy of UF_6 (gas) at 1 atm (see Eq. 42).

The entropy values derived from Eq. 42 agree well with the spectroscopic values (Bigeleisen, Mayer, Stevenson, and Turkevich, 1948) given in Table 13.7. The entropy values derived from spectroscopic data, vapor pressure data (Weinstock and Crist, 1948), and calorimetric measurements of the heats of vaporization of UF_6 (Masi, 1949) are in good agreement.

(c) Velocity of Sound in Gaseous Uranium Hexafluoride and the Ratio C_p/C_v. The velocity of sound in gaseous uranium hexafluoride has been measured by two British observers. One observer reported a mean value of 8,870 cm/sec at 49°C and pressures of 320 and 539 mm Hg (British 18), and the other found the velocity to be 8,990 cm/sec at 49.2°C and 420 mm Hg pressure (British 8).

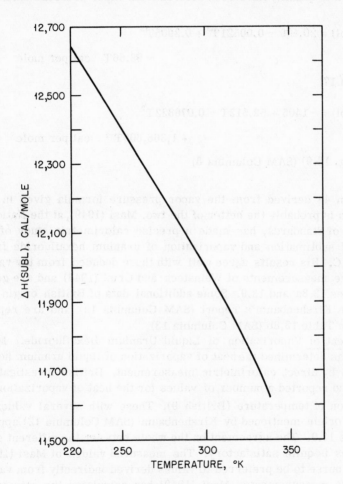

Fig. 13.9—Heat of sublimation of UF_6 (solid) (see Eq. 44).

Assuming deviations from the gas laws to be negligible and using the value v = 8,990 cm/sec, it is found that

$$C_p/C_v = 1.063 \pm 0.005$$

at 49.2°C. This agrees well with the value (1.066) that is obtained from Eq. 40, assuming the perfect gas laws hold.

(d) Heat of Sublimation of Solid Uranium Hexafluoride. A value of 10.36 cal per mole for the heat of sublimation of uranium hexafluoride .was reported by Ruff and Heinzelmann (1911). Two empirical formulas for this quantity have been derived from vapor pressure data.

$$\Delta H \text{ (subl)} = 20{,}400 - 0.00521 T^3 + 0.3905 T^2$$

$$- 98.66T \quad \text{cal per mole} \quad (43)$$

(British 17)

$$\Delta H \text{ (subl)} = -1405 + 52.512T - 0.076822 T^2$$

$$+ 1{,}368{,}600 T^{-1} \quad \text{cal per mole} \quad (44)$$

(see Fig. 13.9) (SAM Columbia 8)

Equation 44 derived from the vapor pressure formula given in Eq. 27a or b is probably the better of the two. Masi (1949), at the National Bureau of Standards, has made a precise calorimetric study of the heats of sublimation and vaporization of uranium hexafluoride from 4 to 90°C. His results agree well with those deduced from the vapor pressure measurements of Weinstock and Crist (1948) and are given in Tables 13.8a and 13.9. Some additional data of British origin are given in Kirshenbaum's report (SAM Columbia 12) and are reproduced in Table 13.8b (SAM Columbia 13).

(e) Heat of Vaporization of Liquid Uranium Hexafluoride. Masi (1949) has determined the heat of vaporization of liquid uranium hexafluoride by direct calorimetric measurement. British investigators have also reported a number of values for the heat of vaporization as a function of temperature (British 9). These with several values of British origin mentioned by Kirshenbaum (SAM Columbia 12) appear in Table 13.9. The agreement on the whole between the different sets of values is quite satisfactory. The measured values of Masi (1949) are of course to be preferred to values derived indirectly from vapor pressure measurements. Masi (1949) has calculated the entropy of UF_6 ideal gas (Tables 13.8a and 13.9) from his data, and these agree well with the spectroscopic values given in Table 13.7. The following equation for the heat of vaporization of liquid uranium hexafluoride .was derived by Kirshenbaum (SAM Columbia 8) from the vapor pressure equation (Eq. 31a or b):

$$\Delta H \text{ (vap)} = 2473.4 + 14.476T - 0.028546T^2$$

$$+ 987,670T^{-1} \quad \text{cal per mole} \quad (45)$$

(see Fig. 13.10)

Table 13.8a—Heat and Entropy of Sublimation of Solid Uranium Hexafluoride[*]

Temp., °K	l, cal/mole[†]	l/T, cal/mole/°C	S°,[‡] cal/mole/°C
273.16	12,023 (12,220)	44.02	87.56
280	11,988	42.81	88.39
290	11,929	41.13	89.57
298.16	11,872 (11,970)	39.82	90.50
300	11,858	39.53	90.70
310	11,772	37.97	91.80
320	11,666	36.46	92.86
330	11,537	34.96	93.87
337.21	11,429	33.89	94.59

[*]After Masi (1949).
[†]Values in parentheses are from Weinstock and Crist (1948).
[‡]Entropy of ideal gas.

Table 13.8b—Heat of Sublimation of Uranium Hexafluoride[*]

Temp., °C	ΔH (sublimation), kcal/mole
−75	12.20 ± 0.10
−50	12.15 ± 0.10
−25	12.08 ± 0.10
0	11.98 ± 0.10
20	11.87 ± 0.10
40	11.73 ± 0.10
60	11.53 ± 0.10
65	11.47 ± 0.10

[*]After British measurements.

(f) Heat of Fusion of Uranium Hexafluoride. The heat of fusion can be determined by direct calorimetric measurements or calculated as the difference between the heat of sublimation of the solid and the heat of evaporation of the liquid. Table 13.10 gives a summary of the results obtained in these two ways.

(g) Thermal Conductivity of Gaseous and Liquid Uranium Hexafluoride. A direct measurement of the thermal conductivity by the "hot wire" method has been made at the Carbide and Carbon laboratory (SAM Carbide and Carbon 3). The values were measured relative to argon and nitrogen, but in corresponding work carried out by the British the reference gas is not named. This may account for

Table 13.9—Heat of Vaporization of Liquid Uranium Hexafluoride

Temp., °K	l, kcal/mole		
	a	b	c
337.21	6.859	...	...
338	...	6.95	6.92 ± 0.1
340	6.817	...	...
343	...	6.89	6.89 ± 0.1
348	...	6.82[d]	...
350	6.671	...	...
353	...	6.74	...
358	...	6.66	...
360	6.533	...	...
363	...	6.57	6.67 ± 0.1
368	...	6.46	6.31 ± 0.1
370	6.404	...	...
403	...	...	5.80 ± 0.1

[a]Masi (1949); the entropy of the ideal gas calculated from the data in this column is 94.59, 94.88, 95.90, 96.91, and 97.95 e.u. at the respective temperatures.
[b]British 9.
[c]British results quoted by Kirshenbaum (SAM Columbia 12).
[d]Weinstock and Crist (1948) give a value of 6,487 cal/mole at this temperature.

the discrepancy between the British and American work. In Table 13.11 the thermal conductivity of UF_6 (gas) at a number of temperatures is given. Included for comparison are values calculated from viscosity measurement by the Eucken equation.

$$K = \frac{1}{4} (9\gamma - 5) \frac{\eta C_v}{M} \tag{46}$$

where γ = ratio of the specific heats at constant volume and pressure, η = viscosity, and the other symbols have their usual meaning. As shown in Table 13.11, the American values obtained directly and by calculation from viscosity data are in excellent agreement. The four sets of data may be summarized analytically.

American:	$K = 1.450 + 0.0067_5 t$	$(10^5$ cal/cm sec °C$)$	(47)
British (old):	$K = 1.542 + 0.0059 t$	$(10^5$ cal/cm sec °C$)$	(48)
British (new):	$K = 1.39 + 0.0052 t$	$(10^5$ cal/cm sec °C$)$	(49)
From viscosity:	$K = 1.454 + 0.0067 t$	$(10^5$ cal/cm sec °C$)$	(50)

These may be averaged to give Eq. 51, which probably represents the best available data for the thermal conductivity of the gas.

$$\overline{K} = 1.459 + 0.00614t \qquad (10^5 \text{ cal/cm sec °C}) \qquad (51)$$

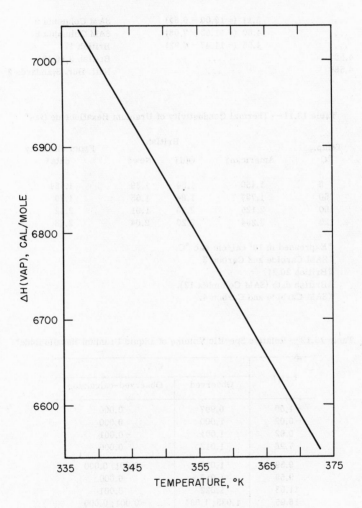

Fig. 13.10—Heat of vaporization of UF_6 (liq) (see Eq. 45).

One value for the thermal conductivity of UF_6 (liq) has been given by Columbia workers (SAM Columbia 15). This is a preliminary value of 3.83 ± 3 per cent $\times 10^{-4}$ cal/cm sec °C at 72°C. No values for solid UF_6 are available.

Table 13.10—Heat of Fusion of Uranium Hexafluoride

ΔH (fusion), kcal/mole

Calorimetric determination	ΔH (fusion) = ΔH (sublimation) −ΔH (vaporization)	Reference
...	5.11 (= 12.00 − 6.89)	SAM Columbia 9
...	4.60 (= 11.63 − 7.03)	SAM Columbia 8
...	4.55 (= 11.47 − 6.92)	British 17
4.58	...	British 9,16,19
4.588	...	Natl. Bur. Standards 2

Table 13.11—Thermal Conductivity of Uranium Hexafluoride Gas*

Temp., °C	American[†]	British Old[‡]	British New[§]	From viscosity data[¶]
0	1.450	1.54	1.39	1.454
50	1.787	1.84	1.65	1.79
100	2.125	2.13	1.91	2.12
125	2.294	2.26	2.04	2.28

*Expressed in 10^5 cal/cm sec °C.
[†]SAM Carbide and Carbon 3.
[‡]British 20,21.
[§]British data (SAM Columbia 12).
[¶]SAM Carbide and Carbon 4.

Table 13.12—Relative Specific Volume of Liquid Uranium Hexafluoride*

$t-t_f$[†]	V/V_f Observed	V/V_f Observed-calculated
−1.60	0.997	0.000
−0.02	1.000	0.000
0.92	1.001	−0.001
7.26	1.013	0.000
9.59	1.076	−0.001; 0.000
9.59	1.017	0.000
11.93	1.022	0.001
18.95	1.033; 1.034	−0.001; 0.000
21.29	1.038	0.000
25.98	1.048	0.001
28.33	1.053; 1.051	0.001; 0.001

*Calculated from Eq. 52.
[†]t_f = 64.052°C.

2.3 <u>Mechanical Properties</u>. (a) Density of Solid UF_6. The density of UF_6 (solid) was reported by Ruff and Heinzelmann (1911) to be 4.68 g/cc at 20.7°C. From x-ray diffraction data, the Cornell group

Table 13.13 — Density of Liquid Uranium Hexafluoride

Temp., °C	Density, g/cc			
	I	II	III*	IV
65	3.624	3.667	3.624	...
68.66	...	...	(3.604)	...
70	3.593	3.636	3.595	...
72.91	...	...	(3.576)	...
75	3.561	3.604	3.565	...
77	...	...	...	3.63 ± 0.18
80	3.530	3.572	3.532	...
85	3.498	3.540	3.502	...
90	3.464	3.508	3.470	...
92.65	...	...	(3.452)	...
95	...	...	3.437	...
100	...	...	3.404	...
112.67	...	...	(3.316)	...
120	...	...	3.263	...
131.12	...	...	(3.183)	...
140	...	...	3.111	...
156	...	...	...	3.11 ± 0.16
160	...	...	2.948	...
162.59	...	...	(2.920)	...
203	...	...	...	2.62 ± 0.13
215	...	...	...	2.32 ± 0.12
225	...	...	...	2.09 ± 0.10
230	...	...	...	1.63 ± 0.08

*The values in parentheses were experimentally determined. For details see Hoge and Wechsler (1949).

first calculated a density of 4.95 g/cc and later a more accurate value of 5.060 ± 0.005 g/cc (Cornell 1). British investigators found a density of 5.09 ± 0.06 g/cc at 20.7°C (British 10). Hoge and Wechsler (1949) found a value of 4.87 g/cc for the density of the solid at 62.5°C.

(b) <u>Density of Liquid UF_6</u>. The earliest measurement was made at Columbia University, where the density of UF_6 (liq) was found to be 3.667 ± 0.05 g/cc at 65.1°C (SAM Columbia 16). At the National Bureau of Standards (Natl. Bur. Standards 4) the ratio of liquid volume at 65.1°C to the solid volume at 63.5°C was found to be 1.343 ± 0.002, and Eq. 52 was given for the ratio of volume at the temperature t to the volume at the triple point, t_f (64.052°C):

$$V/V_f = 1 + 1.727 \times 10^{-3}(t - t_f) + 3.59 \times 10^{-6}(t - t_f)^2 \qquad (52)$$

Table 13.12 shows the measured and the calculated values of this ratio between 62.45 and 92.4°C.

Table 13.13 contains a summary of measured and calculated absolute densities of liquid UF_6. Column I gives values calculated from the expansion data of Table 13.12, using 3.630 g/cc for the density at 64.052°C. Column II was calculated in the same way, using 3.667 g/cc for the density at 65.1°C. Column III contains a set of direct determinations made at the National Bureau of Standards (Natl. Bur. Standards 3; Hoge and Wechsler, 1949) and the values interpolated from these measured points by means of the empirical expression

$$\rho = 3.630 - 5.805 \times 10^{-3}(t - t_f) - 1.36 \times 10^5(t - t_f)^2 \qquad (53)$$

where $t_f = 64.052$°C. This equation represents the density of saturated liquid UF_6 within the experimental error over the temperature range 68.66 to 162.59°C. The values that were experimentally determined are enclosed in parentheses. Column IV contains values that were not available to us in publications but were gathered from various summary reports issued by British and American groups (SAM Columbia 8).

(c) Density of UF_6 (gas). Gaseous UF_6 behaves like a nearly perfect gas. There is no evidence for any association. Ruff and Heinzelmann (1911) found an average molecular weight of 338 in two vapor density determinations at 448°C. Amphlett, Mullinger, and Thomas (1948; British 8,22,23) have redetermined the vapor density of uranium hexafluoride as a function of temperature and pressure. Their results are presented in Tables 13.14a and b. There appears to be no systematic dependence of molecular weight on the temperature. The arithmetic mean observed values of the molecular weight are 355 ± 3 at room temperature and 353.3 ± 1 at 50°C; the calculated molecular weight of UF_6 is 352.8. From the observed variation of the vapor density with pressure, it is found that

$$\left(\frac{\partial \rho}{\partial p}\right)_{t=49.2°C} = 1.32 \pm 0.01 \times 10^{-8}$$

The value computed for this quantity from kinetic theory, assuming a perfect gas of the molecular weight of UF_6, is 1.314×10^{-8}. It appears, therefore, that UF_6 vapor at ordinary temperatures and pressures consists of single unassociated molecules.

Since UF_6 behaves like a perfect gas, its density at various temperatures and pressures can be calculated from Eq. 54 (SAM Columbia 10).

$$\rho \text{ (g/liter)} = 13.28 \times \frac{323.2}{T°K} \times \frac{cm\ Hg}{76.0} \qquad (54)$$

(d) Self-diffusion Coefficient of UF_6 (gas). The Virginia group determined a value of 234 ± 9 micropoises at 30°C for the product ρD,

Table 13.14a—Vapor Density of Uranium Hexafluoride*

Temp., °C	Pressure, mm Hg	Density (H = 1)	Molecular weight (O = 16)
14.1	52.0	176	355
14.8	54.4	178	357
17.4	66.1	174	352
29.7	157.5	174	352
50.0	521.0	176	355
50.0	419.0	175	353
50.0	392.1	174	352

*After Amphlett, Mullinger, and Thomas (1948).

Table 13.14b—Vapor Density of Uranium Hexafluoride as a Function of Pressure at 49.2°C*

Pressure, mm Hg	Vapor density, 10^3 g/cc	Pressure, 10^{-5} dyne/cm^2
523.0	9.016	7.027
470.0	5.295	4.246
424.0	6.482	5.072
378.0	7.240	5.717
316.0	8.119	6.315
259.3	4.278	3.479
213.3	3.627	2.862
186.4	3.130	2.534

*After Amphlett, Mullinger, and Thomas (1948).

where ρ = density and D = self-diffusion coefficient (Virginia 1; Ney and Armistead, 1947). Since according to Eq. 54 the density at 30°C is

$$\rho = 13.28 \times \frac{323.2}{303.2} \times \frac{15.5}{76.0} = 2.887 \text{ g per liter} \qquad (55)$$

it is found that

$$D = \frac{0.000234}{0.002887} = 0.081 \text{ cm}^2/\text{sec} \qquad (56)$$

This makes it possible to calculate a value for ϵ from the equation $\rho D = \epsilon \eta$, where η = viscosity coefficient. According to British data, η (30°C) = 180 micropoises. Columbia University data indicate that η (30°C) = 206 micropoises (see Sec. 2.3f). The value of ϵ for UF_6 therefore is either 1.30 or 1.14. The large difference between these two values indicates the need for renewed study. The kinetic theory of gases shows that ϵ should be 1.200 for hard-sphere molecules and 1.504 for molecules interacting as the inverse fifth power of distance. A theoretical formula for D (UF_6) has been derived, using the value ϵ = 1.50 (SAM Columbia 10).

$$D = (0.0606) \frac{20.0 \text{ cm}}{p_{cm\,Hg}} [1 + 6.29(t \times 10^{-3}) + 9.6(t \times 10^{-3})^2] \text{cm}^2/\text{sec} \qquad (57)$$

The uncertainty of this value, however, restricts the usefulness of the equation.

(e) Mean Free Path. The following formula has been given for the mean path, L, of UF_6 molecules in the gas phase:

$$L = \frac{20.0 \text{ cm}}{p_{cm\,Hg}} \times 6.32 [1 + 4.45(t \times 10^{-3}) + 3.1(t \times 10^{-3})^2] \times 10^{-6} \text{ cm} \qquad (58)$$

where t = °C from 0 to 200°C.

Values for the mean free path calculated from the equation

$$L = \frac{1}{\sqrt{2}\pi N \sigma^2}$$

where σ, the molecular diameter, is derived from the viscosity measurement, check fairly well with those computed from Eq. 58.

(f) Viscosity of UF_6. The Columbia group measured the viscosity of liquid UF_6 from 67.2 to 75.0°C and found the values in Table 13.15 (SAM Columbia 17). The absolute viscosity of UF_6 is a little less than twice as large as that of water (0.422 centipoise at 67.2°C); the kinematic viscosity is about one-half as large $[\eta$ (H_2O) at 67.2°C is 0.431 centipoise]. The results of British measurements, given in parentheses in Table 13.15 (SAM Columbia 12), do not agree well with those obtained at Columbia.

The viscosity of gaseous UF_6 has been measured by British workers (British 20,21,24). For the temperature range 0 to 200°C, the following relation was found to reproduce the British data in Table 13.16 to within ±2 per cent:

$$\eta \times 10^4 \text{ poises} = 1.67 - 0.0044t \qquad t = °C \qquad (59)$$

At Columbia University, however, a determination of the viscosity of gaseous UF_6 gave results that differed by 20 per cent from those of

Table 13.15 — Viscosity of Liquid Uranium Hexafluoride

Temp., °C	Absolute viscosity, centipoises	Kinematic viscosity, centistokes
67.2	0.731	0.200
70	(0.91)	...
72.5	0.692	0.191
72.9	0.685	0.189
73.4	0.679	0.188
74.7	0.669	0.185
75.0	0.663	0.184
80	(0.85)	...
90	(0.80)	...
100	(0.75)	...

Table 13.16 — Viscosity of Gaseous Uranium Hexafluoride

Temp., °C	Viscosity, micropoises	
	British	SAM Carbide and Carbon
0	167	...
20	175	...
40	...	178.7
50	189	...
60	...	189.0
80	...	199.9
100	211	...
110	...	216.1
140	...	231.9
150	233	...
170	...	248.0
200	255	261.1

the British (SAM Columbia 18). A redetermination at the Carbide and Carbon laboratory (SAM Carbide and Carbon 5) has shown that the British values are substantially correct. Sutherland's constant has been computed from the British data and is found to be about 101 (British 10).

(g) Surface Tension of Liquid UF_6. The surface tension of liquid UF_6 in equilibrium with the vapor has been measured by British workers, using the capillary-rise method in the temperature range 70.0 to 100°C (British 14). The results are summarized in Table 13.17.

Table 13.17—Surface Tension of Liquid Uranium Hexafluoride

Temp., °C	Surface tension, dynes/cm
70	16.8 ± 0.3
80	15.6 ± 0.3
90	14.3 ± 0.3
100	13.1 ± 0.3

Table 13.18—Refractive Index of Liquid Uranium Hexafluoride

Temp., °C	Refractive index	
	4360 A	5890 A
70	1.383	1.367
80	1.374	1.358
90	1.365	1.350
100	1.355	1.342

2.4 Optical Properties. (a) Refractive Index. (British 25.) The refractive index of liquid UF_6 has been determined from 70 to 100°C by measurements of the angle of total reflection and is probably accurate to 0.4 per cent in absolute value. The relative values are probably better than this. The results are summarized in Table 13.18.

The molecular refraction, R, was calculated from the well-known expression

$$R = \frac{M}{\rho} \frac{(n^2 - 1)}{(n^2 + 2)}$$ (60)

where M = molecular weight, ρ = density, and n = wavelength of light. These results are

$$R_{4360\ A} = 22.59 \pm 0.08 \qquad t = 85°C$$

$$R_{5890\ A} = 21.83 \pm 0.08 \qquad t = 85°C$$

On the assumption that the molecular refraction of the vapor is the same as that of the liquid, the index of refraction of the gas at 5100 A should be (British 10)

$$n - 1 = 5.2 \times 10^{-4} \, p_{mm \, Hg}/T°K \tag{61}$$

(b) Ultraviolet Absorption Spectrum of UF_6.   The ultraviolet absorption spectrum of UF_6 vapor was first photographed by Lipkin and Weisman (UCRL 1) with a large Hilger instrument (dispersion 5 A per millimeter). A 1-meter layer of the vapor in equilibrium with the solid at room temperature was used. Strong continuous absorption was found to set in at about 3300 A; it was preceded by a region of banded absorption between 3400 and 3800 A. No analysis of the spectrum was made.

Table 13.19 — Absorption Bands of Uranium Hexafluoride Gas

Frequency, cm^{-1}

24,364	25,344	26,573	27,342
24,460	25,478	26,644	27,420
24,504	25,538	26,755	27,562
24,794	25,720	26,935*	27,655
24,860	26,031	27,015†	27,820‡
24,972	26,175	27,135	27,907
25,108	26,227	27,184	28,080
25,211§	26,379	27,234¶	28,415**

*Doubtful.
†Moderately strong absorption (violet edge).
‡Strong absorption (violet edge).
§Somewhat sharper.
¶Strongest absorption (violet edge).
**Center.

Martin and Amphlett (British 26) also examined the ultraviolet absorption of UF_6 vapor at room temperature and found strong selective absorption, with a band head at 3701 A and a minimum at about 3417 A. Amphlett, Mullinger, and Thomas (1948) have more recently reported some preliminary results on the absorption of UF_6 vapor between 2000 and 4000 A. In contradiction to the results obtained at Columbia, they observed continuous absorption, with no structure, and complete extinction below about 3100 A. Since these results are based on much less experimental work than those of Columbia workers, the discrepancy must be resolved in favor of the latter.

The absorption spectrum in the region 3500 to 4000 A was again studied at Columbia (SAM Columbia 19). The UF_6 gas was at a pressure of 36 mm Hg, and a path length of 102 cm was used. The five most prominent bands appear to be separated by about 590 cm^{-1} (Table 13.19). Around each main band there are groups of closely spaced bands, separated by frequencies of the order of 100 cm^{-1}. In addition there are apparently four bands separated from the main bands by about 200 cm^{-1}. It was further reported that a thick layer (12 cm) of clear crystalline solid UF_6 showed complete absorption below 4271 A (SAM Columbia 20). The UF_6 absorption at 42,000 cm^{-1} is about 1,000 times stronger than that in the band region discussed here. It is continuous and extends, with gradually declining intensity, to about 31,000 cm^{-1}. A magnetic field of 19,000 gauss was found to have no effect on the spectrum.

(c) Fluorescence Spectrum of UF_6. (SAM Columbia 21.) Solid UF_6 excited by λ 3660 A shows a bright-violet fluorescence at 77°K which disappears at room temperature. No fluorescence could be observed in UF_6 vapor even with long exposures. The fluorescence spectrum shows diffuse bands with a separation of about 220 cm^{-1} (Table 13.20). The maximum of intensity appears to be near 23,700 cm^{-1}; the range

Table 13.20 — Fluorescence Bands of Solid Uranium Hexafluoride at 77°K

Frequency, cm^{-1}

22,230	23,319
22,450	23,537
22,662	23,739
22,883	23,994
23,098	24,166

covered is 22,000 to 24,700 cm^{-1}. A magnetic field of 25,000 gauss had no effect on the fluorescence spectrum.

(d) Infrared Absorption and Raman Spectra. The infrared spectrum of UF_6 (gas) has been studied by Turkevich (SAM Columbia 8; Bigeleisen, Mayer, Stevenson, and Turkevich, 1948). The results of this study are summarized in Table 13.21.

The Raman spectrum has also been studied. The Columbia group first reported that liquid UF_6 at 70.9°C gives two Raman lines at 603 and 228 cm^{-1} on exposure to the mercury line 4358.34 A (SAM Columbia 22) for 3 hr. The most intense line was at 603 cm^{-1}. Another line at 572 cm^{-1} was predicted but was not found; it was, perhaps, masked by the line at 603 cm^{-1}.

A more detailed study of this spectrum was then made at Columbia (SAM Columbia 14; Bigeleisen, Mayer, Stevenson, and Turkevich, 1948).* A number of experiments were performed using 4358 A excitation. Despite efforts to reduce photodecomposition of liquid UF_6, a sufficient amount of fine fluffy solid formed in 15 min to necessitate stopping the exposure because of the scattering of light by the solid.

Table 13.21 — Infrared Absorption Bands of Gaseous Uranium Hexafluoride

Frequency, cm^{-1}	Band, μ	Intensity
623	16.05	100
675	14.82	10
713−719	14.03−13.91	0.5
825	12.12	2
850	11.68	0.25
1163	8.60	4
1295	7.72	3
2053	4.87	0.5

It was impossible to determine whether or not there were any Raman lines below 300 cm^{-1}. Two Raman lines were found outside this region. The average values obtained from exposures of the liquid were 656 ± 3 and 511 ± 3 cm^{-1}. The high-frequency line was sharp and strong, but the lower frequency line was quite weak and somewhat diffuse. Attempts were made to extend the data by the use of UF_6 solutions in fluorocarbons (i.e., C_7F_{16}). With 4358 A radiation, photodecomposition still took place as with pure hexafluoride. With 5460 A radiation, no decomposition was evident, and a satisfactory Raman spectrum was obtained at 20°C. A strong line at 202 cm^{-1} was found in addition to the line previously found at 656 cm^{-1}. The line at 511 cm^{-1} did not appear, probably because of low intensity, and the 656 cm^{-1} line was shifted to 666 cm^{-1} in the fluorocarbon solution.

The infrared and Raman data were analyzed by Mayer and Bigeleisen (SAM Columbia 14; Bigeleisen, Mayer, Stevenson, and Turkevich, 1948). The infrared data indicate that there may be an accidental degeneracy. It is assumed that the fundamental frequency is about 640 cm^{-1} and that a double overtone falls at about the same point so that the two lines at 623 cm^{-1} and 675 cm^{-1} originate from the interaction of two levels at about 640 cm^{-1}. The frequency assignment given in Table 13.22 was made on this assumption. Calculated and observed infrared active double overtones are compared in Table

*The apparatus used was that of the American Cyanamid Company.

13.23. The number and positions of active lines both in the infrared and Raman spectra are considered strong evidence in favor of a completely symmetrical structure for uranium hexafluoride.

Table 13.22 — Frequency Assignment for Gaseous Uranium Hexafluoride
Based on Raman Spectra*

Designation	Frequency, cm⁻¹	Symmetry	Degeneracy	Activity	Observed frequency, cm⁻¹
ν_1	656	A_{1g}	1	Raman active	656 ± 3
ν_2	511	E_g	2	Raman active	511 ± 3
ν_3	200	T_{2g}	3	Raman active	202 ± 3
ν_4	130	T_{2u}	3	Inactive	...
ν_5	200	T_{1u}	3	Infrared active	...
ν_6	640	T_{1u}	3	Infrared active	623 or larger

*Bigeleisen, Mayer, Stevenson, and Turkevich (1948).

Table 13.23 — Calculated and Observed Infrared Overtones*

	Frequency, cm⁻¹		
Designation	Calculated	Observed	Strength
ν_5	200	...	
$\nu_4 + \nu_5$	300	...	
$\nu_5 + \nu_3$	400	...	
ν_6	640	623	1000
$\nu_4 + \nu_2$	641	675	10
$\nu_5 + \nu_2$	711	713 – 719	0.5
$\nu_6 + \nu_3$	840	825	2
$\nu_5 + \nu_1$	856	850	0.25
$\nu_6 + \nu_2$	1151	1163	4
$\nu_6 + \nu_1$	1296	1295	3
		2053	0.5

*Bigeleisen, Mayer, Stevenson, and Turkevich (1948).

2.5 Electrical and Magnetic Properties. (a) Dielectric Constant.

Martin and Amphlett reported the dielectric constant of UF_6 (gas) at room temperature and atmospheric pressure to be 1.004 (British 26). They later revised this value to 1.0043 ± 0.0001 (British 10). More recently, Amphlett, Mullinger, and Thomas (1948) have reported a value $\epsilon = 1.0038 \pm 0.0001$ at 19.6°C and 760 mm Hg pressure for the gas. The molar polarization computed from this value is 30.7 ± 1.0 cc at the same temperature and pressure. A careful determination at Princeton University (Princeton 1) gave the results listed in Table

13.24. From a knowledge of the dielectric constant ϵ of the gas, the molar polarization P can be calculated, and from this, in turn, by the use of the Clausius-Mosotti equation,

$$P = \frac{M}{\rho} \cdot \frac{\epsilon - 1}{\epsilon + 2} \tag{62}$$

the dielectric constant of the liquid can be obtained. The Princeton group reported P = 27.1 cc and ϵ = 2.18 at 65°C. These values are typical of nonpolar liquids.

Table 13.24—Dielectric Constant of Gaseous Uranium Hexafluoride

Temp., °C	Dielectric constant, ϵ	Molar polarization, P, cc
59.5	1.00297 ± 0.00006	27.0 ± 0.2
67.4	1.00292 ± 0.000003	27.2 ± 0.3
89.0	1.00273 ± 0.00005	27.0 ± 0.2

(b) <u>Dipole Moment</u>. Smyth and Hannay (Princeton 1) have calculated the dipole moment of gaseous UF_6 at various temperatures from the dielectric constant. It was found (see Table 13.24) that the molar polarization does not vary with temperature. P is defined by the well-known Debye relation

$$P = \frac{\epsilon - 1}{\epsilon + 2} \cdot \frac{M}{d} = \frac{4\pi N}{3}\alpha + \frac{4\pi N\mu^2}{9kT} \tag{63}$$

The absence of any significant variation of P with temperature indicates that the dipole moment is probably zero. By the usual criteria it is concluded that, although UF_6 could have a dipole moment of less than 0.5×10^{-18}, it is logical to assume that its moment is zero.

(c) <u>Magnetic Susceptibility</u>. Uranium hexafluoride appears to be paramagnetic. Henkel and Klemm (1935) found the specific susceptibility of solid UF_6 to be $+0.12 \times 10^{-6}$ and the molar susceptibility $+43 \times 10^{-6}$. With the molar susceptibility corrected for diamagnetism (U = 23, F = 7.25; Tilk, 1939), a value of $+106 \times 10^{-6}$ is obtained. The paramagnetism seems to be independent of the temperature.

(d) <u>Ionization Potential</u>. The critical ionization potentials of the univalent ions produced by slow-electron bombardment of UF_6 have been determined (White and Cameron, 1947; CEW-TEC 4) and are given in Table 13.25. The data have a probable error range of from 5 per cent for the UF_5^+ ion to 15 per cent for the U^+ ion.

2.6 <u>Structure of Uranium Hexafluoride Molecule</u>. The structure of uranium hexafluoride has been studied by electron diffraction, x-ray crystallography, Raman spectrum, infrared spectrum, and dipole moment. The evidence available indicates that UF_6 possesses an undistorted octahedral structure.

(a) Electron-diffraction Studies. Electron-diffraction studies of UF_6 were first carried out by Braune and Pinnow (1937). They reported irregular structures for the hexafluorides of uranium, tungsten, and molybdenum, possessing holohedral orthorhombic symmetry, the three pairs of metal-fluorine distances having the ratios

Table 13.25—Critical Ionization Potentials of UF_6 Gas*

Ion	Potential, volts	Ion	Potential, volts
UF_5^+	15.5	UF_2^+	29.9
UF_4^+	20.1	UF_1^+	37.9
UF_3^+	23.5	UF^+	50.3

*White and Cameron (1947).

$1:1.12:1.22$. For the uranium compound this leads to 1.78, 1.99, and 2.17 A for the metal-fluorine distances. Bauer (SAM Columbia 23) investigated the electron diffraction of UF_6 vapor, using 42-kv electrons, and concluded that all six uranium-fluorine distances could not be identical. Two models could be reconciled with his data. (The symmetrical octahedral structure as well as several other possible structures were excluded.) Bauer selected as the most probable structure a deformed octahedron with the fluorine atoms at six corners and the uranium atom in the plane of four fluorine atoms. The interatomic distances for this model are

$$\text{Two adjacent U-F} = 1.87 \pm 0.02 \text{ A}$$
$$\text{Two opposed U-F} = 2.12 \pm 0.03 \text{ A}$$
$$\text{Two adjacent U-F} = 2.22 \pm 0.03 \text{ A}$$

The distortion is in a direction that produces nearly equal F-F separations (about 2.9 A). This structure is similar to that of Braune and Pinnow, but the absolute values of the interatomic distances are larger by approximately 5 per cent.

The model proposed by Bauer has no center of symmetry and must therefore have a small dipole moment. The analogy to the structure of uranium metal found by Jacob and Warren (1937) is clear.

(b) Dipole Moment. Smyth (Princeton 1) pointed out that the structure proposed by Bauer necessitated a dipole moment of not less than 1.5×10^{-18} but probably larger. Measurement of the dielectric constant showed the moment to be no larger than 0.5×10^{-18} and, very likely, zero. This renders Bauer's proposed structure untenable. Smyth further pointed out that one of the models Bauer had discarded fitted the electron-diffraction data quite well. This structure has undistorted octahedral angles with interatomic distances

$$\text{Two opposed U-F} = 1.87 \text{ A}$$
$$\text{Two opposed U-F} = 2.12 \text{ A}$$
$$\text{Two opposed U-F} = 2.22 \text{ A}$$

This structure would have three pairs of mutually opposing and canceling dipoles, which would give zero moment to the molecule as a whole.

(c) Evidence from Raman and Infrared Spectra. (SAM Columbia 14, 20 to 22.) The appearance of only three Raman lines is good evidence for a symmetrical octahedral structure for UF_6, since, for any other model, more lines should appear. If the UF_6 molecule is a regular octahedron it belongs to symmetry group O_h, and selection rules show that only three Raman lines are to be expected; a tetragonal octahedron (D_{4h} or C_{4v}) would give a much larger number of lines. Only a plane hexagon (D_{6h}) would give an equally small number of lines as the regular octahedron. It has further been pointed out that, even if only two opposite uranium-fluorine bonds were elongated, the Raman lines at 511 and 202 cm^{-1} should be split. No such splitting is observed. The infrared spectrum shows only one fundamental frequency between 600 and 2100 cm^{-1}, and this also can be explained on the basis of complete octahedral symmetry.

(d) X-ray Crystal Structure. (Cornell 1.) X-ray analyses of cylindrical single crystals show that UF_6 (solid) possesses orthorhombic holohedry, space group D_{2h}^{16}-Pnma. At 24°C the parameters are

$$a = 9.900 \pm 0.002 \text{ A} \qquad b = 8.962 \pm 0.002 \text{ A} \qquad c = 5.207 \pm 0.002 \text{ A}$$

The unit cell contains four molecules, and the x-ray density of the solid UF_6 at 25°C is 5.060 ± 0.005 g/cc. A comparison between experimentally determined and "ideal" parameters for this structure is given in Table 13.26. The x-ray data do not indicate a completely regular octahedral configuration for the molecule as it exists in the crystal. The fluorine atoms are in approximately closest packing; there appears to be a real but small deviation from ideal symmetry in a slight flattening of the molecule along the normal to the plane of symmetry. This slight distortion may be the result of nonisotropic vibrations in the crystal and is compatible with a perfectly regular configuration of the molecules in the vapor. The configurations favored by Bauer for the vapor are definitely excluded for the crystal.

3. CHEMICAL PROPERTIES OF URANIUM HEXAFLUORIDE

Uranium hexafluoride is a highly reactive substance. Its chemical properties will be discussed under three headings: (1) general chemical behavior, (2) solutions of uranium hexafluoride in various

Table 13.26—Comparison of Experimentally Determined and Ideal Parameters of the Uranium Hexafluoride Structure

Atom	Experimental parameters			Ideal parameters		
	x	y	z	x	y	z
U	0.1295	$\frac{1}{4}$	0.081	$\frac{1}{8}$	$\frac{1}{4}$	$\frac{1}{2}$
F_I	0.003	$\frac{1}{4}$	-0.250	0	$\frac{1}{4}$	$-\frac{1}{4}$
F_{II}	0.250	$\frac{1}{4}$	0.417	$\frac{1}{4}$	$\frac{1}{4}$	$\frac{5}{12}$
F_{III}	0.014	0.093	0.250	0	$\frac{1}{12}$	$\frac{1}{4}$
F_{IV}	0.246	0.093	-0.083	$\frac{1}{4}$	$\frac{1}{12}$	$-\frac{1}{12}$

solvents, and (3) corrosive effects on metals and other materials of construction.

3.1 General Chemical Behavior. (a) Water. UF_6 reacts vigorously with water, forming mainly UO_2F_2 and HF. Considerable heat is evolved in this reaction (see Sec. 2.2a). When working on a large scale, it is advantageous to carry out the hydrolysis by introducing UF_6 into dilute alkali, either as vapor or by liquefying UF_6 vapor and allowing drops to fall into the "drowning vessel." Rough calculations indicate that the heat of solution of UF_6 in dilute sodium hydroxide is about -118 kcal per mole (Du Pont 2).

The reaction of UF_6 with NaOH proceeds as follows:

$$UF_6 + 4OH^- \rightarrow UO_2^{++} + 6F^- + 2H_2O \qquad \text{pH } 7.0-7.5 \qquad (64)$$

$$2UO_2^{++} + 6OH^- + 2Na^+ \rightarrow Na_2U_2O_7 + 3H_2O \qquad \text{pH } 10.0-10.5 \qquad (65)$$

At a partial pressure of 15 mm Hg of UF_6, reaction with water vapor gives a dense white smoke which settles rapidly. At 0.4 to 0.5 mm Hg of UF_6 no smoke is visible (British 27). X-ray studies showed that interaction of UF_6 with H_2O vapor does not give pure UO_2F_2 but rather a complex of UO_2F_2, HF, and H_2O, from which pure UO_2F_2 can be obtained by heating to 180°C or higher (British 28). This statement probably requires verification.

(b) Oxygen, Nitrogen, and Carbon Dioxide. (Ruff, 1911.) UF_6 is completely stable to oxygen, nitrogen, and dry air. No reaction occurs with CO_2, but caution must be exercised since fluorination to carbonyl fluoride and subsequent explosion may occur.

(c) Halogens. (Ruff, 1911.) UF_6 is soluble to a considerable extent in liquid chlorine and bromine; it does not react with gaseous chlorine or bromine either in the cold or on heating. The behavior of iodine has not been studied.

(d) Reduction of Uranium Hexafluoride by Nonmetals. (MP Ames 1.) Ruff and Heinzelmann stated that UF_6 reacts with hydrogen instantaneously at room temperature. However, more recently, repeated attempts to reduce UF_6 with hydrogen have all resulted in failure. Thus the Ames group found no reaction below 390°C. Even above 500°C reaction was only moderately fast and incomplete, leading to a complex mixture of products (MP Ames 2). Addition of a small amount of chlorine did not catalyze the reduction. Similar observations have been made by British workers, who found that the reaction of UF_6 with H_2 required a considerable energy of activation; reaction could be initiated, for example, by ultraviolet light or by electrical discharge. Various materials were tried as catalysts. Among them were H_2O, UF_4, Br_2, I_2, UN_2, Ni, UCl_4, Cl_2, HCl, Hg_2Cl_2, and $FeCl_3$, but only the chlorides had any appreciable effect. Even in their presence reaction is slow, and only impure products have been isolated (British 25,28).

Sulfur dioxide failed to reduce UF_6 when passed through UF_6 either at room temperature or on heating the gaseous mixture in a nickel tube to 150°C (MP Ames 3). Ethylene, under the same conditions, apparently reduced UF_6. The product, however, contained a considerable amount of carbonaceous material (MP Ames 3). It has been found that UF_6 can be reduced to UF_4 with propane, using a torch to carry out the reaction (SAM Carbide and Carbon 6). A mixture of hydrogen and arsine fails to reduce UF_6 at room temperature.

UF_6 reacts with hydrogen chloride at 200°C and much more rapidly at 250 or 300°C (MP Ames 4). At 200°C a mixture of brown, green, and black products was obtained which became green when degassed in vacuum (compare U_2F_9 and U_4F_{17}). Material from the 250°C experiment was predominantly brown, with some green and black product also present. The reaction is best carried out in an all-nickel apparatus. The UF_6 vapor is swept in a stream of nitrogen into the chamber where it is brought into contact with hydrogen chloride just inside the chamber at a temperature of 225°C. The reaction product consists of white and blue needles which rapidly change to green in air and which must be heated in vacuum to obtain pure UF_4. This is probably the best method for reducing UF_6 to UF_4.

With hydrogen bromide the reaction proceeds smoothly at 80°C, giving a light-green fluffy powder. Analysis showed the presence of oxyfluoride, a contamination that was attributed to the water content of the bromine used in preparing the hydrogen bromide.

The following thermodynamic estimates were made for the reduction of UF_6 by hydrogen and the two hydrogen halides (at 25°C):

$$UF_6 \text{ (solid)} + H_2 \text{ (gas)} \rightarrow UF_4 \text{ (solid)} + 2HF \text{ (gas)} \tag{66}$$
$$\Delta F° = -68,900 \text{ cal}$$

$$UF_6 \text{ (solid)} + 2HBr \text{ (gas)} \rightarrow UF_4 \text{ (solid)} + 2HF \text{ (gas)} + Br_2 \tag{67}$$
$$\Delta F° = -43,500 \text{ cal}$$

$$UF_6 \text{ (solid)} + 2HCl \rightarrow UF_4 \text{ (solid)} + 2HF \text{ (gas)} + Cl_2 \text{(gas)} \tag{68}$$
$$\Delta F° = -23,400 \text{ cal}$$

Reduction of UF_6 with hydrogen must have a particularly high activation energy, since this reaction is by no means fast even at 600°C, whereas the other two reactions proceed much more readily.

The Ames group has studied the reduction of UF_6 by ammonia (MP Ames 5). Gaseous ammonia reacts with UF_6 at dry-ice temperature. The reaction has also been carried out in the gas phase at 300°C in a nickel reactor. The product obtained in this way is not pure UF_4. X-ray studies showed the presence of ammonium fluoride. Some new unidentified lines were also present, but there were no lines due to UF_4. Chemical analysis indicated the presence of a double salt corresponding to a mixture of 2 per cent NH_4F and 98 per cent NH_4UF_5.

Experiments in which UF_4 was fused with an excess of acid ammonium fluoride at 450°C until all the HF was expelled yielded a product that had an x-ray pattern practically indistinguishable from the reaction product of UF_6 and NH_3. Analytically, also, the products were practically identical, differing only in ammonium fluoride content. The x-ray diagram indicates the arrangement of uranium atoms in this double salt to be very different from that in UF_4.

The liquid-phase reduction of UF_6 by $SiCl_4$, $SOCl_2$, and PCl_3 has been studied. All these reagents reduce UF_6 at room temperature, and thionyl chloride apparently gives a quantitative yield of UF_4 (CEW-TEC 5).

Amorphous carbon reduces UF_6 to UF_4 on heating, with the formation of CF_4 and other fluorinated hydrocarbons. Silicon, on heating, yields SiF_4, and arsenic yields AsF_3. Phosphorus reacts like silicon and on moderate heating gives PF_3. UF_6 also reacts with sulfur, giving UF_4, US_2, and a sulfur fluoride gas (m.p., $-135°C$; b.p., $-40°C$), which is probably SF_4 (Ruff, 1911).

The reaction of UF_6 with mercuric cyanide has been studied in an attempt to prepare volatile uranium compounds containing the $-NC$ group (which is isosteric with CO). UF_6 was distilled onto dry $Hg(CN)_2$ and allowed to react for 24 hr. However, no volatile products other than UF_6 could be detected, and the $Hg(CN)_2$ appeared to be unaffected (British 29).

3.2 Solutions of Uranium Hexafluoride in Various Solvents. (a) Miscellaneous Organic Compounds. (Ruff, 1911; SAM Columbia 3.) Uranium hexafluoride dissolves in several organic solvents to form practically perfect solutions. Since UF_6 is a strong fluorinating agent, it reacts with most organic compounds to give hydrogen fluoride and carbon fluorides. Solutions in some chlorinated hydrocarbons, however, are quite stable for periods up to several weeks.

Alcohol and ether react rapidly at room temperature with UF_6 to give HF, UO_2F_2, and carbonaceous material. Benzene, toluene, and xylene also react rapidly. UF_6 dissolves in nitrobenzene to give a dark-red solution which fumes in air. Paraffin hydrocarbons (i.e., n-cetane, $C_{16}H_{34}$) do not dissolve UF_6 but react rapidly at room temperatures with evolution of hydrogen fluoride and carbonization. Oils containing unsaturated hydrocarbons react more rapidly. UF_6 is insoluble in carbon disulfide. With dry CS_2 a slow reaction takes place; with moist CS_2, sulfur fluorides very similar to if not identical with those obtained by the action of UF_6 on sulfur are formed.

(b) Halogenated Hydrocarbons. (Ruff, 1911; SAM Columbia 3.) UF_6 is soluble in carbon tetrachloride, chloroform, and symmetrical tetrachloroethane. Of these hydrocarbons, the latter, $Cl_2CHCHCl_2$, forms the most stable solution, extensive reaction occurring only after several days at room temperature. On boiling, the yellow solution of UF_6 in $C_2H_2Cl_4$ becomes colorless; it regains its yellow color on cooling, indicating that the color may be due to complex formation. Some reactions of UF_6-$C_2H_2Cl_4$ solutions have been examined. Nitric oxide, NO, colors such solutions blue-green; gaseous ammonia forms a green color with the simultaneous separation of a flaky material· containing fluorine, ammonium, and quadrivalent uranium. This substance is easily soluble in dilute sulfuric acid, thus differing from UF_4. Its composition is unknown. Arsenic trichloride precipitates a rust-brown solid soluble in excess reagent. These reactions of dissolved UF_6 merit more detailed study. Pentachloroethane, CCl_3CCl_2H, also forms a yellow UF_6 solution which becomes colorless on boiling. This solution deposits good crystals of JF_6 on cooling. Solutions in this solvent are more stable than those in tetrachloroethane with only a small amount of decomposition occurring after several weeks. 1,2-Difluoro-1,1,2,2-tetrachloroethane, FCl_2CCCl_2F, reacts rather rapidly with UF_6. Gas evolution occurs (probably CCl_2F_2), and ultimately all the UF_6 is reduced to UF_4. Trichloroethylene, $Cl_2C{=}CClH$, reacts immediately to form UF_4.

Table 13.27 — Physical Properties of Some Fluorocarbons

Empirical formula	Melting point, °C	Boiling point, °C	d^{20} (liq)
CF_4	−183.5	−128.0	1.91 (−183°C)
C_2F_6	−100.6	−78.2	1.70 (−100°C)
C_3F_8	−183	−38	1.45 (0.2°C)
C_5F_{10}	−12	22	1.67
C_6F_{12}	50	52	1.714
C_7F_{14}		76	1.789
C_7F_{16}		83	1.730
C_7F_{16}	−60	82.2	1.721
C_8F_{16}		100	1.85
C_8F_{18}		104	1.73
C_9F_{18}		123	1.89
C_9F_{20}		123	1.80
$C_{10}F_{18}$	−75	138	1.92
$C_{12}F_{22}$	75	90 (90 mm Hg)	
$C_{16}F_{34}$	114	240	

(c) Fluorocarbons. The completely fluorinated hydrocarbons are extremely inert to most reagents. The only reagent known to react with them at ordinary temperatures is fluorine, which reacts to form CF_4. Metallic sodium or potassium reacts above 400°C. The thermal stability of the fluorocarbons is high; temperatures of 400 to 500°C do not affect them. Nitric acid (96 per cent), fuming sulfuric acid, acid chromate, or permanganate solution are without effect; bromine and iodine do not react although hydrogen (100 atm, 450°C) removes some fluorine as hydrogen fluoride. Dilute or concentrated alkalis do not affect the fluorocarbons up to 100°C. The liquid, saturated fluorocarbons are insoluble in water, alcohols, and hydrocarbons, but they show appreciable solubility and, at higher temperatures, complete miscibility with chlorinated hydrocarbons. The fluorocarbons are completely miscible with ethyl ether and with partly fluorinated hydrocarbons such as benzotrifluoride, $C_6H_5{\cdot}CF_3$. For details of the preparation and other properties of these interesting compounds, the reader is referred to the very extensive work on fluorocarbons summarized in Division VII, Volume 1, and to other volumes of the Series. No attempt can be made here to enumerate the various workers and laboratories responsible for the recent spectacular advances in this field.

The physical properties of a few fluorocarbons are given in Table 13.27.

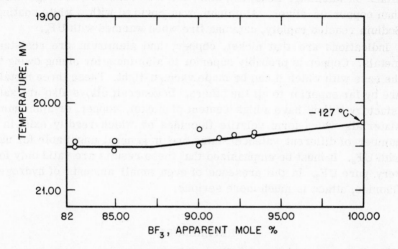

Fig. 13.11—Melting point diagram of the boron trifluoride–uranium hexafluoride system. ○, experimental data. For multijunction thermocouple, 0.16 millivolt = 1°C.

Solutions of UF_6 in fluorocarbons are extremely stable. Reduction, which is occasionally observed, is usually attributable to reaction with the container or to small amounts of hydrofluorocarbons which may be present in the solvent. Pure fluorocarbons (C_6F_{12}, C_7F_{16}) show no reaction.

(d) <u>Solutions of UF_6 in Inorganic Solvents.</u> Qualitative behavior of uranium hexafluoride with a large number of inorganic "solvents" has been studied at the SAM Laboratories of Columbia University (SAM Columbia 24). Quantitative data have been obtained for UF_6-BF_3 systems.

The melting point–composition diagram for the UF_6-BF_3 system (Fig. 13.11) has also been evolved at Columbia (SAM Columbia 25). UF_6 is only very sparingly soluble in BF_3 at about −125°C.

3.3 <u>Reactions of UF_6 with Metals and Other Materials of Construction.</u> Uranium hexafluoride reacts with most metals or other common materials of construction. The problem of finding inert materials has therefore received a great deal of attention.

Metals and Alloys. A variety of methods have been employed in evaluating the corrosion resistance of various materials. Ruff and Heinzelmann (1911) made a number of qualitative observations by exposing various substances to UF_6 vapor. Gold and platinum were reported to show no reaction in the cold but tarnished slightly on heating. Mercury reacted in the cold. Copper and silver showed slight attack on warming. Lead, tin, zinc, and iron reacted more vigorously than copper or silver. Aluminum was covered with a white coating. Sodium reacted rapidly, catching fire when warmed with UF_6.

Indications are that nickel, copper, and aluminum are resistant metals. Copper is probably superior to aluminum for piping owing to the ease with which it can be made vacuum-tight. These three metals are by far superior to all the others. In general, alloys also are satisfactory if they have a high content of nickel, copper, or aluminum. Materials which form volatile fluorides or which readily exist in a number of different valence states are in general unsuitable for use with UF_6. It must be emphasized that these results are valid only for very pure UF_6. In the presence of even small amounts of hydrogen fluoride, attack is much more serious.

REFERENCES

1900 H. Moissan, "Le Fluor," G. Steinheil, Paris.
1909 O. Ruff and A. Heinzelmann, Ber., 42: 495.
1911 O. Ruff and A. Heinzelmann, Z. anorg. Chem., 72: 63-84.
1919 G. Bahlau, dissertation, University of Danzig.
1931 F. W. Aston, Nature, 128: 725.
1932 A. V. Grosse and P. Kronenberg, Z. anorg. u. allgem. Chem., 204: 184.
1935 R. Henkel and W. Klemm, Z. anorg. u. allgem. Chem., 222: 70.
1937 H. Braune and P. Pinnow, Z. physik. Chem., B-35: 239.
1937 G. W. Jacob and B. Warren, J. Am. Chem. Soc., 59: 2588.
1938 A. L. Henne and M. W. Renoll, J. Am. Chem. Soc., 60: 1060.
1939 W. Tilk and W. Klemm, Z. anorg. u. allgem. Chem., 240: 355.
1947 E. P. Ney and F. C. Armistead, Phys. Rev., 71: 14.
1947 J. R. White and A. E. Cameron, Phys. Rev., 71: 907.
1948 C. B. Amphlett, L. W. Mullinger, and L. F. Thomas, Trans. Faraday Soc., 44: 927.
1948 J. Bigeleisen, M. Goeppert Mayer, P. C. Stevenson, and J. Turkevich, J. Chem. Phys., 16: 442.
1948 F. G. Brickwedde, H. J. Hoge, and R. B. Scott, J. Chem. Phys., 16: 429.
1948 B. Weinstock and R. H. Crist, J. Chem. Phys., 16: 436.
1949 H. J. Hoge and M. T. Wechsler, J. Chem. Phys., 17: 617.
1949 J. F. Masi, J. Chem. Phys., 17: 755.

Project Literature

British 1: W. L. Congreve, Report BR-436, June 2, 1944.
British 2: H. A. Skinner, Report BR-423, May 8, 1944.

British 3: M. W. Perrin, Report B-33, September, 1941.
British 4: A. G. Dowson, Report BR-287, Appendix I, Aug. 26, 1943.
British 5: W. J. Skinner, Report BR-347, Nov. 18, 1943.
British 6: M. A. E. Hodgson, Report BR-633, July 18, 1945.
British 7: F. Simon and J. Ferguson, Report B-37, January, 1942.
British 8: W. N. Haworth, Report B-79, June, 1942.
British 9: F. Simon, Report B-3, May, 1942.
British 10: Directorate of Tube Alloys, [B]LRG-64, Mar. 13, 1944.
British 11: A. T. Williamson and Lynch, Report B-18, July 9, 1941.
British 12: C. B. Amphlett and L. F. Thomas, Report B-18, Aug. 26, 1941.
British 13: C. B. Amphlett and L. F. Thomas, Report B-27, Oct. 21, 1941.
British 14: Directorate of Tube Alloys, [B]LRG-19, April, 1943.
British 15: J. H. Awbery, Report BR-290, August, 1943.
British 16: Arms, Report B-36, December, 1941.
British 17: F. Simon, Report B-55, March, 1942.
British 18: W. N. Haworth, Report B-38, February, 1942.
British 19: Report B-32, May, 1942.
British 20: D. R. Llewellyn and Swaine, Report B-100, September, 1942.
British 21: D. R. Llewellyn and Swaine, Report BR-427, 1944; L. E. J. Roberts, Report BR-468, 1944; cited in Report [A]M-2511.
British 22: C. H. Johnson, Report B-19, July 7, 1941.
British 23: C. H. Johnson, Report B-97, July, 1942.
British 24: L. E. J. Roberts, Report BR-468, July, 1944.
British 25: Directorate of Tube Alloys, [B]LRG-20, May, 1943.
British 26: A. E. Martin and C. B. Amphlett, Report B-117, November, 1942.
British 27: Gordon, Report B-99, August, 1942.
British 28: Directorate of Tube Alloys, [B]LRG-22, July, 1943.
British 29: Directorate of Tube Alloys, [B]LRG-17, February, 1943.

Brown 1: C. A. Kraus, Report CC-1717, July 29, 1944, p. 4.

CEW-TEC 1: F. Smith, K. Grandy, and F. W. Tober, Report CD-4101, Jan. 16, 1945.
CEW-TEC 2: G. P. Happ and A. E. Cameron, Report CD-569, Mar. 12, 1945.
CEW-TEC 3: A. J. Miller, K. B. Brown, H. R. Grady, and H. A. Young, Report CD-453, July 19, 1944; J. Whitney, F. Smith, and A. J. Miller, Report CD-2.355.1, Oct. 15, 1945.
CEW-TEC 4: A. E. Cameron and J. R. White, Report CD-B-6.460.9, Jan 17, 1946.
CEW-TEC 5: V. P. Calkins, Report CD-0.350.4, Sept. 25, 1945.

Cornell 1: J. L. Hoard and J. D. Stroupe, Reports A-1242, Mar. 1, 1944, and A-1296, June 30, 1944.

Du Pont 1: F. B. Downing, A. F. Benning, and A. L. Linch, Report A-728, May 17, 1943.
Du Pont 2: D. X. Klein, Reports A-1055, June 3, 1944, and A-1058, Sept. 16, 1944.

Johns Hopkins 1: R. D. Fowler, The Preparation of Uranium Hexafluoride from Oxides, in National Nuclear Energy Series, Division VIII, Volume 6.

MP Ames 1: I. B. Johns, A. D. Tevebaugh, E. Gladrow, K. Walsh, P. Chiotti, B. Ayers, F. Vaslow, and R. W. Fisher, The Reduction of Uranium Hexafluoride to Uranium Tetrafluoride, in National Nuclear Energy Series, Division VIII, Volume 6.
MP Ames 2: A. D. Tevebaugh, K. Walsh, and I. B. Johns, Report CC-1245, Jan. 8, 1944.

MP Ames 3: A. D. Tevebaugh, F. Vaslow, and R. W. Fisher, Report CK-1516, Mar. 10, 1944, p. 6.

MP Ames 4: E. M. Gladrow and P. Chiotti, Report CK-1498, May 10, 1944, p. 11.

MP Ames 5: B. Ayers, Report CC-1504, Aug. 10, 1944, p. 6.

MP Chicago 1: H. S. Brown, Report CC-408, Jan. 8, 1943.

MP Chicago 2: S. Fried and N. R. Davidson, Report N-1722, Nov. 14, 1944; The Reaction of UF_4 with Dry O_2; a New Synthesis of UF_6, in National Nuclear Energy Series, Division VIII, Volume 6.

Natl. Bur. Standards 1: P. Abelson, Report A-58.

Natl. Bur. Standards 2: F. G. Brickwedde, H. J. Hoge, and R. B. Scott, Report A-607.

Natl. Bur. Standards 3: H. J. Hoge and M. T. Wechsler, Report A-1591, Dec. 18, 1943.

Natl. Bur. Standards 4: M. T. Wechsler and H. J. Hoge, Report A-456, Feb. 25, 1943.

Princeton 1: C. P. Smyth and N. B. Hannay, Report A-2130, Oct. 1, 1944.

SAM Carbide and Carbon 1: A. V. Grosse, Report A-3389.

SAM Carbide and Carbon 2: W. K. Ergen, Report A-3387, July 29, 1942.

SAM Carbide and Carbon 3: A. H. Taylor and P. A. Agron, to be published.

SAM Carbide and Carbon 4: R. D. Present, Report [A]M-2511, June 8, 1945.

SAM Carbide and Carbon 5: A. L. Meyerson and J. H. Eicker, Report A-3825, Dec. 26, 1945.

SAM Carbide and Carbon 6: H. A. Bernhardt, D. J. Voss, and J. P. Brusie, Report A-3618, Dec. 6, 1945.

SAM Columbia 1: D. S. McClure, Report A-1224, Oct. 28, 1943.

SAM Columbia 2: B. Weinstock, Report A-569, Mar. 15, 1943.

SAM Columbia 3: A. V. Grosse, Report A-83, June 25, 1941.

SAM Columbia 4: N. R. Davidson and E. A. Long, Report A-154, Apr. 13, 1942.

SAM Columbia 5: H. F. Priest, Report A-121, Feb. 20, 1942.

SAM Columbia 6: A. Turkevich, N. Metropolis, and W. F. Libby, Report A-180, June, 1942.

SAM Columbia 7: J. F. G. Hicks, L. Weil, C. Coleman, and T. Yurman, Report [A]M-1136, Aug. 8, 1944.

SAM Columbia 8: I. Kirshenbaum, Report A-753, July 1, 1943.

SAM Columbia 9: R. H. Crist and B. Weinstock, Report A-571, Apr. 10, 1943.

SAM Columbia 10: Extracts from Report A-763, July 19, 1943.

SAM Columbia 11: E. A. Long and N. R. Davidson, quoted by W. F. Libby, Report A-1228x, Aug. 29, 1944.

SAM Columbia 12: I. Kirshenbaum, Report A-753, Addendum, Oct. 26, 1943.

SAM Columbia 13: E. T. Booth, D. Callihan, E. Haggsstrom, and N. Nordsick, Report A-87, Dec. 27, 1941.

SAM Columbia 14: M. Goeppert Mayer and J. Bigeleisen, Report A-1269, May 4, 1944.

SAM Columbia 15: H. F. Priest, Report A-2142, October, 1942.

SAM Columbia 16: H. F. Priest, Report A-139, Mar. 29, 1942.

SAM Columbia 17: A. D. Kirshenbaum, Report A-732, June 10, 1943.

SAM Columbia 18: R. D. Fowler, H. C. Anderson, and C. E. Weber, Report A-1398, Nov. 15, 1942.

SAM Columbia 19: A. B. F. Duncan, Report A-565, Mar. 15, 1943.

SAM Columbia 20: A. B. F. Duncan, Report A-584, Apr. 15, 1943.

SAM Columbia 21: H. C. Urey, Report A-750, July 10, 1943.

SAM Columbia 22: A. B. F. Duncan and G. M. Murphy, Report A-580, Apr. 5, 1943.

SAM Columbia 23: S. H. Bauer, Report A-1209, Aug. 10, 1943.

SAM Columbia 24: I. Kirshenbaum, Report A-711, May 14, 1944.

SAM Columbia 25: I. Kirshenbaum, Report [A]M-709, Feb. 16, 1944.

UCRL 1: D. Lipkin and S. Weisman, Report A-520, Dec. 17, 1942.

Virginia 1: E. P. Ney and F. C. Armistead, quoted by R. S. Mulliken, Report CC-567, Mar. 25, 1943.

Yale 1: M. Fytelson, H. S. Lowenhaupt, and D. Wasserman, Report A-540, Feb. 11, 1943.

Chapter 14

URANIUM-CHLORINE COMPOUNDS

Four binary uranium-chlorine compounds are known: UCl_3, UCl_4, UCl_5, and UCl_6. The first three were prepared and studied by the earliest workers in uranium chemistry. The preparation of the fourth, UCl_6, has only recently been effected at the University of California Radiation Laboratory. Since this compound and uranium hexafluoride are the only known oxygen-free sexivalent uranium compounds, they are of special interest. Considerable new information on uranium chlorides has been acquired since 1942, and this material furnishes most of the subject matter of this chapter.

For convenience a number of the physical properties of the uranium-chlorine compounds are collected in Table 14.1.

1. URANIUM TRICHLORIDE

Uranium trichloride was first prepared by reduction of UCl_4 with hydrogen (Péligot, 1842). This is still the most convenient method of preparation, particularly if no pure uranium metal is available. The unusual stability of uranium trichloride in air has stimulated considerable interest in the compound.

1.1 <u>Preparation of Uranium Trichloride</u>. In addition to the reduction of UCl_4 with hydrogen, uranium trichloride can also be prepared by reaction of uranium hydride with hydrogen chloride. Some other reactions also leading to UCl_3 are of no practical importance.

(a) <u>UCl_4 + Hydrogen.</u> (Brown 1 to 6; MP Ames 1.) The reduction of UCl_4 by hydrogen,

$$UCl_4 + \tfrac{1}{2}H_2 \rightleftharpoons UCl_3 + HCl \tag{1}$$

does not proceed rapidly below 500°C. At 525 to 550°C the rate is still so slow as to require about 5 hr for the reaction to go to completion (see Sec. 2.3d). It is essential, however, to keep the temperature below 590°C so as to avoid formation of liquid UCl_4, since the molten material easily becomes covered by a layer of UCl_3 which prevents further reduction. It is, therefore, advisable to convert most of the UCl_4 to UCl_3 below 575°C; the reduction can then be completed at temperatures up to 650°C. The necessity of using carefully purified hydrogen does not need to be stressed.

Hydrogenation under pressure is much more expeditious (Brown 4,6). Under 7 atm at 525 to 550°C the reaction proceeds four times as fast as at atmospheric pressure. To prevent sintering, the reaction mixture is first kept at 525°C for 30 min and then slowly heated to

Table 14.1 — Some Physical Properties of Uranium-Chlorine Compounds

Compound	M.p., °C	Density,* g/cc	Vapor pressure $\log p_{mm\ Hg} = A - \dfrac{B}{T}$		ΔH (sublimation), kcal/mole	$F^{\circ}_{298°K}$, kcal/mole	$S_{298°K}$, e.u.
			A	B			
UCl_3	842 ± 5	5.51	10.0	12,000	55.0	−196.2	40.0
UCl_4	590 ± 1	4.87	13.2995	10,427	47.7	−229.0	47.14
UCl_5	300†	3.81‡	...	...	...	−235.7	57.0†
UCl_6	177.5†	3.59	6.634	2,422	11.12	−241.4	68.26

*From x-ray data.
†These quantities are estimated.
‡Direct measurement by immersion in benzene.

550°C. The reaction may be carried out in a stainless-steel vessel equipped with vents for exhausting the gases slowly. Uranium trichloride produced in this way is a dark-green crystalline substance.

(b) UH₃ + HCl. (MP Ames 2,3.) When finely divided uranium metal or uranium hydride is heated with hydrogen chloride at 250 to 300°C, olive-green UCl_3 is obtained.

$$UH_3 + 3HCl \rightarrow UCl_3 + 3H_2 \tag{2}$$

The reaction proceeds smoothly and, when metallic uranium is available, this is a very convenient method. A suitable apparatus is the one used for preparing UBr_3 (MP Ames 4) (see Chap. 15). Hydrogen chloride is strongly absorbed by the product and can only be removed in vacuum at 150°C.

(c) Miscellaneous Preparations. Uranium trichloride can be reduced by ammonia at elevated temperatures, but the product contains nitrogen (Rammelsberg, 1842). Hydrogen iodide may also be used.

$$UCl_4 + HI \xrightarrow{300\text{-}350°C} UCl_3 + HCl + \frac{1}{2}I_2 \qquad (3)$$

In this case the product may be freed from iodine by a stream of dry nitrogen at 200 to 300°C (CEW-TEC 1).

It has been observed that UCl_4 vapors react with heated metals according to the equation

$$xUCl_4 + M \rightarrow MCl_x + xUCl_3 \qquad (4)$$

If the metal M forms a volatile chloride, very good yields of crystalline UCl_3 can be obtained. Metallic zinc is particularly suitable (CEW-TEC 2,32).

When uranium mononitride, UN, is heated in UCl_4 vapor at 500°C for several hours, a product thought to be UCl_3 is obtained. The reacted mass is found to dissolve in water with the evolution of considerable gas and the formation of a purple solution (MP Ames 5). This reaction is of no importance as a method of preparation.

Solutions of UCl_3 in concentrated hydrochloric acid are more stable than in pure water, and such solutions can be prepared by electrolytic reduction of a 15 per cent solution of UO_3 in HCl (sp. gr. = 1.12) (Rosenheim, 1908). The electrolysis is carried out under carbon dioxide in a cell equipped with a porous diaphragm and a mercury cathode. The reduction is performed at 110 volts and 1.5 to 2.0 amp, with concentrated hydrochloric acid as the anolyte. After the solution, which rapidly turns green at first, has acquired a dirty brown-green color, the current is reduced to 0.75 to 1.0 amp. Further electrolysis is carried out at 0°C with periodic additions of hydrochloric acid until the characteristic red color of uranium(III) is developed. Solutions of uranium(III) can also be prepared by reduction of UO_2Cl_2 with zinc and hydrochloric acid (Zimmermann, 1882).

(d) Purification of UCl_3. Uranium trichloride can be freed from volatile materials (such as UCl_4) by heating to 770°C in vacuum for several hours (Brown 7). An ingenious purification method is based on the conversion of UCl_3 to UCl_3I by distillation in a stream of iodine vapor.

$$UCl_3 + \frac{1}{2}I_2 \rightarrow UCl_3I \qquad (5)$$

UCl_3I is volatile; on cooling, decomposition to UCl_3 and iodine again occurs. The net result is the distillation of the UCl_3 at temperatures far below those required by its own volatility (UCRL 1).

1.2 Physical Properties of Uranium Trichloride. At room temperature UCl_3 is olive-green in color. At 300°C it develops a reddish-brown color, and above 450°C it develops a dark-purple color. On cooling to room temperature the olive-green color returns. Material prepared by reduction of UCl_4 with H_2 is usually red.

(a) Melting Point. (Brown 5,8.) The melting point of UCl_3 is 842 ± 5°C as determined by thermal analysis.

(b) Volatility. The vapor pressure of UCl_3 has been measured by the molecular effusion method. The upper limit of the volatility is in the vicinity of 10^{-2} mm Hg at 820°C and 3.5×10^{-3} mm Hg at 750°C (UCRL 2). An empirical equation has been devised which fits the vapor pressure data in the temperature range 600 to 1000°C (UCRL 3).

$$\log p_{mm\ Hg} = -\frac{12,000}{T} + 10.00 \tag{6}$$

A heat of sublimation of 55,000 cal per mole is deduced from this equation.

(c) Crystal Structure of UCl_3. (MP Chicago 1; Zachariasen, 1948a.) Uranium trichloride has a hexagonal unit cell containing two molecules with the lattice constants

$$a_1 = 7.428 \pm 0.003\ A \qquad a_3 = 4.312 \pm 0.003\ A$$

The space group is $C6_3/m(C_{6h}^2)$, and the atomic positions are

$$2\ U\ \text{in}\ \pm(\tfrac{1}{3}\ \tfrac{2}{3}\ \tfrac{1}{4})$$
$$6\ Cl\ \text{in}\ \pm(x\ y\ \tfrac{1}{4})\ (\bar{y},\ x-y,\ \tfrac{1}{4})\ (y-x,\ \bar{x},\ \tfrac{1}{4})$$

The parameter values are found to be

$$x = 0.375 \pm 0.014 \qquad y = 0.292 \pm 0.014$$

Each metal atom thus has nine halogen neighbors, three at a distance U-Cl of 2.95 A and six at a distance U-Cl of 2.96 A. The closest distance of approach of two chlorine atoms is Cl-Cl = 3.45 A. The UCl_3 structure is of coordination type, and the U-Cl bonds are probably of predominantly ionic character. UCl_3 is isomorphous with $CeCl_3$, $LaBr_3$, and $CeBr_3$, and a number of rare earth hydroxides.

The density of UCl_3 is computed to be 5.51 g/cc from the x-ray data. By direct displacement in CCl_4 a value of 5.35 g/cc was obtained. The density of UCl_3 is therefore higher than that of UCl_4 (MP Ames 1).

(d) Thermochemical Data. (Natl. Bur. Standards 1.) The specific heat, entropy, and enthalpy of UCl_3 have been determined for the temperature range 15 to 380°K (Natl. Bur. Standards 1) and from 273 to 1000°K (Natl. Bur. Standards 6; Ginnings and Corruccini, 1947). The results are probably accurate to within a few tenths of 1 per cent; they are given in Tables 14.2a and 14.2b. The values in Table 14.2b are to be preferred.

The heat of formation of UCl_3 has recently been redetermined at the University of California Radiation Laboratory (UCRL 4) by the method of Biltz and Fendius (1928a) (see Sec. 2.2f). The result was

$$U \text{ (solid)} + \tfrac{3}{2}Cl_2 \text{ (gas)} \rightarrow UCl_3 \text{ (solid)}$$

$$\Delta H_{273°K} = -212.1 \pm 0.6 \text{ kcal per mole} \tag{7}$$

For details of the calculation the reader is referred to the original report.

The equilibrium

$$UCl_4 \text{ (solid)} + \tfrac{1}{2}H_2 \text{ (gas)} \rightleftarrows UCl_3 \text{ (solid)} + HCl \text{ (gas)} \quad \text{(Eq. 1)}$$

has been studied at the University of California Radiation Laboratory (UCRL 5,6), and the results have been critically evaluated by MacWood (UCRL 7). The equilibrium constants for this reaction are given in Table 14.3 (p. 459). The change in heat capacity in Eq. 1 can be calculated as follows:

UCl_3 (solid):	$C_p = 23.42 + 4.132 \times 10^{-3}T$
HCl (gas):	$C_p = 6.7 + 0.84 \times 10^{-3}T$
$-UCl_4$ (solid):	$-C_p = -27.56 - 4.412 \times 10^{-3}T$
$-\tfrac{1}{2}H_2$ (gas):	$-\tfrac{1}{2}C_p = -3.31 - 0.405 \times 10^{-3}T$
	$\Delta C_p = -0.75 + 1.55 \times 10^{-4}T$

This leads to the expression

$$\Delta H_T = 17.270 \text{ kcal} - 0.75T + 7.79 \times 10^{-5}T^2$$

for the heat of reaction (Eq. 1). From the equilibrium constants a value of $\Delta S_{298°K} = 23.00$ is obtained. Unfortunately this value differs by 3.1 units from that derived from third-law considerations ($\Delta S_{298°K} = 19.92$). The reason for the discrepancy is unknown. If it is assumed that the value of the entropy of UCl_4 at 298.16°C derived from the third law, 47.14 e.u., is correct, then the entropy of UCl_3 at 298.16°C would be 40.5 e.u.

Using the latter value, MacWood concludes that the best available data for the reaction

$$U \text{ (solid)} + \tfrac{3}{2}Cl_2 \text{ (gas)} \rightarrow UCl_3 \text{ (solid)} \quad \text{(Eq. 7)}$$

are

$\Delta H_{298°K} = -212.0 \text{ kcal per mole}$

$\Delta F°_{298°K} = -196.7 \text{ kcal per mole}$

$\Delta S_{298°K} = -51.5 \text{ e.u.}$

$$\Delta H = -213,700 + 5.70T - 7.79 \times 10^{-4}T^2 \tag{8}$$

$$\Delta F° = -213,700 - 13.13T \log T + 7.79 \times 10^{-4}T^2 + 89.23T \tag{9}$$

1.3 Chemical Properties of Uranium Trichloride.

In this section emphasis will be placed on the reactions of solid UCl_3. UCl_3 is a typical trivalent halide; it resembles the rare earth trihalides in its physical and chemical properties. The principal difference is that UCl_3, both as a solid and in solution, is a strong reducing agent.

(a) Water. Uranium trichloride is hygroscopic but to a much smaller degree than the other uranium chlorides. No absorption occurs if $p_{H_2O} = 2.4$ mm Hg; slight absorption occurs at $p_{H_2O} = 3.4$ mm Hg. In a nitrogen stream containing water vapor at a $p_{H_2O} = 9$ mm Hg, UCl_3 deliquesces to form a purple solution which slowly evolves hydrogen. In moist air ($p_{H_2O} = 9$ mm Hg) UCl_3 does not exhibit deliquescence but instead is rapidly hydrolyzed and simultaneously oxidized to $UOCl_2$. The product is a dry cake (Brown 9). UCl_3 gains about 1 per cent by weight per minute on exposure to moist air (UCRL 2).

In water, uranium trichloride dissolves to give a fugitive purple color which soon turns to the dirty green indicative of uranium(IV). Uranium trichloride is soluble in water, forming a solution between 3M and 4.8M at 0°C. A 0.02M solution of UCl_3 in water has a pH of about 2.4 (UCRL 8). This is a very rough value, but it gives an indication of the extent of hydrolysis.

When UCl_3 is dissolved in water, the amount of hydrogen evolved corresponds within a few per cent to the oxidation from the trivalent to the quadrivalent state (UCRL 8). The evolution of hydrogen is about one-fourth as rapid at 0°C as at 24°C. At 0°C, UCl_3 can be kept in a saturated solution in contact with the solid salt with little oxidation for considerable periods of time. The oxidation has been followed by the change in absorption spectra. As the oxidation proceeds, the char-

Table 14.2a —Specific Heat, Entropy, and Enthalpy of UCl_3 from 0 to 380°K

Temp., °K	C_p, int. joules/mole/°C	S, int. joules/mole/°C	$H - E_0^s$, int. joules/mole
0	0.0	0.0	0.0
5	0.476*	0.159	0.597
10	3.15*	1.181	8.73
15	6.64	3.13	33.39
20	9.30	5.43	73.67
25	12.38	7.84	127.8
30	15.51	10.37	197.6
35	18.94	13.01	283.4
40	23.07	15.80	388.1
45	27.2	18.76	513.8
50	31.6	21.85	660.8
55	36.7	25.10	831.6
60	41.8	28.51	1,028
65	46.4	32.04	1,248
70	50.9	35.64	1,492
75	55.2	39.30	1,757
80	59.1	42.99	2,043
85	62.9	46.69	2,348
90	66.6	50.39	2,672
95	70.3	54.10	3,014
100	73.1	57.78	3,373
105	75.4	61.40	3,744
110	77.8	64.96	4,128
115	79.9	68.46	4,522
120	82.1	71.91	4,927
125	84.0	75.30	5,342
130	85.6	78.63	5,766
135	87.2	81.89	6,198
140	88.6	85.09	6,638
145	89.9	88.22	7,084
150	91.2	91.29	7,537
155	92.4	94.30	7,996
160	93.4	97.25	8,460
165	94.3	100.14	8,930
170	95.1	102.97	9,403
175	95.8	105.73	9,881
180	96.4	108.44	10,361
185	97.0	111.09	10,845
190	97.4	113.68	11,331
195	97.8	116.22	11,819
200	98.1	118.70	12,309
205	98.4	121.12	12,800
210	98.8	123.50	13,293
215	99.2	125.83	13,788
220	99.6	128.11	14,285

Table 14.2a — (Continued)

Temp., °K	C_p, int. joules/mole/°C	S, int. joules/mole/°C	$H - E_0^s$, int. joules/mole
225	100.1	130.35	14,784
230	100.5	132.56	15,286
235	100.8	134.72	15,789
240	101 1	136.85	16,294
245	101.4	138.94	16,800
250	101.7	140.49	17,308
255	102.0	143.01	17,817
260	102.2	144.99	18,328
265	102.6	146.94	18,839
270	102.7	148.85	19,352
275	102.8	150.73	19,865
280	102.9	152.58	20,379
285	103	154.40	20,893
290	103	156.19	21,408
295	103	157.95	21,923
298.16	103	159.04	22,248
300	103	159.68	22,438
305	103	161.38	22,953
310	103	163.06	23,468
315	103	164.71	23,983
320	103	166.33	24,498
325	103	167.93	25,013
330	103	169.50	25,528
335	103	171.05	26,043
340	103	172.57	26,558
345	103	174.08	27,073
350	103	175.56	27,588
355	103	177.02	28,103
360	103	178.46	28,618
365	103	179.88	29,133
370	103	181.28	29,648
375	103	182.67	30,163
380	103	184.03	30,678

*Extrapolated using Debye function, 15.665 D $\left(\dfrac{68.4}{T}\right)$.

acteristic bands of the uranium(III) ion (a wide band in the green and two sharp adjacent bands in the red) fade, but the strong band in the yellow-green, characteristic of uranium(IV), does not appear immediately.

The rate of solution of UCl_3 in hydrochloric acid decreases with increasing concentration; the same weight of crystals that will dissolve completely in 0.5 hr at 0°C in 1M HCl is not completely dissolved in 1.0 hr in 6M HCl. Solutions of UCl_3 in concentrated hydrochloric acid are considerably more stable than in water. The heat of solution of UCl_3 in hydrochloric acid (1 mole of HCl to 8.808 moles of H_2O) is

Table 14.2b — Specific Heat, Entropy, and Enthalpy of UCl_3 from 0 to 727°C*

Temp., °C	C_p, int. joules/mole/°C	S, int. joules/mole/°C	$H - E_0^s$, int. joules/mole
0	101.7	149.99	19,674
50	102.5	167.15	24,778
100	103.3	181.95	29,921
150	104.1	194.99	35,103
200	104.9	206.66	40,329
250	105.9	217.23	45,601
300	106.8	226.93	50,921
350	107.9	235.90	56,290
400	109.0	244.28	61,714
450	110.3	252.13	67,194
500	111.7	259.55	72,741
550	113.5	266.61	78,369
600	115.8	273.38	84,101
650	119.0	279.92	89,970
700	123.5	286.29	96,028
726.84 (1000°K)	126.7	289.69	99,379

*4.1833 int. joules = 1 cal; 0°C = 273.16°K.

40.6 kcal per mole (Biltz, 1928a). It has been reported that the addition of sulfuric acid to an aqueous solution of UCl_3 results in the precipitation of a uranium(III) sulfate; but the addition of ammonia, phosphoric acid, oxalic acid, or hydrofluoric acid gives only uranium(IV) salts (Rosenheim, 1908) (see Chap. 12).

(b) Other Solvents. (MP Ames 6,7; Ephraim, 1933.) Uranium trichloride dissolves in glacial acetic acid to a red solution. Methanol or formamide reacts when treated with UCl_3. Uranium trichloride is insoluble in carbon tetrachloride, acetone, chloroform, pyridine, hydrocarbons, or nonpolar solvents. In polar solvents the solubility of UCl_3 is less than that of UCl_4 or UBr_4. Uranium trichloride is insoluble in inorganic solvents such as PCl_3 and $SnCl_4$. With PCl_3 a green emulsion is formed.

(c) Thermal Decomposition of UCl_3. When a sample of UCl_3 is heated in vacuum to 840°C, several different substances volatilize and collect in the cooler parts of the apparatus. The material deposited farthest from the furnace is brown UCl_5; nearest the furnace is a light-tan deposit of UCl_3; and in between is a green deposit of UCl_4. Deposition of uranium tetrachloride can first be observed when the temperature reaches 500°C; UCl_5 appears at 720°C; and unchanged UCl_3 sublimes first at 840°C. The residue from such sublimation consists of bright, shining, metallic-appearing lumps (UCRL 2). These

Table 14.3 — Equilibrium Constants for the Reduction of UCl_4*

Temperature		$\frac{1}{T} \times 10^3$	K†	K‡
°C	°K			
400	673	1.486	0.295	0.386
420	693	1.443	0.438	0.549 (425°C)
453	726	1.377	0.789	0.762 (450°C)
477	750	1.333	1.100	1.014 (475°C)
502	775	1.290	1.550	1.277 (500°C)

$$*K = \frac{p_{HCl}}{p_{H_2}^{1/2}}.$$

†UCRL 6.
‡UCRL 5.

observations have been interpreted as indicating a disproportionation of UCl_3.

$$2UCl_3 \rightarrow UCl_2 + UCl_4 \qquad (10)$$

The evidence for the formation of UCl_2 is, however, far from conclusive (MP Ames 3; MP Chicago 2; see also Chap. 15).

(d) Oxidizing Agents. When UCl_3 is heated to 700°C in air and then subjected to vacuum distillation, a yellow solid, presumably UO_2Cl_2, sublimes. The probable reaction is (UCRL 2)

$$2UCl_3 + O_2 \rightarrow UO_2Cl_2 + UCl_4 \qquad (11)$$

Uranium trichloride begins to react with pure oxygen at about 150°C. The product is probably UO_2Cl_2, with some unreacted UCl_3 and a small amount of UO_2 (MP Ames 8). Chlorine reacts with UCl_3 smoothly at 250°C to give UCl_4.

$$UCl_3 + \frac{1}{2}Cl_2 \xrightarrow{250°C} UCl_4 \qquad (12)$$

Little heat is developed, and melting does not occur (MP Ames 2). At higher temperatures, however, the reaction is very vigorous. Thus when UCl_3 is treated with chlorine at 300°C, the UCl_3 burns, and the reaction mixture reaches an estimated temperature of 800°C. Brown vapors are evolved in this process. When more chlorine is passed over the now solidified reaction mass for 6 hr at 420°C, a second

Table 14.4 — Phase Relations of Some Uranium Trichloride — Metal Halide Systems

System	Compound	M.p. of compound, °C	Eutectic	M.p. of eutectic, °C
UCl_3-KCl	$2KCl \cdot UCl_3$	625 ± 5	$KCl-K_2UCl_5$ 20 mole % UCl_3	590 ± 5
	$3KCl \cdot UCl_3$	At 530 ± 5 transformed to $2KCl \cdot UCl_3$ and KCl	$2K_2UCl_5-UCl_3$ 50 mole % UCl_3	545 ± 5
UCl_3-NaCl	None		$NaCl-UCl_3$ 33 mole % UCl_3	520 ± 5
UCl_3-$BaCl_2$	None		One undetermined eutectic compound	635 − 640
UCl_3-UCl_4	None		Compounds immiscible in liquid phase and only slightly soluble in the solid phase	

volatile brown product can be obtained. The first brown sublimate has the composition $UCl_{3.67}$; the second sublimate has the composition $UCl_{5.33}$. A small amount of green crystals is also obtained; they condense just outside the furnace and have the composition $UCl_{4.83}$. Although the analytical results are not particularly good, these observations are reported here because of their bearing on the possible existence of additional uranium-chlorine compounds, similar to U_2F_9 and U_4F_{17} (MP Ames 1) (see Chap. 12).

Bromine and iodine react with UCl_3 to form mixed halides (MP Ames 9) (see Chap. 15). Anhydrous hydrogen fluoride oxidizes UCl_3 at elevated temperatures to UF_4 (CEW-TEC 1).

$$UCl_3 + 4HF \xrightarrow{450°C} UF_4 + 3HCl + \frac{1}{2}H_2 \qquad (13)$$

Liquid anhydrous hydrogen fluoride reacts with UCl_3 at 25°C with the evolution of hydrogen chloride; the product is presumed to be UF_3 (Brown 10).

(e) Reducing Agents. Hydrogen has been reported to reduce UCl_3 with the formation of HCl and UCl_2.

$$UCl_3 + \tfrac{1}{2}H_2 \rightarrow UCl_2 + HCl \qquad (14)$$

However, no experimental details or analytical results are available to substantiate this conclusion (MP Chicago 2). UCl_3 was found to be unattacked by atomic hydrogen at 700°C (MP Chicago 3).

(f) Complex Compounds of UCl_3. A number of complexes containing UCl_3 together with metal halides have been examined by thermal analysis (Brown 8). The results are summarized in Table 14.4.

2. URANIUM TETRACHLORIDE

Uranium tetrachloride was first prepared by Péligot (1842). Since then the compound has been studied by many workers.

Table 14.5 — Free Energies for Some Preparations of Uranium Tetrachloride

Equation No.	Equation*	$\Delta H_{298°K}$,[†] kcal	$\Delta S_{298°K}$,[†] e.u.	$\Delta F_{300°K}$, kcal	$\Delta F_{800°K}$, kcal
15	$UO_2 + 4HCl \rightarrow UCl_4 + 2H_2O$	−8.0	−59.9	+10.0	+39.9
16	$UO_2 + C + 2Cl_2 \rightarrow UCl_4 + CO_2$	−75.1	−28.4	−66.5	−52.3
17	$UO_2 + CCl_4 \rightarrow UCl_4 + CO_2$	−49.4	+5.4	−51.0	−53.7
18	$UO_2 + 2CCl_4 \rightarrow UCl_4 + 2COCl_2$	−36.2	+14.5	−40.6	−47.8
19	$UO_2 + 2COCl_2 \rightarrow UCl_4 + 2CO_2$	−62.5	−3.7	−61.4	−59.5
20	$UO_2 + 2SOCl_2 \rightarrow UCl_4 + 2SO_2$	−37.0	−0.1‡	−36.7	−36.9
21	$UO_2 + \tfrac{1}{2}S_2 \text{ (gas)} + 2Cl_2 \rightarrow UCl_4 + SO_2$§	−51.5	−46.1	...	−14.7
22	$UO_2 + 2PCl_5 \rightarrow UCl_4 + 2POCl_3$	−75.4	+8.3‡	−77.9	...
23	$UO_2 + 2PCl_3 + 2Cl_2 \rightarrow UCl_4 + 2POCl_3$	−117.4	−72.3‡	...	−59.5

*All substances except UCl_4, UO_2, and C are taken in the vapor state. For UO_2: $S_{298°K} = 18.63$ e.u.; $\Delta H_{298°K} = -270.4$ kcal (see Chap. 11). For UCl_4: $S_{298°K} = 47.14$ e.u.; $\Delta H_{298°K} = -251.0$ kcal (see Sec. 2.2f).

†Heats of formation from F.R. Bichowsky and F. D. Rossini, "The Thermochemistry of the Chemical Substances," Reinhold Publishing Corporation, New York, 1936; entropies from K. K. Kelley, "IX. The Entropies of Inorganic Substances," U. S. Bur. Mines Bull. 434, 1941.

‡Entropy of $SOCl_2$ and $POCl_3$ from D. M. Yost and H. Russell, Jr., "Systematic Inorganic Chemistry," Prentice-Hall, Inc., New York, 1944.

§This represents the reaction when S_2Cl_2 is used as chlorinating agent.

A number of halogenation reactions have received study in connection with the preparation of uranium tetrachloride. The heats and free energies of some of them are listed in Table 14.5. The entropies of halogenation are rather large, and this makes it essential to consider free energies rather than heats of reaction in any attempt to predict whether an active chlorinating agent is likely to prove useful.

Table 14.5 lists free energies at 300 and 800°K, the latter because temperatures of this order of magnitude are necessary to make most halogenation reactions proceed at a practically useful rate.

Parallelism between the free energy liberated and the rate of establishment of a reaction equilibrium is often observed in a series of analogous reactions but is, of course, not a general rule. Therefore it is not certain that the relative ease with which the various agents listed in Table 14.5 react with UO_2 corresponds exactly to the order of the ΔF values of halogenation. It can be predicted, however, that halogenation reactions which occur without marked decrease in standard free energy will be useless for practical purposes. Even if reactions of this kind, e.g., that with HCl, would proceed rapidly toward equilibrium, the small equilibrium concentration of the reaction products would make it impossible to remove them at a rate allowing conversion to be completed within a reasonable time.

A point to be kept in mind is that the free energies listed in Table 14.5 are average values for two chlorination steps, e.g.,

$$UO_2 + SOCl_2 \rightarrow UOCl_2 + SO_2 \tag{24}$$

$$UOCl_2 + SOCl_2 \rightarrow UCl_4 + SO_2 \tag{25}$$

In all likelihood the first step liberates more energy than the second; therefore, if the chlorinating agent is weak, the reaction may stop at the $UOCl_2$ stage.

All the other reagents listed can be expected to convert UO_2 to UCl_4 more or less readily. Carbon tetrachloride and phosgene (Eq. 19) appear to be powerful chlorinating agents, as is thionyl chloride (Eq. 20). Phosphorus pentachloride (Eq. 22) is especially useful at low temperatures where it does not dissociate into PCl_3 and Cl_2.

Considerations other than thermodynamics, however, enter into the choice of a reagent. It is essential to obtain pure UCl_4, free of the chlorinating reagent; this reduces, for example, the usefulness of sulfur chloride (Eq. 21). Corrosion resistance, ease of manipulation, and the nature of the other reaction products formed are other factors involved in the selection of a reagent.

The most important starting materials for the preparation of UCl_4 are the uranium oxides. The free metal, hydride, carbide, nitride, and sulfide can all be used, but these materials are less readily available. Uranyl compounds, on the other hand, are easily obtainable but insufficiently reactive to be of much value.

An important factor for all chlorination reactions with uranium oxides as starting material is the reactivity of the oxide used. Reactive uranium dioxide can be obtained, for example, by reduction of the higher oxides with methane, ammonia, or ethanol below 750°C, or by thermal decomposition of uranium(IV) compounds, such as the sulfate, oxalate, or benzoate. Uranium dioxide obtained in this way is a gray-black pyrophoric powder which oxidizes rapidly in air. The coffee-brown dioxide prepared by reduction or thermal decomposition above 800°C is much more inert.

The reasons for the variations in reactivity of the oxides are at present not entirely clear, particularly in the case of the dioxide. Recent investigations have shown that uranium trioxide can exist in at least four different crystal forms which differ considerably in thermal stability (see Chap. 11). More detailed examination may perhaps reveal a polymorphism also in the case of the dioxide. The reactivity of the oxides may also depend on the exact U/O ratio in the product (according to Chap. 11). This ratio can vary in the case of the lower oxide between 1.75 and 2.30 without change in crystal structure.

Halogenation reactions can be carried out either by gas-phase reactions in which the vaporized reagent is passed over the uranium compound at elevated temperatures or by liquid-phase reactions at more moderate temperatures. Many of the halogenating agents can be utilized in either of these two ways.

Preparative methods for the production of UCl_4 have been critically reviewed by Wagner (CEW-TEC 4); this discussion follows Wagner's review in many respects.

2.1 Preparation of Uranium Tetrachloride. (a) Uranium Dioxide and Hydrogen Chloride. Thermodynamic data (cf. Table 14.5) indicate that at elevated temperatures the equilibrium of conversion of uranium dioxide to UCl_4 by hydrogen chloride is not favorable for halogenation. The equilibrium concentrations of UCl_4 and H_2O, calculated from the relation $-\Delta F = RT \ln K$, are so small that these products cannot be removed rapidly enough to ensure a reasonable rate of conversion. It has in fact been found by the Purdue group (Purdue 1) that practically no reaction occurs when UO_2 [prepared by decomposition of uranium(IV) oxalate] is treated with hydrogen chloride in the vapor phase at 400°C. At low temperatures the thermodynamic conditions become less prohibitive. Experiments showed that, if a solvent such as water or aqueous alcohol is used, chlorination does occur; however, as anticipated above, the oxide is converted only to $UOCl_2$ rather than to UCl_4 (UCRL 9). Uranium trioxide also reacts with hydrogen chloride in ethanol or carbon tetrachloride, but, again, only UO_2Cl_2 and no UCl_4 is produced. Active uranium dioxide has been reported to react with hydrogen chloride in anhydrous ethanol, but the formation of UCl_4 in such solutions has not been proved (Purdue 1).

(b) Uranium Oxide, Carbon, and Chlorine. Uranium tetrachloride was first prepared by the reaction of chlorine with an intimate mixture of uranium dioxide and carbon (Péligot, 1842); this reaction has since been utilized by many investigators (Zimmermann, 1882; Fischer, 1913; Voigt, 1924; Hönigschmid, 1928). Carbon tetrachloride is probably the actual chlorinating agent. In order to obtain an intimate and reactive mixture of oxide and carbon, sugar can be mixed with oxide and ignited. Colani (1907) suggested that sugar charcoal or lampblack is particularly suitable, but other workers have found that carbon obtained in various other ways also is quite satisfactory. Under the conditions of the experiment, involving the passage of a large excess of chlorine, considerable quantities of the uranium pentachloride are also produced (Roscoe, 1874). Biltz and Fendius (1928b) tried to minimize the contamination by the pentachloride by using a large excess of charcoal in the mixture, but this device seems to be of little value. The best ratio of oxide to graphite is about 0.6 (CEW-TEC 5). Any oxide or oxyhalide of uranium can be used instead of the dioxide, but, other things being equal, the dioxide will give the least amount of pentachloride. With a sodium uranate–lampblack mixture the reaction proceeds satisfactorily at 600°C (Brown 11).

The principal advantage of this method is readily accessible starting materials. The disadvantages, which seem to outweigh the advantages, are the production of comparatively large amounts of UCl_5 (which must be eliminated by subsequent treatment), the high temperature of reaction which imposes a severe strain on the equipment, and the presence of considerable quantities of phosgene in the waste gases.

Recently this method has again been studied at the Tennessee Eastman laboratories (CEW-TEC 6). Pellets made by compressing a mixture of uranium trioxide and carbon were supported in a vertical reactor on a perforated graphite disk, and chlorine was conducted through the mixture from the bottom. The reaction zone was maintained at 850°C. At this temperature uranium tetrachloride is liquid; it therefore trickles through the bed and collects in the bottom of the reactor. However, large amounts of UCl_5 were formed and swept out as dust in the gas stream. On the whole, the results did not seem too encouraging.

(c) Uranium Oxide and Carbon Tetrachloride Vapor. Carbon tetrachloride was first introduced by Watts and Bell (1878) as an agent for the conversion of metal oxides to chlorides. The utility of this reagent was rediscovered by Demarcay (1887) and by Lothar Meyer (1887), who examined its reaction with a large number of oxides. Carbon tetrachloride was first applied to the halogenation of the oxides of uranium by Colani (1907), who found that both U_3O_8 and UO_2 reacted at red heat with gaseous CCl_4. He noticed that the amount of UCl_5 formed concurrently with UCl_4 was much smaller with UO_2 than with U_3O_8. Camboulives (1910), who treated some thirty metal oxides with CCl_4 vapor, also noted that U_3O_8 reacted at 300°C to form a mixture of UCl_4 and UCl_5. These early workers all utilized gaseous carbon tetrachloride at elevated temperatures. Michael and Murphy (1910) were the first to study the liquid-phase halogenation. They used liquid CCl_4 as well as solutions of chlorine in CCl_4, and their work forms the basis of the present-day methods of conversion of uranium oxides to UCl_4, UCl_5, and UCl_6 by liquid-phase chlorination.

The course of reaction of uranium dioxide with carbon tetrachloride was studied by measuring and analyzing the gaseous products. At temperatures of 450 ± 20°C (the optimum temperature for the reaction with "inactive" UO_2) the principal reaction was found to be (UCRL 16) (cf. Eq. 17)

$$UO_2 + CCl_4 \rightarrow UCl_4 + CO_2$$

Small amounts of phosgene, carbon monoxide, and chlorine found in the exit gases (UCRL 14) indicate the occurrence of competitive reactions, such as

$$UO_2 + CCl_4 \rightarrow UOCl_2 + COCl_2 \tag{26}$$

$$UOCl_2 + CCl_4 \rightarrow UCl_4 + COCl_2 \tag{27}$$

$$UO_2 + 2COCl_2 \rightarrow UCl_4 + 2CO_2 \quad \text{(Eq. 19)}$$

$$COCl_2 \rightarrow CO + Cl_2 \tag{28}$$

$$UCl_4 + \tfrac{1}{2}Cl_2 \rightarrow UCl_5 \tag{29}$$

The principal reaction (Eq. 17) is exothermic to the extent of about 50 kcal per mole, and local overheating may cause partial fusion and thus lead to incomplete conversion. Attempts have been made to run the reaction at temperatures high enough to sublime the UCl_4 out of the reaction zone, but this procedure appears unsatisfactory since considerable amounts of uranium pentachloride and hexachloroethane are then formed.

The reaction of uranium trioxide with CCl_4 vapor has also received considerable study since direct conversion of UO_3 to UCl_4, without preliminary preparation of UO_2, often appears desirable. The reaction of UO_3 and CCl_4 vapor proceeds under much the same conditions as does that of UO_2, but, because of the higher valence of the oxide, the amount of UCl_5 formed is greater. As the temperature is increased from 300 to 350°C, the UCl_4 production increases more than that of UCl_5 (UCRL 17), but the proportion of UCl_5 in the product becomes higher (see Table 14.6). With "active" UO_3 from calcined $UO_4 \cdot 2H_2O$, complete chlorination was obtained at 350°C.

The mechanism of this reaction has been studied by methods similar to those employed in the corresponding reaction with UO_2. There

Table 14.6 — Formation of UCl_5 from UO_3 and CCl_4 Vapor

Temp., °C	Conversion, %	UCl_5 formed, %	UCl_4 formed, %	UCl_5 in UCl_4, %
300	79.6	8.3	71.3	11.7
350	92.2	11.7	80.5	14.5

appears to be a rapid initial reaction in which carbon dioxide and chlorine are formed and the trioxide is converted to UO_2Cl_2 (UCRL 18). The second stage of the reaction, the conversion of the oxychloride to tetrachloride, is slow at 400°C. The main stoichiometric reaction at 400°C is considered to be (UCRL 18,19)

$$2UO_3 + 3CCl_4 \rightarrow 2UCl_4 + 3CO_2 + 2Cl_2 \qquad (30)$$

The following individual reactions have been postulated to account for the presence of $COCl_2$, CO, Cl_2, and UCl_5 among the reaction products:

$$UO_3 + CCl_4 \rightarrow UO_2Cl_2 + COCl_2 \qquad (31)$$

$$UO_3 + COCl_2 \rightarrow UO_2Cl_2 + CO_2 \qquad (32)$$

$$2UO_2Cl_2 + 2CCl_4 \rightarrow 2UCl_6 + 2CO_2 \qquad (33)$$

$$2UCl_6 \rightarrow 2UCl_5 + Cl_2 \qquad (34)$$

$$2UCl_5 \rightarrow 2UCl_4 + Cl_2 \qquad (35)$$

$$COCl_2 \rightarrow CO + Cl_2 \quad (Eq.\ 28)$$

The results obtained with U_3O_8 are intermediate between those found with UO_2 and UO_3. This oxide can be chlorinated at about 450°C with the formation of 8 to 12 per cent UCl_5.

Attempts have been made to control the amount of UCl_5 produced from UO_3 by introducing CO, $CHCl_3$, or other compounds to react with the liberated chlorine,

$$CO + Cl_2 \rightarrow COCl_2 \tag{36}$$

$$CHCl_3 + Cl_2 \rightarrow CCl_4 + HCl \tag{37}$$

thereby preventing the formation of UCl_5 (UCRL 20,21). It has been claimed that the presence of CO greatly diminishes the formation of UCl_5 (UCRL 22), but the consensus is that this expedient is on the whole unsatisfactory. The addition of CH_4, SO_2, $CHCl_3$, and various other chlorinated hydrocarbons was found ineffective. The use of active UO_2 and a low temperature of reaction remain the best prescription for the production of UCl_5-free tetrachloride.

(d) Liquid-phase Reaction of UO_3 and CCl_4. Decomposition of UCl_5. It was found by the Brown University group that UO_2 or UO_3 reacts slowly with liquid CCl_4 at the boiling point (77°C) to give UCl_4 (Brown 2). Higher temperatures and pressures greatly accelerate the liquid-phase reaction. Thus uranium dioxide reacts rapidly with liquid CCl_4 at 250°C.

The higher uranium oxides react more easily, i.e., at lower temperatures and pressures, than does UO_2; consequently the greatest attention has been paid to the reaction of liquid CCl_4 with UO_3.

At 115 to 185°C the primary reaction product of UO_3 and liquid CCl_4 probably is UCl_5.

$$2UO_3 + 6CCl_4 \rightleftharpoons 2UCl_5 + 6COCl_2 + Cl_2 \tag{38}$$

Both UCl_5 and UCl_4 may be prepared by this reaction under the proper conditions.

Uranium pentachloride undergoes decomposition to the tetrachloride when heated in a stream of nitrogen, carbon dioxide, or even chlorine (Brown 3,4,6,14 to 17; MP Ames 10).

$$2UCl_5 \rightleftharpoons 2UCl_4 + Cl_2 \quad \text{(Eq. 35)}$$

Decomposition begins below 100°C and is complete at 250°C. The volatility and poor heat conductivity of the UCl_5 cause some complications in the practical dechlorination procedure. The dechlorination can be smoothly effected by placing the UCl_5 in a narrow, vertical glass cylinder which is then slowly lowered into a furnace maintained at 500°C. A stream of carbon dioxide or nitrogen is played on the surface of the UCl_5 to help condense the material that may otherwise be lost by sublimation.

Dechlorination of UCl_5 can also be accomplished by refluxing with S_2Cl_2, $SnCl_4$, C_3Cl_6, etc. (CEW-TEC 9). It is usually difficult, however, to free the product from the solvent. Thionyl chloride works better;

with this reagent, both dechlorination and removal of the solvent can be achieved without much difficulty (CEW-TEC 10).

(e) Chlorination in the Vapor Phase with Other Organic Halogen Compounds. A large number of halogenated organic compounds have been examined as reagents that might convert uranium oxides to UCl_4 at low temperatures without formation of phosgene or UCl_5. These compounds are listed in Table 14.7. In general, compounds with more than four carbon atoms are unsatisfactory because of excessive carbon formation. Chloroform and $CHCl_3$-CCl_4 mixtures also produce considerable carbonization. Chloroform itself reacts slowly; conversion is incomplete, and the UCl_4 is contaminated with pyrolytic products (UCRL 21). Hexachloropropene and hexachlorocyclopentadiene, unsaturated compounds that have chlorine atoms in the alpha position relative to a double-bonded carbon atom, are fairly satisfactory. However, none of these compounds has any advantages over carbon tetrachloride as far as vapor-phase chlorination is concerned.

(f) Miscellaneous Liquid-phase Halogenations. (CEW-TEC 12,15, 33; Purdue 3.) A large number of organic halogen compounds have been examined for utility in liquid-phase halogenation (CEW-TEC 11, 33) (see Table 14.7). The saturated chlorine compounds must have a much higher boiling point than CCl_4 in order to show appreciable reaction and are, on the whole, not very effective. Compounds with a vinylic chlorine (i.e., a $C=C-Cl$ group) have little chlorinating capacity, as may be expected from the general inertness of such chlorine atoms in substitution reactions. Trichloroacrylyl chloride, however, reacts with UO_3 to form a water-insoluble compound, which is soluble in acetone or benzene and appears to be a partly chlorinated oxide. Compounds with an allylic chlorine (a $C=C-CCl$ group) are in general more effective. Thus hexachloropropene, hexachlorocyclopentadiene (C_5Cl_6), and benzotrichloride ($C_6H_5CCl_3$), among others, react vigorously with many oxides, but it is not always possible to isolate pure products from these reaction mixtures.

It has been found, for example, at the Tennessee Eastman laboratories that $MoCl_4$, WCl_6, $WOCl_4$, $CoCl_3$, $CbCl_5$, $TaCl_5$, VCl_3, and $HgCl_2$ can be prepared by refluxing the corresponding oxide with halogenated hydrocarbons such as hexachloropropene (CEW-TEC 14). In general, little or no reaction is observed with oxides of the first four groups of the periodic table. Refluxing for a short time with hexachloropropene converts UO_3, U_3O_8, $UO_2 + 2UO_3$, $UO_4 \cdot 2H_2O$, UO_2Cl_2, UO_2SO_4, $UO_2(NO_3)_2$, and other uranyl compounds to UCl_4. Pure uranium dioxide and $(NH_4)_2U_2O_7$ do not appear to undergo this reaction (Purdue 4; CEW-TEC 9,13,33). With UO_3, a rapid and exothermic reaction starts below 100°C. Uranium pentachloride, which is easily soluble in the reaction mixture, forms first, and it is then decomposed by further heating to 165°C. Reaction is incomplete if the mole ratio of hexachloropropene to uranium trioxide falls below 5/1; the UCl_4 should be

Table 14.7 — Some Organic Compounds That Have Been Investigated As Halogenating Agents for Uranium Trioxide*

Compound	Formula	Melting point, °C	Boiling point,† °C
Carbon tetrachloride	CCl_4	−22.6	76.8
Hexachloroethane	CCl_3CCl_3	186.9−187.4	185.6 (777)
Tetrachloroethylene	$CCl_2 = CCl_2$	−19	120.8
Trichloroacrylyl chloride	$Cl_2C = CCl - COCl$		158
Hexachlorobutadiene	$CCl_2 = CCl - CCl = CCl_2$	39	283−284
		32	268−269
Trichloroacetyl chloride	CCl_3COCl		118
Hexachloropropene	$CCl_2 = CClCCl_3$	−70	210 (760)
			98 (11)
Hexachlorocyclopentadiene	C_5Cl_6		
Dichlorodibromomethane	CBr_2Cl_2	22	135
Trichlorobromomethane	$CBrCl_3$	−21	104.1
Trichloroacetonitrile	CCl_3CN		83−84
Methyl chloride	CH_3Cl	−97.7	−24
Methylene chloride	CH_2Cl_2	−96.7	40−41
Chloroform	$CHCl_3$	−63.5	61.2
1,1,1-Trichloroethane	CH_3CCl_3		74.1
sym-Tetrachloroethane	$CHCl_2CHCl_2$	−36	146.3
1,2,3-Trichloropropane	$CH_2ClCHClCH_2Cl$	−14.7	158
Trichloroethylene	$CCl_2 = CHCl$	−73	87.2
Trichlorocumene	$(CH_3)_2CHC_6H_2Cl_3$		
Chloral	CCl_3CHO	−57	97.6 (768)
Allyl chloride	$CH_2 = CHCH_2Cl$	−136.4	44.6
2,3-Dichloropropylene-1	$H_2C = CCl - CH_2Cl$		94
Benzotrichloride	$C_6H_5CCl_3$	−4.75	220.7
3,3-Dichloropentanedione-2,4	$CH_3COCCl_2COCH_3$		87 (18−20)
			79 (11)
1,1-Dichloroethane	CH_3CHCl_2	−96.7	57.3
1,1,2-Trichloroethane	$CH_2ClCHCl_2$	−36.7	113.5
Ethyl trichloroacetate	$CCl_3COOCH_2CH_3$		168
2,2-Dichloropropane	$CH_3CCl_2CH_3$	−34.6	69.7

*UCRL 24; CEW-TEC 11,33; Purdue 2.
†Numbers in parentheses indicate pressures expressed in millimeters of mercury.

washed with CCl_4 (after filtration) to remove traces of organic matter. The principal organic product of this reaction is trichloroacrylyl chloride, $CCl_2 = CClCOCl$ (b.p. 158°C) (Purdue 5; CEW-TEC 13,33).

Since this compound tends to hold uranium in solution, it may be necessary to remove it by distillation before attempting the isolation of the reaction products.

A number of other chlorinated hydrocarbons have been studied as halogenating agents for UO_3, using sealed glass bomb tubes (CEW-TEC 16). CBr_2Cl_2 and $CBrCl_3$ produce pure UCl_4. At 130°C, the compounds

I	II
$CH_3COCCl_2COCH_3$	$CCl_3COOCH_2CH_3$
CCl_3CHO	$CHCl=CCl_2$
CCl_3COCl	CCl_3CN
CH_3CHCl_2	CH_3CCl_3
$CCl_2=CCl_2$	$CH_3CCl_2CH_3$
$CH_2ClCHCl_2$	

are less satisfactory. These compounds are arranged approximately in order of decreasing reactivity, starting at the head of column I. Chloroform does not react even in the presence of appreciable quantities of UCl_5.

(g) Uranium Oxide and Phosgene. Phosgene was introduced as a halogenating agent by Chauvenet (1911), who found it to be a powerful reagent of wide applicability. Reaction occurs not only with uranium oxides but also with sulfides and phosphates. Thus the mineral autunite reacts at 800°C.

$$2UO_2 \cdot CaO \cdot P_2O_5 + 8COCl_2 \rightarrow 2UCl_4 + 2POCl_3 + 8CO_2 + CaCl_2 \quad (39)$$

Chauvenet found the reaction of phosgene with U_3O_8 to proceed at 450°C; this was confirmed by Rosenheim and Kelmy (1932). Active UO_2 reacts at slightly lower temperatures. In recent work, phosgene has not been utilized extensively, presumably because of its toxic properties. British workers have observed complete conversion of U_3O_8 to UCl_4 by $COCl_2$ at 900 to 940°C (British 1). Phosgene has been used on a commercial scale for the conversion of thorium dioxide to anhydrous chloride, and there is little doubt of its general efficacy in the preparation of anhydrous chlorides.

(h) Uranium Oxide and Thionyl Chloride. (UCRL 24; Purdue 1,2,4, 6.) Thionyl chloride is an effective reagent for converting the uranium oxides to UCl_4. In the vapor phase, UO_3 (obtained by decomposition of the peroxide) is rapidly converted to UCl_4 at 350°C. U_3O_8 reacts less rapidly. Both $UO_4 \cdot 2H_2O$ and $(NH_4)_2U_2O_7$ can be chlorinated at 350°C. The most satisfactory oxide from the point of view of the least UCl_5 contamination is active UO_2 (prepared from $UO_4 \cdot 2H_2O$ or UO_3 by reduction with alcohol vapor) (UCRL 5).

In the liquid phase active UO_3 can be converted to UCl_5 by refluxing with $SOCl_2$. The addition of UCl_5 aids the reaction. Difficulties are encountered in isolating UCl_5 from the reaction mixture because of the high solubility of UCl_5 in $SOCl_2$ (Table 14.8). A reddish compound, $UCl_5 \cdot xSOCl_2$, crystallizes from such solutions; this complex may be decomposed at 100°C in vacuum. Solutions of UCl_5 in $SOCl_2$ can be simultaneously dechlorinated and desolvated by spray-drying at 300°C (CEW-TEC 10).

Table 14.8 — Solubility of Some Uranium Compounds
in Liquid Thionyl Chloride*

Compound	Solubility in $SOCl_2$, g/ml
UCl_4	3×10^{-4}
UCl_5	1.29
UCl_6	0.17
UO_2Cl_2	3×10^{-4}

*CEW-TEC 17.

Chlorination of UO_3 with liquid $SOCl_2$ has been carried out at 100 to 130°C and 75 to 150 psi (Purdue 4; CEW-TEC 8). The reaction may be directed so as to produce either UCl_4 or UCl_5, but the results are rather unpredictable.

Sulfuryl chloride, SO_2Cl_2, does not chlorinate either UO_3 or U_3O_8.

(i) Uranium Oxide and Sulfur Chloride. Sulfur monochloride, S_2Cl_2, was first suggested as a chlorinating agent by Matignon and Bourion (1905), and the reaction was later applied to uranium oxides by a number of workers. Colani (1907) found that a mixture of chlorine and S_2Cl_2 reacts with U_3O_8 at red heat but that, unless the UCl_4 is sublimed away, the reaction is incomplete. Considerable amounts of UCl_5 are formed. Lely and Hamburger (1914) found that, if S_2Cl_2 was used without chlorine, conversion to UCl_4 occurred without much UCl_5 formation (see also Hülsmann, 1934). Moore (1923) has given a detailed description of the preparation of UCl_4 by reaction of S_2Cl_2 with U_3O_8. Goggin (1926) studied the reaction of U_3O_8 with S_2Cl_2 and chlorine at 900°C. The study was conducted in such a way that reaction and sublimation could proceed simultaneously.

"Active" oxides begin to react at 250°C, and reaction is usually complete below 450°C. At elevated temperatures the reagent is actually dissociated into sulfur and chlorine.

Thermodynamically, the process

$$UO_2 + 2S_2Cl_2 \rightarrow UCl_4 + SO_2 + 3S \tag{40}$$

is more favorable than

$$2UO_2 + S_2Cl_2 + 3Cl_2 \rightarrow 2UCl_4 + 2SO_2 \tag{41}$$

The use of chlorine as a carrier gas is advisable, however, since contamination of the product by sulfur should be reduced by its use. The reaction proceeds at much lower temperatures in the liquid phase. A 95 per cent conversion can be effected by refluxing U_3O_8 with S_2Cl_2 (138°C) for 7 hr at atmospheric pressure (Brown 2,14; UCRL 24). All the oxides of uranium undergo this reaction. Higher pressures facilitate the conversion. The product is usually heavily contaminated with sulfur, which is difficult to remove. It is practically impossible to remove it by vacuum sublimation (British 1,2; Natl. Bur. Standards 2). It has been proposed to purify UCl_4 containing sulfur by treatment with chlorine, which was expected to convert sulfur into volatile SCl_4 (UCRL 26), but large amounts of UCl_5 may be formed under these conditions.

(j) Chlorination with Other Anhydrous Inorganic Chlorides. (CEW-TEC 19,20; UCRL 26.) The chlorination of UO_2 by a large number of inorganic halides has been studied, particularly at the Tennessee Eastman laboratories. Thermodynamically, reaction can be expected only with the halides of metals that form very stable oxides, since UO_2 itself is very stable. None of the reagents studied proved to be particularly useful, except perhaps aluminum trichloride. Ammonium chloride, which is widely used for the conversion of rare earth oxides to chlorides, fails in the case of uranium (Booth, 1939). The results of UO_2−metal halide reactions are summarized in Table 14.9.

Uranium tetrafluoride has also been studied with respect to its behavior toward many of the above-mentioned reagents. It can be readily converted to UCl_4 with $AlCl_3$ and BCl_3 in the vapor phase (CEW-TEC 21). In the liquid phase, UF_4 can be converted to UCl_4 to the extent of 5 per cent by refluxing with $SiCl_4$ (1 hr) and to the extent of 10 to 20 per cent by refluxing with $SbCl_3$ (2 hr); 80 per cent conversion can be achieved by refluxing with $SnCl_4$ (2 hr), and a 90 per cent yield by refluxing with $TaCl_5$ (1 hr). As could be anticipated from the heats of formation of the fluorides, fusion of UF_4 with NaCl, KCl, CuCl, $NiCl_2$, $MnCl_2$, $PbCl_2$, and $SnCl_2$ yields no UCl_4; but $BeCl_2$, $MgCl_2$, $CaCl_2$, $BaCl_2$, and $NaAlCl_4$ convert UF_4 to UCl_4 at 700 to 800°C (CEW-TEC 22). It is difficult to recover UCl_4 free from the chlorinating agent from the products of these reactions.

Uranium(IV) sulfate and the eutectic mixture NaCl + CaCl$_2$ (51.5 per cent CaCl$_2$) yield a sublimate of UCl$_4$ when heated to 750°C in vacuum; 10 per cent conversion can be obtained after several hours. With BaCl$_2$ at 580°C, only a small amount of green sublimate is obtained after 13 hr (Purdue 1).

Table 14.9 — Reaction of UO$_2$* with Inorganic Halides

Halide	Conditions†	Result
NH$_4$Cl, CuCl, KCl, Hg$_2$Cl$_2$	Fusion at melting point of salt	No reaction
CaCl$_2$	700–800°C	5% UCl$_4$
BeCl$_2$	500°C	Complete conversion to UCl$_4$ in 4 hr
MgCl$_2$	700–800°C	15% UCl$_4$
SnCl$_2$, ZnCl$_2$	500°C	Small amount of green sublimate
NiCl$_2$, CuCl$_2$, FeCl$_2$, CoCl$_2$, CdCl$_2$, PbCl$_2$, HgCl$_2$, BaCl$_2$, MnCl$_2$	500°C	No reaction
AlCl$_3$	250°C	Reacts readily
SbCl$_3$	Reflux	No chlorination
PCl$_3$	Reflux	5% UCl$_4$
FeCl$_3$	300°C	An unidentified uranium compound (not UCl$_4$)
SiCl$_4$	Reflux	No reaction
SnCl$_4$	Reflux	No reaction
PCl$_5$	300°C	20% conversion
TaCl$_5$	Reflux	10–20% UCl$_4$

*Mallinckrodt UO$_2$, which is very unreactive, was used in these experiments.

†Where the temperature is above the boiling point of the halide, the reaction was run in a bomb tube.

(k) Preparation of UCl$_4$ in Aqueous Solutions. Aqueous solutions of UCl$_4$ can be readily prepared by reduction of uranyl chloride solutions. Reduction may be effected electrolytically; photochemically; by amalgams of silver, cadmium, or bismuth; by metallic copper; or by sodium hyposulfite, Na$_2$S$_2$O$_4$ (Zimmermann, 1882; Kohlschütter, 1900; Giolitti, 1905; Aloy, 1922). Treatment with air may be required to oxidize any uranium(III) that is formed.

Attempts to isolate anhydrous UCl$_4$ from such solutions have been only partly successful. Rosenheim and Kelmy (1932) reported the preparation of the hydrate UCl$_4$·10H$_2$O by evaporation of the aqueous solution to a syrup, followed by treatment with absolute alcohol saturated with hydrogen chloride. The Ames group was unable to confirm this result. The product they obtained by dehydration in a stream of hydrogen chloride was a mixture of UO_2 and $UOCl_2$ (MP Ames 11). The Ames results were later confirmed by other investigators.

The direct dehydration of uranium(IV) solutions according to the procedure of Kleinheksel and Kremers (1928), which is very effective for rare earths, does not give anhydrous UCl_4. However, the $UOCl_2$ which can be obtained in this way may be converted, at least in part, to UCl_4 by thermal disproportionation (UCRL 9) (see Chap. 16).

Promising results have been obtained by dehydration of aqueous UCl_4 solutions by azeotropic distillation (CEW-TEC 23). Benzene, carbon tetrachloride, chloroform, etc., have been added to the aqueous solution of UCl_4, the azeotrope being distilled until all the water has been removed. Removal of the dehydrating solvent itself leaves solid green products from which 10 to 70 per cent of UCl_4 can then be obtained by sublimation.

(1) Miscellaneous Preparations of UCl_4. Halogenation of Various Uranium Compounds. Uranium metal readily reacts with many halogenation agents to give the tetrachloride (Péligot, 1842; Zimmermann, 1882; Moissan, 1896).

$$U + 2Cl_2 \xrightarrow{250°C} UCl_4 \tag{42}$$

Hydride can be used instead of the metal (see Chap. 8).

$$UH_3 + 2COCl_2 \xrightarrow{250°C} UCl_4 + 2CO + \tfrac{3}{2}H_2 \tag{43}$$

$$UH_3 + \tfrac{7}{2}Cl_2 \longrightarrow UCl_4 + 3HCl \tag{44}$$

At higher temperatures uranium metal also reacts with CH_3Cl or HCl. The direct reaction of UH_3 with chlorine is not suitable for laboratory use since, once it has started at 200°C, it cannot be controlled. The reaction sequence

$$U + \tfrac{3}{2}H_2 \rightarrow UH_3 \tag{45}$$

$$UH_3 + 3HCl \rightarrow UCl_3 + 3H_2 \quad (Eq. 2)$$

$$UCl_3 + \tfrac{1}{2}Cl_2 \rightarrow UCl_4 \quad (Eq. 12)$$

is much more convenient when uranium metal is to be converted to UCl_4 (MP Ames 2).

Uranium sulfide (UCRL 26) and carbide (Moissan, 1897) can be chlorinated with chlorine gas at 400°C; the carbide also reacts with hydrogen chloride at elevated temperatures. Uranium nitrides, UN and $UN_{1.75}$, react readily with hydrogen chloride at 400 to 500°C to yield a mixture of ammonium and uranium chlorides (MP Ames 12) (see Chap. 10). Most uranyl salts are susceptible to chlorination in the vapor phase but react more slowly than the oxides. Uranates and diuranates are partially chlorinated in the liquid phase by carbon tetrachloride, but the products have not yet been well characterized (CEW-TEC 22).

Attempts have been made to convert organic uranium(IV) compounds to UCl_4 by double decomposition. Thus uranium(IV) acetyl acetonate has been treated, in benzene or ethyl ether solution, with anhydrous hydrogen chloride. Black tars resulted and no UCl_4 could be isolated (UCRL 27). As is also the case with thorium, the white solid that first precipitates when the solution is treated with hydrogen chloride still contains organic matter. Uranium(IV) benzoate, $U(OOCC_6H_5)_4 \cdot H_2O$, suspended in ethyl ether and treated with anhydrous hydrogen chloride yields green compounds with chlorine/uranium ratios between 2.5 and 3.8, which contain 20 to 40 per cent organic matter (CEW-TEC 24). The green product dissolves in water to a green solution, but no UCl_4 can be obtained from it by sublimation.

(m) <u>Purification of Uranium Tetrachloride</u>. Uranium tetrachloride can be purified either by direct sublimation in vacuum or by sublimation as a higher chloride in a stream of chlorine (275 to 700°C) and subsequent reconversion to UCl_4 (Brown 14). Sublimation can also be performed in a stream of inert gas, such as nitrogen or helium. In nitrogen, sublimation is rapid at 600 to 650°C at a pressure of 3 to 5 mm Hg; the UCl_4 sublimes without melting. If the sublimation is performed at atmospheric pressure, the nonvolatile residue can be greatly reduced by adding CCl_4 to the nitrogen stream (Brown 3). Helium with a small admixture of chlorine has been used for the same purpose.

The quality of purified UCl_4 can be evaluated by several criteria, the most important being solubility in certain selected solvents and volatility (Purdue 7,8; CEW-TEC 26).

2.2 <u>Physical Properties of UCl_4</u>. (a) <u>Melting Point</u>. The most reliable value of the melting point of UCl_4 that has been determined is 590 ± 1°C (CEW-TEC 27,35). The unweighted average of a number of determinations made by various other observers is 589°C (MP Ames 13; Brown 3,5,8; UCRL 32 to 34). The published value of 564°C (Voigt, 1924) is definitely too low.

(b) <u>Transition Points</u>. A study of the cooling curve of UCl_4 showed solid-solid transitions at 212, 506 to 522, and 543 to 565°C. Only small heat effects (approximately 1 kcal per mole) appeared to be associated with these transitions (UCRL 31). Observers at the National Bureau of Standards (Natl. Bur. Standards 3) were unable to confirm the phase transition at 211°C. The reality of all the reported phase transitions therefore remains uncertain.

(c) Boiling Point. The best direct measurement of the boiling point of UCl_4 (made at the Tennessee Eastman laboratories) gave 792°C at 760 mm Hg pressure (CEW-TEC 27,35). There appears to be no appreciable decomposition at this temperature. Several other determinations of the boiling point are of doubtful accuracy (MP Ames 14, 15; UCRL 32).

Table 14.10 — Vapor Pressure of Solid Uranium Tetrachloride

Temp., °C	Pressure, mm Hg	Temp., °C	Pressure, mm Hg
350	0.00037	430	0.030
360	0.00067	440	0.047
370	0.0012	450	0.076
380	0.0022	460	0.12
390	0.0038	470	0.18
400	0.0064	480	0.28
410	0.011	490	0.43
420	0.018	500	0.64
		505	0.80

(d) Vapor Pressure. The vapor pressure of solid UCl_4 has been determined by two methods — by effusion at Berkeley and by transpiration at Ames. The results disagree. The reason for this is not obvious, but the effusion data probably are more reliable. According to a critical survey of the effusion measurements (UCRL 35), the vapor pressure of the solid UCl_4 can be expressed by the equation

$$\log p_{mm\ Hg} = -\frac{10,427}{T} + 13.2995 \quad (350-505°C) \qquad (46)$$

Values calculated from this equation are given in Table 14.10. The heat of sublimation is 47.7 kcal per mole, and the free energy of sublimation is given by the equation

$$\Delta F° = 60,606 + 36.85T \log T - 170.9T \quad \text{cal per mole} \qquad (47)$$

The vapor pressure of liquid UCl_4 has been determined at the Tennessee Eastman laboratories (CEW-TEC 27,35); the results obtained there supersede the older values (UCRL 32). The results can be expressed by the equation

$$\log p_{\text{mm Hg}} = -\frac{7205}{T} + 9.65 \quad (590-790°C) \quad (48)$$

Experimental values are given in Table 14.11 and plotted in Fig. 14.1. The heat of vaporization calculated from these data is 33 kcal per mole. The latent heat of fusion is then $47.7 - 33 = 14.7$ kcal per mole,

Table 14.11 — Vapor Pressure of Liquid Uranium Tetrachloride

Temp., °C	Pressure, cm Hg	Temp., °C	Pressure, cm Hg
789.5	74.63	714	23.0
788	74.55	697.5	17.2
791	74.26	692	15.0
769	54.6	660	9.0
770	54.1	648	6.1
763.5	54.0	639	6.0
761	47.2	632	4.7
757	46.0	628	4.6
747	39.5	611	3.2
742	37.4	603	2.7
733	30.9	597	2.3
727.5	28.0	591	1.9

and Trouton's constant is 31. Liquid UCl_4 thus appears to be a highly associated liquid.

Extrapolation of the vapor pressure curves of liquid UCl_4 gives $p = 19.5 \pm 1$ mm Hg for the vapor pressure at the melting point, $590 \pm 1°C$. This gives the best available estimate for the triple-point coordinates.

The vapor pressure measurements made at Ames by the transpiration method (MP Ames 15) gave results of an altogether different order of magnitude (much higher than those listed in Table 14.11). They must contain a systematic error and will not be further discussed here.

(e) Crystal Structure and Density. Uranium tetrachloride forms metallic, dark-green octahedral crystals. X-ray study (MP Chicago 4,5; Mooney, 1949) shows that they have tetragonal symmetry, with the lattice constants

$$a_1 = 8.296 \pm 0.009 \text{ A} \qquad a_3 = 7.487 \pm 0.009 \text{ A}$$

There are four UCl_4 molecules per unit cell, and the calculated density is 4.87 g/cc. The symmetry group is I4/amd, and the positions of the atoms are as follows:

4 U in 4 b

16 Cl in 16 h

with x = 0.281 and z = 0.398. These Chicago results have later been confirmed elsewhere (Johns Hopkins 1, 2; MP Ames 16).

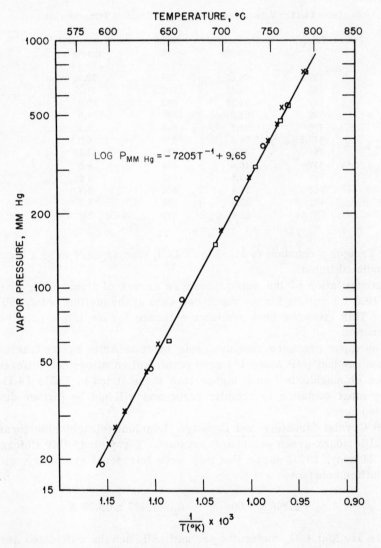

Fig. 14.1—Vapor pressure of liquid uranium tetrachloride. ×, first lot; □, second lot; ○, third lot. Normal boiling point is 792°C.

The density has been determined also by direct measurement. The older values were 4.854 (Biltz, 1928b) and 4.725 (Hönigschmid, 1928). The density has been determined by gas displacement with hydrogen from room temperature down to liquid-air temperatures (Hülsmann and Biltz, 1934). These workers obtained 4.950 at $-193°C$ and 4.890 at $-78°C$. Their value at 19°C agrees well with the x-ray density. The Ames group finds a density of 4.88 g/cc at 24°C which also agrees very well with the x-ray value (MP Ames 10).

The density of UCl_4 vapor has also been measured (Zimmermann, 1881). The average value is 13.31 (referred to air); it shows that no association takes place. An ebullioscopic determination of the molecular weight in bismuth trichloride as solvent indicates that UCl_4 is present in the form of simple molecules in solutions also—at least, in dilute solutions at high temperatures (Rugheimer, 1908).

(f) Thermochemical Data. Observers of the National Bureau of Standards (Natl. Bur. Standards 4) have determined the specific heat,[*] entropy, and enthalpy of UCl_4 (solid) from 15 to 355°K (Table 14.12). Measurements were made in an atmosphere of dry helium to avoid oxidation and hydrolysis. The values of S and the function $H - E_0^s$ were obtained by tabular integration, using Simpson's rule. The entropy of UCl_4 (solid) at 298.16°K is 47.14 e.u. These results have been extended in a second series of determinations (Natl. Bur. Standards 3; Ginnings and Corruccini, 1947), which covered the temperature range 0 to 427°C (Table 14.13). The differences between the two sets of values in the range where overlapping occurs is about 0.7 per cent; the high-temperature values (Table 14.13) are probably more reliable since the UCl_4 used was of better quality.

The following equations[†] fit the experimental data in Table 14.13 with the indicated precision. They should not be used for temperatures below 0°C.

$$C_p = 0.07608 + 2.88 \times 10^{-5}t - \frac{1.3}{(t + 41)^2} \quad \text{cal/g/°C} \quad (49)$$

(Precision: ± 0.35 per cent, 0 to 427°C)

where t is in degrees centigrade.

---

[*]The sample used contained less than 0.02 per cent nonvolatile impurities.

[†]Alternative equations for the same data are given by Ginnings and Corruccini (1947).

$$H - H_{0°C} = 0.07608t + 1.44 \times 10^{-5}t^2 - 0.032$$

$$+ \frac{1.3}{t + 41} \quad \text{cal per gram} \quad (50)$$

(Precision: ±0.2 per cent, 0 to 427°C)

where t is in degrees centigrade.

$$S - S_{0°C} = 0.15706 \log T + 2.88 \times 10^{-5}T - 0.39062 + \frac{0.0056}{T - 232}$$

$$+ 5.6 \times 10^{-5} \log \left(\frac{T - 232}{T}\right) \quad \text{cal/g/°C} \quad (51)$$

(Precision: ±0.2 per cent, 0 to 427°C)

where $T = t + 273.16$.

At temperatures above 100°C, the last terms in Eqs. 49 and 50 and the last two terms in Eq. 51 are negligible.

To obtain values of enthalpy and entropy referred to 0°K, the values 14.64_4 cal per gram and 0.1174_8 cal/g/°C, respectively, taken from Table 14.12, can be added to those given in Table 14.13 and Eqs. 50 and 51.

The heat of formation of UCl_4 was first determined by Biltz and Fendius (1928a); this determination was repeated at Berkeley (UCRL 7). The method consisted in dissolving uranium metal and UCl_4 separately in 12N HCl containing 10 per cent ferric chloride and comparing the heats of solution (measured in an all-glass bunsen-type ice calorimeter). The acid and the ferric salt were present in such large excess that heat effects caused by concentration changes could be neglected. In the reaction of uranium metal with Fe(III) the liberated hydrogen may reduce some of the Fe(III) ions; this effect can be taken into account by determining the ferrous iron with ceric sulfate after the dissolution has been completed.

The heats of solution in combination with other known thermochemical data permit the calculation of the heat of formation of UCl_4.

$$U \text{ (solid)} + 2Cl_2 \text{ (gas)} \rightarrow UCl_4 \text{ (solid)} \quad (Eq. 42)$$

$$\Delta H_{273°K} = -250.9 \pm 0.6 \text{ kcal per mole}$$

The earlier determination by Biltz and Fendius gave a very similar result.

Table 14.12 — Specific Heat, Entropy, and Enthalpy of Solid Uranium Tetrachloride from 0 to 355°K

Temp., °K	C_p, int. joules/mole/°C	S, int. joules/mole/°C	$H - E_0^s$, int. joules/mole
0	0.0	0.0	0.0
5	0.255*	0.0843	0.318
10	2.035*	0.6793	5.098
15	6.473	2.251	25.22
20	13.01	4.977	73.55
25	20.50	8.673	156.9
30	28.17	13.09	278.8
35	35.27	17.98	437.6
40	41.64	23.11	630.2
45	47.26	28.35	852.5
50	52.24	33.59	1,102.0
55	56.68	38.79	1,374
60	60.78	43.88	1,668
65	64.73	48.93	1,982
70	68.34	53.83	2,314
75	71.72	58.66	2,665
80	74.88	63.40	3,032
85	77.80	68.04	3,413
90	80.61	72.56	3,809
95	83.12	76.97	4,219
100	85.48	81.30	4,640
105	87.72	85.55	5,074
110	89.85	89.66	5,517
115	91.82	93.72	5,972
120	93.72	97.63	6,435
125	95.47	101.5	6,908
130	97.10	105.3	7,390
135	98.66	109.0	7,879
140	100.1	112.6	8,376
145	101.5	116.1	8,880
150	102.8	119.6	9,391
155	103.9	123.0	9,908
160	105.0	126.3	10,430
165	106.1	129.5	10,959
170	107.1	132.7	11,491
175	108.0	135.9	12,029
180	108.9	138.9	12,571
185	109.7	141.9	13,118
190	110.5	144.9	13,668
195	111.2	147.7	14,223

*Extrapolation using Debye function, $0.15396D\left(\dfrac{130.7}{T}\right)$.

Table 14.12 — (Continued)

Temp., °K	C_p, int. joules/mole/°C	S, int. joules/mole/°C	$H - E_0^s$, int. joules/mole
200	111.9	150.6	14,781
205	112.6	153.3	15,342
210	113.2	156.0	15,906
215	113.8	158.7	16,474
220	114.4	161.3	17,045
225	115.0	163.9	17,618
230	115.5	166.4	18,194
235	116.1	168.9	18,773
240	116.6	171.4	19,355
245	117.0	173.8	19,939
250	117.5	176.2	20,525
255	118.0	178.5	21,114
260	118.5	180.8	21,705
265	118.9	183.0	22,298
270	119.4	185.3	22,894
275	119.8	187.5	23,492
280	120.3	189.6	24,092
285	120.7	191.8	24,695
290	121.2	193.9	25,299
295	121.6	196.0	25,906
298.16	121.8	197.2	26,291
300	122.0	198.0	26,515
305	122.4	200.0	27,126
310	122.7	202.0	27,739
315	123.0	204.0	28,353
320	123.4	205.9	28,970
325	123.7	207.8	29,587
330	124.0	209.7	30,206
335	124.3	211.6	30,827
340	124.5	213.4	31,449
345	124.8	215.2	32,072
350	125.1	217.0	32,697
355	125.3	218.8	33,323

The reduction equilibrium

$$UCl_4 + \tfrac{1}{2}H_2 \rightleftharpoons UCl_3 + HCl$$

has been discussed in Sec. 1 of this chapter (cf. Eq. 1). The following values for the heat and free energy of formation of UCl_4 are derived from the experimental results given there:

$$\Delta H_{298^\circ K} = -251.0 \text{ kcal per mole}$$

$$\Delta F_{298^\circ K}^\circ = -229.65 \text{ kcal per mole}$$

$$\Delta S_{298^\circ K} = -71.5 \text{ e.u.}$$

$$\Delta H = -252,600 + 5.70T - 7.79 \times 10^{-4}T^2$$

$$\Delta F^\circ = -252,600 - 13.13T \log T + 7.79 \times 10^{-4}T^2 + 109.25T$$

The heat of formation agrees well with that derived above by the method of Biltz and Fendius. The formation of uranium tetrachloride

Table 14.13 — Specific Heat, Enthalpy, and Entropy of Solid Uranium Tetrachloride from 0 to 427°C*

Temp., °C	C_p, cal/g/°C	$H - H_{0^\circ C}$, cal/g	$S - S_{0^\circ C}$, cal/g/°C
0	0.07516	0	0
50	0.07735	3.815	0.012830
100	0.07917	7.728	0.024083
150	0.08063	11.726	0.034132
200	0.08196	15.790	0.043204
250	0.08328	19.920	0.051501
300	0.08472	24.120	0.059164
350	0.08616	28.392	0.066309
400	0.08760	32.736	0.073011
426.84 (700°K)	0.08838	35.099	0.076450

*The data in this table are from Ginnings and Corruccini (Natl. Bur. of Standards 3). Recalculated values which differ slightly have been published by Ginnings and Corruccini (1947).

from UCl_3 and $\frac{1}{2}Cl_2$ is endothermal; therefore, with increasing temperature, UCl_4 becomes more and more stable with respect to dissociation into UCl_3 and chlorine. Calculations show that UCl_4 (gas) cannot be decomposed to UCl_3 at any attainable temperature under less than atmospheric pressure.

(g) Optical Properties. The red vapor of UCl_4 absorbs light between 500 and 220 mμ. Absorption becomes strong around 310 mμ and rises toward shorter wavelengths. As the concentration of the vapor increases (t = 430°C and higher), absorption spreads throughout the visible spectrum. Pressures of UCl_4 vapor as low as 3×10^{-3} mm Hg could be detected optically in a 37.5-cm tube (UCRL 36).

The absorption spectrum of aqueous solutions of UCl_4 also has been studied (MP Chicago 6). Maxima occur at 429.5, 485, 495, 549, 649, and 672 mμ. Beer's law is obeyed at 649 mμ over the concentration range from 0.007M to 0.020M.

(h) Electrical and Magnetic Properties.   Magnetic Susceptibility. The molal susceptibility of UCl_4 (solid) has been determined as a function of the temperature by Sucksmith (1932) (Table 14.14). Curie's law is obeyed at low temperatures. Since Sucksmith stated that UCl_4

Table 14.14 — Molal Susceptibility of Uranium Tetrachloride*

Temp., °K	χ (molal) × 10^5	χ (molal) × T
90	1,058	0.952
198	488	0.965
273	413	1.130
289	410	1.185

*Sucksmith, 1932.

is unstable above room temperature, it must be concluded that no adequate precautions were taken to exclude moisture and air during the measurements. Consequently these results are of doubtful value and are cited here only in the absence of better data.

Electrical Conductivity.   Molten UCl_4 was found by Hampe (1888) to be a good conductor of electricity. The specific conductivity as a function of temperature was studied by Voigt and Biltz (1924). Although the measurements seem good in themselves, doubt is cast on

Table 14.15 — Electrical Conductivity of Uranium Tetrachloride

Temp., °C	Conductivity, reciprocal ohms
570	0.34
598	0.42
620	0.48

the validity of the results by the fact that the boiling point of the UCl_4 given by them differed considerably from the value currently accepted for pure tetrachloride. Their data are presented, however, in Table 14.15 since no other measurements are available. Electrolytic reduction of molten UCl_4 has been utilized in the preparation of uranium metal (see Chap. 4).

2.3 Chemical Properties of Uranium Tetrachloride. Uranium tetrachloride attracts water and oxidizes in air and should therefore be handled in a dry inert atmosphere.

(a) Water. Uranium tetrachloride is very hygroscopic. Although less sensitive to moisture than UCl_5 or UCl_6, the tetrachloride will react when water vapor is present at a partial pressure of more than 2 mm Hg. The water absorption is not limited to the formation of a UCl_4 hydrate but proceeds further by chemical reaction with the formation of $UOCl_2$ and hydrogen chloride until it results in complete dissolution and decomposition of the tetrachloride (UCRL 37). Brown University observers found, however, that very little hydrogen chloride is evolved until sufficient water has been absorbed to form $UCl_4 \cdot 2H_2O$ (Brown 8).

Uranium tetrachloride dissolves in water with the evolution of heat to form a green solution which exhibits the characteristic properties of the uranium(IV) ion (Zimmermann, 1883). The heat of solution in aqueous hydrochloric acid (1 mole of HCl per 8.808 moles of H_2O) has been determined as 39.4 ± 0.2 kcal per mole (Biltz, 1928a). Since aqueous solutions of UCl_4 always have an acid reaction, considerable hydrolysis must occur. When an attempt is made to concentrate the aqueous solution by evaporation, hydrogen chloride is evolved, and a complex mixture of hydrated uranium oxides precipitates. [Consideration of the ionic species present in aqueous solutions of UCl_4 is deferred until the general discussion of the properties of uranium(IV) solutions in Part II.]

Action of steam on UCl_4 at 600°C leads to U_3O_8.

(b) Nonaqueous Solvents. UCl_4 is generally soluble in strongly polar solvents and insoluble in nonpolar solvents, such as hydrocarbons, benzene, chloroform, and ether. It dissolves in certain oxygen-containing organic solvents, but solvolysis makes it impossible to recover the UCl_4 from such solutions. For example, alcoholysis occurs when UCl_4 is dissolved in methanol or ethanol (Fischer, 1913). Small amounts of water can apparently be present in alcoholic solutions without causing hydrolysis (Purdue 2). Solutions of UCl_4 in alcohol conduct electricity.

Uranium tetrachloride dissolves in acetone, probably also with chemical reaction, forming a greenish-yellow solution (Renz, 1903). Pierlé (1919) found that this solution has good electrical conductivity. Pyridine, ethyl acetate, and ethyl benzoate also dissolve UCl_4 with chemical reaction.

Uranium tetrachloride is insoluble at temperatures up to 70°C in inorganic solvents such as PCl_3, $SnCl_4$, and $POCl_3$ (MP Ames 17) and also does not dissolve in liquid anhydrous hydrogen fluoride (British 3).

(c) Oxidizing Agents. The reaction of UCl_4 with oxygen at elevated temperatures has been studied at the Tennessee Eastman laboratories (CEW-TEC 28; Van Wazer and John, 1948). The results differed from those obtained in the earlier work at Berkeley (UCRL 38). The reaction of oxygen and UCl_4 proceeds, according to Van Wazer and John, in two steps:

$$UCl_4 + O_2 \rightarrow UO_2Cl_2 + Cl_2 \tag{52}$$

$$3UO_2Cl_2 + O_2 \rightarrow U_3O_8 + 3Cl_2 \tag{53}$$

The lowest temperature at which measurable reaction occurs is $230 \pm 5°C$ for reaction 52 and $250 \pm 10°C$ for reaction 53. No analytical data were obtained to show that the "black compound" obtained by oxidation actually was U_3O_8, although this was postulated to be the case. The energies of activation for the two reactions were estimated to be 12 and 45 kcal, respectively. An examination of the kinetic data indicates that the rate-determining step is the oxidation itself and not the diffusion of the gas through the solid.

Chlorine reacts with UCl_4 at elevated temperatures to form mixtures of UCl_5 and UCl_6. (This reaction is discussed in connection with the preparation of UCl_5 in Sec. 3.1.) Uranium tetrachloride may also be chlorinated to a mixture of UCl_5 and UCl_6 at 115 to 125°C by means of chlorine solution in CCl_4. Fluorine converts UCl_4, as it does all other uranium compounds, to UF_6. The reaction may be dangerous because of the formation of explosive mixtures of chlorine and fluorine.

(d) Reducing Agents. Hydrogen reduces UCl_4 to UCl_3 under atmospheric pressure at a temperature of 525 to 550°C (Brown 5) (see Sec. 1.1a). Contrary to statements in the older literature, hydrogen bromide and hydrogen iodide also can reduce UCl_4 to the trivalent state (CEW-TEC 1) (instead of merely exchanging the halogen atoms).

$$UCl_4 + 4HBr \xrightarrow{300-350°C} UBr_3 + 4HCl + \tfrac{1}{2}Br_2 \tag{54}$$

$$UCl_4 + HI \xrightarrow{350-420°C} UCl_3 + HCl + \tfrac{1}{2}I_2 \tag{Eq. 3}$$

At higher temperatures, however, HBr forms UBr_4, and HI gives mixed iodochlorides. Hydrogen fluoride always gives UF_4. Alkali metals (lithium, sodium, potassium, calcium, magnesium, and aluminum) reduce UCl_4 to metallic uranium (see Chap. 4).

(e) Double Decomposition Reactions. Sulfides and selenides react with UCl_4 to give the corresponding uranium(IV) sulfide or selenide (Colani, 1907).

$$UCl_4 + 2H_2S \xrightarrow{550°C} US_2 + 4HCl \qquad (55)$$

$$UCl_4 + 2H_2Se \xrightarrow{550°C} USe_2 + 4HCl \qquad (56)$$

In general, UCl_4 reacts at elevated temperatures with most oxygen-containing compounds to form UO_2. Silicon dioxide (quartz), SiO_2, was said not to react with molten UCl_4 (Hönigschmid, 1928), but glass has been reported to be attacked severely (Biltz, 1928b). Uranium dioxide itself reacts with molten UCl_4 to give $UOCl_2$.

(f) Complex Compounds of UCl_4. Uranium tetrachloride is coordinatively unsaturated and consequently forms numerous complex compounds. Gaseous ammonia is absorbed by UCl_4 at moderate temperatures to give compounds containing 1, 2, or 3 moles of ammonia. Part of the ammonia taken up in this way is retained even in vacuum. Liquid ammonia forms a greenish-white compound, $UCl_4 \cdot 12NH_3$.

$$UCl_4 \text{ (solid)} + 12NH_3 \rightarrow UCl_4 \cdot 12NH_3 + 183 \text{ kcal} \qquad (57)$$

At 100°C enough ammonia is lost by the latter compound to form $UCl_4 \cdot 8NH_3$. Both these compounds are unstable in air and react with water to form hydrated uranium(IV) hydroxide. At high temperatures, a mixture of oxide and nitride results (Beck, 1932). Quinoline and pyridine form similar addition compounds with UCl_4 by reaction in acetone solution. The complex compounds separate as yellow crystals, which may be recrystallized from boiling absolute alcohol (Renz, 1903).

Double salts of the type M_2UCl_6 have been prepared, where M may be Li, Na, K, or R_4N. Some of these compounds found application in the electrochemical preparation of uranium metal (see Chap. 4). Because of their greater stability in air, they are sometimes used also in chemical preparative work in lieu of anhydrous UCl_4. Compounds of this type cannot be prepared by precipitation from aqueous solution. The potassium compound, K_2UCl_6, may be prepared by the reaction of UCl_4 vapors with KCl at red heat (Aloy, 1899). It is a dark-green crystalline material, somewhat hygroscopic in air, melting at 350°C. It can be heated to 550°C without evolution of vapors. It dissolves in water with decomposition, is soluble in alcohol and ethyl acetate, and is practically insoluble in ether. The sodium and rubidium compounds are prepared in the same way and have similar properties. The existence of a lithium compound of this type is still doubtful (see Table 14.16).

The alkaline earths form similar double chlorides with UCl_4. Two such compounds have been prepared, namely, $BaUCl_6$ and $SrUCl_6$, both resembling the potassium double salt closely. Pyridinium and quinolinium salts, $(C_5H_6N)_2UCl_6$ and $(C_9H_8N)_2UCl_6$, are also known.

Table 14.16 — Phase Relations of Uranium Tetrachloride and Various Halides

System	Compound	M.p. of compound, °C	Eutectic	M.p. of eutectic, °C
UCl_4-KCl	K_2UCl_6	650 ± 5[†]	KCl-K_2UCl_6 (25 mole % UCl_4)	560 ± 5
			K_2UCl_6-UCl_4 (53 mole % UCl_4)	350 ± 5
UCl_4-NaCl	Na_2UCl_6	Transformed to NaCl + UCl_4 at 430 ± 5	$2Na_2UCl_6$-UCl_4 (50 mole % UCl_4)	370 ± 5
UCl_4-LiCl	None		?	410 and 430 ± 5
UCl_4-$CaCl_2$	None		?	485 ± 5
UCl_4-$BaCl_2$	Ba_2UCl_8		?	515 and 425 ± 5
UCl_4-UCl_3[*]	None		Components insoluble in liquid phase; slightly soluble in solid phase	

*The Ames group reports the melting point of the UCl_3-UCl_4 system containing 11.7 mole % UCl_3 to be lower by 14°C than the melting point of UCl_4.

†We are unable to account for the discrepancy between this value and the one given on p. 487.

These can be prepared by adding pyridine or quinoline to a solution of UCl_4 in absolute alcohol saturated with hydrogen chloride (Rosenheim, 1932).

The phase relations in systems containing UCl_4, together with a number of other halides, have been studied by thermal analysis (Brown 8). The results are summarized in Table 14.16. No x-ray investigations of these systems have as yet been made (see Chap. 12).

3. URANIUM PENTACHLORIDE

Uranium pentachloride, UCl_5, was discovered by Roscoe in 1874 as a by-product of the preparation of UCl_4. Subsequently, many investigators observed that higher uranium chlorides are formed when UCl_4, in the process of its preparation, is permitted to come in contact with chlorine at elevated temperatures, but not much attention was paid to these by-products. Ruff and Heinzelmann (1911) attempted unsuccessfully to use UCl_5 as an intermediate in the preparation of UF_6, but, in general, the difficulties of manipulation of the pentachloride discouraged the further investigation of this compound.

The interest in UCl_5 was reawakened when it was discovered in 1943 that this halide undergoes thermal disproportionation into tetrachloride and the formerly unknown hexachloride (UCRL 42).

$$2UCl_5 \rightarrow UCl_4 + UCl_6 \qquad (58)$$

Analysis of UCl_5 is usually performed by hydrolysis and determination of uranium(IV) and uranium(VI) in the resultant solution. If $UOCl_4$ or UO_2Cl_2 are present in the analyzed material, the value obtained for the uranium(VI) content will be too high.

3.1 Preparation of Uranium Pentachloride. There are two practical methods for the preparation of the pentachloride. The first consists in the treatment of the lower uranium chlorides with chlorine gas at high temperatures; the second, based on the work of Michael and Murphy (1910), is the liquid-phase chlorination of uranium trioxide by carbon tetrachloride.

(a) Reaction of UCl_4 with Chlorine and CCl_4 Vapor. Uranium tetrachloride can be converted to UCl_5 by the action of chlorine at 520 to 555°C (UCRL 43).

$$2UCl_4 + Cl_2 \rightarrow 2UCl_5 \qquad (59)$$

The addition of small amounts of CCl_4 vapor to the chlorine stream has been advocated but does not appear to be particularly useful, especially if the starting material (UCl_4) is sufficiently pure (UCRL 44).

The Cl/U ratio in the product obtained by reaction 59 depends on the rate of cooling and may be as high as 5.5, indicating a considerably greater content of chlorine than that corresponding to the formula UCl_5. It has been suggested that compounds of the type $xUCl_6 \cdot yUCl_4$ are formed, where the ratio x/y may have values greater than 1. However, it seems more likely that in these intermediary compounds, and in UCl_5 as well, all uranium atoms are equivalent and bear the same charge [cf. the structure of the uranium fluorides (Chaps. 12 and 13) and of U_2O_5 (Chap. 11)].

Mixtures of UCl_4 and UCl_5 can be converted to pure UCl_5 by the action of liquid chlorine at elevated temperature and pressure (Ruff, 1911). Bubbling of chlorine through UCl_4 suspended in boiling CCl_4 (76°C) does not produce UCl_5 (Brown 5).

Attempts have been made to prepare UCl_5 by reaction of uranium oxide with carbon tetrachloride vapor at 550°C in a stream of air, but the product was found to contain oxygen (as UO_2Cl_2 or $UOCl_4$) (UCRL 45). The existence of fairly stable $UOCl_4$ has been made likely by experiments on the low-temperature chlorination of UO_2Cl_2 by CCl_4 (cf. UCRL 46). This point will be discussed in greater detail in Sec. 3.1b.

(b) Liquid-phase Reaction of Uranium Oxides with CCl_4. As pointed out in the section on the preparation of UCl_4, liquid-phase halogenation of uranium oxides with CCl_4 can be made to yield either the tetrachloride or the pentachloride. Here we shall be concerned with the specific conditions under which UCl_5 is obtained.

The reaction of uranium oxides with liquid carbon tetrachloride was first studied by Michael and Murphy (1910) who found that at 250°C, UO_3 or U_3O_8 rapidly reacted with liquid CCl_4 in a sealed tube to give UCl_5 (cf. Brown 1). More recently the behavior of boiling CCl_4 at atmospheric pressure has also been investigated. Under these conditions the reaction is much slower, 24 hr of refluxing resulting in only 10 to 30 per cent conversion. If even a small amount of UCl_5 is added at the beginning, only a short heating period is required to obtain completely water-soluble products from either UO_3 or U_3O_8. However, these products are mainly oxychlorides, and their conversion to the pentachloride proceeds only very slowly.

Uranium trioxide can be converted completely to UCl_5 at the boiling point of CCl_4 after many hours if a large excess of carbon tetrachloride is used; considerable amounts of UCl_5 must be added. Chlorine must be bubbled continuously through the boiling mixture (Brown 4,5, 19). A difficulty of the liquid-phase reaction at atmospheric pressure is that minute traces of acid seem to inhibit the reaction quite strongly (UCRL 47).

Mechanism of Conversion. (Brown 22,23.) In the interpretation of the mechanism of the liquid-phase halogenation of uranium oxide with carbon tetrachloride, the following three facts must be explained: (1) the reaction can be made to yield either UCl_4, UCl_5, or UCl_6; (2) UCl_5 acts as a catalyst; and (3) chlorine and phosgene are found among the products.

The over-all reactions by which CCl_4 converts UO_3 or U_3O_8 to UCl_5 with the liberation of chlorine and phosgene can be represented by the following equations:

$$2UO_3 + 6CCl_4 \rightarrow 2UCl_5 + 6COCl_2 + Cl_2 \quad \text{(Eq. 38)}$$

$$2U_3O_8 + 16CCl_4 \rightarrow 6UCl_5 + 16COCl_2 + Cl_2 \quad \text{(60)}$$

At higher temperatures UCl_5 decomposes into UCl_4 and chlorine:

$$2UCl_5 \rightleftharpoons 2UCl_4 + Cl_2 \tag{61}$$

Under these conditions the product may contain much UCl_4 or may even be pure UCl_4 (see Sec. 2). When extra chlorine is added to the system, not only will equilibrium (Eq. 61) be shifted back to the left, but chlorination may even proceed further, leading to the formation of UCl_6.

$$2UCl_5 + Cl_2 \rightleftharpoons 2UCl_6 \tag{62}$$

Even if no extra chlorine is added and the solvent is removed below its boiling point, a few per cent of UCl_6 is always found in the UCl_5. This may be interpreted as indicating the liberation of enough chlorine by reaction 61 to produce a small amount of UCl_6 by reaction 62. Alternatively, the same may result from direct disproportionation of two molecules of UCl_5 into UCl_4 and UCl_6.

In Eqs. 38 and 60 it was assumed that the primary product is UCl_5. A more reasonable hypothesis, however, is that the primary product (at least with UO_3) is UCl_6

$$UO_3 + 3CCl_4 \rightarrow UCl_6 + 3COCl_2 \tag{63}$$

and that all the other uranium halides arise by thermal decomposition of the UCl_6.

A detailed investigation has been carried out at Brown University to determine the mechanism of the autocatalytic action of UCl_5 on the reaction of UO_3 with liquid CCl_4. Although UO_3 and CCl_4 alone do not react appreciably at 65°C even after 10 hr, a water-soluble product is formed within 3 hr if UCl_5 ($2UCl_5/1UO_3$) is added. A study of the products isolated after 2 to 16 hr of reaction shows a slowly progressing chlorination of the uranium. The completely water-soluble solid obtained by removal of the solvent after 15 hr of interaction was divided into four fractions on the basis of decreasing solubility in CCl_4; each fraction was analyzed by determining the Cl/U ratio and the oxygen content. (Sublimation in vacuum at 750°C leaves a residue of UO_2, the amount of which is proportional to the oxygen content of the investigated fraction.) The results showed fairly conclusively that the water-soluble solid consisted of UCl_5 (soluble in CCl_4), $UOCl_4$ (somewhat soluble), $UOCl_3$ (less soluble), and UO_2Cl_2 (insoluble).

The intermediate formation of the oxychlorides having thus been established, the autocatalytic effect of UCl_5 can be explained by the assumption that reaction first occurs between UO_3 and UCl_5, which gives oxychlorides. The latter then undergo chlorination by CCl_4 more readily than does UO_3 itself. A likely mechanism for the first step is

$$UO_3 + UCl_5 \rightarrow UOCl_3 + UO_2Cl_2 \tag{64}$$

$$UO_2Cl_2 + UCl_5 \rightarrow UOCl_3 + UOCl_4 \tag{65}$$

$$UO_3 + 2UCl_5 \rightarrow 2UOCl_3 + UOCl_4 \tag{66}$$

Chlorination of the intermediates by CCl_4 can then occur according to the equations

$$2UOCl_3 + 2CCl_4 \rightarrow 2UCl_5 + 2COCl_2 \tag{67}$$

$$UOCl_4 + CCl_4 \rightarrow UCl_6 + COCl_2 \tag{68}$$

The over-all reaction will then be

$$UO_3 + 2UCl_5 + 3CCl_4 \rightarrow 2UCl_5 + UCl_6 + 3COCl_2 \tag{69}$$

One mole of UCl_6 should thus be present in the final product for each two moles of UCl_5 added, unless the conditions are such that UCl_6 decomposes into UCl_5 and chlorine. The conditions specified for the preparation of UCl_5 usually assure complete decomposition of UCl_6 (or of its precursor, $UOCl_4$); conditions for obtaining UCl_6 as final product will be discussed in Sec. 4.1. This mechanism indicates why the ratio $2UCl_5/1UO_3$ is necessary to obtain complete conversion at the fastest possible rate.

The thermal stability of the above-postulated intermediate oxychlorides has been investigated in a qualitative way. The mixture of water-soluble reaction intermediates was found to evolve chlorine readily at 130°C, and a measurable amount of chlorine was obtained even at temperatures as low as 88°C. Since UCl_5, as well as UCl_6, gives off only traces of chlorine at 130°C (under atmospheric pressure), the source of the chlorine must be the decomposition of the intermediates. However, the $UOCl_3$ fraction obtained from the crude mixture by elution with CCl_4 was found to be surprisingly stable; its decomposition occurred only above 375°C. The low-temperature decomposition is therefore best explained by the reaction

$$UOCl_4 \rightarrow UOCl_3 + \tfrac{1}{2}Cl_2 \tag{70}$$

The high thermal stability of $UOCl_3$ stops the decomposition of $UOCl_4$ before $UOCl_2$ is reached.

The chlorination of UO_3 by liquid $SOCl_2$ and UCl_5 probably follows the same course as the reaction with CCl_4.

3.2 Physical Properties of UCl_5. Few of the physical properties of this compound have been studied. As prepared by reaction of UCl_4 with chlorine, it usually is a red-brown microcrystalline powder. High-pressure autoclave preparations are reddish-black crystals with a metallic luster. Attempts have been made to recrystallize the red-brown microcrystals of UCl_5 from CCl_4 solution in an autoclave, but these attempts have been unsuccessful (UCRL 53).

(a) Density. A density determination by direct displacement in benzene gave a value of 3.81 g/cc (MP Ames 18). The packing density of autoclave-prepared material is 1.95 g/cc (UCRL 54).

(b) Melting Point and Volatility. Since UCl_5 undergoes both thermal decomposition and disproportionation, no accurate values for the melting point or volatility are available. The vapor pressure has been estimated to be 10^{-7} mm Hg at 50°C; because of disproportionation, doubts have been expressed concerning the possibility of UCl_5 existing as a vapor above 60°C (UCRL 55). It has been found difficult if not impossible to separate UCl_5 and UCl_6 by sublimation, and this was interpreted as indicating a rather high volatility of the penta-chloride. UCl_5 can be distilled in an atmosphere of chlorine containing some CCl_4, but the sublimate is enriched in UCl_6 (UCRL 56).

(c) Molecular Weight. (CEW-TEC 29.) The molecular weight of UCl_5 has been determined by ebullioscopy in carbon tetrachloride solution. It was found to correspond quite closely to a dimer, $(UCl_5)_2$. It may thus be suggested that UCl_5 should be formulated as U_2Cl_{10} in solution.

(d) Thermochemistry. (UCRL 4.) The heat of formation of UCl_5 has been determined at Berkeley by measuring the difference in the heats of solution of UCl_4 and UCl_5 in water containing excess $FeCl_3$.

$$UCl_4 \text{ (solid)} + 2FeCl_3 \text{ (aq)} + 2H_2O \text{ (liq)} \rightarrow UO_2Cl_2 \text{ (aq)}$$
$$+ 4HCl \text{ (aq)} + 2FeCl_2 \text{ (aq)} \quad (71)$$

$$\Delta H = -24.1 \text{ kcal}$$

$$UCl_5 \text{ (solid)} + FeCl_3 \text{ (aq)} + 2H_2O \text{ (liq)} \rightarrow UO_2Cl_2 \text{ (aq)}$$
$$+ 4HCl \text{ (aq)} + FeCl_2 \text{ (aq)} \quad (72)$$

$$\Delta H = -33.9 \text{ kcal}$$

$$FeCl_2 \text{ (aq)} + \tfrac{1}{2}Cl_2 \text{ (aq)} \rightarrow FeCl_3 \text{ (aq)} \qquad (73)$$

$$\Delta H = -21.0 \text{ kcal}$$

Therefore

$$UCl_4 \text{ (solid)} + \tfrac{1}{2}Cl_2 \text{ (aq)} \rightarrow UCl_5 \text{ (solid)} \qquad (74a)$$

$$\Delta H = -71.2 \text{ kcal per mole}$$

and, using Eq. 42,

$$U \text{ (solid)} + \tfrac{5}{2}Cl_2 \text{ (gas)} \rightarrow UCl_5 \text{ (solid)} \qquad (74b)$$

$$\Delta H_{273°K} = -262.1 \pm 0.6 \text{ kcal per mole}$$

The entropy of UCl_5 has been estimated to be $S_{298°K} = 57.0$ e.u.; ΔS for Eq. 74a then becomes -88.3 e.u., and $\Delta F°_{298°K} = -235.7 \pm 2.0$ kcal per mole (UCRL 57).

(e) Dielectric Constant. The dielectric constant of solid UCl_5 is 1.9 and does not vary appreciably with the frequency (CEW-TEC 30).

(f) Absorption Spectrum of UCl_5 in Solution. Sterett and Calkins (1949) have examined the absorption spectrum of UCl_5 in carbon tetrachloride and thionyl chloride solutions. UCl_5 shows a rather broad peak at 850 mμ in both solvents, with evidence of another peak at 650 mμ in carbon tetrachloride and 950 mμ in thionyl chloride. UCl_6, according to these authors, exhibits peaks at the same wavelengths, and UCl_4 shows no absorption in this region. Although there are distinct differences in the UCl_5 and UCl_6 spectra, Sterett and Calkins conclude from the fact that the peaks occur at approximately the same positions that UCl_5 should be written $UCl_4 \cdot UCl_6$. The evidence for this conclusion is rather tenuous. However, until x-ray data are available on the solid, judgment must be deferred.

3.3 Chemical Properties of UCl_5. (a) Effect of Solvents. Uranium pentachloride is very hygroscopic. It is sensitive to moisture when the partial pressure of water is 0.007 mm Hg or more (Brown 18). Liquid water decomposes it immediately.

$$2UCl_5 + 2H_2O \rightarrow UCl_4 + UO_2Cl_2 + 4HCl \qquad (75)$$

To avoid loss of hydrogen chloride when this reaction is used for analytical purposes, it is advisable to add the UCl_5 to dilute sulfuric acid cooled to $-78°C$ and to allow the reactants to warm up slowly (Brown 24).

The solubility of UCl_5 in a large number of organic solvents has been examined. The solubility in CCl_4 is given in Table 14.17 as a function of temperature. Table 14.17 shows that the addition of Cl_2 decreases the solubility of UCl_5 in CCl_4.

A number of miscellaneous observations on solubility are summarized in Table 14.18.

Dimethyl ether reacts vigorously with UCl_5 to form an orange water-soluble substance. Diethyl ether, acetone, ethyl acetate, formamide, dioxane, and alcohols also react immediately; chloroform,

Table 14.17 — Solubility of Uranium Pentachloride in Carbon Tetrachloride*

Temp., °C	Grams of UCl_5 per 100 g of CCl_4	Grams of UCl_5 per 100 g of $CCl_4 + Cl_2$[†]
−20	...	0.26
−16	0.29	...
0.0	0.45	0.43
22.8	0.94	...
25.0	1.04	0.82
70	5.3	
160	20 (estimated)	

*Brown 9,19,23.
[†]87.5% CCl_4 + 12.5% Cl_2.

acetophenone, and chlorobenzene react more slowly. No reaction is observed with isopropyl ether (MP Ames 1; Brown 23). Uranium pentachloride is highly soluble in carbon disulfide and thionyl chloride.

(b) Oxidation Reactions. Oxygen reacts with UCl_5 to give a mixture of uranium oxyhalides. Chlorine partly converts UCl_5 to UCl_6 at elevated temperatures. The reaction

$$2UCl_5 + Cl_2 \rightleftharpoons 2UCl_6 \quad \text{(Eq. 62)}$$

is reversible, but practically complete conversion of UCl_5 to UCl_6 can be obtained by treatment with chlorine in CCl_4 vapor (Brown 6). This reaction will be discussed in Sec. 4. Fluorine gas converts UCl_5 to UF_6.

(c) Double Decomposition Reactions. Liquid anhydrous hydrogen fluoride converts UCl_5 to UF_5 (see Chap. 12).

$$UCl_5 + 5HF \rightarrow UF_5 + 5HCl \quad (76)$$

(d) Reduction Reactions. Sodium metal reduces UCl_5 to metallic uranium. Uranium pentachloride is not, however, a particularly suitable source of uranium metal (Hunter, 1923). Chloroform reduces UCl_5 to UCl_4 (UCRL 59). This reduction can be carried out at 140°C in an autoclave but is of little practical value since the UCl_4 obtained is contaminated with hexachloroethane. Uranium pentachloride can

Table 14.18 — Solubility of Uranium Pentachloride in Various Solvents*

Solvent	Solubility	Remarks
Thionyl chloride	Soluble	Reddish-brown solution
Pyridine	Slightly soluble	Appears to form a complex
Benzotrichloride	Soluble	Black solid forms on standing or refluxing
β,β'-Dichloroethyl ether	Somewhat soluble	Green solution
Tetralin	Slightly soluble	
Tetrachloroethane	Soluble	Precipitate forms on refluxing
Tetrachloroethylene	Soluble: 9% at 22°C, 25% at 120°C	Dark-brown solution
Benzene, xylene	No solution	No reaction
Freon 11, Freon 21	Rapid reaction with Cl_2 evolution	
Freon 113	Soluble	Precipitate forms on boiling

*UCRL 58.

be dechlorinated by refluxing with S_2Cl_2, $SnCl_4$, hexachloropropene, etc., but it is usually difficult to free the product from the solvent.

(e) Disproportionation and Thermal Decomposition. When UCl_5 is heated, it undergoes the decomposition (Roscoe, 1874)

$$2UCl_5 \rightleftharpoons 2UCl_4 + Cl_2 \quad \text{(Eq. 61)}$$

Some decomposition occurs below 100°C; at 250°C it is rapid, and its rate is about the same in atmospheres of chlorine or pure nitrogen (Brown 18), indicating that the reaction is not markedly reversible under these conditions. At the Tennessee Eastman laboratories it has been reported that UCl_5 evolves noticeable amounts of Cl_2 at 75°C and that the rate of decomposition increases rapidly at 175°C (CEW-TEC 31).

The thermal stability of UCl_5 has been studied by Martin and Eldau (1943). These workers were apparently unaware of the existence of UCl_6, and an examination of their experimental procedure indicates that their results probably refer to a mixture of UCl_5 and UCl_6 rather than to pure UCl_5. The kinetics of the thermal decomposition have been investigated at Berkeley (UCRL 55). The reaction appears to be first order with respect to UCl_5. The rate constant of reaction 61 was found to be of the order of 10^{-3} per min for the temperature range 100 to 152°C, and the activation energy was computed to be 28 kcal per mole.

In addition to thermal decomposition, UCl_5 also undergoes disproportionation.

$$2UCl_5 \rightarrow UCl_4 + UCl_6 \quad (Eq. 58)$$

at 100 to 175°C in a high vacuum. Uranium pentachloride should not be formulated $UCl_4 \cdot UCl_6$ because of this reaction. Very likely, as in the case of UF_5, all the uranium atoms in the solid are equivalent. The details of the thermal disproportionation of UCl_5 to UCl_4 and UCl_6 will be discussed in Sec. 4.

(f) Complex Compounds of UCl_5. It has been reported that a double salt, $\overline{UCl_5 \cdot PCl_5}$, is formed by reactions of UO_3 and PCl_5 at 180 to 190°C in a sealed tube (Cronander, 1873). After removal of $POCl_3$, the double salt remains as a yellow-red amorphous mass which melts without decomposition. It is reduced by hydrogen to HCl and PH_3; fusion with KCl gives UCl_4, PCl_5, and chlorine. Water precipitates uranium(IV) phosphate and forms UO_2^{++}. A closer study of this complex would be of interest.

4. URANIUM HEXACHLORIDE

Before the preparation of uranium hexachloride, uranium hexafluoride was the only known oxygen-free sexivalent uranium compound. Uranium hexachloride was discovered as a result of the observation that initially pure UCl_5 could not be completely sublimed unless the temperature was high enough to cause the sublimation of UCl_4 as well; when sublimation was carried out below 550°C, a residue with a composition close to UCl_4 always remained (UCRL 42). The sublimate could be resublimed repeatedly in vacuum or in an atmosphere of nitrogen or chlorine without leaving any further residue. Analysis of the sublimate showed it to be UCl_6, and it became clear that it must arise by disproportionation of UCl_5.

$$2UCl_5 \rightarrow UCl_4 + UCl_6 \quad (Eq. 58)$$

The rather unexpected stability of the hexachloride, coupled with its marked volatility, has aroused considerable interest. Methods of preparation have been intensively studied, and some progress has been made in the examination of its physical and chemical properties.

4.1 Preparation of Uranium Hexachloride. Three methods are known for the preparation of UCl_6: (1) The above-mentioned disproportionation of UCl_5 in vacuum. The maximum amount of UCl_6 obtainable in this way is 54.6 per cent of the weight of UCl_5. (2) Reaction of UCl_4 or UCl_5 with chlorine gas at elevated temperatures. Products with a chlorine/uranium ratio up to 5.8/1 can be obtained in this way. They can be used as material for the preparation of pure UCl_6 by fractionation. (3) Reaction of a lower chloride (usually UCl_5) or of UO_3 with a mixture of chlorine and carbon tetrachloride in the liquid phase. The last method also does not yield pure UCl_6 directly; further treatment is necessary to remove the UCl_5 with which the product is mixed.

(a) Preparation of UCl_6 by Disproportionation of UCl_5. (MP Ames 4.) The preparation of UCl_6 by sublimation of UCl_5 in vacuum can be carried out in a glass or a stainless-steel vessel equipped with a condenser cooled with solid CO_2. The purity of the UCl_6 obtained depends to a considerable extent on the composition of the starting material. The higher the chlorine/uranium ratio in the material used, the fewer fractional sublimations will be required to obtain pure UCl_6 (MP Ames 1,19). The necessity for repeated fractionation arises from the fact that UCl_5 itself possesses an appreciable vapor pressure (Brown 25 to 27; UCRL 60).

Various workers have utilized disproportionation temperatures ranging from 80 to 240°C (UCRL 54). Although the rate of sublimation increases with increasing temperature of the charge, temperatures above 150°C are better avoided in order to minimize the codistillation of UCl_5. The disproportionation is best carried out at 120 to 150°C under pressures of the order of 10^{-4} mm Hg (UCRL 61). The closer the composition of the starting charge corresponds to UCl_5, the lower must the temperature be kept to avoid losses of the pentachloride. Since low operating pressures favor the disproportionation, a good vacuum is essential. A large-capacity pumping system, capable of maintaining a high vacuum even when gases are continuously evolved by the reacting mass, should be used.

(b) Preparation of UCl_5-UCl_6 Mixtures by High-temperature Vapor-phase Chlorination. When UCl_4 is treated with chlorine at elevated temperatures, mixtures of UCl_5 and UCl_6 result.

$$UCl_4 + \tfrac{1}{2}Cl_2 \rightarrow UCl_5 \quad \text{(see Eq. 29)}$$

$$UCl_5 + \tfrac{1}{2}Cl_2 \rightarrow UCl_6 \quad \text{(see Eq. 62)}$$

$$UCl_4 + Cl_2 \rightarrow UCl_6 \tag{77}$$

These reactions proceed at temperatures above 350°C, and the composition of the product depends to a considerable extent on the rapidity with which the reaction products are cooled (Brown 18). The more rapid the cooling, the higher the UCl_6 content. The chlorine/uranium ratio in the product is higher at 350°C than at 450°C (Brown 20). Experiments showed that the sometimes suggested addition of CCl_4 vapor to the chlorine is not particularly beneficial. At the above-given temperatures a product consisting of 30 per cent UCl_6 and 70 per cent UCl_5 can be obtained.

(c) Purification of UCl_6. Pure anhydrous UCl_6 can be sublimed at 75 to 100°C in a vacuum of 10^{-4} mm Hg. UCl_6 is ordinarily purified by repeated vacuum sublimation, but it can also be distilled in a stream of inert gas of low pressure. The product obtained by subliming UCl_6 in a stream of argon at 1 to 2 cm Hg pressure is crystalline and has been claimed to be purer than the product obtained by vacuum sublimation (MP Ames 10). Sublimation in chlorine or nitrogen at 350°C, however, appears to be unsatisfactory. When chlorine is used, large amounts of chlorine are absorbed which cannot be removed even by prolonged passage of inert gas over the product (UCRL 64). The effect of traces of CCl_4 on the sublimation of UCl_6 also has been studied (UCRL 65). After solution in CCl_4 or exposure to CCl_4 vapor the sublimation of UCl_6 is never complete, and a residue that cannot be sublimed at 100°C always remains.

4.2 Physical Properties of UCl_6. Uranium hexachloride forms fine black or dark-green crystals. The exact appearance depends on the method of purification and the rate at which the crystals are grown.

(a) Melting Point. (UCRL 66.) Despite thermal instability, it has been found possible to determine the melting point of UCl_6. The measurement was performed by immersing thin-walled tubes containing UCl_6 in an oil bath preheated to various temperatures. At atmospheric pressure a liquid phase is formed momentarily at 177.5 ± 2.5°C. The equilibrium pressure of chlorine caused by the decomposition

$$UCl_6 \rightarrow UCl_4 + Cl_2 \tag{78}$$

must be quite high at that temperature, as judged by the very rapid disappearance of liquid UCl_6.

(b) X-ray Structure and Density. (MP Chicago 5,8; Zachariasen, 1948b.) Crystalline UCl_6 has been examined by x-ray methods and found to possess a hexagonal unit cell

$$a_1 = 10.90 \pm 0.02 \text{ A} \qquad a_3 = 6.03 \pm 0.01 \text{ A}$$

There are three molecules per unit cell, and the space group is $C\bar{3}m(D_{3d}^3)$. The atomic positions are

$$1 \text{ U}_I \text{ at } (0\ 0\ 0)$$
$$2 \text{ U}_{II} \text{ at } \pm(\tfrac{1}{3}\ \tfrac{2}{3}\ z) \quad \text{with } z = \tfrac{1}{2}$$

There are three sets of chlorine atoms in $\pm(x\ 2x\ z)\ (2\bar{x}\ \bar{x}\ z)\ (x\ \bar{x}\ z)$, with the parameters

	x	z
Cl_I	0.10	0.25
Cl_{II}	0.43	0.25
Cl_{III}	0.77	0.25

UCl_6 has a typical molecular structure composed of individual UCl_6 molecules in a three-dimensional array, which is consistent with the observed high volatility of the compound. The six chlorine atoms about each uranium form a nearly perfect octahedron with a U-Cl distance of 2.42 A. From the normal single-bond radius of chlorine of 0.99 A, the single-bond radius of sexivalent uranium is calculated to be 1.43 A. The distance of closest approach between chlorine atoms of different UCl_6 molecules is 3.85 A.

The density computed from the x-ray data is 3.59 g/cc. A direct measurement by immersion in benzene has given a value of 1.56 g/cc (MP Ames 19), but measurements in tetrahydronaphthalene gave 3.36 ± 0.17 g/cc (UCRL 67). The packing density is about 1.75 g/cc (UCRL 68).

(c) Vapor Pressure of UCl_6. The vapor pressure of UCl_6 has been determined by two independent methods. The results obtained are not in good agreement. One set of measurements was made with an all-glass clicker gauge (UCRL 69); the vapor pressure and free energy of sublimation could be expressed to within 1 per cent or better by the equations

$$\log p = -\frac{2422}{T} + 6.633_7 \quad (0 \text{ to } 200°C) \tag{79}$$

$$\Delta H \text{ (subl)} = 11,120 \text{ cal per mole} \tag{80}$$

$$\Delta F° = 17,500 + 36.85T \log T - 129.0T \tag{81}$$

Values of the vapor pressure computed from Eq. 79 are given in Table 14.19.

The vapor pressure of UCl_6 has also been determined by a transpiration method, using purified helium gas (MP Ames 4,20). The results were corrected for evolution of chlorine arising from decomposition of the UCl_6. The vapor pressures are considerably lower than those obtained by the clicker gauge method (Table 14.20).

Table 14.19—Vapor Pressure of Uranium Hexachloride, Clicker Gauge Method

Temp., °C	Pressure, mm Hg	Temp., °C	Pressure, mm Hg
0	0.005	120	2.95
20	0.023	140	5.75
40	0.08	160	11.07
60	0.22	180	18.60
80	0.59	200	32.35
100	1.35		

Table 14.20—Vapor Pressure of Uranium Hexachloride, Transpiration Method

Temp., °C	Pressure, mm Hg
75.5	0.06
98.5	0.24
120.0	0.89
138.0	1.75

The experimental data from the transpiration method can be represented by the following equation:

$$\log p = -\frac{3788}{T} + 9.52 \tag{82}$$

$$\Delta H \text{ (subl)} = -17,300 \text{ cal per mole} \tag{83}$$

The values obtained by the transpiration method are probably more accurate in principle than the clicker gauge values. In the latter case no correction was made for thermal decomposition of the UCl_6, which will necessarily give rise to high results. However, the workers using the clicker gauge reported only negligible decomposition below 130°C, whereas the transpiration-method workers found decomposition at practically all temperatures. Before any final conclusions can be drawn, it will be necessary to obtain comparative measurements on the same sample.

Table 14.21 — Specific Heat, Entropy, and Enthalpy of Uranium Hexachloride*
from 0 to 350°K

Temp., °K	C_p, int. joules/mole/°C	S, int. joules/mole/°C	$H - E_0^s$, int. joules/mole
0	0.0	0.0	0.0
5	0.793†	0.264	0.991
10	6.01	2.078	15.52
15	15.66	6.26	68.78
20	25.18	12.08	171.0
25	34.35	18.68	319.9
30	43.51	25.76	514.6
35	52.18	33.12	754.1
40	60.33	40.63	1,036
45	67.73	48.17	1,356
50	74.53	55.66	1,712
55	80.74	63.06	2,100
60	86.37	70.33	2,518
65	91.46	77.45	2,963
70	96.01	84.40	3,432
75	100.2	91.16	3,923
80	104.6	97.77	4,435
85	108.6	104.23	4,968
90	112.0	110.54	5,519
95	115.3	116.69	6,088
100	118.4	122.68	6,672
105	121.3	128.53	7,271
110	124.2	134.24	7,885
115	126.9	139.82	8,513
120	129.6	145.28	9,154
125	132.1	150.62	9,808
130	134.4	155.85	10,475
135	136.8	160.96	11,153
140	139.0	165.98	11,842
145	141.0	170.89	12,542
150	143.0	175.71	13,252
155	144.8	180.42	13,972
160	146.5	185.05	14,700
165	148.1	189.58	15,437
170	149.6	194.03	16,181
175	151.1	198.39	16,933
180	152.5	202.66	17,692
185	153.8	206.86	18,458
190	155.1	210.98	19,230
195	156.4	215.02	20,009
200	157.6	219.00	20,794
205	158.7	222.90	21,585
210	159.8	226.74	22,381
215	160.9	230.51	23,183
220	161.9	234.23	23,990

Table 14.21 — (Continued)

Temp., °K	C_p, int. joules/mole/°C	S, int. joules/mole/°C	$H - E_0^s$, int. joules/mole
225	163.0	237.88	24,802
230	164.0	241.47	25,619
235	164.9	245.01	26,442
240	165.9	248.49	27,269
245	166.8	251.92	28,100
250	167.7	255.30	28,936
255	168.6	258.62	29,777
260	169.5	261.91	30,622
265	170.4	265.14	31,472
270	171.2	268.34	32,326
275	172.1	271.48	33,184
280	172.9	274.59	34,046
285	173.7	277.66	34,912
290	174.5	280.69	35,783
295	175.2	283.68	36,657
298.16	175.7	285.54	37,212
300	176.0	286.63	37,535
305	176.7	289.54	38,417
310	177.5	292.42	39,303
315	178.2	295.27	40,192
320	178.9	298.08	41,084
325	179.7	300.86	41,981
330	180.5	303.61	42,882
335	181.3	306.33	43,786
340	182.1	309.02	44,694
345	182.9	311.69	45,607
350	183.8	314.33	46,524

*The UCl_6 contained 6.3 per cent UCl_4; the appropriate corrections were made.

†Extrapolated using Debye function, $58.17D\left(\dfrac{89.4}{T}\right)$.

(d) <u>Thermochemical Data.</u> The specific heat, entropy, and enthalpy of UCl_6 from 0 to 350°K have been determined by the National Bureau of Standards (Natl. Bur. Standards 5) (Table 14.21). $S_{298°K}$ is found to be 68.28 e.u.

The heat of formation of UCl_6 has been determined at the University of California Radiation Laboratory (UCRL 4).

$$UCl_4 \text{ (solid)} + 2FeCl_3 \text{ (aq)} + 2H_2O \text{ (liq)} \rightarrow UO_2Cl_2 \text{ (aq)}$$
$$+ 4HCl \text{ (aq)} + 2FeCl_2 \text{ (aq)} \quad (84)$$

$$\Delta H = -24.1 \text{ kcal}$$

$$UCl_6 \text{ (solid)} + 2H_2O \text{ (liq)} \rightarrow UO_2Cl_2 \text{ (aq)} + 4HCl \text{ (aq)} \qquad (85)$$

$$\Delta H = -44.7 \text{ kcal}$$

$$2FeCl_2 \text{ (aq)} + Cl_2 \text{ (gas)} \rightarrow 2FeCl_3 \qquad (86)$$

$$\Delta H = -42.0 \text{ kcal}$$

Therefore,

$$U \text{ (solid)} + 3Cl_2 \text{ (gas)} \rightarrow UCl_6 \text{ (solid)} \qquad (87)$$

$$\Delta H_{273°K} = -272.3 \pm 0.7 \text{ kcal per mole}$$

$$\Delta F°_{298°K} = -241.4 \pm 1.0 \text{ kcal per mole}$$

(e) Optical Properties. (SAM Columbia 1,2.) The absorption spectrum of uranium hexachloride vapor is difficult to study because of

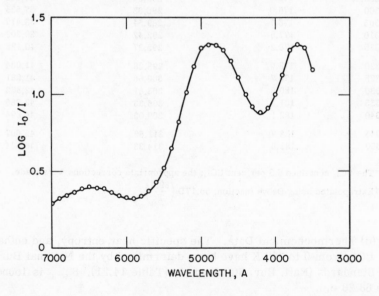

Fig. 14.2 — Absorption spectrum of uranium hexachloride in perfluoroheptane.

the low vapor pressure. The absorption spectrum has been examined at 76.5 and 100°C, using a tube 50 cm long. A weak continuous absorption band at 20,000 cm^{-1} was found. A much stronger band appeared at 27,000 cm^{-1}, obviously corresponding to the absorption band of UF_6. The absorption spectrum of solid UCl_6 films at 77°K was also investigated. As would be expected, the band structure is more pronounced at the lower temperature. A sequence of rather diffuse bands is found at 20,000 cm^{-1}; below 17,000 cm^{-1} the bands become weaker but sharper. The ultraviolet spectrum below 30,000 cm^{-1} was not investigated.

The absorption spectrum of UCl_6 in a fluorocarbon solution is shown in Fig. 14.2 and appears qualitatively to resemble the absorption spectrum of UCl_6 vapor.

4.3 Chemical Properties of UCl_6. Little is known about the chemical properties of UCl_6. It is extremely unstable in the presence of moist air and must therefore be handled either in a vacuum apparatus or in a dry box. Uranium hexachloride reacts violently with liquid water to form UO_2Cl_2.

(a) Solubility. Uranium hexachloride dissolves in CCl_4 to give a stable brown solution. The solubility in CCl_4 and in a number of other solvents has been determined (Table 14.22). Uranium hexachloride

Table 14.22 — Solubility of Uranium Hexachloride in Various Solvents*

Solvent	Temp., °C	Grams of UCl_6 per 100 g of solution
CCl_4	−18	2.64
CCl_4	0	4.9
CCl_4	20	7.8
6.6% Cl_2 − 93.4% CCl_4	−20	2.4
12.5% Cl_2 − 87.5% CCl_4	−20	2.23
12.5% Cl_2 − 87.5% CCl_4	0	3.98
Liquid chlorine	−33	2.20
Benzene	80	Insoluble
CH_3Cl	−24	1.16†
Freon 113	45	1.83†

*Brown 23.
†No apparent reaction between solvent and solute.

appears to react with tetrachloroethylene (UCRL 58) and naphthenic hydrocarbons (SAM Columbia 1). It is slightly soluble in the fluorocarbon C_7F_{16} and in isobutyl bromide.

(b) Thermal Decomposition. No exact statements are possible at this time on the thermal decomposition of UCl_6. The compound is probably stable up to temperatures of 120 to 150°C. The kinetics of the thermal decomposition of UCl_6 have been studied and found to be complicated (UCRL 55). When the rate of decomposition is plotted against time, the curve is found to have a point of inflection; this has been interpreted as indicating a solid-phase transition from one crystal form of UCl_6 to another more stable form. A well-defined induction period, decreasing from 30 to 5 min as the temperature is raised from 140 to 170°C, is observed. The induction period may well be a function of the previous history of the sample. The decomposition is also markedly affected by catalysis. Stainless steel and monel metal are active catalysts for thermal decomposition of UCl_6; gold- and nickel-plated copper wire are quite inert.

The rate of decomposition of UCl_6 in the vapor phase is negligible by comparison to that in the solid phase. In the range 130 to 180°C, the rate constant, k, of the reaction

$$2UCl_6 \text{ (gas)} \rightarrow 2UCl_5 \text{ (solid)} + Cl_2 \text{ (gas)} \tag{88}$$

is between 10^{-3} and 10^{-1} per min. These wide limits reflect the uncertainty of the vapor pressure data for UCl_6. An activation energy of approximately 40 kcal per mole is required for reaction 88.

(c) <u>Reaction of UCl_6 with Hydrogen Fluoride</u>. (Brown 26,27.) When UCl_6 is treated with purified anhydrous liquid hydrogen fluoride at room temperature (in a platinum-lined reactor), UF_5 rather than UF_6 is obtained.

$$2UCl_6 + 10HF \rightarrow 2UF_5 + 10HCl + Cl_2 \tag{89}$$

REFERENCES

1842   E. Péligot, Ann. chim. et phys., (3) 5: 5-20.
1842   C. Rammelsberg, Pogg. Ann., 55: 322; 56: 129.
1873   W. Cronander, Bull. soc. chim. Paris, (2) 19: 500.
1874   H. Roscoe, Ber., 7: 1131-1133.
1878   C. W. Watts and C. A. Bell, J. Chem. Soc., 33: 442.
1881   C. Zimmermann, Ber., 14: 1934-1939.
1882   C. Zimmermann, Ann., 213: 301, 304.
1882   C. Zimmermann, Ber., 15: 849.
1883   C. Zimmermann, Ann., 216: 9, 10.
1887   E. Demarcay, Compt. rend., 104: 111.
1887   L. Meyer, Ber., 20: 681.
1888   W. Hampe, Chem. Ztg., 12: 106.
1896   H. Moissan, Compt. rend., 122: 1092.
1897   H. Moissan, Bull. soc. chim. Paris, (3) 17: 266.
1899   J. Aloy, Bull. soc. chim. Paris, (3) 21: 264.
1900   V. Kohlschütter and M. Rossi, Ber., 34: 1473.
1903   C. Renz, Z. anorg. Chem., 36: 110.
1904   V. Pimmer, dissertation, University of Zürich, from J. W. Mellor, "A Comprehensive Treatise on Inorganic and Theoretical Chemistry," vol. XII, p. 84, Longmans, Green & Co., Inc., New York, 1932.
1905   F. Giolitti and G. Agamennone, Atti accad. nazl. Lincei, (5) 14: 165.
1905   C. Matignon and F. Bourion, Ann. chim. et phys., (8) 5: 127-136.
1907   A. Colani, Ann. chim. et phys., (8) 12: 66-80.

1908   A. Rosenheim and H. Loebel, Z. anorg. Chem., 57: 235.
1908   L. Rugheimer and L. Gonder, Ann., 364: 45.
1910   P. Camboulives, Compt. rend., 150: 175.
1910   C. Matignon and F. Bourion, Ann. chim. et phys., (8) 21: 49-131.
1910   A. Michael and A. Murphy, Jr., Am. Chem. J., 44: 365-384.
1911   E. Chauvenet, Compt. rend., 152: 87, 1250.
1911   O. Ruff and A. Heinzelmann, Z. anorg. Chem., 72: 65.
1913   A. Fischer, Z. anorg. Chem., 81: 177, 190.
1914   O. Lely, Jr., and L. Hamburger, Z. anorg. Chem., 87: 209-228.
1919   C. A. Pierlé, J. Phys. Chem., 23: 541.
1922   J. Aloy and E. Rodier, Bull. soc. chim. France, (4) 31: 247.
1923   M. A. Hunter and A. Jones, Trans. Am. Electrochem. Soc., 44: 27.
1923   R. W. Moore, Trans. Am. Electrochem. Soc., 43: 317-323.
1924   A. Voigt and W. Biltz, Z. anorg. u. allgem. Chem., 133: 277-305.
1926   J. F. Goggin, J. J. Cronin, H. C. Fogg, and C. James, Ind. Eng. Chem., 18: 114-116.
1928a   W. Biltz and C. Fendius, Z. anorg. u. allgem. Chem., 176: 49-63.
1928b   W. Biltz and C. Fendius, Z. anorg. u. allgem. Chem., 172: 386-391.
1928   O. Hönigschmid and W. E. Schilz, Z. anorg. u. allgem. Chem., 170: 145-160.
1928   J. Kleinheksel and H. Kremers, J. Am. Chem. Soc., 50: 959.
1932   G. Beck, Z. anorg. u. allgem. Chem., 206: 416, 421.
1932   A. Rosenheim and M. Kelmy, Z. anorg. u. allgem. Chem., 206: 31, 32.
1932   W. Sucksmith, Phil. Mag., (7) 14: 1121.
1933   F. Ephraim and M. Mezener, Helv. Chim. Acta, 16: 1259.
1934   O. Hülsmann and W. Biltz, Z. anorg. u. allgem. Chem., 219: 357-366.
1939   H. S. Booth, editor, "Inorganic Syntheses," vol. I, p. 28, McGraw-Hill Book Company, Inc., New York.
1943   H. Martin and K. H. Eldau, Z. anorg. u. allgem. Chem., 251: 295-304.
1947   D. C. Ginnings and R. J. Corruccini, J. Research Natl. Bur. Standards, 39: 309-316.
1948   J. R. Van Wazer and G. S. John, J. Am. Chem. Soc., 70: 1207.
1948a   W. H. Zachariasen, J. Chem. Phys., 16: 254.
1948b   W. H. Zachariasen, Acta Crystallographica, 1: 285-287.
1949   R. C. L. Mooney, Acta Crystallographica, 2: 189-191.
1949   C. C. Sterett and V. P. Calkins, Report AECD-2443.

Project Literature

British 1: T. K. Wood, Report [B]LRG-4, Dec. 22, 1942.
British 2: W. Haworth, Report B-20.
British 3: Directorate of Tube Alloys, Report B-55, March, 1942.

Brown 1: C. A. Kraus, Report [A]M-1060, Nov. 2, 1942.
Brown 2: C. A. Kraus, Report CC-342, Nov. 15, 1942.
Brown 3: C. A. Kraus, Report CC-365, Nov. 17, 1942.
Brown 4: C. A. Kraus, Report A-505, Jan. 9, 1943.
Brown 5: C. A. Kraus, Report A-522, Feb. 1, 1943.
Brown 6: C. A. Kraus, Report A-726, June 1, 1943.
Brown 7: C. A. Kraus, Report CC-1715, July 6, 1944.
Brown 8: C. A. Kraus, Report [A]M-251, July 1, 1943.
Brown 9: C. A. Kraus, Report N-29, Apr. 15, 1943.
Brown 10: C. A. Kraus, Report CC-1772, February, 1944.
Brown 11: D. F. Stedman and A. G. Brown, Report A-7, Apr. 1, 1941.

Brown 12: C. A. Kraus, Reports A-2313, July 12, 1945; A-2315, July 23, 1945; A-2321, Sept. 17, 1945.
Brown 13: C. A. Kraus, Report A-2300, October, 1944.
Brown 14: C. A. Kraus, Report [A]M-885, Oct. 7, 1942.
Brown 15: C. A. Kraus, Report CC-402, Dec. 1, 1942.
Brown 16: C. A. Kraus, Report A-1088, Nov. 3, 1943.
Brown 17: C. A. Kraus, Report A-1089, Nov. 1, 1943.
Brown 18: C. A. Kraus, Report A-557, Mar. 1, 1943.
Brown 19: C. A. Kraus, Report N-12, Dec. 15, 1942.
Brown 20: C. A. Kraus, Report N-18, Apr. 1, 1943.
Brown 21: E. C. Evers and C. A. Kraus, Report A-2329, May 1, 1946.
Brown 22: C. A. Kraus, Report A-1085, Aug. 1, 1943.
Brown 23: C. A. Kraus, Report A-1087, Oct. 1, 1943.
Brown 24: C. A. Kraus, Report [A]-BM-4, Jan. 3, 1944.
Brown 25: C. A. Kraus, Report CC-1685, Mar. 10, 1944.
Brown 26: C. A. Kraus, Report A-1090, Dec. 1, 1943.
Brown 27: C. A. Kraus, Report A-1091, December, 1943.
Brown 28: C. A. Kraus, Report CC-1716, July 7, 1944.
Brown 29: C. A. Kraus, Report A-1086, Sept. 1, 1943.
Brown 30: C. A. Kraus, Report N-1789, July 1, 1943.
Brown 31: C. A. Kraus, Report CC-1717, March, 1944.
Brown 32: C. A. Kraus, Report CC-1519, Feb. 22, 1944.

CEW-TEC 1: J. W. Gates and L. J. Andrews, Report CD-477, Nov. 9, 1944.
CEW-TEC 2: H. S. Young, E. L. Wagner, and A. J. Miller, Report CD-2.350.5, Nov. 23, 1945.
CEW-TEC 3: F. R. Diakiw, R. E. Gluyas, C. Tanford, R. L. Tichenor, and C. E. Larson, Report CD-3.385.7, Sept. 26, 1945.
CEW-TEC 4: E. L. Wagner, Report CD-0.350.9, February, 1946.
CEW-TEC 5: E. L. Wagner and H. A. Gottschall, Report CD-MPR-13, February, 1945.
CEW-TEC 6: G. S. Parsons, C. L. McCabe, G. F. Engle, and C. E. Larson, Reports CD-564, Apr. 2, 1945; CD-541; and CD-5701, May, 1945; G. S. Parsons and C. E. Larson, Reports CD-5715, June, 1945, and CD-5718, June, 1945.
CEW-TEC 7: B. M. Pitt, E. L. Wagner, and A. J. Miller, Report CD-2.355.2, December, 1945.
CEW-TEC 8: J. V. Hubbard, H. A. Perlmutter, H. L. Goren, and G. H. Clewett, Report CD-5.350.3, Aug. 28, 1945; R. S. Lowrie, J. V. Hubbard, and G. H. Clewett, Report CD-5.350.6, Dec. 20, 1945; H. A. Perlmutter, J. H. Coobs, R. S. Lowrie, and A. J. Miller, Report CD-2.350.9, Mar. 12, 1946.
CEW-TEC 9: B. M. Pitt, W. F. Curran, J. M. Schmitt, E. L. Wagner, and A. J. Miller, Report CD-2.350.3, July, 1945.
CEW-TEC 10: G. S. Parsons, C. L. McCabe, G. F. Engle, and C. E. Larson, Report CD-5701, May, 1945.
CEW-TEC 11: L. B. Dean, E. L. Wagner, and A. J. Miller, Report CD-4107, June 12, 1945; E. L. Wagner, B. M. Pitt, W. F. Curran, J. M. Schmitt, and A. J. Miller, Report CD-2.350.1, July 10, 1945.
CEW-TEC 12: E. L. Wagner and B. M. Pitt, Report CD-MPR-14, March, 1945, reported by A. J. Miller and H. A. Young, CEW-TEC Part II, p. 12.
CEW-TEC 13: E. L. Wagner, B. M. Pitt, L. B. Dean, and A. J. Miller, Report CD-4106, April, 1945.
CEW-TEC 14: B. M. Pitt, E. L. Wagner, and A. J. Miller, Report CD-2.355.3, Jan. 14, 1946.

CEW-TEC 15: H. A. Gottschall, J. F. Manneschmidt, J. M. Schmitt, E. L. Wagner, and A. J. Miller, Report CD-2.350.6, Jan. 29, 1946.

CEW-TEC 16: G. Gavlin and J. V. Hubbard, Report CD-5.350.5, Dec. 12, 1945.

CEW-TEC 17: C. C. Sterett and J. R. Van Wazer, Report CD-3.360.4, Sept. 21, 1945.

CEW-TEC 18: J. V. Hubbard, H. A. Perlmutter, M. Castillo, and G. H. Clewett, Report CD-5.381.12, January, 1945.

CEW-TEC 19: V. P. Calkins and C. E. Larson, Report CD-0.350.4, September, 1945.

CEW-TEC 20: A. J. Miller and E. L. Wagner, Report CD-0.100.25, January, 1946.

CEW-TEC 21: V. P. Calkins and C. E. Larson, Report CD-0.350.5, September, 1945.

CEW-TEC 22: E. L. Wagner and H. A. Gottschall, Report CD-0.100.25, January, 1946.

CEW-TEC 23: W. A. Gregory, H. J. Crogg, P. F. Grieger, and J. W. Gates, Report CD-1.350.1, Aug. 1, 1945.

CEW-TEC 24: E. L. Wagner and B. M. Pitt, Report CD-MPR-13, Sec. III, February, 1945.

CEW-TEC 25: E. L. Wagner, H. S. Young, S. S. Marsden, W. J. McLean, and A. J. Miller, Report CD-2.350.2, July 28, 1945.

CEW-TEC 26: C. C. Sterett and J. R. Van Wazer, Report CD-GS-37, Apr. 26, 1945.

CEW-TEC 27: E. L. Wagner, H. F. Grady, and A. J. Miller, Report CD-2.350.4, Sept. 18, 1945.

CEW-TEC 28: G. S. John and J. R. Van Wazer, Report CD-3.385.4, Aug. 22, 1945.

CEW-TEC 29: H. L. Goren, R. S. Lowrie, and J. V. Hubbard, Report CD-5.350.8, Feb. 28, 1946.

CEW-TEC 30: C. C. Sterett, I. Halpern, and J. R. Van Wazer, Report CD-GS-28, Mar. 16, 1945.

CEW-TEC 31: G. Gavlin, J. V. Hubbard, and G. H. Clewett, Report CD-PPO-14, Apr. 3, 1945.

CEW-TEC 32: H. S. Young, Preparation of Uranium Trichloride by Reduction of the Tetrachloride with Metals, in National Nuclear Energy Series, Division VIII, Volume 6.

CEW-TEC 33: A. J. Miller and L. B. Dean, Preparation of Uranium Tetrachloride with Hexachloropropylene and Other Chlorinated Organic Compounds, in National Nuclear Energy Series, Division VIII, Volume 6.

CEW-TEC 34: H. S. Young, Purification of Uranium Tetrachloride by Fractional Condensation, in National Nuclear Energy Series, Division VIII, Volume 6.

CEW-TEC 35: H. S. Young and H. F. Grady, Physical Constants of Uranium Tetrachloride, in National Nuclear Energy Series, Division VIII, Volume 6.

Johns Hopkins 1: W. B. Burford III, Report [A]M-2121, February, 1945.

Johns Hopkins 2: W. B. Burford III, Report [A]M-2122, March, 1945.

MP Ames 1: O. Johnson and T. Butler, Report CC-1496, May 11, 1944.

MP Ames 2: A. S. Newton, O. Johnson, A. Kant, and R. W. Nottorf, Report CC-705, June 7, 1943.

MP Ames 3: T. Butler, Report CC-1500, June 17, 1944.

MP Ames 4: O. Johnson, T. Butler, and A. S. Newton, The Preparation, Purification, and Properties of Anhydrous Uranium Chlorides, in National Nuclear Energy Series, Division VIII, Volume 6.

MP Ames 5: O. Johnson, Report A-3190, Nov. 25, 1944.

MP Ames 6: O. Johnson and T. Butler, Report CC-1974, Nov. 18, 1944.

MP Ames 7: V. P. Calkins, Report CC-1500, June 17, 1944.

MP Ames 8: O. Johnson, Report CC-1500, June 17, 1944.

MP Ames 9: F. Edwards and J. C. Warf, Report CC-1496, May 11, 1944.

MP Ames 10: O. Johnson and T. Butler, Report CC-1524, March, 1944.

MP Ames 11: C. Neher, H. Lipkind, W. Lyon, and W. Sleight, Report CT-1780, Aug. 9, 1944; F. Spedding, Report CT-1780, July 10, 1944.

MP Ames 12: F. Spedding, Report CC-1504, June 10, 1944.

MP Ames 13: T. Butler and A. S. Newton, Report CC-1500, June 17, 1944.

MP Ames 14: T. Butler and R. W. Nottorf, Report CC-1975, Sept. 10, 1944, p. 7.

MP Ames 15: O. Johnson and T. Butler, Report CC-1778, Aug. 18, 1944.

MP Ames 16: N. C. Baenziger and R. E. Rundle, Report CC-1500, June 17, 1944.

MP Ames 17: V. P. Calkins, Reports CC-1496, May 11, 1944, and CC-1500, June 17, 1944.

MP Ames 18: O. Johnson and T. Butler, Report CC-1500, June 17, 1944.

MP Ames 19: T. Butler, Report CC-1778, Aug. 18, 1944.

MP Ames 20: T. Butler and O. Johnson, Report CC-1975, Sept. 10, 1944.

MP Chicago 1: W. H. Zachariasen, Report CK-1487, Mar. 27, 1944.

MP Chicago 2: G. T. Seaborg, Report N-930, Apr. 5, 1944.

MP Chicago 3: J. Karle, Report CK-1512, Apr. 14, 1944.

MP Chicago 4: R. C. L. Mooney, Reports CP-1533, Mar. 31, 1944, and CP-1507, Apr. 3, 1944.

MP Chicago 5: W. H. Zachariasen, Report CC-2768, Mar. 12, 1945.

MP Chicago 6: J. J. Howland, Jr., Absorption Spectra of Uranium(III) and Uranium(IV) in Molar Hydrochloric Acid, in National Nuclear Energy Series, Division VIII, Volume 6.

MP Chicago 7: S. Fried and N. R. Davidson, The Ignition of U_3O_8 in Oxygen at High Pressures and the Crystallization of UO_3, in National Nuclear Energy Series, Division VIII, Volume 6.

MP Chicago 8: W. H. Zachariasen, Reports N-1574, Feb. 28, 1945, and CF-2796, Mar. 15, 1945.

Natl. Bur. Standards 1: W. J. Ferguson and J. L. Prather, Report A-3143, Nov. 18, 1944.

Natl. Bur. Standards 2: C. J. Rodden, Report A-36, March, 1942.

Natl. Bur. Standards 3: D. C. Ginnings and R. J. Corruccini, Report A-3948, July, 1946.

Natl. Bur. Standards 4: W. J. Ferguson, J. L. Prather, and R. B. Scott, Report A-1920, May 3, 1944.

Natl. Bur. Standards 5: W. J. Ferguson and R. D. Rand, Report A-3357, Mar. 28, 1945.

Natl. Bur. Standards 6: D. C. Ginnings and R. J. Corruccini, Report A-4120, date received, Nov. 20, 1946.

Princeton 1: J. E. White, R. W. Thompson, and A. Schelberg, Report A-178, May 23, 1942.

Purdue 1: E. T. McBee, Report [A]M-2102, Mar. 21, 1945.

Purdue 2: E. T. McBee, Report [A]M-2101, Feb. 21, 1945.

Purdue 3: E. T. McBee, Report [A]M-2113, January, 1946.

Purdue 4: E. T. McBee, Report [A]M-2103, March, 1945.

Purdue 5: E. T. McBee and L. R. Evans, Report A-2704, Dec. 14, 1945.

Purdue 6: E. T. McBee, D. W. Pearce, and R. Mezey, Report A-2706, Feb. 5, 1946.

Purdue 7: R. E. Burns and T. DeVries, Report A-2714, Mar. 12, 1946.

Purdue 8: E. T. McBee, T. DeVries, and G. M. Rothrock, Report A-2705, Dec. 14, 1944.

SAM Columbia 1: H. C. Urey, Report A-750, July 10, 1943.

SAM Columbia 2: A. B. F. Duncan, Report A-584, Apr. 15, 1943.

UCRL 1: C. H. Barkelew, Report RL-4.6.906, Jan. 18, 1945.

UCRL 2: D. Altman, Report RL-4.6.156, June 11, 1943.

UCRL 3: G. E. MacWood and D. Altman, Report RL-4.7.600, Oct. 24, 1944.

UCRL 4: C. H. Barkelew, Report RL-4.6.929, Aug. 28, 1945.

UCRL 5: D. Altman, Report RL-4.6.276, Aug. 14, 1945.

UCRL 6: N. W. Gregory, Report RL-4.6.928, Aug. 24, 1945.

UCRL 7: G. E. MacWood, Thermodynamic Properties of Some Uranium Compounds, in National Nuclear Energy Series, Division VIII, Volume 6.

UCRL 8: D. Altman, Report RL-4.6.166, June 25, 1943.

UCRL 9: P. H. Davidson and M. E. Mueller, Report RL-4.6.250, Feb. 26, 1944.

UCRL 10: O. H. Cook and R. C. Feber, Report RL-4.6.926, July 3, 1945.

UCRL 11: H. G. Reiber, Report RL-4.6.169, July 3, 1943.

UCRL 12: H. G. Reiber, Report RL-4.6.187, July 31, 1943.

UCRL 13: S. Rosenfeld, Report RL-4.6.58.

UCRL 14: H. G. Reiber, Report RL-4.7.600, Sec. D, Oct. 24, 1944.

UCRL 15: M. J. Polissar, Report RL-4.6.35, Nov. 3, 1942; S. Rosenfeld, Report RL-4.6.52, Mar. 22, 1943.

UCRL 16: H. G. Reiber and H. A. Young, Report RL-4.6.54.

UCRL 17: G. Steele, Report RL-4.6.113, Apr. 13, 1943.

UCRL 18: H. G. Reiber, Report RL-4.6.70, Mar. 13, 1943.

UCRL 19: H. G. Reiber, Report RL-4.6.105, Apr. 10, 1943.

UCRL 20: H. G. Reiber, Report RL-4.6.67.

UCRL 21: S. Rosenfeld, Report RL-4.6.119, Aug. 28, 1944.

UCRL 22: S. H. Babcock, Report RL-4.6.61, July 13, 1943.

UCRL 23: W. C. Wood, H. Bradley, and H. S. Carroll, Report RL-4.6.287, August, 1944.

UCRL 24: M. Kamen, Report RL-4.6.8, Nov. 30, 1942.

UCRL 25: R. L. Kinderman, M. B. Leboeuf, and H. Walter, Report RL-4.6.99, July 27, 1945.

UCRL 26: F. A. Jenkins, Report RL-4.6.232, Dec. 15, 1943.

UCRL 27: G. D. Dorough, Report RL-4.6.283, Aug. 18, 1944.

UCRL 28: F. A. Jenkins, O. E. Anderson, and A. DeHaan, Report RL-4.6.1.

UCRL 29: S. Rosenfeld, Report RL-4.6.66, Mar. 23, 1943.

UCRL 30: R. Schmidt, Report RL-4.6.117, Apr. 24, 1943.

UCRL 31: N. W. Gregory, Report RL-4.6.923, Aug. 1, 1943.

UCRL 32: F. A. Jenkins, Report RL-4.6.3.

UCRL 33: M. J. Polissar, Report RL-4.6.36, Oct. 23, 1942.

UCRL 34: D. Altman, Report RL-4.6.276, Aug. 14, 1944.

UCRL 35: M. E. Mueller, Report RL-4.6.934, Sept. 18, 1945.

UCRL 36: F. L. Reynolds and P. H. Davidson, Reports RL-4.6.231, Nov. 30, 1943, and RL-4.6.236, Dec. 28, 1943.

UCRL 37: M. E. Mueller, Report RL-4.6.191, Aug. 19, 1943.

UCRL 38: J. Booker and A. DeHaan, Report RL-4.6.6, Nov. 22, 1942.

UCRL 39: R. Cummings, Report RL-4.6.176, July 27, 1943.

UCRL 40: P. H. Davidson and I. Streeter, Report RL-4.6.289, Sept. 5, 1944.

UCRL 41: R. Cummings and C. Hollingshead, Reports RL-4.6.89, Mar. 26, 1943, and RL-4.6.92, Apr. 3, 1943; R. Cummings, Report RL-4.6.137, May 20, 1943.

UCRL 42: J. F. Carter, F. A. Jenkins, M. Kamen, C. E. Larson, D. Lipkind, and S. Weissman, Report RL-4.6.13, Jan. 17, 1943.

UCRL 43: A. D. Webb, Reports RL-4.6.102 and RL-4.6.87, Mar. 25, 1943; H. P. Kyle, Report RL-4.6.50, Aug. 7, 1944; A. D. Webb and H. P. Kyle, Report RL-4.6.51, March, 1943.

UCRL 44: H. G. Reiber, Report RL-4.6.116, Apr. 23, 1943.

UCRL 45: C. D. Wilder, Reports RL-4.6.65, RL-4.6.68, RL-4.6.73, RL-4.6.103, RL-4.6.106, and RL-4.6.129, Apr. 26, 1943; H. A. Young, Report RL-4.6.111, Apr. 20, 1943.

UCRL 46: M. E. Mueller, Report RL-4.6.88, Mar. 28, 1943.

UCRL 47: F. L. Reynolds, Report RL-4.6.80, Feb. 13, 1943.

UCRL 48: W. C. Wood and H. Bradley, Report RL-4.6.263, May 20, 1944; W. C. Wood, H. Bradley, and H. S. Carroll, Report RL-4.6.287, Aug. 29, 1944.

UCRL 49: W. C. Wood, Report RL-4.7.600, E-1, Aug. 9, 1944.

UCRL 50: F. A. Jenkins, Report RL-4.6.97, Apr. 6, 1943.

UCRL 51: A. Bell, Report RL-4.6.525, Aug. 9, 1943.

UCRL 52: J. V. Hubbard, Report RL-4.6.168, June 30, 1943.

UCRL 53: G. D. Steele, Report RL-4.6.99, Mar. 26, 1943.

UCRL 54: F. A. Jenkins, Reports RL-4.6.85A, Feb. 22, 1943, and RL-4.6.47, Mar. 9, 1943.

UCRL 55: D. Altman, Report RL-4.6.161, June 21, 1943.

UCRL 56: H. G. Reiber, Report RL-4.6.60.

UCRL 57: G. E. MacWood, Report RL-4.7.602, Dec. 18, 1945.

UCRL 58: M. B. Allen, Report RL-4.6.19, Feb. 13, 1943.

UCRL 59: W. C. Wood and H. Bradley, Report RL-4.6.288, Sept. 2, 1944.

UCRL 60: R. Krohn, Report RL-4.6.238, Jan. 4, 1944.

UCRL 61: P. H. Davidson, Report RL-4.6.86, Mar. 23, 1943.

UCRL 62: H. G. Reiber, Reports RL-4.6.233, Dec. 15, 1943, and RL-4.6.125, May 3, 1943.

UCRL 63: H. G. Reiber, Report RL-4.6.131, May 8, 1943.

UCRL 64: M. B. Allen, Report RL-4.6.77, Feb. 6, 1943.

UCRL 65: F. L. Reynolds, Report RL-4.6.78, Feb. 7, 1943.

UCRL 66: F. L. Reynolds, Report RL-4.6.188, May 2, 1943.

UCRL 67: P. H. Davidson and J. A. Holmes, Report RL-4.6.182, July 29, 1943.

UCRL 68: F. A. Jenkins, Report RL-4.6.47, Mar. 9, 1943.

UCRL 69: D. Altman, D. Lipkin, and S. Weissman, Report RL-4.6.22, Feb. 27, 1943.

UCRL 70: D. Lipkin and S. Weissman, Reports RL-4.6.18, Feb. 3, 1943; RL-4.6.112, Apr. 27, 1943; and RL-4.6.13, Jan. 17, 1943.

Chapter 15

BROMIDES, IODIDES, AND PSEUDOHALIDES OF URANIUM

1. URANIUM TRIBROMIDE, UBr_3 (MP AMES 1)

Uranium tribromide resembles uranium trichloride in many respects. It is a somewhat more hygroscopic compound but in a general way undergoes the same reactions and possesses physical properties similar to those of UCl_3. The methods of preparation for UBr_3 and UCl_3 are also very similar. The resemblance is emphasized here because the chemistry of UCl_3 is considerably better explored than that of UBr_3, and much qualitative information about UBr_3 may be obtained by analogy from UCl_3.

1.1 Preparation of UBr_3. The oldest method of preparation of UBr_3 is the reduction of UBr_4 with hydrogen (Alibegoff, 1886). This reaction, however, is rather unsatisfactory, and procedures that give a purer product have been developed.

(a) Preparation of UBr_3 from $UH_3 + HBr$. (MP Ames 2 to 6; MP Chicago 1; UCRL 1.) Hydrogen bromide reacts readily and smoothly with uranium hydride at 300°C to form UBr_3.

$$2U + 3H_2 \xrightarrow{250°C} 2UH_3 \tag{1}$$

$$UH_3 + 3HBr \xrightarrow{300°C} UBr_3 + 3H_2 \tag{2}$$

Suitable apparatus is described in the National Nuclear Energy Series, Division VIII, Volume 6 (MP Ames 1) (also see Fig. 15.1). It is necessary to use pure reagents; traces of oxygen must be removed from the hydrogen, and the bromine used in preparing the hydrogen bromide must be dried over P_2O_5. If uranium metal is available, this is an excellent method of preparation.

(b) Reaction of Metallic Uranium with Bromine. (MP Berkeley 1 to 3.) Metallic uranium and bromine vapors react to yield UBr_3.

$$2U + 3Br_2 \rightarrow 2UBr_3 \tag{3}$$

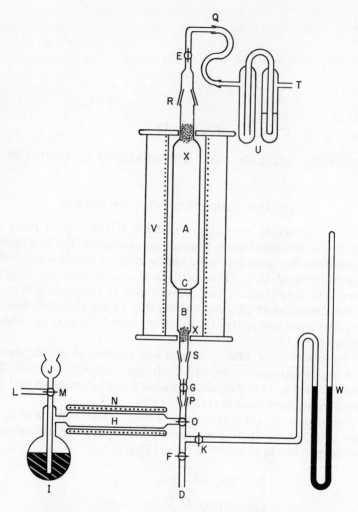

Fig. 15.1—Apparatus for the preparation of anhydrous uranium tribromide.

A, reaction chamber
B, scrubbing chamber
C, sintered glass disk
D, outlet for flushing system
E, stopcock
F, stopcock
G, stopcock
H, platinized asbestos
I, bromine bubbler
J, thistle tube
K, stopcock
L, gas inlet

M, three-way stopcock
N, resistance furnace
O, three-way stopcock
P, standard taper joint
Q, rubber tubing
R, standard taper joint
S, standard taper joint
T, outlet to hood
U, sulfuric acid bubbler and trap
V, resistance furnace
W, manometer
X, glass wool plug

To obtain UBr$_3$, stoichiometric quantities are used. A large pyrex tube with a side arm for the bromine is a suitable reaction vessel. The bromine is kept frozen in the reservoir until the metal is heated to 300 to 500°C; the bromine is then liquefied, and the vapors are allowed to react with the metal. The uranium may advantageously be powdered by forming the hydride and subsequently decomposing it. When the bromine is all consumed, the entire system is heated to 500 to 550°C. A product which is shown by analysis to be close to UBr$_3$ is obtainable in this way.

(c) Reduction of UBr$_4$ with Hydrogen. Historically the preparation of UBr$_3$ was first accomplished by hydrogen reduction at temperatures near the melting point of UBr$_4$ (Alibegoff, 1886; Zimmermann, 1882).

$$2UBr_4 + H_2 \rightarrow 2UBr_3 + 2HBr \qquad (4)$$

It has been found, however, that reduction does not necessarily stop at the tribromide stage and that products with a uranium/bromine ratio significantly less than 3 are sometimes obtained (MP Ames 6; MP Berkeley 4,5). This observation has been confirmed both by analysis of the product and by measurement of the amount of hydrogen bromide evolved. It is not clear whether reduction to a lower uranium bromide occurs or whether traces of moisture or oxygen are responsible. The reduction reaction appears to start at 430°C, but the rate is extremely slow below 550°C. At 700°C the rate is rapid, but considerable attack on the quartz reactors is experienced. The uncertainty of the composition of the product, coupled with the fact that the reaction is slow, renders the hydrogen reduction method of doubtful utility.

1.2 Physical Properties of UBr$_3$. Uranium bromide resembles UCl$_3$ and the rare earth tribromides LaBr$_3$ and NdBr$_3$ in its general physical properties. Like these, it is a high-melting nonvolatile crystalline substance.

(a) Melting Point. A number of values for the melting point of UBr$_3$ are available (MP Ames 7; MP Berkeley 2,6,7; MP Chicago 2; UCRL 1). A value of 755°C found at Ames is not based on a direct determination of the melting point but is derived from phase-rule studies of the U-UBr$_3$ and UBr$_3$-UBr$_4$ systems. At the University of California Radiation Laboratory a value of 730°C was found by direct measurement and is probably more accurate.

(b) Crystal Structure and Density. (MP Chicago 3; Zachariasen, 1948.) X-ray methods indicate UBr$_3$ to have a hexagonal unit cell, with

$$a_1 = 7.926 \pm 0.002 \text{ A} \qquad a_3 = 4.432 \pm 0.002 \text{ A}$$

There are two molecules per unit cell, with nine bromine atoms arranged about every uranium atom at an average U-Br distance of 3.12 A. Uranium tribromide is isomorphous with UCl_3, $LaCl_3$, $NdCl_3$, $LaBr_3$, $PrBr_3$, $La(OH)_3$, and $Nd(OH)_3$ (MP Ames 8,9; MP Chicago 4). The space group is $C6_3/m$ (C^2_{6h}). For details of this structure type, see Chap. 14, Sec. 1.2c.

The density of UBr_3 determined from the x-ray data is 6.53 g/cc. A direct pycnometric determination (in xylene) gave an average value of 5.98 g/cc (MP Ames 10).

Table 15.1 — Vapor Pressure of Uranium Tribromide
(Effusion Method, Mo Capsule)

Temperature, °C	Vapor pressure, mm Hg
600	2.0×10^{-5}
650	1.6×10^{-4}
700	1.3×10^{-3}
750	8.0×10^{-3}
800	3.2×10^{-2}
850	1.6×10^{-1}
900	5.0×10^{-1}
950	1.6
1000	5.0

(c) Vapor Pressure of UBr_3. The determination of the vapor pressure of UBr_3 has been difficult. The measurements must be made at elevated temperatures, where attack on the apparatus is serious. Further, the composition of the residue changes slowly from a uranium/bromine ratio of 1/3 to values approaching 1/2. The ratio in the distillate also varies (from 1/3.2 to 1/3.8), and, although an appreciable fraction of the distillate is thought to be UBr_3, some UBr_4 is almost certainly present, particularly at temperatures above 900°C. The vapor pressure of UBr_3 has been measured by an effusion method, using molybdenum capsules (UCRL 1). The data given in Table 15.1 are probably the best available for UBr_3. The vapor pressure data may be represented by the equation

$$\log p_{mm\ Hg} = -\frac{15,000}{T} + 12.5 \tag{5}$$

This leads to a heat of vaporization of 68 kcal per mole.

A gas saturation (transpiration) method has also been used (MP Berkeley 1). To suppress disproportionation a mixture of UBr_3 and uranium is used. The liquid phase approximated UBr_2 in composition, whereas the distillate corresponded to UBr_3 (Table 15.2). The agreement between the two sets of measurements is poor, and it is questionable whether the same phenomenon was being measured. It may be that one or both series of the vapor pressure data actually are the measure of disproportionation rather than of true volatility. The need for further work is indicated.

(d) Thermochemical Data. (UCRL 2.) The heat of formation of UBr_3 was determined by measuring its heat of solution in ferric chlo-

Table 15.2 — Vapor Pressure of Uranium Tribromide
(Transpiration Method)

Temperature, °C	Vapor pressure, mm Hg
915	0.30
1014	1.9
1062	3.1
1124	6.6

ride solution. Combination of thermochemical data for other uranium halides with pertinent data from the literature allows the heat and free-energy formation of UBr_3 to be computed from this measurement.

$$U \text{ (crystal)} + \tfrac{3}{2}Br_2 \text{ (gas)} \rightarrow UBr_3 \text{ (crystal)} \tag{6}$$

$$\Delta H_{273°K} = -181.6 \pm 0.6 \text{ kcal per mole}$$

MacWood (UCRL 3) gives 50.6 e.u. as the value of S_{298} for UBr_3. This leads to the value -49.4 e.u. for the entropy of formation and -166.9 ± 1.5 for $\Delta F°_{298°K}$. The origin of these entropy values is obscure, and they doubtlessly are based, at least in part, on estimates. No heat capacity data for UBr_3 are available.

No phase transitions have been detected in the temperature range 100 to 500°C (UCRL 4).

1.3 Chemical Properties of Uranium Tribromide. Uranium tribromide is a dark-brown crystalline substance. It is very hygroscopic, much more so than uranium trichloride. At room temperature there is appreciable reaction with the oxygen in air.

(a) Water. Uranium tribromide dissolves in water, with vigorous evolution of gas, to give a clear, dark red-violet solution, which changes its color in 30 to 60 sec to the clear brilliant green characteristic of uranium(IV) solutions (MP Berkeley 4). The evolved hydrogen has been measured, and the amount corresponds within a few per cent to that calculated for the oxidation from the uranium(III) to the uranium(IV) state (MP Berkeley 8).

(b) Other Solvents. (MP Berkeley 4,5,9.) Uranium tribromide, like UBr_4, is almost insoluble in nonpolar solvents. For example, the solubility of UBr_3 in benzene and ethyl bromide is much less than 0.5 g per 100 ml of solvent. Anhydrous crystals of UBr_3 undergo no color change in p-xylene. In polar solvents, UBr_3 is in general less soluble than UBr_4. Dissolution of UBr_3 in polar solvents usually results in extensive reaction. Thus, in formamide UBr_3 forms a deep-red solution, which in the course of several minutes turns green with the evolution of gas. In formamide—hydrogen bromide (32 g of HBr per 100 ml of formamide), UBr_3 decomposes immediately with vigorous gas evolution to a green solution of uranium(IV). Addition of UBr_3 to acetamide at 110 to 115°C results in the formation of a very deep-red solution, which in a few seconds turns green with simultaneous gas evolution. In ethanol UBr_3 decomposes rapidly in either the absence or presence of atmospheric oxygen, and hydrogen is evolved. Uranium tribromide is insoluble in inorganic solvents such as PCl_3 and $SnCl_4$ (MP Ames 11).

(c) Reduction of UBr_3. (See Sec. 1.3e.) Calcium metal reduces UBr_3 to metallic uranium in good yield. The customary technique can be used (MP Ames 4,12; see also Chap. 4). The electrochemical reduction of UBr_3 in molten salt baths has been studied (MP Berkeley 1,10,11). Solution of UBr_3 (4 to 10 per cent) in molten $BaBr_2$ at 870 to 910°C gave unsatisfactory, spongy, nonadherent deposits of uranium metal. However, a 10 to 20 mole % solution in $SrBr_2$ at 700°C can be used to deposit uranium on an iron, molybdenum, or uranium cathode. The bath operates with about 50 per cent current efficiency. Molten UBr_3 itself is not a satisfactory electrolyte for the deposition of uranium.

(d) Oxidation of UBr_3. The oxidation of UBr_3 to UBr_4 by bromine is discussed in Sec. 2.1. Oxygen reacts with UBr_3 even at room temperature, but the oxidation products have not been identified. In a qualitative way UBr_3 seems to undergo oxidation more readily than UCl_3.

(e) Disproportionation of UBr_3. Phase Relations. In connection with the discussion of vapor pressure it was pointed out that the composition of the volatilized material could be interpreted as indicating a partial disproportionation at temperatures above 900°C.

$$2UBr_3 \rightarrow UBr_2 + UBr_4 \tag{7}$$

The question of the existence of a bivalent halide of uranium naturally aroused interest, particularly after it was observed that metallic

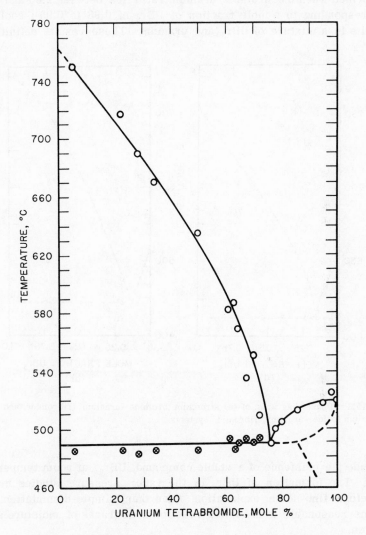

Fig. 15.2 — Uranium tribromide – uranium tetrabromide phase diagram.

uranium could be dissolved in molten UBr$_3$ to give, on cooling, solid products with empirical formulas closely approximating UBr$_2$. The problem was attacked by a phase-rule study of the U–UBr$_3$ system

(MP Berkeley 2). The results clearly indicate that no compound is formed by the solution of uranium metal in molten UBr_3. A eutectic is formed when the bromine/uranium ratio lies between 2.85 and 2.9 (corresponding to a mole fraction of UBr_3 of 0.98 to 0.95); cooling results in a mixture of UBr_3 and uranium. These results definitely

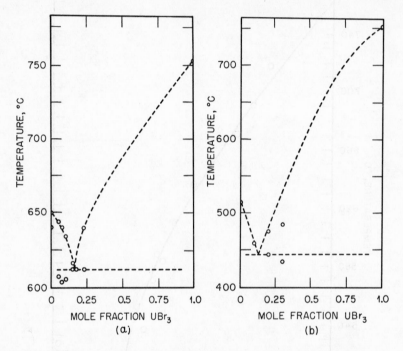

Fig. 15.3 — Phase diagrams of (a) strontium bromide – uranium tribromide and (b) strontium iodide – uranium tribromide systems.

exclude the existence of a stable compound, UBr_2, at room temperature. The formation of UBr_4 in the vapor pressure studies must therefore find some explanation other than simple dismutation. It seems reasonable to assume a reaction with traces of moisture and oxygen.

The phase diagram for the UBr_3-UBr_4 system has been obtained at Ames (MP Ames 7) (Fig. 15.2). A simple eutectic system is indicated with negligible solubility in the liquid phase but with appreciable solubility of UBr_3 in solid UBr_4. The eutectic occurs at 76 mole % UBr_3 and melts at 490°C.

The SrI$_2$-UBr$_3$ and SrBr$_2$-UBr$_3$ systems have also been studied (MP Berkeley 7). Preliminary indications are that compound formation does not occur in these systems (Fig. 15.3).

(f) Complex Compounds. Miscellaneous Reactions. Uranium tribromide undergoes reaction on exposure to ammonia gas. The characteristic brown color of UBr$_3$ disappears, and a very bulky green material, which has the composition UBr$_3$·6NH$_3$, forms. Treatment of UBr$_3$ with liquid ammonia and drying in vacuum gave a solid that contained between three and four NH$_3$ groups per UBr$_3$ (MP Berkeley 9).

When UBr$_3$ is heated in quartz until no further distillate is formed, it is found that the residue is free from bromine. Examination shows it to consist of a water-soluble fraction and a black insoluble portion. The reaction

$$12UBr_3 + 2SiO_2 \rightarrow USi_2 + 2UO_2 + 9UBr_4 \tag{8}$$

has been proposed to explain these observations. Incidentally, this indicates the inadvisability of long-continued heating of UBr$_3$ in quartz or glass at elevated temperatures. A few metals have been examined for corrosion resistance to UBr$_3$; molybdenum is found to be quite resistant and is superior to stainless steel (UCRL 1).

2. URANIUM TETRABROMIDE, UBr$_4$ (MP AMES 1)

Uranium tetrabromide resembles UCl$_4$ quite closely in its physical and chemical properties. It is a less stable substance, as evidenced by a smaller free energy of formation.

$$U \text{ (crystal)} + 2Cl_2 \text{ (gas)} \rightarrow UCl_4 \text{ (crystal)} \tag{9}$$

$$\Delta F^\circ_{298°K} = -229.0 \pm 1.0 \text{ kcal per mole}$$

$$U \text{ (crystal)} + 2Br_2 \text{ (gas)} \rightarrow UBr_4 \text{ (crystal)} \tag{10}$$

$$\Delta F^\circ_{298°K} = -190.2 \pm 1.5 \text{ kcal per mole}$$

Methods of preparation for uranium tetrabromide are less varied than for uranium tetrachloride. The only known reagent that converts the uranium oxides directly to the tetrabromide is a mixture of carbon and bromine at elevated temperature. Unlike carbon tetrachloride, which is so useful for the preparation of UCl$_4$, carbon tetrabromide does not come into consideration as a brominating agent, since it completely decomposes at the high temperatures required for reaction. Hydrogen bromide does not convert the uranium oxides at a sufficiently high rate at any temperature despite the fact that the thermodynamic conditions are more favorable for this reaction than for the chlorination of uranium oxides by hydrogen chloride. Since the reaction of oxides with carbon and bromine at elevated temperatures is difficult experimentally, UBr_4 must be prepared from uranium compounds other than the oxides. Free bromine is a sufficiently powerful brominating agent to convert uranium metal and uranium tribromide as well as uranium carbide, nitride, or sulfide to UBr_4. Most of these reactions have been studied at Ames.

2.1 Preparation of Uranium Tetrabromide. Since both bromine and UBr_4 are corrosive at elevated temperatures, preparations are usually carried out in glass or quartz. Nickel and gold are fairly resistant, but stainless steel, monel, and copper are unsatisfactory (MP Ames 6).

(a) UO_2 + C + Br_2. As mentioned above, this is the original, but inconvenient, method for the preparation of UBr_4. It was first employed by Hermann (1861) and followed in principle by all subsequent workers (Zimmermann, 1882) until very recently. The efforts of Colani (1907) to improve the procedure were unsuccessful. (He also made the claim that hydrogen bromide would convert UCl_4 to UBr_4 at 590°C, which is doubtful.) Preparation and purification of UBr_4 by the carbon-bromine reaction were discussed in considerable detail by Richards (1902) and Hönigschmid (1915) who used the purified UBr_4 for atomic weight determinations.

The reaction is carried out by mixing UO_2 or U_3O_8 with excess carbon and treating this mixture with bromine at bright red heat. The bromine is best carried in an inert gas such as nitrogen or CO_2. The UBr_4 condenses in the cooler portions of the apparatus.

(b) Uranium Metal and Bromine. (Zimmermann, 1882; MP Ames 13; MP Chicago 5.) This reaction is very suitable for the preparation of UBr_4. Bromine (carried in helium gas) acts smoothly with uranium metal turnings at 650°C. The operation is carried out in a quartz apparatus which permits distillation of the product as it forms. The apparatus developed at Ames is shown in Fig. 15.4.

(c) Uranium Metal and Bromine (by Way of Uranium Hydride). (MP Ames 2.) The conversion of uranium metal to the tetrabromide may also be effected by intermediary formation of uranium hydride.

$$2U + 3H_2 \rightarrow 2UH_3 \quad (Eq.\ 1)$$

$$2UH_3 + 7Br_2 \rightarrow 2UBr_4 + 6HBr \tag{11}$$

This is not so satisfactory as the direct conversion of the metal. The reaction is more difficult to control, and local hot spots are apt to form which eventually clog the apparatus.

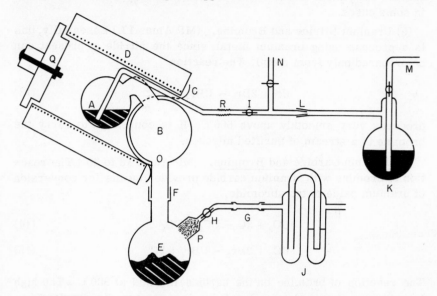

Fig. 15.4—Apparatus for the preparation of anhydrous uranium tetrabromide.

A, quartz reaction flask containing uranium tribromide or uranium metal
B, 1-liter quartz flask used as condensing chamber
C, aluminum foil reflector used for insulation
D, resistance furnace
E, 1-liter pyrex flask used as collector
F, quartz-to-pyrex joint sealed with Apiezon W
G, rubber tubing
H, stopcock
I, stopcock
J, sulfuric acid bubbler and trap
K, bromine reservoir
L, standard taper joint
M, gas inlet
N, outlet to hood
O, opening (which occasionally plugs up)
P, glass wool plug to retain fine product
Q, removable end for observation
R, quartz-to-pyrex seal

(d) Uranium Tribromide and Bromine. (MP Ames 2,4,10,14 to 16.) This reaction is excellent for preparing UBr₄.

$$2UBr_3 + Br_2 \xrightarrow{300°C} 2UBr_4 \qquad (12)$$

The bromine is carried by a stream of nitrogen (rigorously purified over copper oxide at 600°C, ascarite, anhydrone, and, finally, uranium nitride at 600°C). The reaction proceeds smoothly at 300°C, but the apparatus requires occasional shaking to prevent packing and channeling. Below 200°C reaction is incomplete. The reaction can be carried out without distillation of the UBr_4. This may be advantageous in some cases.

(e) Uranium Nitride and Bromine. (MP Ames 17.) Essentially, this is a process using uranium metal, since the nitride at present can be prepared only from metal. The reaction

$$UN + 2Br_2 \rightarrow UBr_4 + \tfrac{1}{2}N_2 \tag{13}$$

proceeds very smoothly above 600°C. It is convenient to carry the bromine in a stream of purified nitrogen.

(f) Uranium Carbide and Bromine. (MP Ames 18 to 20.) The reaction of bromine with uranium carbide provides means for conversion of uranium oxide to tetrabromide.

$$UO_2 + 3C \rightarrow UC_2 + CO_2 \tag{14}$$

$$UC_2 + 2Br_2 \rightarrow UBr_4 + 2C \tag{15}$$

The reaction of bromine on the carbide is rapid at 800°C. The high temperature required (see Sec. 2.1a) is a serious disadvantage because of the difficult corrosion conditions created. Another drawback is that some of the finely divided carbon produced is carried along by the gas stream and contaminates the product.

(g) Uranium Sulfide and Bromine or Hydrogen Bromide. (MP Ames 20,21.) Uranium sulfide is converted by bromine or hydrogen bromide to UBr_4.

$$US_2 + 3Br_2 \rightarrow UBr_4 + S_2Br_2 \tag{16}$$

$$US_2 + 4HBr \rightarrow UBr_4 + 2H_2S \tag{17}$$

Reaction with bromine occurs at room temperature; hydrogen bromide requires somewhat higher temperatures. The sulfur bromide can be distilled away below 450°C, and the reaction is usually completed by distilling the UBr_4 at 600°C. These reactions do not have much practical significance because of the difficulties of obtaining uranium sulfide.

(h) Uranium Dioxide, CS_2, and Bromine ($U_3O_2S_4$ and Br_2). (MP Ames 13,15,17.) It is well-known that many oxides (particularly of bivalent and trivalent metals) are converted to sulfides by treatment with carbon disulfide at elevated temperatures. With uranium oxide the reaction proceeds, however, to give a mixed uranium(IV) oxysulfide (see Chap. 11).

$$3UO_2 + 4CS_2 \xrightarrow{900°C} U_3O_2S_4 + 4COS \qquad (18)$$

When this compound is treated with bromine, reaction occurs at 50 to 100°C, and S_2Br_2 forms, which can be removed at 400°C. The residue, a light-brown powder, decomposes at 600°C with the formation of UBr_4. The residue from this treatment is $UOBr_2$, which in turn decomposes to UO_2 and UBr_4 at 800°C.

$$U_3O_2S_4 + 6Br_2 \xrightarrow{50-100°C} 2UOBr_2 + UBr_4 + 2S_2Br_2 \qquad (19)$$

$$2UOBr_2 \xrightarrow{800-1000°C} UBr_4 + UO_2 \qquad (20)$$

$$U_3O_2S_4 + 6Br_2 \xrightarrow{800°C} 2UBr_4 + UO_2 + 2S_2Br_2 \qquad (21)$$

If the reaction is completed at 800°C, the net result will be the conversion of 3 moles of UO_2 to 2 moles of UBr_4. The residual mole of UO_2 will require reprocessing. This interesting cycle merits further study.

(i) Uranium Dioxide, Sulfur, and Bromine. Carter (1949) prepared uranium tetrabromide by the liquid-phase reaction of uranium dioxide with either sulfur monobromide or a mixture of sulfur and bromine. An excess of sulfur and bromine is refluxed for 12 hr at 170°C; excess reagent is then distilled away, and the product washed with carbon tetrachloride. It may be anticipated that considerable difficulty will be encountered in obtaining a sulfur-free product.

(j) Summary of Preparation Methods. Table 15.3 summarizes the reactions discussed above by which uranium tetrabromide can be produced from various uranium compounds and free bromine.

(k) Preparation of UBr$_4$ from Solution. Since aqueous solutions of UBr_4 may be readily prepared, attempts have been made to prepare anhydrous UBr_4 from them. Solutions of UBr_4 can be made by electrochemical reduction of uranyl bromide, by chemical reduction with agents such as uranium metal, or by dissolving hydrated uranium(IV) hydroxide in hydrobromic acid. It has been found that such solutions cannot be dehydrated without very considerable hydrolysis. In this respect aqueous solutions of UBr_4 resemble those of UCl_4, with the difference that hydrolysis occurs even more readily in the case of UBr_4. Hydrolysis can take place even in a stream of hydrogen bromide. This is not surprising when it is recalled that hydrogen bromide is incapable of converting UO_2 to UBr_4. Considerable oxidation occurs when UBr_4 solutions are concentrated under reduced pressure either at room temperature or at 180°C even in a hydrogen bromide atmosphere (MP Ames 22). It has been reported that UBr_4 is formed on photochemical reduction of a solution of UO_3 in alcoholic hydrogen bromide, but this requires verification (Aloy, 1922).

Table 15.3 — Preparation of Uranium Tetrabromide

Reaction No.	Reactants (+ bromine)	Temp., °C	Carrier gas	By-products
22*	UO_2, C	700−900	N_2, CO_2	CO
23	U_3O_8, C	700−900	N_2, CO_2	CO
10*	U	650	He	None
11	UH_3	?	He	HBr
12*	UBr_3	300	N_2	None
13	UN	600	N_2	N_2
15	UC_2	800	N_2	C
16	US_2	450	N_2	S_2Br_2
21	$U_3O_2S_4$	800	N_2	UO_2, S_2Br_2
	UO_2, S	170		

*These reactions are considered most practicable from the point of view of accessibility of starting materials as well as ease of manipulation.

(1) Purification of UBr_4. (MP Ames 6,15,23; Princeton 1.) Uranium tetrabromide is best purified by vacuum distillation or by sublimation in a stream of inert gas containing a small amount of bromine. Highly purified nitrogen or helium is a suitable carrier gas.

2.2 Physical Properties of Uranium Tetrabromide. The physical properties of UBr_4 have not been investigated as systematically as those of UCl_4, and the information available is rather fragmentary.

(a) Melting Point. (MP Ames 17,24; UCRL 5.) The melting point of UBr_4 has been determined at Ames to be 519 ± 2°C. A value of 516°C was obtained from thermal analysis (UCRL 4).

(b) Boiling Point. (MP Ames 17,25.) Direct measurement yields a value of 765°C for the boiling point (with decomposition) under 740 mm Hg pressure. Extrapolation from vapor pressure data (transpiration method) gives a value of 761°C at 760 mm Hg, which is in satisfactory agreement.

(c) Density. The older values (Richards, 1902) for the density are probably in error. A direct determination, using benzene and xylene as the pycnometer fluids, gave a value of 5.35 g/cc at 26°C (MP Ames 10). There are no x-ray crystallographic data from which the density of UBr$_4$ can be calculated. The vapor density determined by Zimmermann (1882) shows the vapor to exist as a monatomic gas.

(d) <u>Vapor Pressure</u>. The vapor pressure of UBr$_4$ has been studied by a molecular effusion technique and by a transpiration method. Since different temperature ranges were explored, the results cannot be readily compared except for one point (450°C), at which the agreement is satisfactory. The results obtained by the effusion method are given in Table 15.4.

Table 15.4 — Vapor Pressure of Uranium Tetrabromide
(Effusion Method)*

Temperature, °C	Vapor pressure, mm Hg
300	1.08×10^{-4}
330	$8.45, 9.55 \times 10^{-4}$
350	$3.08, 3.24, 3.33 \times 10^{-3}$
365	5.61×10^{-3}
375	1.27×10^{-2}
400	$2.25, 6.23 \times 10^{-2}$
450	2.9×10^{-1}

*Princeton 1,2.

The results of the vapor pressure determination by the transpiration method are given in Table 15.5 (MP Ames 1). From these data the following equations have been derived:

For sublimation: $\log p_{\text{mm Hg}} = -\dfrac{10,900}{T} + 14.56$ (24)

For vaporization: $\log p_{\text{mm Hg}} = -\dfrac{7,060}{T} + 9.71$ (25)

ΔH (subl)   = 50,000 cal per mole

ΔH (vap)   = 32,000 cal per mole

ΔH (fusion) = 18,000 cal per mole

The data in Table 15.4 lead to a value of ΔH (sublimation) of 48,700 cal per mole. Trouton's constant ($\Delta H/T$) is computed to be 31.0, and Hildebrand's constant is computed to be 32.8. These values indicate considerable association in the liquid state. The second set of data appear to be more reliable than those obtained by the effusion method;

the thermodynamic data based on the transpiration-method vapor pressure data are therefore probably more accurate.

(e) <u>Thermochemical Data</u>. The heat of solution of UBr_4 in water has been found to be -33.1 ± 0.1 kcal per mole. By combination with

Table 15.5 — Vapor Pressure of Uranium Tetrabromide
(Transpiration Method)*

Temperature, °C	Vapor pressure, mm Hg
450	0.30
475	0.95
500	2.85
525	9.15
550	13.5
575	24.0
600	40.8
625	70.1

*MP Ames 17.

other appropriate data the heat of formation of UBr_4 can be calculated (UCRL 2) (Eq. 10).

$$U \text{ (crystal)} + 2Br_2 \text{ (gas)} \rightarrow UBr_4 \text{ (crystal)}$$

$$\Delta H^\circ_{298^\circ K} = -211.3 \pm 0.6 \text{ kcal per mole}$$

MacWood (UCRL 3) gives a value of $S_{298^\circ K} = 58.5$ e.u. for the entropy of UBr_4, from which the following can be deduced:

$$\Delta S_{298^\circ K} = -70.8 \text{ e.u.}$$

and

$$\Delta F^\circ_{298^\circ K} = -190.2 \pm 1.5 \text{ kcal per mole}$$

The equilibrium constant for the reaction

$$UBr_4 \text{ (crystal)} + \tfrac{1}{2}H_2 \text{ (gas)} \rightarrow UBr_3 \text{ (crystal)} + HBr \text{ (gas)} \quad (26)$$

has been determined in the temperature range 375 to 525°C at the University of California Radiation Laboratory (UCRL 5 to 7) (Table 15.6). The more recent results of Gregory (UCRL 13) are probably more precise. Assuming a value for ΔC_p for reaction 26 of -7.59 at 500°C, ΔH_0 is found to be 19.913 kcal per mole. The data in Table 15.6 also lead to $\Delta S_{298°K} = 23.96$ e.u. for the same equation.

2.3 Chemical Properties of Uranium Tetrabromide. Uranium tetrabromide is a crystalline substance that varies in color from light

Table 15.6 — Equilibria for the Reduction of Uranium Tetrabromide with Hydrogen*

Temp., °C	Temp., °K	$\frac{1}{T} \times 10^3$	K (UCRL 7)	K (UCRL 5)
375	648	1.543	0.080	...
400	673	1.486	0.124	0.141
425	698	1.433	0.188	0.170 (420°C)
450	723	1.383	0.256	0.250
475	748	1.337	0.365	0.320
500	773	1.294	...	0.440
525	798	1.253	...	0.570

*$K = p_{HBr}/p_{H_2}^{1/2}$.

tan to dark brown, depending on the crystal size. It is very hygroscopic and is rapidly oxidized in moist air to uranyl bromide. Consequently, UBr$_4$ must be handled in a dry inert atmosphere.

(a) Solubility. (MP Ames 10,26; MP Berkeley 5,9,12.) Uranium tetrabromide dissolves readily in water to form a green solution with the characteristic properties of uranium(IV) solutions. The solution is markedly hydrolyzed.

The behavior of UBr$_4$ in a great number of organic solvents has been studied. Nonpolar solvents, such as n-heptane, benzene, toluene, p-xylene, bromobenzene, tetralin, and carbon tetrachloride, neither dissolve uranium tetrabromide nor react with it. Nonpolar inorganic liquids, such as PCl$_3$, POCl$_3$, Br$_2$, and SnCl$_4$, likewise fail to dissolve UBr$_4$ at temperatures up to 70°C. On the other hand, uranium tetrabromide is soluble in polar solvents. Solvents such as acetic acid, methanol, ethanol, phenol, and aniline react with UBr$_4$ to evolve hydrogen bromide. Polar solvents with oxidizing properties, for example, nitrobenzene, nitromethane, or benzaldehyde, convert UBr$_4$ to uranyl bromide. Solvents containing a basic oxygen grouping, such as dioxane, acetone, diethyl ether, ethyl acetate, and amyl acetate, react with UBr$_4$ to form stable solvates from which UBr$_4$ cannot be easily regenerated. These latter solutions are green by reflected light and red by transmitted light. Carbonyl-containing compounds are much better solvents than those with only an ether linkage. Nitriles such as acetonitrile and benzonitrile dissolve the tetrabromide to the extent of perhaps 5 g of UBr_4 per 100 ml of solvent and form insoluble solvates on standing.

Uranium tetrabromide is insoluble in liquid ammonia but forms insoluble ammoniated salts. Other amines, i.e., aniline, butyl amine, triethyl amine, diethyl aniline, and pyridine, scarcely dissolve UBr_4

Table 15.7—Solubility of Uranium Tetrabromide in Organic Solvents at 25°C*

Solvent	Density of saturated solution, g/ml	Grams of UBr_4 per 100 g of solvent (by U analysis)
Acetic acid	1.573	64.81
Methyl alcohol	1.841	185.2
Ethyl acetate	1.523	91.3
Amyl acetate	1.126	35.3
Dioxane	1.057	2.84
Nitrobenzene	...	17.86
Nitromethane	1.271	14.5
Benzene†	0.869	0.0
Carbon tetrachloride	1.583	0.0
n-Heptane	0.678	0.0

*MP Ames 26.
†UBr_4 is also insoluble in toluene, p-xylene, and tetralin.

but do form insoluble addition products. Formamide dissolves UBr_4 with the evolution of considerable heat. The formamide solution appears stable at room temperature, but at temperatures above 100°C some reaction occurs as evidenced by the formation of a precipitate. Uranium tetrabromide is very soluble in molten acetamide (50 g of UBr_4 per 100 ml of acetamide). The solution appears to be perfectly stable at 145°C for several days and is identical in color and absorption spectrum with the formamide solution. Some quantitative data on the solubility of UBr_4 in various solvents are summarized in Tables 15.7 and 15.8.

The solubility of UBr_4 in some molten salts has also been determined. At 900°C a 10 per cent UBr_4 solution in $BaBr_2$ can be prepared. Uranium tetrabromide is also soluble to an unknown extent in a molten NaCl-KCl mixture at 725°C (MP Berkeley 10).

(b) Oxidation Reactions. Oxygen converts UBr_4 to UO_2Br_2. Chlorine converts UBr_4 to UCl_4. Bromine (liquid) does not react with UBr_4 even at 230°C (Zimmermann, 1882). No evidence for the formation of UBr_5 or higher bromides is found when UBr_4 is distilled with bromine vapor at 750°C.

(c) Reduction Reactions. Hydrogen reduces UBr$_4$ to UBr$_3$ at 470 to 700°C.

$$UBr_4 + \frac{1}{2}H_2 \rightarrow UBr_3 + HBr \quad (Eq. 26)$$

The evolution of HBr drops sharply when reduction to the UBr$_3$ stage is achieved (MP Berkeley 5) (see preparation of UBr$_3$, Sec. 1.1).

Table 15.8 — Solubility of Uranium Tetrabromide in Various Solvents*

Solvent†	Solubility, g of UBr$_4$ per 100 ml of solvent	Color
Ethyl bromide‡	Very slightly soluble	Highly colored
Ethyl bromide (SO$_2$)‡	Very slightly soluble	Highly colored
Bromobenzene	Very slightly soluble	Highly colored
Diethyl ether	Slightly soluble	Reddish orange
Dioxane	Slightly soluble	Reddish orange
Acetone	15	Dark red-brown
Acetone (SO$_2$)	30	Dark red-brown
Acetic acid	55	Dark red-brown
Nitrobenzene	3	Dark violet-red
Formamide	55	Very dark green
Formamide (HBr)	5	Dark red-brown

*MP Berkeley 12.

†(SO$_2$) indicates the solvent to be saturated with SO$_2$, and (HBr) indicates the presence of about 40 g of HBr per 100 ml of formamide.

‡5 per cent benzene by volume added.

Calcium and magnesium reduce UBr$_4$ to uranium metal. Owing to the volatility of UBr$_4$, a pressure bomb is required for the reaction. Uranium tetrabromide can be reduced (1) electrochemically, (2) from molten salt solution (MP Berkeley 10), or (3) from solutions of UBr$_4$ in certain organic solvents (MP Berkeley 5,9,12). Extensive, albeit inconclusive, investigations have been carried out at Berkeley on the latter. Solutions of UBr$_4$ in benzene, ethyl bromide, bromobenzene, diethyl ether, dioxane, acetone, acetic acid, and nitrobenzene give complex nonmetallic deposits at the cathode. Acetonitrile, benzonitrile, water, and liquid ammonia solutions also fail to give metal. From solutions in formamide, acetamide, and ethyl alcohol, however, deposits are obtained (on platinum cathodes) which have been claimed to be metallic uranium. The identification of the deposit was made on the basis of its insolubility in water and solubility in dilute hydrochloric acid with evolution of gas. Unfortunately, certain uranium compounds other than the metal may behave in this way, and consequently the identity of the deposit remains in doubt. The efficiency of the electrolysis is very low (about 5 per cent) even under the most favorable conditions attained. The procedures, therefore, have little or no practical significance. Nevertheless, further studies on the complex phenomena associated with the electrolytic behavior of these organic solutions may be of interest.

(d) Thermal Decomposition of UBr$_4$. Distillation of UBr$_4$ in a stream of nitrogen leads to partial decomposition (Hönigschmid, 1915).

$$UBr_4 \rightarrow UBr_3 + \tfrac{1}{2}Br_2 \tag{27}$$

Uranium tetrabromide vapor is reduced to metallic uranium on a hot filament in an evacuated tube (see Chap. 4 for discussion of the production of metallic uranium by thermal decomposition).

(e) Coordination Compounds. Uranium tetrabromide absorbs ammonia gas at room temperature and atmospheric pressure with slight evolution of heat. Treatment of UBr$_4$ for 1 hr followed by pumping for 5 min gives a gray solid, UBr$_4 \cdot 4NH_3$. Liquid ammonia gives solids varying in composition from UBr$_3 \cdot 5NH_3$ to UBr$_4 \cdot 6NH_3$. The hexaammoniate is white with a greenish tinge (MP Berkeley 9).

It has been reported (Aloy, 1899) that UBr$_4$ vapor reacts with alkali metal bromides at elevated temperatures to give double salts of the type M$_2$UBr$_6$. The reaction is performed in the same way as with the corresponding chlorine compound but is more difficult. The product obtained from KBr and UBr$_4$ vapor is a green substance, easily soluble in water. A similar sodium compound has also been prepared. Data relating to these compounds are meager, and their existence and composition require verification.

3. URANIUM(III) AND URANIUM(IV) IODIDES, UI$_3$ AND UI$_4$

Uranium triiodide and tetraiodide are best discussed together because of their ready interconversion. The tetraiodide differs from the other uranium(IV) halides in being thermally unstable. We recall that uranium tetrachloride is not decomposed to UCl$_3$ at any conveniently attainable temperature (at atmospheric pressure) and that uranium tetrabromide undergoes only slight decomposition to tribromide at temperatures of the order of 700 to 800°C. Uranium tetraiodide, on the other hand, so readily decomposes to UI$_3$ and I$_2$ that, even at comparatively low temperatures, it is necessary to specify the partial pressure of iodine in the system

$$UI_4 \rightleftharpoons UI_3 + \frac{1}{2}I_2 \qquad (28)$$

in order to know which of the uranium iodides will be stable.

The only uranium iodide mentioned in the older literature was UI_4 (Gmelin, 1936). Guichard (1907), in an investigation based on an observation of Moissan (1896), treated uranium metal at 550°C in a sealed tube with iodine vapor at 1 atm pressure and obtained a product whose analysis showed it to be UI_4. However, Guichard described this compound as melting above 500°C and being practically nonvolatile, which are characteristics of UI_3 rather than UI_4.

Uranium carbide appears to react with iodine, but it has not been possible to obtain pure products this way (MP Ames 27); however, more recent results (UCRL 8) make it likely that this is a potentially useful reaction.

Early attempts to prepare uranium tetraiodide by the action of iodine on a mixture of uranium dioxide and carbon failed (Hermann, 1861). An examination of the pertinent thermodynamic data indicates that the equilibrium of this reaction is unfavorable for iodination. Attempts to prepare uranium iodides by iodination of UO_2 with AlI_3, SbI_3, and BiI_3 have not hitherto been successful, despite the fact that a reaction is known to occur with these compounds (UCRL 8). Some UI_4 is apparently produced when uranium tetrachloride is treated with hydrogen iodide, but the product is heavily contaminated with chlorides (Colani, 1907). Preparations in aqueous solution have also been investigated. It was found that aqueous solutions of uranium iodide can be prepared by dissolving hydrated uranium(IV) hydroxide in hydriodic acid. Attempts to dehydrate this solution were unsuccessful. Products containing considerable amounts of free iodine were obtained after removal of the solvent (Sendtner, 1879). Recently this work was repeated at the Metallurgical Laboratory (MP Chicago 6,7). The dehydration was effected either in vacuum or in a stream of hydrogen and hydrogen iodide, and analysis of the products showed a U/I ratio of 2.70 to 2.73. In view of the difficulties experienced in dehydrating solutions of the other uranium(IV) halides (Sec. 2.1k), it is not surprising that compounds of such low iodine content — presumably containing oxide — were obtained.

3.1 Preparation of UI_3 and UI_4 by Reaction of Uranium with Iodine. The most satisfactory method for the preparation of both iodides remains the direct combination of the elements. For the preparation of UI_3, the stoichiometric quantity of iodine is slowly distilled into an evacuated flask containing finely divided uranium at 350°C. When all the iodine has been absorbed, the reaction vessel is sealed off and

TO VACUUM LINE VIA LIQUID-AIR TRAP

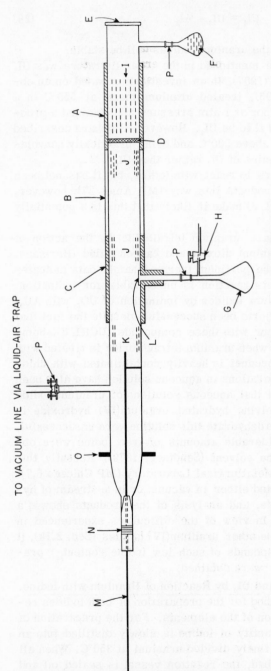

Fig. 15.5 — Apparatus for the preparation of uranium triiodide.

A, furnace reaction zone, 525°C (approx. 10 in. long)

B, furnace condensation zone, 400°C (approx. 10 in. long)

C, furnace condensation zone, 350°C (approx. 10 in. long)

D, perforated alundum disk

E, ground-glass seal (sealing wax)

F, iodine

G, uranium triiodide collector

H, stopcock for evacuation of collector after removal from tube

I, uranium metal charge, introduced at E

J, position of maximum uranium triiodide deposition

K, location of uranium tetraiodide deposition

L, cover preventing contamination of collector

M, stainless-steel rake for transfer of uranium triiodide deposit

N, Wilson seal attached to sylphon bellows

O, ground-glass seal (sealing wax)

P, ground-glass joints

heated, first at 130°C for several hours and then at 570°C for 15 to 20 hr. The product is practically pure UI_3 with very little UI_4 (MP Berkeley 11). The same procedure may be used to prepare UI_4, but in this case it is well to attach to the system an iodine reservoir so that the partial pressure of iodine may be maintained close to 1 atm during the entire course of the reaction.

The reaction may also be carried out in a flow system, which is preferable when large quantities are to be prepared. Figure 15.5 shows the apparatus used at the University of California Radiation Laboratory. A horizontal reactor is employed which is connected to a vacuum line. A reservoir of iodine is attached to the reactor inlet, and iodine vapor is pumped through metallic uranium heated to 525°C. The collector is separated from the reactor by a sintered glass disk and is maintained at 400°C. The desired partial pressure of the iodine is obtained by immersion of the iodine reservoir in a bath at 30°C. The reaction is slow, but good conversion yield is obtained. The product is 98 to 99 per cent pure UI_3 (UCRL 8).

To prepare UI_4 the same system may be employed. The only difference is that the partial pressure of iodine must be high enough to prevent the equilibrium

$$UI_4 \rightleftharpoons UI_3 + \tfrac{1}{2}I_2 \quad \text{(Eq. 28)}$$

from being displaced to the right. A pressure of 100 to 200 mm Hg (i.e., an iodine reservoir temperature of 120 to 140°C) is adequate for this purpose. To prevent the iodine from being pumped through the system too rapidly, connection to the vacuum line is made through a capillary tubing. The proper temperature of the collector is 300°C. The iodine pressure in the system must not be allowed to drop until the collector tube is cold.

In preliminary work on this method of preparation, the products obtained when iodine was passed over uranium metal at 500 to 530°C were deposited on collectors having a thermal gradient. The collector was then cut into sections, and the composition of the sublimate was determined as a function of the temperature of the surface (UCRL 9) (Table 15.9). The results indicate the optimum condensation ranges for UI_3 and UI_4 and also provide evidence for the existence of a higher iodide. Further investigation of this point would be valuable.

A satisfactory procedure for UI_4 preparation is iodination of UI_3. A vertical reactor is used, with liquid iodine in the bottom and UI_3 in a constricted tube above. The UI_4 formed melts and falls into the pool of excess iodine.

Another iodination reaction which has been investigated is that of uranium hydride with methyl iodide at 275 to 300°C. The reaction is rapid and yields UI_3, but no information on the purity of the product is available (MP Ames 28).

3.2 Physical Properties of UI_3 and UI_4. The absence of reliable data on the physical properties of the iodides is a serious gap in our present knowledge of uranium halide chemistry.

(a) Melting and Boiling Points. (UCRL 13.) The melting point of UI_3 has been estimated as 680°C. The melting point of UI_4 is 506°C

Table 15.9—Iodine/Uranium Ratio as a Function of Condensation Temperature

Temp. range, °C	Iodine/uranium ratio
310−386	3.05−2.99
278−354	3.65−3.19
250−290	4.54−4.20
235−253	5.69

(measured in an atmosphere of iodine); its boiling point was estimated as 762°C.

(b) Crystal Structure. (MP Chicago 8.) Uranium triiodide and uranium tetraiodide are both black needlelike crystalline substances. Uranium triiodide is orthorhombic with four molecules per unit cell.

$$a_1 = 13.98 \pm 0.02 \text{ A} \qquad a_2 = 4.33 \pm 0.02 \text{ A} \qquad a_3 = 9.99 \pm 0.02 \text{ A}$$

The calculated density is 6.76 g/cc. The substance is isomorphous with LaI_3, $NdBr_3$, and $SmBr_3$, and the space group is Ccmm (D_{2h}^{17}). The atomic positions are

$$4 \text{ U in } \pm(x_1 \ 0 \ \tfrac{1}{4}) \quad x_1 = 0.250$$

$$4 \text{ I in } \pm(x_2 \ 0 \ \tfrac{1}{4}) \quad x_2 = 0.069$$

$$8 \text{ I in } \pm(x_3 \ 0 \ z_3) \ (x_3 \ 0 \ \tfrac{1}{2}-z_3) \quad x_3 = 0.361, \ z_3 = -0.050$$

Each uranium atom is bonded to eight iodine atoms, with an average U-I interatomic distance of 3.35 A.

There are no crystallographic data available for UI_4.

(c) Vapor Pressure of UI_4. (Princeton 1.) Uranium triiodide is relatively nonvolatile, thereby resembling UCl_3. Uranium tetraiodide is more volatile, and a number of determinations of its vapor pressure have been made. These measurements are complicated by the dissociation of UI_4 to UI_3 and iodine, and consequently it is difficult to assess the precision of the results. As a rough indication of the volatility of UI_4, the data given in Table 15.10 are probably satisfactory. The heat of sublimation of UI_4 is then ΔH_s = 52,800 cal per mole, a value that seems high in comparison with those found for other uranium(IV) halides. More recent work at the University of California Radiation Laboratory confirmed these values roughly; the

Table 15.10 — Vapor Pressure of Uranium Tetraiodide (Effusion Method)

Temperature, °C	Vapor pressure, mm Hg
300	2.4×10^{-5}
330	2.32×10^{-4}
350	9.48×10^{-4}
360	1.98×10^{-3}
380	7.04×10^{-3}
410	4.25×10^{-2}

vapor pressure data found there could be represented by the following equation (cf. MP Ames 28; UCRL 10):

$$\log p = 15.53 - \frac{11.52 \times 10^3}{T} \tag{29}$$

(d) Thermochemical Data. (UCRL 3.) The following heats and free energy of formation of UI_3 and UI_4 have been determined at the University of California Radiation Laboratory:

$$U \text{ (solid)} + \tfrac{3}{2}I_2 \text{ (gas)} \rightarrow UI_3 \text{ (solid)} \tag{30}$$

$$\Delta H_{298°K} = -136.9 \pm 0.6 \text{ kcal per mole}$$

$$U \text{ (solid)} + 2I_2 \text{ (gas)} \rightarrow UI_4 \text{ (solid)} \tag{31}$$

$$\Delta H_{298°K} = -156.7 \pm 0.6 \text{ kcal per mole}$$

The entropies of UI_3 (solid) and UI_4 (solid) have been estimated to be $S_{298°K}$ = 59.5 e.u. (UI_3) and $S_{298°K}$ = 68.0 e.u. (UI_4). This gives for the free energies: $\Delta F^°_{298°K}$ = -123.2 ± 1.5 kcal per mole for UI_3 and $\Delta F^°_{298°K}$ = -136.2 ± 1.5 kcal per mole for UI_4.

3.3 Chemical Properties of UI_3 and UI_4. (a) Uranium Iodides and Water. The uranium iodides are very hygroscopic and dissolve readily in water. Uranium triiodide forms a beautiful dark-red solution; UI_4 solutions have the characteristic green color of the uranium(IV) ion (UCRL 9). Solutions of UI_4 in water are extensively hydrolyzed as indicated by their strong acid reaction.

The dissolution of UI_3 in water often is accompanied by vigorous reaction. Freshly prepared, the substance reacts quite violently with water (or even moist air), giving dense clouds of iodine. Preliminary heating to 800°C greatly moderates these reactions (MP Ames 28).

(b) Reduction of Iodides. Hydrogen reduces UI_4 at moderately elevated temperatures to UI_3 and HI. Reduction of UI_3 may be effected electrochemically, metallic uranium being produced by electrolysis of a solution of UI_3 in molten SrI_2 (MP Berkeley 11). A tungsten anode and molybdenum cathode can be used; the bath is maintained at 540°C.

(c) Oxidation of Iodides. Uranium triiodide is moderately stable in acid oxygen-free solutions but is rapidly oxidized in alkaline solution (MP Chicago 9,10). Chlorine reacts with UI_4 at room temperature with evolution of heat. Oxygen or dry air converts UI_3 and UI_4 to U_3O_8 at elevated temperatures and to UO_2I_2 at room temperature. Formation of intermediates such as UOI_2 has been suspected in the latter reaction, but no definite results are available (MP Chicago 6).

(d) Thermal Decomposition of UI_4. As mentioned before (cf. Eq. 28), an equilibrium exists between UI_3, UI_4, and I_2. The dissociation pressures of iodine in equilibrium with solid UI_3 and UI_4 have been measured (UCRL 11) (Table 15.11). Assuming a value of $\Delta C_p = -1.54$ for Eq. 28, the following relations can be calculated (UCRL 3):

$$\Delta H = 18,720 - 1.54T$$

$$\Delta H_{298°K} = 18.3 \text{ kcal per mole} \qquad \Delta F^\circ_{298°K} = 11.51 \text{ kcal per mole}$$

$$\Delta S_{298°K} = 22.64 \text{ e.u.}$$

These data may be used to calculate the partial pressure of iodine required to prepare either UI_3 or UI_4 by the reaction of uranium and iodine.

The thermal decomposition of uranium tetraiodide finds practical application in the "hot-wire" method for preparing uranium metal (UCRL 12). The process depends on the thermal dissociation of UI_4 at sufficiently high temperatures.

$$UI_4 \text{ (gas)} \rightarrow U \text{ (crystal)} + 2I_2 \text{ (gas)} \qquad \text{(Eq. 31)}$$

The decomposition is accomplished by means of an incandescent filament in an evacuated system. Uranium tetraiodide is thermodynamically unstable with respect to its elements at temperatures over 1300°K. A difficulty is introduced by the decomposition of UI_4 to I_2 and the nonvolatile UI_3 (according to Eq. 28). The amount of gaseous iodine present must be very carefully controlled to achieve a favorable relation between the desired reaction (Eq. 31) and the undesirable reaction (Eq. 28). These two decomposition equilibria have therefore

Table 15.11 — Dissociation Pressure of Iodine in Equilibrium with Uranium Triiodide and Uranium Tetraiodide

Temperature, °K	Iodine pressure, mm Hg
523	8.82×10^{-8}
588	3.82×10^{-6}
625	1.84×10^{-5}
644.2	4.17×10^{-5}
666	8.05×10^{-5}

been studied in some detail (cf. Chap. 4 on preparation of metallic uranium).

(e) <u>Miscellaneous Reactions</u>. At elevated temperatures UI_3 is a rather corrosive substance; glass and porcelain are attacked with the formation of silicon at 800°C. Of the metals, nickel and silver are more resistant than platinum (MP Ames 28).

4. URANIUM(III) AND URANIUM(IV) MIXED HALIDES*

The crystal structures of the trivalent and quadrivalent uranium halides are such that in any particular compound it is possible to replace some of the halogen atoms by others over a wide range of compositions. For certain stoichiometric proportions the solid solutions possess maximum stability and behave like true compounds. Six classes containing 18 individual compounds of the type UX_nY_{4-n} (X and Y being two different halogens) can exist. At least one compound of each class has been prepared; a number of trivalent mixed halides are also known.

The methods of preparation of these compounds cannot be based on displacement of one halogen atom by another, since such reactions fail to occur. The two methods for the preparation of uranium(IV) mixed halides are (1) the treatment of a trivalent uranium halide with a halogen of higher atomic number and (2) the reaction of two different tetrahalides at high temperatures. The mixed uranium(III) halides

*UCRL 13; MP Ames 29.

Table 15.12 — Uranium(III) and Uranium(IV) Mixed Halides

Compound	Method of preparation	B.p., °C	M.p.,* °C	Temp. at which v.p. = 10^{-3} mm Hg,* °C	Stability	$\Delta F^\circ_{298°K}$, kcal/mole	$\Delta H_{298°K}$, kcal/mole	$S_{298°K}$, e.u.
UClF$_3$	UF$_3$ + ½Cl$_2$				Stable but cannot be volatilized			
UCl$_2$F$_2$	UO$_2$F$_2$ + 2CCl$_4$	(460)			Disproportionates to UCl$_4$ and UF$_4$			
UBrF$_3$	UF$_3$ + ½Br$_2$				Stable to 450°C			
UIF$_3$	UF$_3$ + ½I$_2$				Stable below 100°C			
UBrCl$_3$	UCl$_3$ + ½Br$_2$	784	521	356	Sublimes without decomposition	−220.3 ± 1.5	−241.6	50.0
UBr$_2$Cl$_2$	UCl$_4$ + UBr$_4$	780	510	340	Sublimes without decomposition	−210.5 ± 1.5	−230.6	57.7
UBr$_3$Cl	UCl$_4$ + 3UBr$_4$	771	502	341	Sublimes without decomposition	−200.5 ± 1.5	−220.1	60.7
UICl$_3$	UCl$_3$ + ½I$_2$	<490		(340)	Evolves I$_2$ above 250°C	−206.4 ± 1.5	−227.3	53.1
UI$_2$Cl$_2$	UI$_4$ + UCl$_4$	<500		(340)	Evolves I$_2$ above 250°C			
UI$_3$Cl	UI$_3$ + UCl$_4$	<500		(340)	Sublimes without decomposition at low I$_2$ pressure in system			
UIBr$_3$	UBr$_3$ + ½I$_2$	478		330	Sublimes without decomposition at low I$_2$ pressure in system	−177.5 ± 1.5	−195.4	70.9
UI$_2$Br$_2$	UI$_4$ + UBr$_4$	<500		(330)	Evolves I$_2$ at 300°C			
UI$_3$Br	3UI$_4$ + UBr$_4$	<500		(330)	Evolves I$_2$ at 250°C; less stable than UI$_3$Cl			
UBrCl$_2$	2UCl$_3$ + UBr$_3$	(800)		(700)	Stable	−187.2 ± 1.5	−202.1	44.8
UBr$_2$Cl	UBr$_3$Cl + H$_2$ →	(775)		(700)	Stable	−178.0 ± 1.5	−191.9	50.6
UICl$_2$	UICl$_2$I$_2$ → UCl$_3$ + ½I$_2$	(750)		(700)	Stable			
UI$_2$Cl	UCl$_3$ + 2UI$_3$	(725)		(700)	Stable			
UIBr$_2$	UI$_2$Br$_2$ →	(700)		(700)	Stable			
UI$_2$Br	UIBr$_2$ + ½I$_2$ → UI$_3$Br → UI$_2$Br + ½I$_2$	(690)		(700)	Stable			

*Figures in parentheses are estimated values.

can be prepared (1) by thermal decomposition of mixed uranium(IV) halides, (2) by hydrogen reduction of mixed uranium(IV) compounds, and (3) by fusion of two trivalent halides.

Table 15.12 lists all the mixed halides known at present. Their preparation and most of their properties will be discussed individually in the order in which the compounds appear in the table; but, for the sake of convenience, the vapor pressures and thermochemical data of all mixed halides will be presented together in review form in Secs. 4.6 and 4.7.

The most extensive research in this field has been carried out at the University of California Radiation Laboratory. For details the papers by Gregory in Division VIII, Volume 6, of the National Nuclear Energy Series should be consulted.

4.1 Uranium(IV) Chloro-, Bromo-, and Iodofluorides. (a) Uranium Monochlorotrifluoride, $UClF_3$. (MP Ames 30,31.) This compound has been prepared by the reaction

$$UF_3 + \tfrac{1}{2}Cl_2 \rightarrow UClF_3 \qquad (32)$$

At 315°C the reaction is slow, and many hours are required for its completion. Treatment of UF_3 with HCl does not give any $UClF_3$.

An x-ray diffraction pattern of an impure sample indicated that $UClF_3$ has a cubic structure with $a_1 = 8.64$ A. The compound is rather stable and gives off no chlorine when heated in argon to 400°C. It is, however, nonvolatile.

(b) Uranium Dichlorodifluoride, UCl_2F_2. This compound was first prepared (CEW-TEC 1,2) by the reaction

$$UO_2F_2 + 2CCl_4 \rightarrow UCl_2F_2 + 2COCl_2 + Cl_2 \qquad (33)$$

The reaction may be effected either in the liquid phase under pressure at 130°C or in the vapor phase at 450°C. Of these the vapor-phase reaction is more convenient. It is interesting to note that the liquid-phase reaction produces a small amount of carbon tetrachloride — soluble material, and it has been suggested that it is a quinquevalent mixed halide, such as UCl_3F_2 or UCl_4F. However, compounds of this type have not as yet been definitely identified. UCl_2F_2 dissolves in water to give chloride and fluoride ions, indicating that the compound is not a mechanical mixture of UCl_4 and UF_4, since UF_4 is insoluble in water. UCl_2F_2 undergoes disproportionation on heating.

$$2UCl_2F_2 \rightarrow UF_4 + UCl_4 \qquad (34)$$

The UCl_4 can be removed by vacuum distillation (CEW-TEC 1,2). The mixed halide as such cannot be distilled.

Several attempts (MP Ames 31; UCRL 14) have been made to prepare UCl_2F_2 by the reaction

$$UCl_4 + UF_4 \rightarrow 2UCl_2F_2 \tag{35}$$

On the basis of fragmentary observations, it seems that UCl_2F_2 does form and that it melts at about 460°C. However, the product obtained in this way is less pure than that obtained by reaction of UO_2F_2 with CCl_4, probably because of the higher temperature used.

(c) Uranium Monobromotrifluoride, $UBrF_3$. (MP Ames 31.) $UBrF_3$ can be prepared in the same way as the corresponding chlorine compound.

$$UF_3 + \tfrac{1}{2}Br_2 \rightarrow UBrF_3 \tag{36}$$

Thus bromine vapor may be carried in a stream of helium over UF_3 at 250°C. Usually mixtures of $UBrF_3$ and UF_3 result unless the reaction is continued for very long periods. The compound $UBrF_3$ is stable in helium to 450°C; hydrogen at 200 to 400°C slowly reduces it to UF_3.

(d) Uranium Monoiodotrifluoride, UIF_3. (MP Ames 31.) Iodine vapor (carried in helium) reacts at 180°C with UF_3 to give mixtures of UF_3 and UIF_3. The mixture is brownish black in color, hygroscopic, and evolves iodine at room temperature in air. It appears to be stable to 100°C in the absence of air or moisture. This compound is imperfectly identified.

None of the chloro-, bromo-, or iodofluorides can be volatilized without decomposition.

4.2 Uranium(IV) Bromochlorides and Iodochlorides. These compounds have been studied in more detail than any of the mixed halides of uranium. Data are now available on the vapor pressures, heats of formation, and crystal chemistry of some of the members of this group. Their identity has been well established, but more extensive x-ray data would be desirable.

(a) Uranium Monobromotrichloride, $UBrCl_3$. (UCRL 14 to 16.) This compound is best prepared by bromination of UCl_3.

$$UCl_3 + \tfrac{1}{2}Br_2 \rightarrow UBrCl_3 \tag{37}$$

The reaction is carried out by heating UCl_3 in a closed system to 500°C in contact with bromine vapors. Since the rate of bromination is strongly dependent on the partial pressure of bromine, the latter is maintained at its equilibrium value over liquid bromine at room temperature. The product is purified by sublimation.

With liquid bromine the rate of reaction 37 at 125°C is too slow to be of practical use.

A number of other methods have also been explored. Fairly good results can be obtained by fusion of UBr_4 and UCl_4.

$$UBr_4 + 3UCl_4 \rightarrow 4UBrCl_3 \qquad (38)$$

However, the product of sublimation contains as much as 20 per cent UCl_4. The same type of reaction has been applied to UCl_3 and UBr_4.

$$UCl_3 + UBr_4 \rightleftharpoons UBrCl_3 + UBr_3 \qquad (39)$$

Reaction occurs at 550°C, but apparently an equilibrium is established; the product obtained by subliming the reaction product is a mixture of $UBrCl_3$ and UBr_4. This reaction is therefore unsuitable for the preparation of pure $UBrCl_3$.

Another approach, which has been partly explored, is the decomposition of complex compounds of UCl_4 and $NaBr$. On heating $NaBr$ and UCl_4 to 500°C, a compound, probably $Na_2UCl_4Br_2$, is formed. Subliming this compound in vacuum gives a mixture of UCl_4 (25 per cent) and $UBrCl_3$ (75 per cent). When $NaBr$ and UCl_4 are fused at 700°C and sublimed in vacuum, a mixture of 73 per cent $UBrCl_3$ and 27 per cent UBr_2Cl_2 is obtained. The following reactions are postulated to explain these results:

$$UCl_4 + 2NaBr \rightleftharpoons Na_2UBr_2Cl_4 \qquad (40)$$

$$Na_2UBr_2Cl_4 \rightarrow UBrCl_3 + NaBr + NaCl \qquad (41)$$

$$UBrCl_3 + 2NaBr \rightleftharpoons Na_2UBr_3Cl_3 \qquad (42)$$

$$Na_2UBr_3Cl_3 \rightarrow UBr_2Cl_2 + NaBr + NaCl \qquad (43)$$

The compound $Na_2UBr_4Cl_2$ can be prepared in a similar fashion, but the method yields only a mixture of UBr_3Cl and UBr_2Cl_2 on sublimation.

$UBrCl_3$ is a hygroscopic, dark-green crystalline substance, which resembles UCl_4 in most respects. It dissolves readily in water to give a green solution from which bromine may be displaced by addition of HNO_3. The compound melts at 521°C with an estimated heat of fusion of 11.6 kcal per mole. Treatment with hydrogen at 400 to

500°C results in the formation of approximately 70 mole % $UBrCl_2$ and 30 mole % UCl_3. The thermodynamics of this equilibrium is discussed below.

The compound is thermally stable and can be sublimed in vacuum without decomposition. It is stable in dry air, but it reacts very readily with water vapor to form oxyhalides.

A preliminary examination of the crystal structure of $UBrCl_3$ has been made (MP Ames 9,32), but unfortunately the work was confined to preparations of rather indefinite composition. $UBrCl_3$ apparently possesses a tetragonal structure very similar to UCl_4, with $a_1 = 8.434$ A, $a_3 = 7.690$ A, and a unit cell slightly larger than pure UCl_4. The pattern suggests that the mixed halide is not truly body-centered and that the replacement of chlorine in the lattice is not random.

(b) Uranium(IV) Dibromodichloride, UBr_2Cl_2. (UCRL 16.) UBr_2Cl_2 is best prepared by fusion (at 590°C) of an equimolar mixture of UCl_4 and UBr_4 in a quartz tube for about three days. The reaction is carried out in an atmosphere of helium, and, on completion of the heating period, about three-quarters of the charge is usually found to have sublimed. The product is a dark-green crystalline substance and has the composition UBr_2Cl_2 (containing approximately 1 per cent excess UCl_4). The composition of the sublimate indicates a compound rather than a mixture of UCl_4 and UBr_4, since these two substances possess very different vapor pressures and sublimate from an equimolar mixture would be much richer in bromine than is actually the case.

In its appearance, UBr_2Cl_2 resembles UCl_4 more closely than UBr_4. It melts at 510°C with an estimated heat of fusion of 12.3 kcal per mole. The boiling point (calculated from vapor pressure data; see below) is 780°C. Reduction with hydrogen (350 to 450°C) yields a mixture of roughly 60 per cent UBr_2Cl and 40 per cent $UBrCl_2$; the thermodynamics of the reduction is discussed in Sec. 4.7a. UBr_2Cl_2 is thermally stable and can be sublimed in vacuum without decomposition. It is stable in dry air but must be preserved from traces of moisture since it is very hygroscopic.

(c) Uranium(IV) Tribromomonochloride, UBr_3Cl. Heating a mixture of UCl_4 and UBr_4 (mole ratio 1 to 3) to 590°C for one day in a helium atmosphere yields a sublimation product of the composition UBr_3Cl. Direct chlorination of UBr_3 leads only to UCl_4. As expected, UBr_3Cl resembles UBr_4 more closely than UCl_4. It is a light-brown substance melting at 502°C with a heat of fusion of 13.6 kcal per mole (UCRL 13). UBr_3Cl may be sublimed in vacuum without decomposition. Reaction with hydrogen at 350 to 450°C results in the formation of 40 per cent UBr_3 and 60 per cent UBr_2Cl.

(d) Uranium(IV) Monoiodotrichloride, $UICl_3$. (UCRL 15.) This compound is rather difficult to prepare in pure form because of its thermal instability. It can be obtained by iodination of UCl_3.

$$UCl_3 + \frac{1}{2}I_2 \xrightleftharpoons{500°C} UICl_3 \qquad (44)$$

$UICl_3$ appears to be stable in the vapor phase even at 500°C. Decomposition occurs, however, when solid is deposited; it is essential, therefore, to maintain the partial pressure of the iodine in the system above the dissociation pressure of $UICl_3$.

$UICl_3$ is a reddish-brown to black substance which loses iodine even at 225°C in vacuum. The decomposition pressures range from 5×10^{-4} mm Hg at 200°C to 1.34×10^{-1} mm Hg at 302°C. Decomposition is too extensive to permit vapor pressure measurements, but the indications are that the compound has a vapor pressure greater than that of UCl_4 and probably quite similar to that of UI_4.

The compound is extraordinarily hygroscopic. Moist air liberates iodine and forms oxyhalide. $UICl_3$ must, therefore, be preserved from all contact with air or moisture.

(e) Uranium(IV) Diiododichloride, UI_2Cl_2. This compound can be prepared in the same way as UBr_2Cl_2, e.g., by fusion of an equimolar mixture of UCl_4 and UI_4 (in the presence of a small amount of iodine to minimize decomposition of the UI_4) (UCRL 13).

$$UCl_4 + UI_4 \rightarrow 2UI_2Cl_2 \qquad (45)$$

UI_2Cl_2 is a black crystalline substance. It is distinctly more stable than $UICl_3$. The vapor pressure and melting point of UI_2Cl_2 and $UICl_3$ appear to be quite similar.

(f) Uranium(IV) Triiodomonochloride, UI_3Cl. This compound has been prepared by fusion of UCl_4 and UI_3 above the melting point of UCl_4 (UCRL 14).

$$UI_3 + UCl_4 \xrightarrow{600°C} UI_3Cl + UCl_3 \qquad (46)$$

After sublimation, material containing over 90 per cent UI_3Cl is obtained, with UCl_3 as the most persistent contaminant. The presence of the latter in the sublimate is due to a small amount of free iodine, which is formed through unavoidable exposure of the starting material (UI_3) to air and moisture. This iodine converts UCl_3 to $UICl_3$, which sublimes together with UI_3Cl but decomposes back into UCl_3 and I_2 upon condensation.

UI_3Cl is a black crystalline substance with a red cast in the finely divided state. It is much more stable than $UICl_3$ or any of the other iodochlorides and can be sublimed without apparent decomposition with a low partial pressure (0.1 mm Hg) of iodine vapor in the system. It appears to be slightly more volatile than UBr_4.

4.3 <u>Uranium(IV) Iodobromides</u>. All three possible members of this class are known. They are relatively unstable compounds, which

Table 15.13 — Thermal Stability of Iodochlorides and Iodobromides

Decreasing stability	Iodochlorides	Iodobromides
↓	UI_3Cl	$UIBr_3$
	UI_2Cl_2	UI_2Br_2
	$UICl_3$	UI_3Br

lose iodine rather easily. $UIBr_3$ is the only member of the group which can be sublimed without decomposition. A comparison of the order of thermal stability of the iodobromides with that of the iodochlorides is of interest (Table 15.13).

All the iodobromides are extremely hygroscopic and must be prepared and stored in the complete absence of air or moisture. They are all black crystalline substances.

(a) <u>Uranium(IV) Monoiodotribromide, $UIBr_3$</u>. This compound is best prepared by the reaction

$$UBr_3 + \frac{1}{2}I_2 \xrightarrow{500°C} UIBr_3 \tag{47}$$

A partial pressure of 0.5 mm Hg of iodine is maintained during cooling. This method succeeds because $UIBr_3$ is much more stable than the corresponding chlorine compound, $UICl_3$ (for whose preparation this method is quite inappropriate).

$UIBr_3$ is volatile and has an appreciable vapor pressure $(1 \times 10^{-2}$ mm Hg) at 375°C; it may be sublimed without decomposition in a system in which $p_{I_2} = 0.1$ mm Hg. Decomposition to UBr_3 and iodine occurs at 400°C in vacuum. $UIBr_3$ melts at 478°C.

(b) <u>Uranium(IV) Diiododibromide, UI_2Br_2</u>. The most convenient method of preparation is the fusion of equivalent quantities of UI_4 and UBr_4.

$$UI_4 + UBr_4 \rightarrow 2UI_2Br_2 \tag{48}$$

Qualitatively, this compound resembles $UIBr_3$ closely except for a much lower thermal stability.

(c) Uranium(IV) Triiodomonobromide, UI_3Br. Fusion of a mixture of UI_4 and UBr_4 in appropriate proportions leads to the formation of this compound.

$$UBr_4 + 3UI_4 \rightarrow 4UI_3Br \tag{49}$$

The product is sublimed from the reaction zone. The compound is thermally unstable and undergoes ready conversion to UI_2Br and iodine at 300°C. Its thermal stability is comparable to that of UI_2Br_2.

4.4 Uranium(IV) Halides Containing Three Different Halogens. The compounds $UIBrCl_2$ and $UIBr_2Cl$ have not been prepared in pure form. However, mixtures of the two have been obtained by the following reactions:

$$2UBr_2Cl_2 + H_2 \rightarrow UBrCl_2 + UBr_2Cl + HBr + HCl \tag{50}$$

Iodination of the product then gives

$$UBrCl_2 + UBr_2Cl + I_2 \rightarrow UIBrCl_2 + UIBr_2Cl \tag{51}$$

There is reason to believe that $UIBr_2Cl$ is more stable than $UIBrCl_2$ with respect to loss of iodine.

4.5 Uranium(III) Mixed Halides. Mixed halides of trivalent uranium can be prepared by (1) thermal decomposition of mixed uranium(IV) halides, (2) reduction of mixed uranium(IV) halides by hydrogen, and (3) fusion of a mixture of two uranium(III) halides.

All mixed uranium(III) halides are black crystalline substances that resemble UCl_3 and UI_3 with respect to thermal stability, volatility, etc. Little is known of their physical or chemical properties. The reduction of the uranium(IV) halides with hydrogen, however, has permitted the calculation of thermodynamic constants for $UBrCl_2$ and UBr_2Cl.

(a) Uranium(III) Monobromodichloride, $UBrCl_2$. The reduction of $UBrCl_3$ with hydrogen produces a mixture of $UBrCl_2$ and UCl_3. Iodination of this mixture removes the UCl_3 (as volatile $UICl_3$). The best method for preparing $UBrCl_2$, however, is fusion of UCl_3 and UBr_3.

$$2UCl_3 + UBr_3 \xrightarrow{850°C} 3UBrCl_2 \tag{52}$$

The product can be purified from unreacted trichloride by treatment with iodine vapor and sublimation of the $UICl_3$ formed.

(b) Uranium(III) Dibromomonochloride, UBr_2Cl. This compound has been prepared by reduction of UBr_3Cl with hydrogen. The mixture of UBr_2Cl and UBr_3 which is produced in this way is treated with iodine; the $UIBr_3$ and $UIBr_2Cl$ vapors are then passed through a tube heated to 385°C in a stream of iodine vapor of 0.1 mm Hg pressure. The $UIBr_2Cl$ decomposes under these conditions to UBr_2Cl, and the more stable $UIBr_3$ sublimes away.

Fusion of UBr_3 and UCl_3 (1 to 2) should also yield UBr_2Cl.

Table 15.14 — Vapor Pressure Equations* for Some Uranium(IV) Mixed Halides†

Compound	A	B	Temp. range, °C
$UBrCl_3$	13.852	10,526	320 – 430
UBr_2Cl_2	13.149	9,901	330 – 404
UBr_3Cl	13.280	10,000	320 – 420
$UIBr_3$	13.416	9,901	315 – 382

$$*\log p_{mm\ Hg} = A - \frac{B}{T}.$$

†UCRL 3.

Table 15.15 — Interpolated Vapor Pressure (in Millimeters of Mercury) of Some Uranium(IV) Mixed Halides

Temp., °C	$UBrCl_3$	UBr_2Cl_2	UBr_3Cl	$UIBr_3$
320	1.25×10^{-4}	2.85×10^{-4}	2.57×10^{-4}	5.25×10^{-4}
330	2.48×10^{-4}	5.42×10^{-4}	4.95×10^{-4}	9.93×10^{-4}
340	4.75×10^{-4}	1.00×10^{-3}	9.30×10^{-4}	1.84×10^{-3}
350	9.00×10^{-4}	1.82×10^{-3}	1.70×10^{-3}	3.34×10^{-3}
360	1.66×10^{-3}	3.23×10^{-3}	3.02×10^{-3}	5.96×10^{-3}
370	3.00×10^{-3}	5.70×10^{-3}	5.40×10^{-3}	1.04×10^{-2}
380	5.40×10^{-3}	9.80×10^{-3}	9.40×10^{-3}	1.79×10^{-2}
390	9.50×10^{-3}	1.67×10^{-2}	1.61×10^{-2}	
400	1.64×10^{-2}	2.77×10^{-2}	2.69×10^{-2}	
410	2.80×10^{-2}	4.62×10^{-2}	4.50×10^{-2}	
420	4.70×10^{-2}		7.30×10^{-2}	
430	7.90×10^{-2}			

(c) Uranium(III) Monoiododichloride, $UICl_2$. Thermal decomposition of UI_2Cl_2 in vacuum yields pure $UICl_2$.

(d) Uranium(III) Monoiododibromide, $UIBr_2$. This compound is obtainable by the thermal decomposition of UI_2Br_2 in vacuum.

(e) Uranium(III) Diiodomonobromide, UI_2Br. This compound may be readily prepared by the thermal decomposition of UI_3Br in vacuum.

All these compounds dissolve in water to give a red solution. They are hygroscopic but not to the same extent as the uranium(IV) compounds.

4.6   Vapor Pressure Measurements.   The vapor pressures of UBrCl$_3$ (UCRL 17), UBr$_3$Cl (UCRL 18), UBr$_2$Cl$_2$ (UCRL 19), and UIBr$_3$ (UCRL 20) have been studied by a molecular effusion method. Tables 15.14 and 15.15 summarize the experimental data. The heats and free energies of sublimation are given in Table 15.16.

Table 15.16 — Heats and Free Energies of Sublimation of Some Uranium(IV) Mixed Halides

Compound	ΔH (sublimation), kcal/mole	ΔF (sublimation),* cal/mole		
		ΔH_0,	A	B
UBrCl$_3$	48.5	57,692	36.85	168.5
UBr$_2$Cl$_2$	45.5	56,000	36.85	167.1
UBr$_3$Cl	46.3	54,474	34.55	158.2
UIBr$_3$	43.0	51,417	34.55	154.7

*ΔF (sublimation) = ΔH_0 + AT log T $-$ BT.

Table 15.17 — Equilibria of Reduction of Mixed Halides with Hydrogen

Reaction No.	Reaction	$\Delta H_{700°K}$, kcal/mole	$\Delta F^0_{700°K}$, kcal/mole	$\Delta S_{700°K}$, e.u.
53	UBrCl$_3$ (solid) + ½H$_2$ (gas) ⇌ UBrCl$_2$ (solid) + HCl (gas)	16.2	1.73	20.67
54	UBrCl$_3$ (solid) + ½H$_2$ (gas) ⇌ UCl$_3$ (solid) + HBr (gas)	16.3	2.98	20.03
55	UBr$_2$Cl$_2$ (solid) + ½H$_2$ (gas) ⇌ UBr$_2$Cl (solid) + HCl (gas)	17.6	2.68	21.62
56	UBr$_2$Cl$_2$ (solid) + ½H$_2$ (gas) ⇌ UBrCl$_2$ (solid) + HBr (gas)	16.9	3.18	20.03
57	UBr$_3$Cl (solid) + ½H$_2$ (gas) ⇌ UBr$_3$ (solid) + HCl (gas)	17.1	3.61	19.84
58	UBr$_3$Cl (solid) + ½H$_2$ (gas) ⇌ UBr$_2$Cl (solid) + HBr (gas)	16.9	2.98	20.47

4.7   Thermochemistry.   Heats of solution, dissociation pressures, and reduction equilibria with hydrogen have been measured for seven mixed halides; these measurements permit calculation of the heat and free energy of formation and entropy of these compounds. No specific heat data are available, and the consequent uncertainties in the assumed ΔC_p for the reduction equilibria are reflected in the final values.

(a) Reduction Equilibria.   (UCRL 21.) The results of measurements of the reduction equilibria are summarized in Table 15.17.

(b) Decomposition Pressures of Iodine over UIBr$_3$ (UCRL 22) and UICl$_3$. (UCRL 20.) The decomposition pressure of iodine as a function of temperature has been measured for these two compounds. The results are summarized in Table 15.18.

(c) Thermodynamic Constants. These constants have been determined for seven mixed halides. The results are given in Table 15.12. For details of the calculations, we refer to the paper by MacWood in the National Nuclear Energy Series, Division VIII, Volume 6.

Table 15.18 — Iodine Pressure above $UICl_3$ and $UIBr_3$

UICl₃		UIBr₃	
Temp., °C	Iodine pressure, mm Hg	Temp., °C	Iodine pressure, mm Hg
198.6	4.73×10^{-4}	277.3	5.44×10^{-4}
221.3	2.09×10^{-3}	293.2	1.20×10^{-3}
241.0	6.64×10^{-3}	314.9	3.05×10^{-3}
261.7	1.70×10^{-2}	331.1	6.18×10^{-3}
280.0	5.07×10^{-2}	349.7	1.49×10^{-2}
302.0	1.35×10^{-1}	368.2	2.44×10^{-2}
		381.4	3.75×10^{-2}

5. URANIUM(IV) BOROHYDRIDE, $U(BH_4)_4$*

The metal borohydrides (Univ. Chicago Dept. Chem. 1,2) are a relatively new class of compounds. They can be considered as "pseudo-halides," with the radical BH_4 playing the part of a halide, and this justifies their treatment in this chapter.

Prior to the discovery of uranium borohydride the borohydrides of lithium, beryllium, and aluminum were known (Schlesinger, 1939, 1940; Burg, 1940). These compounds were first made by the action of diborane on an alkyl compound of the metal.

$$2LiC_2H_5 + 2B_2H_6 \rightarrow 2LiBH_4 + (C_2H_5)_2B_2H_4 \qquad (59)$$

$$Be(C_2H_5)_2 + 2B_2H_6 \rightarrow Be(BH_4)_2 + (C_2H_5)_2B_2H_4 \qquad (60)$$

$$2Al(C_2H_5)_3 + 6B_2H_6 \rightarrow 2Al(BH_4)_3 + 3(C_2H_5)_2B_2H_4 \qquad (61)$$

The reactions are more complex than represented by the equations since trialkyl boron compounds are produced in the early stages and complex mixtures of alkyl diboranes in the later stages. Lithium and beryllium borohydrides are saltlike, but aluminum borohydride is a typical nonpolar substance. This is a volatile liquid (b.p. 44.5°C), soluble in benzene. Since beryllium borohydride and aluminum borohydride are the most volatile of all known compounds of these metals,

*Univ. Chicago Dept. Chem. 1,2.

it seemed worth while to try to prepare the corresponding compound of uranium.

5.1 <u>Preparation of Intermediates.</u> Unlike the other borohydrides, uranium(IV) borohydride cannot be prepared by the reaction of metal alkyls with diborane, since uranium alkyls are unknown. The best method of preparation found so far is the reaction of aluminum borohydride with uranium tetrafluoride. The method of preparing $Al(BH_4)_3$ described in the literature is very tedious and suitable only for the preparation of very small amounts. It was necessary, therefore, to devise more practical methods for the preparation of $Al(BH_4)_3$. The detailed results of this interesting work are to appear elsewhere. The following reactions may be employed in preparing aluminum borohydride:

$$NaH + B(OCH_3)_3 \rightarrow NaBH(OCH_3)_3 \tag{62}$$

$$2NaBH(OCH_3)_3 + B_2H_6 \rightarrow 2NaBH_4 + 2B(OCH_3)_3 \tag{63}$$

$$3NaBH_4 + AlCl_3 \rightarrow Al(BH_4)_3 + 3NaCl \tag{64}$$

Diborane can be prepared (MP Chicago 11) by one of the following two reactions:

$$6LiH + 8(C_2H_5)_2O:BF_3 \rightarrow B_2H_6 + 6LiBF_4 + 8(C_2H_5)_2O \tag{65}$$

or

$$6NaBH(OCH_3)_3 + 8(C_2H_5)_2O:BF_3 \rightarrow B_2H_6 + 6NaBF_4$$
$$+ 6B(OCH_3)_3 + 8(C_2H_5)_2O \tag{66}$$

5.2 <u>Preparation of Uranium(IV) Borohydride.</u> Uranium(IV) borohydride is formed by treating UF_4 with $Al(BH_4)_3$. Reaction is spontaneous (and exothermal) at room temperature. It proceeds according to the equation

$$UF_4 + 2Al(BH_4)_3 \rightarrow U(BH_4)_4 + 2Al(BH_4)F_2 \tag{67}$$

The uranium tetrafluoride in this reaction is best prepared by dehydration of $2UF_4 \cdot 5H_2O$ (cf. Chap. 12). Final dehydration is accomplished just before use: Uranium tetrafluoride monohydrate is attached to a vacuum line and dehydrated at 350°C at 10^{-5} mm Hg; aluminum borohydride is distilled into the reaction tube, which is then sealed off from the vacuum line. Since considerable heat is evolved in reaction 67, it is advisable to place the reaction tube in an ice-salt mixture for several hours and then to allow the reaction to go to completion at room temperature. The reaction tube is then joined again to the vacuum line through a breaker seal, and $U(BH_4)_4$ and excess $Al(BH_4)_3$ are distilled away. The $U(BH_4)_4$ is trapped in a U tube at $-20°C$ and the $Al(BH_4)_3$ in a U tube immersed in liquid nitrogen. With large quantities (25 g) the sublimation may require many hours, the necessary time depending greatly on the diameter of the tubing.* Eighty to 90 per cent yields of $U(BH_4)_4$ (calculated in relation to the UF_4 used) can be attained.

Efforts have been made to devise procedures that would not require the use of the highly inflammable $Al(BH_4)_3$. The only procedure that so far offers any promise is the reaction of UF_4 with $LiBH_4$ in ether. Evidence for the production of $U(BH_4)_4$ by this reaction was obtained, but neither the yield nor the purity of the product was satisfactory. Further research is indicated. A number of other reactions were investigated but proved unsuccessful. Several such unsuccessful reactions are listed below.

(a) Action of $Al(BH_4)_3$ on UCl_4. A little $U(BH_4)_4$ was obtained, but the bulk of the uranium appeared to have been reduced to the trivalent state.

(b) Action of $LiBH_4$, $NaBH_4$, and KBH_4 on UF_4 or UCl_4 at Temperatures up to 125°C in the Absence of Solvents. No $U(BH_4)_4$ was obtained.

(c) Action of Diborane or Borine Carbonyl, $BH_3 \cdot CO$ (Burg, 1937), on Uranium Metal or Uranium Hydride. No reaction resulted under a wide range of temperatures and pressures. It is certain, however, that both the uranium metal and the hydride were impure and reexamination is desirable.

The formula of uranium(IV) borohydride was established by hydrolysis.

$$U(BH_4)_4 + 16H_2O \rightarrow U(OH)_4 + 4H_3BO_3 + 16H_2 \qquad (68)$$

The hydrogen was measured with a Toepler pump, the boron was determined as methyl borate, after distillation, by titration with $Ba(OH)_2$ and mannitol, and the uranium was determined as U_3O_8. The experimental results lead to the formula $U_{1.0}B_{4.0}H_{16.1}$ or $U(BH_4)_4$.

---

*The nonvolatile residue from the reaction is spontaneously and violently inflammable in air; the residue must be hydrolyzed by slowly admitting water vapor before air is allowed to enter the reaction vessel.

5.3 Preparation of Alkyl Derivatives of Uranium(IV) Borohydride. Treatment of U(BH$_4$)$_4$ with boron trimethyl in an evacuated sealed vessel at 50 to 70°C results in the formation of a green, moderately viscous liquid together with a small amount of a brown nonvolatile solid. This reaction proceeds slowly below 50°C. From the mixture two volatile uranium compounds have been isolated. The less volatile compound is a crystalline solid, colorless in thin layers, lavender to almost black in thicker crystals. The more volatile compound, after purification by fractional sublimation, is a green solid. The difference in volatility is sufficiently great to permit partial separation by fractional distillation and sublimation, as the green substance passes through a trap maintained at −10°C where the lavender material is condensed. Separation is incomplete, but treatment of the lavender substance with excess boron trimethyl removes the last traces of the green material. The green substance can be obtained in the pure state by treating the impure product with excess diborane at 30 to 40°C. The rationale of these treatments is apparent from the structures of these two substances, which were established by analysis and by reaction with water and hydrogen chloride (see Sec. 5.5c). The green substance gave with hydrogen chloride both diborane and methyl boron chlorides (chiefly CH$_3$BCl$_2$); the lavender compound gave no diborane. Neither compound produced any methane. Therefore the compounds clearly contain carbon as methyl groups. It is reasonable to conclude that in both the methyl groups are attached to boron and not to uranium, that in the lavender compound there are no BH$_4$ groups, and that each carbon is attached to a different boron atom. In conjunction with the analytical data and molecular weight determinations, the results strongly suggest that the green compound is monomethyl uranium borohydride, U(BH$_4$)$_3$(BH$_3$CH$_3$), and the lavender compound is tetramethyl uranium borohydride, U(BH$_3$CH$_3$)$_4$. The methods of purification mentioned above were based on the premise that B(CH$_3$)$_3$ will alkylate traces of the monomethyl compound present in tetramethyl borohydride and that diborane will dealkylate traces of tetramethyl compound which may be present in the monomethyl derivative.

Boron triethyl reacts with U(BH$_4$)$_4$ in the same way as B(CH$_3$)$_3$, but it has not been possible to isolate pure compounds from the reaction mixture. The same is true of the higher boron alkyls which also react with U(BH$_4$)$_4$ but which yield reaction mixtures hitherto defying separation. The volatility of the ethyl or higher alkyl uranium borohydrides apparently is much less than that of the methyl compounds.

5.4 Physical Properties of Uranium(IV) Borohydrides. Vapor pressures were determined by means of a glass Bourdon gauge in an all-glass system. Vapor pressure data are summarized in Tables

15.19 and 15.20. The same apparatus was used for determining the molecular weight. The molecular weight determinations showed the compounds to be monomers.

Table 15.19—Values of Constants for Equations* Reproducing the Observed Vapor Pressures of Uranium Borohydrides

Substance	A	B
$U(BH_4)_4$	4,264.6	13.354
$U(BH_4)_3(BH_3CH_3)$	3,150.0	10.679
$U(BH_3CH_3)_4$	2,970.0	8.820

$$*\ln p_{mm\ Hg} = B - \frac{A}{T}.$$

Table 15.20—Interpolated Vapor Pressure of $U(BH_4)_4$ and Its Derivatives

Temp., °C	Vapor pressure, mm Hg		
	$U(BH_4)_4$	$U(BH_4)_3(BH_3CH_3)$	$U(BH_3CH_3)_4$
30	0.19	1.92	0.10
35	0.33	2.83	0.15
40	0.54	4.12	0.21
45	0.89	5.93	0.30
50	1.43	8.45	0.42
55	2.27	11.89	0.58
60	3.56	16.48	0.80
65	5.51	22.91	1.08
70*	8.40	31.26	1.45
75*	12.68	42.36	1.93

*Values at this temperature represent considerable extrapolation.

$U(BH_4)_4$ sublimes without melting; it melts at 126°C with extensive decomposition. The melting point of $U(BH_3CH_3)_4$ is approximately 72 to 74°C. The melting point of the monomethyl compound is in doubt. A melting point of 95°C has been observed, but this was on a sample in which disproportionation to $U(BH_4)_4$, and more highly methylated products, may have occurred.

Photochemical Properties. (SAM Columbia 1,2.) $U(BH_4)_4$ vapor decomposes both photochemically and thermally; therefore, the absorption spectrum could be studied only at very low pressures. A continuous band at 21,000 cm^{-1} was found (p = 4 mm Hg, d = 50 cm).

5.5 Chemical Properties of $U(BH_4)_4$ and Its Alkyl Derivatives. Uranium(IV) borohydride is a lustrous green crystalline material, which in large aggregates appears almost black. As mentioned before, the monomethyl compound is also green, but the tetramethyl compound forms lavender crystals.

(a) Solubility. Water and alcohol react vigorously with $U(BH_4)_4$. $U(BH_4)_4$ is soluble in ethyl ether to the extent of 2 g per 100 ml at room temperature. The ether cannot be completely removed without decomposition; at $-80°C$ a stable monoetherate, $U(BH_4)_4 \cdot (C_2H_5)_2O$, is formed. Nonpolar solvents, such as n-heptane, dissolve $U(BH_4)_4$ slowly and to only a slight extent; the borohydride can be recovered unchanged from such solvents. Highly purified benzene does not react with $U(BH_4)_4$, but it is difficult to remove the benzene without decomposition of borohydride. The solubility in benzene is too small for practical purposes. The solubility of the alkyl derivatives has not been determined.

(b) Air. $U(BH_4)_4$ reacts with dry air slowly. Yellow nonvolatile products not further investigated are formed. The compound can be exposed to dry air for short periods without hazard.

(c) Hydrolytic Reactions.

$$U(BH_4)_4 + 16H_2O \rightarrow U(OH)_4 + 4H_3BO_3 + 16H_2 \qquad (69)$$

$$U(BH_4)_4 + 4HCl \rightarrow UCl_4 + 2B_2H_6 + 4H_2 \qquad (70)$$

$$U(BH_4)_4 + 16CH_3OH \rightarrow U(OCH_3)_4 + 4B(OCH_3)_3 + 16H_2 \qquad (71)$$

The alkylated borohydrides react in similar fashion:

$$U(BH_4)_3(BH_3CH_3) + 6HCl \rightarrow UCl_4 + \tfrac{3}{2}B_2H_6 + CH_3BCl_2 + 6H_2 \qquad (72)$$

$$U(BH_3CH_3)_4 + 12HCl \rightarrow UCl_4 + 4CH_3BCl_2 + 12H_2 \qquad (73)$$

(d) Diborane. At room temperature B_2H_6 apparently does not react with $U(BH_4)_4$.

$$U(BH_4)_4 + B_2H_6 \rightarrow \text{no action} \qquad (74)$$

The alkyl borohydrides are converted to the simple borohydride.

$$U(BH_4)_3(BH_3CH_3) + B_2H_6 \rightarrow U(BH_4)_4 + CH_3BH_2BH_3 \qquad (75)$$

$$U(BH_3CH_3)_4 + B_2H_6 \rightarrow U(BH_4)_4 + (CH_3)_4B_2H_2 \qquad (76)$$

(e) <u>Methyl Borate</u>. Reaction occurs when $U(BH_4)_4$ is treated with $B(OCH_3)_3$ at 70°C for 1 hr. The products are distinctly less volatile than $U(BH_4)_4$. The reaction is complex, as H_2, B_2H_6, and dimethoxyborine, $BH(OCH_3)_2$, are obtained in addition to the main solid product. Analysis suggests the white solid product to be a monomethoxy derivative, $U(BH_4)_3(BH_3OCH_3)$.

(f) <u>Boron Alkyls and Other Metal Alkyls</u>. The reaction of $U(BH_4)_4$ and $B(CH_3)_3$ to give methyl derivatives of the simple borohydride has already been discussed. Zinc methyl, $Zn(CH_3)_2$, and aluminum methyl, $Al(CH_3)_3$, cause extensive decomposition of $U(BH_4)_4$ at 50°C. The black solid that forms has not been further investigated.

(g) <u>Lithium Ethyl</u>. Since it is known that the reaction

$$3LiC_2H_5 + Al(BH_4)_3 \rightarrow 3LiBH_4 + Al(C_2H_5)_3 \qquad (77)$$

takes place readily, it appeared reasonable to apply this reaction to the problem of preparing uranium alkyl derivatives.

$$U(BH_4)_4 + xLiC_2H_5 \rightarrow U(BH_4)_{4-x}(C_2H_5)_x + xLiBH_4 \qquad (78)$$

In benzene solution, however, a black nonvolatile material forms, apparently as a result of reduction, and no evidence for the formation of U-C bonds has been found.

(h) <u>Carbon Monoxide</u>. (Ethyl Corp. 1.) Carbon monoxide does not react with $U(BH_4)_4$ at 50°C. At 80°C reaction occurs to yield an unknown product considerably less volatile than the original borohydride.

(i) <u>Thermal Stability of the Borohydrides</u>. The decomposition of the borohydrides of uranium occurs most easily in the vapor phase, particularly when the vapors come in contact with solid surfaces. Pyrex glass walls markedly catalyze the decomposition. Metals such as copper or nickel are also very effective in causing decomposition, whereas aluminum and silver are probably less destructive than pyrex.

The kinetics of the decomposition on copper have been studied (SAM Columbia 3). In the presence of freshly reduced copper, decomposition occurs at the rate of 7.6×10^{-9} mole/cm^2/hr at 55°C and 10^{-4} mm Hg. The reactions that may occur are

$$U(BH_4)_4 \rightarrow U(BH_4)_3 + \tfrac{1}{2}B_2H_6 + \tfrac{1}{2}H_2 \qquad (79)$$

$$U(BH_4)_4 \rightarrow U + 2B_2H_6 + 2H_2 \qquad (80)$$

The amounts of hydrogen and diborane are equal, but this, of course, does not permit differentiation between these two possible reactions.

Decomposition to uranium boride occurs when vapors of $U(BH_4)_4$ are passed through a heated tube, and beautiful mirrors of uranium boride can be deposited on glass in this way.

5.6 Attempts to Prepare Borohydrides of Sexivalent Uranium.

Uranium hexafluoride, uranium hexachloride, uranium penta- and hexamethoxides, uranium penta- and hexaethoxides, and uranium pentamethoxide monoethoxide fail to give a sexivalent borohydride when treated with diborane, borine carbonyl, or lithium borohydride. Reduction to quadrivalent uranium occurs, usually to form uranium(IV) borohydride admixed with other uranium(IV) compounds. This result is to be expected in view of the strong reducing properties of these boron compounds. Uranium trioxide failed to react with any of the above reagents.

6. ATTEMPTED PREPARATIONS OF A URANIUM CARBONYL

By analogy to chromium, molybdenum, and tungsten, which are members of the same group in the periodic table and which form the volatile compounds $Cr(CO)_6$, $Mo(CO)_6$, and $W(CO)_6$, uranium might be expected to form the compound $U(CO)_6$. The hexacarbonyls of chromium, molybdenum, and tungsten are readily prepared in good yield by a variety of methods. These carbonyls are volatile and surprisingly stable.

In consideration of the possible existence of a uranium hexacarbonyl the "rule" of Sidgwick and Bailey (1934) has been quoted. This rule is based on the hypothesis that each CO group in a metal carbonyl donates two electrons to the central metal atom and that only those metal carbonyls are stable in which the central metal atom acquires in this way the electronic configuration of a rare gas. At least twenty-six electrons would have to enter the electronic sphere of uranium for the latter to acquire the electronic configuration of the hypothetical rare gas that follows it in the periodic table. Steric considerations alone would seem to rule out the existence of the resulting compound, $U(CO)_{13}$. This appears as a theoretical argument against the existence of a uranium carbonyl. However, the Sidgwick and Bailey rule has no theoretical basis. The acquisition of a rare gas configuration cannot play a decisive role in determining the stability of molecules other than those formed by elements closely preceding or following a rare gas in the periodic system.

Some of the literature pertaining to the metal carbonyls has been reviewed by Hieber (1942), Blanchard (1940), and Trout (1937). These surveys may be consulted for details of the methods of preparation, structure, and properties of the metal carbonyls.

6.1   Reaction of Uranium Metal with Carbon Monoxide. (Ethyl Corp. 1.) Repeated efforts have been made to induce reaction of uranium metal and carbon monoxide under a wide variety of conditions since all the known metal carbonyls (except that of chromium) have been prepared by direct combination of the metal with carbon monoxide. The following attempted reactions have been reported:

1. Westinghouse powdered uranium metal (98.6 per cent) was heated with carbon monoxide under a pressure of 2,950 psi at 165°C in dry hexane.

2. The same reaction mixture given in No. 1 was used at 30°C and 2,400 psi.

3. "Activated" uranium metal (prepared by decomposition of the hydride) was treated with CO at a maximum temperature of 114°C and a maximum pressure of 7,100 psi.

4. "Activated" metal (prepared by heating Westinghouse metal to 250°C in vacuum) was reacted with carbon monoxide at 266°C and 8,650 psi.

5. Uranium metal (96 per cent uranium) was heated in a copper-lined steel cylinder at temperatures up to 200°C and pressures up to 250 atm (British 1).

No formation of a volatile uranium carbonyl was observed in any of these experiments. It must be pointed out that the metal used was not of a high degree of purity. The finely powdered metal, used by the Ethyl Corporation workers, almost certainly had a surface film that could conceivably prevent the reaction. The carbon monoxide contained as much as 0.5 per cent hydrogen and 2.4 per cent nitrogen; that these impurities could have interfered with the desired reaction, especially at the elevated temperatures used, is not impossible.

6.2   Uranium Hydride with Carbon Monoxide (Ethyl Corp. 1) and Nickel Carbonyl. The uranium hydride used in these experiments was prepared from Westinghouse metal. The decomposition pressures of the hydride (3 mm Hg at 225°C, 150 mm Hg at 360°C) reported by the Ethyl Corporation group are considerably higher than those reported by other observers (cf. Chap. 8), casting doubt on the identity of the substance used. The hydride was treated with carbon monoxide at temperatures from −180 to 300°C and pressures up to 8,500 psi, with no signs of reaction. Treatment of the "uranium hydride" with $Ni(CO)_4$ [in the presence of $AlCl_3$ or $(CH_3)_2AlCl$] also gave no volatile uranium compound.

6.3   Uranium Halides with Carbon Monoxide and Reducing Agents. A new method for preparing metal carbonyls was discovered by Job and Cassal (1927). It consists of the treatment of the metal halide with a reducing agent in the presence of carbon monoxide. Job and Cassal employed Grignard reagents as reducing agents. The method has been extended by the use of finely divided metals such as silver, copper, cadmium, or zinc as reductants (Hieber, 1935; Kocheshkov, 1940). Repeated efforts have been made to obtain a volatile uranium carbonyl by these methods.

Uranium tetrachloride was treated with ethyl magnesium bromide and carbon monoxide in an ether-benzene solution (Princeton 3). The same reaction was also performed with zinc dust instead of the Grignard reagent. In both cases some carbon monoxide was taken up, but no volatile uranium product could be isolated. It is of interest to note that in both cases the reaction mixture was worked up by hydrolysis with water or dilute sulfuric acid, on the apparent presumption that uranium carbonyl, if it exists at all, must not be decomposed by water.

The above reaction was extended to UO_2Cl_2, UCl_3, UCl_4, $UCl_4 + UCl_5$ mixture, UCl_6, and $U(CH_3COCHCOCH_3)_4$ (British 1); magnesium methyl iodide, magnesium ethyl bromide, magnesium phenyl bromide, magnesium α-naphthyl bromide, and lithium phenyl bromide were used as Grignard reagents. No volatile uranium compound could be detected. Dimethyl zinc, diethyl zinc, and lithium butyl were also used without success (Ethyl Corp. 1).

A number of reactions were carried out in which UCl_4 and UBr_4 were treated with carbon monoxide and zinc dust in acetone solutions (Ethyl Corp. 1). This procedure gave excellent results with $MoCl_5$ but failed to give the desired uranium compound. Apparently reduction to the trivalent state is all that occurs.

6.4 Uranium Halides with Carbon Monoxide and Carriers. (Ethyl Corp. 1.) A useful procedure for preparing carbonyls is the treatment of an aqueous solution of a metal halide with carbon monoxide in the presence of alkaline potassium cyanide or sodium sulfide (Manchot, 1929; Windsor, 1933; Blanchard, 1934; Hieber, 1939). A few experiments with uranium halides have been performed in which KCN was used as a "carrier" for carbon monoxide, but no reaction occurred. Cysteine has also been tried as "carrier" with UCl_3 and CO (Schubert, 1933; Coleman, 1936). Some reaction did occur, but the products were not identified. Alkaline hydrogen sulfide or uranyl salts in the presence of carbon monoxide also failed to give a volatile uranium compound (British 1).

6.5 Conclusions. It can be seen from the foregoing discussion that the successful preparation of a volatile uranium carbonyl, or even proof for the existence of such a compound, will not be an easy task. A survey of the recent work shows that most, if not all, of the tested methods for preparing metal carbonyls have been tried. With respect to the work done on metallic uranium it would be very desirable to repeat these experiments, using the very pure metal now available and using pure carbon monoxide. The use of alloys such as nickel-uranium alloy has also been suggested and seems worthy of trial (MP Chicago 12). In all experiments made so far, whenever a reaction occurred (as it frequently did) no attempt was made to ascertain its course. All that was done was a quick examination to determine whether a volatile uranium compound had been formed. An understanding of just what happens with uranium in some of these reactions might be helpful in planning future experiments.

The methods of working up the reaction mixtures were based on the preparation of $Cr(CO)_6$, a compound which is not sensitive to moisture. It is possible that $U(CO)_6$, if it exists, is sensitive to water. The ability of uranium(IV) to function as an oxidizing or reducing agent has not been given sufficient weight in the selection of solvents. Finally, it would perhaps be desirable to reinvestigate some of the reactions studied, using UCl_5 and UCl_6 rather than UCl_4.

No final conclusions as to the stability of a uranium carbonyl can be drawn from the lack of success hitherto encountered in attempts to prepare such compounds. It must be pointed out, however, that uranium is a strongly electropositive element and that no carbonyls of elements of the same type, such as beryllium or magnesium, are known.

REFERENCES

1861   H. Hermann, dissertation, University of Göttingen, p. 29.
1879   R. Sendtner, Ann., 195: 331.
1882   C. Zimmermann, Ann., 216: 2-7.
1886   G. Alibegoff, Ann., 233: 119.
1896   H. Moissan, Compt. rend., 122: 1092.
1899   J. Aloy, Bull. soc. chim. Paris, (3) 21: 265.
1902   T. W. Richards and B. S. Merigold, Z. anorg. Chem., 31: 235.
1907   A. Colani, Ann. chim. et phys., (8) 12: 59-74.
1907   M. Guichard, Compt. rend., 145: 921; or Bull. soc. chim. Paris, (4) 3: (1908).
1915   O. Hönigschmid, Monatsh., 36: 51-68.
1922   J. Aloy and E. Rodier, Bull. soc. chim. France, (4) 31: 247.
1927   A. Job and A. Cassal, Bull. soc. chim. France, 41: 1041.
1929   W. Manchot and H. Gall, Ber., 62: 678.
1933   M. Schubert, J. Am. Chem. Soc., 55: 4563.
1933   M. M. Windsor and A. A. Blanchard, J. Am. Chem. Soc., 55: 1877.
1934   A. A. Blanchard, J. R. Rafter, and W. B. Adams, Jr., J. Am. Chem. Soc., 56: 18.
1934   N. V. Sidgwick and R. W. Bailey, Proc. Roy. Soc. London, A144: 521.
1935   W. Hieber and E. Romberg, Z. anorg. u. allgem. Chem., 221: 322.
1936   G. W. Coleman and A. A. Blanchard, J. Am. Chem. Soc., 58: 2160.
1936   "Gmelins Handbuch der anorganischen Chemie," System No. 55, p. 140, Verlag Chemie, Berlin.
1937   A. B. Burg and H. I. Schlesinger, J. Am. Chem. Soc., 59: 780.
1937   W. E. Trout, J. Chem. Education, 14: 453, 575; 15: 77, 113 (1938).
1939   W. Hieber, H. Schulten, and R. Marin, Z. anorg. u. allgem. Chem., 240: 264.
1939   H. I. Schlesinger, R. T. Sanderson, and A. B. Burg, J. Am. Chem. Soc., 61: 536; 62: 3421 (1940).

1940 A. A. Blanchard, Chem. Revs., 26: 409; 21: 3 (1937).
1940 A. B. Burg and H. I. Schlesinger, J. Am. Chem. Soc., 62: 3425.
1940 K. A. Kocheshkov, A. N. Nesmeyanov, M. M. Nad, I. M. Rossinskaya, and L. M. Borisova, Compt. rend. acad. sci. U.R.S.S., 26: 54-59.
1940 H. I. Schlesinger and H. C. Brown, J. Am. Chem. Soc., 62: 3429.
1942 W. Hieber, Die Chemie, 55: 7, 24.
1948 W. H. Zachariasen, J. Chem. Phys., 16: 254.
1949 J. M. Carter, U. S. Patent 2469916, May 10, 1949.

Project Literature

British 1: Report B-22.

CEW-TEC 1: J. W. Gates, Jr., L. J. Andrews, B. P. Block, and H. A. Young, Report CD-460, Aug. 26, 1944.
CEW-TEC 2: J. W. Gates, G. H. Clewett, and H. A. Young, Report CD-454, July 20, 1945.

Ethyl Corp. 1: G. Calingaert, H. Soroos, F. J. Dykstra, V. Hnizda, and J. V. Capinjola, Report A-542, Feb. 16, 1943.

MP Ames 1: F. H. Spedding, A. S. Newton, R. W. Nottorf, J. Powell, and V. P. Calkins, The Preparation and Some Properties of UBr_3, UBr_4, $UOBr_2$, and UO_2Br_2, in National Nuclear Energy Series, Division VIII, Volume 6.
MP Ames 2: A. S. Newton, D. Johnson, A. Kant, and R. W. Nottorf, Report CC-705, June 7, 1943.
MP Ames 3: A. S. Newton, D. Johnson, A. Kant, and R. W. Nottorf, Report CC-725, June 15, 1943.
MP Ames 4: W. Lyon, J. Illif, and H. Lipkind, Report CK-1526, Apr. 15, 1944.
MP Ames 5: W. Lyon, J. Illif, and H. Lipkind, Report CK-1494, May 4, 1944.
MP Ames 6: R. W. Nottorf and J. Powell, Report CC-1524, Mar. 10, 1944.
MP Ames 7: V. P. Calkins and R. W. Nottorf, Report CC-1975, Oct. 24, 1944.
MP Ames 8: N. C. Baenziger and R. E. Rundle, Report CC-1524, Apr. 14, 1944.
MP Ames 9: N. C. Baenziger and R. E. Rundle, Report CN-1495, Apr. 15, 1944.
MP Ames 10: R. W. Nottorf, J. Powell, and V. P. Calkins, Report CC-1524, Apr. 14, 1944.
MP Ames 11: V. P. Calkins, Report CC-1500, June 17, 1944.
MP Ames 12: W. Lyon, J. Illif, and H. Lipkind, Report CT-1180, Dec. 25, 1943.
MP Ames 13: R. W. Nottorf and J. Powell, Report CC-1781, Aug. 10, 1944.
MP Ames 14: R. W. Nottorf and J. Powell, Report CC-1496, Apr. 10, 1944.
MP Ames 15: R. W. Nottorf and J. Powell, Report CC-1778, July 10, 1944.
MP Ames 16: W. H. Keller, Report CT-1270B, Jan. 29, 1944.
MP Ames 17: R. W. Nottorf and J. Powell, Report CC-1504, June 10, 1944.
MP Ames 18: D. H. Ahmann, Report CC-298, Oct. 15, 1942.
MP Ames 19: W. Lyon, J. Illif, H. Lipkind, and C. Neher, Report CK-1498, June 17, 1944.
MP Ames 20: J. Powell, Report CC-1778, Aug. 18, 1944.
MP Ames 21: J. Powell, Report CC-1504, Aug. 10, 1944.
MP Ames 22: C. Neher, H. Lipkind, W. Lyon, and N. Sleight, Report CT-1780, Aug. 9, 1944.
MP Ames 23: J. Powell, Report CC-1781, Oct. 18, 1944.
MP Ames 24: R. W. Nottorf, J. Powell, and A. S. Newton, Report CC-1500, June 17, 1944.

MP Ames 25: R. W. Nottorf and T. A. Butler, Report CC-1975, Oct. 24, 1944.
MP Ames 26: V. P. Calkins, Report CC-1496, May 11, 1944.
MP Ames 27: A. H. Daane, Report CC-298, Oct. 15, 1942.
MP Ames 28: J. A. Ayres, Report CN-1243, Jan. 8, 1944.
MP Ames 29: J. C. Warf and N. C. Baenziger, Preparation and Properties of Some Mixed Uranium Halides, in National Nuclear Energy Series, Division VIII, Volume 6.
MP Ames 30: J. C. Warf and F. Edwards, Report CC-1496, Apr. 10, 1944.
MP Ames 31: J. C. Warf and coworkers, Report CC-1785, Sept. 10, 1944.
MP Ames 32: R. E. Rundle and N. C. Baenziger, Report CC-1500, May 10, 1944.

MP Berkeley 1: R. A. Webster, Report CK-873, Aug. 20, 1943.
MP Berkeley 2: C. D. Thurmond, Report CC-2522, Jan. 15, 1945.
MP Berkeley 3: E. D. Eastman, B. J. Fontana, and R. A. Webster, Observations on the Preparation of Tribromide and Triiodide of Uranium, in National Nuclear Energy Series, Division VIII, Volume 6.
MP Berkeley 4: B. J. Fontana, Report CC-586, Apr. 21, 1943.
MP Berkeley 5: B. J. Fontana, Report CK-677, May 15, 1943.
MP Berkeley 6: A. E. Stickland, Report CC-939, September, 1943.
MP Berkeley 7: C. D. Thurmond and A. E. Stickland, Report CC-1739, June 1, 1944.
MP Berkeley 8: C. D. Thurmond, Report CC-938, September, 1943.
MP Berkeley 9: B. J. Fontana, Report CK-810, July 20, 1943.
MP Berkeley 10: R. A. Webster, Report CK-671, May 15, 1943.
MP Berkeley 11: R. A. Webster, Report CC-2105, Aug. 15, 1944.
MP Berkeley 12: B. J. Fontana, Report CC-585, Apr. 21, 1943.

MP Chicago 1: J. J. Katz and N. R. Davidson, Report CK-1221, Jan. 5, 1944.
MP Chicago 2: J. C. Warner, Report CK-1396, Feb. 17, 1944.
MP Chicago 3: W. H. Zachariasen, The Crystal Structure of Trichlorides, Tribromides, and Trihydroxides of Uranium and of Rare Earth Elements, in National Nuclear Energy Series, Division VIII, Volume 6.
MP Chicago 4: W. H. Zachariasen, Reports CP-1576, Apr. 27, 1944, and CC-2768, Mar. 12, 1945.
MP Chicago 5: C. Smith and B. B. Cunningham, Report CK-932, Sept. 11, 1943.
MP Chicago 6: L. S. Foster and J. Tvrzicky, Report CK-897, Aug. 28, 1943.
MP Chicago 7: L. S. Foster and J. Tvrzicky, Memorandum 123, Sept. 18, 1943; Report CT-963, Oct. 11, 1943.
MP Chicago 8: W. H. Zachariasen, Reports CP-1811, July 4, 1944, and CP-1954, Aug. 10, 1944.
MP Chicago 9: G. T. Seaborg, Report N-930, Apr. 5, 1944.
MP Chicago 10: T. Magel and L. S. Foster, Report CT-963, Oct. 11, 1943.
MP Chicago 11: H. I. Schlesinger, G. W. Schaeffer, and G. D. Barbaras, Reports CP-1558, May 24, 1944, and CK-2737, Mar. 5, 1945.
MP Chicago 12: N. Sugarman, N. Elliott, and C. D. Coryell, Report CC-144.

Princeton 1: A. Schelberg and R. W. Thompson, Report A-809, Oct. 10, 1942.
Princeton 2: R. W. Thompson and A. Schelberg, Report A-179, May 4, 1942.
Princeton 3: B. McDuffie, Report A-806, Oct. 5, 1942.

SAM Columbia 1: H. C. Urey, Report A-750, July 10, 1943.
SAM Columbia 2: A. B. F. Duncan, Report A-584, Apr. 15, 1943.
SAM Columbia 3: A. Turkevich, H. C. Brown, V. Lewinsohn, and W. F. Libby, Report A-383, Oct. 19, 1942.

UCRL 1: D. Altman, Report RL-4.6.202, Sept. 18, 1943.
UCRL 2: C. H. Barkelew, Report RL-4.6.329, Aug. 28, 1945.
UCRL 3: G. E. MacWood, Report RL-4.7.602, Dec. 18, 1945.
UCRL 4: N. W. Gregory, Report RL-4.6.923, Aug. 1, 1945.
UCRL 5: D. Altman, Report RL-4.6.276, Aug. 14, 1944.
UCRL 6: G. E. MacWood and D. Altman, Report RL-4.6.600 (A-1).
UCRL 7: N. W. Gregory, Report RL-4.6.928, Aug. 24, 1945.
UCRL 8: N. W. Gregory, Report RL-4.6.272, July 20, 1944.
UCRL 9: J. A. Holmes, Report RL-4.6.252, Mar. 7, 1944.
UCRL 10: D. Altman, M. E. Mueller, and C. H. Prescott, Report RL-4.6.273, July 27, 1944.
UCRL 11: G. E. MacWood, M. E. Mueller, and D. Altman, Report RL-4.6.277, Dec. 8, 1944.
UCRL 12: C. H. Prescott, Jr., F. L. Reynolds, and J. A. Holmes, The Preparation of Uranium Metal by Thermal Dissociation of the Iodide, in National Nuclear Energy Series, Division VIII, Volume 6.
UCRL 13: N. W. Gregory, Preparation and Properties of the Uranium Halides, in National Nuclear Energy Series, Division VIII, Volume 6.
UCRL 14: N. W. Gregory, Report RL-4.6.905, Jan. 8, 1945.
UCRL 15: N. W. Gregory, Report RL-4.6.275, Aug. 7, 1944.
UCRL 16: N. W. Gregory, Report RL-4.6.295, Oct. 7, 1944.
UCRL 17: M. Estabrook, P. H. Davidson, and I. Streeter, Report RL-4.6.902, Dec. 6, 1944.
UCRL 18: M. Estabrook, P. H. Davidson, and I. Streeter, Report RL-4.6.903, Dec. 29, 1944.
UCRL 19: P. H. Davidson, I. Streeter, and M. Estabrook, Report RL-4.6.908, Jan. 23, 1945.
UCRL 20: P. H. Davidson, G. E. MacWood, and I. Streeter, Report RL-4.6.932, Aug. 30, 1945.
UCRL 21: N. W. Gregory, Report RL-4.6.936, Sept. 9, 1945.
UCRL 22: P. H. Davidson, G. E. MacWood, and I. Streeter, Report RL-4.6.933, Aug. 31, 1945.

Univ. Chicago Dept. Chem. 1: H. I. Schlesinger and H. C. Brown, Report A-1221, July 29, 1943.
Univ. Chicago Dept. Chem. 2: H. I. Schlesinger and H. C. Brown, Uranium Boro-hydride and Its Alkyl Derivatives, in National Nuclear Energy Series, Division VIII, Volume 6.

Chapter 16

URANIUM OXYHALIDES

The oxyhalides of sexivalent uranium of the type of UO_2^{++} were among the earliest compounds of uranium to be studied. Despite this, few physical constants have been determined with any degree of accuracy, and the chemistry of these compounds is still fragmentary. Until 1940, oxyhalides of uranium(IV) were unknown; but now the compounds $UOCl_2$ and $UOBr_2$ are readily available, and a considerable amount of information about them has been accumulated. Evidence for the probable existence of oxyhalides of uranium(VI) of the type of UO^{+4} and of uranium(V) of the type of UOX_3 has already been discussed in Chap. 14.

Here, attention will be directed to the uranyl halides and to the uranium(IV) oxyhalides. Questions relating to ionic equilibria and the physical properties of aqueous solutions of these compounds are discussed in the second portion of this volume.

1. URANYL FLUORIDE, UO_2F_2

Uranyl fluoride was first prepared by reaction of hydrofluoric acid and uranium oxide. Berzelius (1824) treated UO_3 with hydrofluoric acid and obtained a white amorphous solid on evaporation. Bolton (1866a) prepared solutions of uranyl fluoride by treating U_3O_8 with hydrofluoric acid.

$$U_3O_8 + 8HF \rightarrow UF_4 + 2UO_2F_2 + 4H_2O \tag{1}$$

He was unable to obtain crystalline uranyl fluoride from the solution. Smithells (1883), on repeating the work of Bolton, found that uranyl fluoride could be obtained as a bright-yellow soapy substance by evaporation of the aqueous solution; on one occasion a partially crystalline material with a mother-of-pearl luster was obtained. Smithells designated the hydrated amorphous material, obtained by evaporation of aqueous UO$_2$F$_2$ solutions, as the β form, to distinguish it from the compound that he obtained in very small yield by heating uranium tetrafluoride in a closed crucible and designated as the α form. The latter probably is anhydrous UO$_2$F$_2$, formed by the reaction

$$2UF_4 + O_2 \rightarrow UF_6 + UO_2F_2 \tag{2}$$

Ditte (1884) erroneously formulated the compound obtained by ignition of uranium tetrafluoride in a limited supply of air as UOF$_4$.

1.1 Preparation of Uranyl Fluoride from Aqueous Solution. Uranyl fluoride has a tendency to form acid and basic salts of ill-defined composition. To prepare pure neutral uranyl fluoride (Montreal 1) a known weight of oxide (UO$_3$ or UO$_4\cdot$2H$_2$O) is dissolved in the calculated quantity of hydrofluoric acid, and the solution is then evaporated to dryness. The product is usually found to contain a higher percentage of uranium than calculated for the neutral salt. An amount of hydrofluoric acid equivalent to the excess uranium is then added, with enough water to dissolve the salt, and the resulting product is allowed to crystallize from solution. Crystals obtained in this way appear to be the dihydrate UO$_2$F$_2\cdot$2H$_2$O.

Uranyl fluoride is difficult to crystallize, very viscous syrups being formed on concentration. Even if uranyl fluoride dihydrate crystals can be induced to form, separation of the crystals from the mother liquor is usually very difficult. The crystals of uranyl fluoride dihydrate are soft thin plates, pale yellow, and so hygroscopic that it is difficult to dry them. Consequently, solutions are often evaporated to dryness at 150 to 200°C without attempting to achieve crystallization. At 200°C products with a composition corresponding to that of the neutral salt and containing less than 0.1 per cent water can be obtained.

Anhydrous uranyl fluoride has also been prepared by reaction of uranyl acetate with hydrofluoric acid (von Unruh, 1909). The acetate is repeatedly evaporated with hydrofluoric acid in a platinum crucible on a water bath to volatilize acetic acid. The residue is then evaporated to dryness to remove water and excess hydrofluoric acid. On prolonged drying in a vacuum desiccator, an anhydrous product is obtained. This method was used by British workers to prepare UO$_2$F$_2$ (British 1).

1.2 Preparation of UO$_2$F$_2$ by High-temperature Hydrofluorination. Anhydrous uranyl fluoride is readily prepared by reaction of uranium oxide with HF vapor:

$$UO_3 + 2HF \rightarrow UO_2F_2 + H_2O \qquad (3)$$

A temperature of 350 to 500°C appears suitable (MP Chicago 1). On a commercial scale, temperatures as high as 550°C have been recommended (Du Pont 1). Other workers prefer temperatures in the vicinity of 400°C (British 2; SAM Columbia 1) to avoid formation of U_3O_8 from the UO_3 and consequent contamination of the product with uranium tetrafluoride. The type of UO_3 used will determine the optimum temperature. This is probably the most convenient method for preparing anhydrous uranyl fluoride.

Although not of particular value as a method of preparation, the following reaction can be mentioned as also yielding anhydrous uranyl fluoride:

$$UO_2 \cdot HPO_4 \cdot xH_2O + 2HF \xrightarrow{350-500°C} UO_2F_2 + H_3PO_4 + xH_2O \qquad (4)$$

Above 500°C reduction occurs and uranium tetrafluoride begins to form (CEW-TEC 1).

Uranyl fluoride can be readily prepared by the reaction of anhydrous uranyl chloride with liquid anhydrous hydrogen fluoride at room temperature (Brown 1). The reaction is best carried out in a platinum-lined nickel reactor; the reaction mixture is allowed to stand overnight, and the hydrogen chloride and fluoride are removed by distillation in vacuum at 450°C. A completely water-soluble chlorine-free product results.

Miscellaneous Reactions Yielding Uranyl Fluoride as a Product. Although not especially desirable as methods of preparation, the following reactions give uranyl fluoride as a product (MP Chicago 1):

$$UO_3 + 2F_2 \xrightarrow{350°C} UO_2F_2 + OF_2 \, (?) \qquad (5)$$

$$UO_2 + F_2 \xrightarrow{350°C} UO_2F_2 \qquad (6)$$

$$U_3O_8 + 5F_2 \xrightarrow{350°C} 3UO_2F_2 + 2OF_2 \, (?) \qquad (7)$$

Uranyl fluoride is one of the products of the reaction (MP Chicago 2)

$$2UF_4 + O_2 \rightarrow UF_6 + UO_2F_2 \qquad (8)$$

It has been reported that when uranium hexafluoride is treated with water vapor a complex compound of uranyl fluoride, hydrogen fluoride, and water results, from which uranyl fluoride can be prepared by heating to 180°C (British 3). The existence of this complex requires confirmation.

1.3 Physical Properties of Uranyl Fluoride. Anhydrous uranyl fluoride is a pale-yellow solid; no other physical constants have been determined. Since uranyl fluoride undergoes thermal decomposition, the melting point is unknown.

(a) Crystal Structures of Anhydrous Uranyl Fluoride. The crystal structure of UO_2F_2 has been studied by Zachariasen (MP Chicago 3; Zachariasen, 1948). The ideal structure of UO_2F_2 is rhombohedral with one molecule per unit cell. The unit cell has the dimensions

$$a = 5.755 \pm 0.003 \text{ A} \qquad \alpha = 42° \ 47' \pm 3'$$

The space group is $R\bar{3}m$, with the atomic positions

$$
\begin{array}{lll}
1 \text{ U in } (0\ 0\ 0) & & \\
2 \text{ O in } \pm(u\ u\ u) & \text{with } u = 0.122 \\
2 \text{ F in } \pm(v\ v\ v) & \text{with } v = 0.294
\end{array}
$$

To each uranium atom are bonded two oxygen atoms at a distance U-O = 1.91 A and six fluorine atoms at a distance U-F = 2.50 A. The structure consists of layers of uranium atoms 5.22 A apart; the axes of the uranyl groups are perpendicular to these planes. The fluorine atoms are 0.61 A above and below the uranium atom planes, and weak O-O and O-F bonds hold the layers together. Actual samples of UO_2F_2 all show varying degrees of stacking disorder with the planes displaced relative to each other.

The density as computed from the x-ray data is 6.37 g/cc. A direct measurement by displacement in benzene gave 5.8 g/cc, but this value was thought to be low owing to entrapped air (SAM Columbia 2). British workers report a pouring density of 2.95 g/cc and a packing density of 2.55 g/cc (British 2).

(b) Optical and Photochemical Properties of Uranyl Fluoride. The absorption and fluorescence spectra of uranyl fluoride will be discussed in the second portion of this volume.

(c) Specific Heat, Enthalpy, and Entropy of Uranyl Fluoride. (Natl. Bur. Standards 1; Wacker and Cheney, 1947.) The heat capacity of uranyl fluoride was measured from 13 to 418°K. The enthalpy $H° - H_0°$ was calculated to be 63.96, 77.62, and 108.15 int. joules per gram at 298.16, 338.16, and 423.16°K, respectively, and the entropy was calculated to be 0.4400, 0.4830, and 0.5635 int. joules per degree-gram at the same temperatures. No evidence of a transition was found. The values of the specific heat, enthalpy, entropy, and free energy are tabulated at temperature intervals of 5°K in Table 16.1. The probable error is of the order of a few tenths of 1 per cent.

Table 16.1—Heat Capacity, Enthalpy, Entropy, and Free Energy of Uranyl Fluoride

(In international joules)

Temp., °K	$C°$, j/g °C	$H° - H_0°$, j/g	$S°$, j/g °C	$-(F° - H_0°)$, j/g	Temp., °K	$C°$, j/g °C	$H° - H_0°$, j/g	$S°$, j/g °C	$-(F° - H_0°)$, j/g
0	0	0	0	0	215	0.2949	37.626	0.33687	34.802
5	0.00051	0.00064	0.00017	0.00021	220	0.2979	39.107	0.34368	36.503
10	0.00403	0.01014	0.00135	0.00339	225	0.3008	40.604	0.35041	38.238
15	0.01231	0.0492	0.00440	0.01690	230	0.3037	42.116	0.35705	40.006
20	0.02333	0.1378	0.00943	0.0508	235	0.3064	43.641	0.36361	41.808
25	0.03386	0.2803	0.01575	0.1133	240	0.3090	45.179	0.37009	43.643
30	0.04478	0.4768	0.02288	0.2096	245	0.3116	46.731	0.37649	45.510
35	0.05567	0.7280	0.03060	0.3431	250	0.3142	48.295	0.38281	47.408
40	0.06637	1.0331	0.03874	0.5163	255	0.3166	49.872	0.38905	49.337
45	0.07747	1.3925	0.04719	0.7310	260	0.3190	51.462	0.39523	51.298
50	0.08909	1.8086	0.05595	0.9887	265	0.3214	53.063	0.40133	53.290
55	0.10191	2.2855	0.06503	1.2910	270	0.3237	54.676	0.40735	55.310
60	0.11549	2.8292	0.07448	1.6396	275	0.3259	56.300	0.41331	57.362
65	0.12713	3.4371	0.08420	2.0363	280	0.3280	57.934	0.41920	59.443
70	0.13573	4.0951	0.09395	2.4817	285	0.3300	59.579	0.42503	61.555
75	0.14362	4.7933	0.10358	2.9756	290	0.3320	61.235	0.43079	63.694
80	0.15261	5.5333	0.11313	3.5174	295	0.3339	62.900	0.43648	65.862
85	0.16310	6.3220	0.12269	4.1070	300	0.3357	64.574	0.44210	68.058
90	0.17378	7.1647	0.13232	4.7443	305	0.3374	66.256	0.44767	70.284
95	0.18226	8.0560	0.14196	5.4302	310	0.3390	67.947	0.45317	72.536
100	0.18847	8.9834	0.15147	6.1638	315	0.3406	69.647	0.45860	74.814
105	0.19439	9.9404	0.16081	6.9446	320	0.3422	71.354	0.46398	77.121
110	0.2007	10.928	0.16999	7.7713	325	0.3438	73.069	0.46930	79.454
115	0.2070	11.947	0.17905	8.6440	330	0.3454	74.792	0.47456	81.814
120	0.2132	12.998	0.18800	9.5620	335	0.3470	76.523	0.47976	84.199
125	0.2193	14.079	0.19682	10.524	340	0.3486	78.262	0.48492	86.612
130	0.2252	15.190	0.20554	11.530	345	0.3502	80.009	0.49002	89.049
135	0.2308	16.331	0.21415	12.579	350	0.3517	81.764	0.49507	91.512
140	0.2362	17.498	0.22264	13.671	355	0.3532	83.526	0.50007	94.000
145	0.2413	18.692	0.23102	14.806	360	0.3546	85.296	0.50502	96.512
150	0.2463	19.911	0.23928	15.981	365	0.3560	87.072	0.50992	99.050
155	0.2510	21.154	0.24743	17.198	370	0.3572	88.855	0.51477	101.611
160	0.2556	22.421	0.25547	18.455	375	0.3584	90.645	0.51957	104.196
165	0.2598	23.710	0.26340	19.752	380	0.3596	92.440	0.52433	106.806
170	0.2640	25.019	0.27122	21.089	385	0.3607	94.241	0.52904	109.440
175	0.2679	26.349	0.27893	22.464	390	0.3617	96.046	0.53370	112.098
180	0.2717	27.698	0.28653	23.878	395	0.3627	97.858	0.53831	114.776
185	0.2753	29.066	0.29403	25.330	400	0.3638	99.674	0.54288	117.480
190	0.2788	30.451	0.30141	26.818	405	0.3648	101.495	0.54741	120.206
195	0.2822	31.854	0.30870	28.343	410	0.3658	103.321	0.55189	122.954
200	0.2854	33.273	0.31589	29.906	415	0.3668	105.153	0.55633	125.725
205	0.2886	34.708	0.32297	31.502	420	0.3678	106.989	0.56073	128.518
210	0.2917	36.159	0.32997	33.135	425	0.3688	108.831	0.56508	131.330

1.4 Chemical Properties of Uranyl Fluoride. (a) Water. Anhydrous uranyl fluoride is soluble in water, methanol, and ethanol; unlike the other uranyl halides it is insoluble in ether or amyl alcohol (von Unruh, 1909). The solubility in water has been studied at various temperatures (SAM Columbia 2); the data are given in Table 16.2.

The solubility of uranyl fluoride has also been determined at the SAM Carbide and Carbon laboratory. Their data are given in Table 16.3 (SAM Carbide and Carbon 1). By plotting the mole fraction of

Table 16.2 — Solubility of Uranyl Fluoride in Water

Temp., °C	UO_2F_2, wt. %	Molality	ρ, g/cc
25.0	67.3	5.18	2.405
75.0	69.6	5.59	2.472
99.9	72.4	6.08	2.588

Table 16.3 — Solubility of Uranyl Fluoride at Various Temperatures

Temp., °C	UO_2F_2, %	UO_2F_2, mole fraction
1	61.4	0.0862
25	65.6	0.100
60	71.0	0.125
100	74.1	0.143

uranyl fluoride vs. the reciprocal of the absolute temperature, the heat of the reaction

$$UO_2F_2 \cdot H_2O \text{ (solid)} + H_2O \rightarrow UO_2F_2 \text{ (saturated solution)} \qquad (9)$$

is found to be 1.0 kcal per mole. The solubility of uranyl fluoride in water increases with temperature, but the solubility of uranyl fluoride in hydrofluoric acid has a negative temperature coefficient so that crystallization of uranyl fluoride occurs on heating.

Uranyl fluoride prepared at low temperatures is very hygroscopic, but uranyl fluoride obtained by high-temperature hydrofluorination exhibits no deliquescence even on long standing in air. Hydrated uranyl fluoride can be dehydrated at 120°C without serious decomposition (British 2; SAM Columbia 3). Hydrates of uranyl fluoride generally appear to be unstable above 100°C. Uranyl fluoride containing an excess of UO_3 requires higher temperatures for complete dehydration (up to 250°C). At 250°C uranyl fluoride undergoes a color change; it becomes pink but does not appear to be changed chemically (Montreal 1).

Aqueous solutions of neutral uranyl fluoride do not attack glass even at the boiling point. Solutions of 2M uranyl fluoride have been heated in pyrex tubes to 200°C for ten days without formation of any precipitate (Montreal 1). Superheated steam at 900°C removes all the fluorine, leaving a residue of U_3O_8. At 450°C the residue is largely UO_3 (MP Ames 1).

(b) The UO_2F_2-HF-H_2O System. (SAM Carbide and Carbon 1.) A phase study of this system has been made by the solubility method.

Table 16.4—Composition of the Liquid and Solid Phases in Equilibrium in the UO_2F_2-HF-H_2O System at 25°C

Liquid phase				Solid phase		
HF, %	UO_2F_2, %	H_2O, %	D_{25}^{25}	HF, %	UO_2F_2, %	H_2O, %
0.00	65.55	34.45	2.224	...	...	...
11.88	31.88	56.24	1.440	8.52	51.95	39.53
20.70	22.99	57.01	...	18.20	32.50	49.30
25.75	18.19	56.01	...	14.51	51.25	34.24
32.51	1.35	56.14	1.231	12.06	64.47	23.47
				3.50*	83.76*	12.74*
41.70	6.10	52.20	...	8.77	75.2	16.03

*Partially dried.

(The uranyl fluoride used was prepared by reaction of UO_3 with hydrogen fluoride.) The various mixtures were allowed to come to equilibrium at 25°C in two to three weeks' time. The results obtained are presented in Table 16.4.

The data indicate a marked decrease in solubility with increasing concentration of hydrofluoric acid. In the region investigated, only one solid phase appears. On extrapolating the tie lines joining points corresponding to the compositions of the solution phases and the equilibrium solids, the composition of the equilibrium solid phase is found to be $UO_2F_2 \cdot H_2O$ (Fig. 16.1).

(c) Thermal Decomposition of Uranyl Fluoride. The compound appears to be stable in air to 300°C. Above 300°C decomposition to U_3O_8 occurs (CEW-TEC 2; British 4). The thermal decomposition has been studied in some detail at Ames (MP Ames 2). Decomposition is found to occur even at 200°C, but most of it occurs at 850 to 900°C (in vacuum). The substance does not melt. A black cokelike residue of U_3O_8 remains in the furnace; a yellow substance, UO_2F_2(?), and a green substance, impure UF_4, distill out of the furnace. There is also evidence for the evolution of fluorine (or perhaps UF_6), since a deposit of mercurous fluoride is usually found in the mercury vapor pump. The Ames workers postulated a number of concurrent reactions to account for the observed products of thermal decomposition.

$$2UO_2F_2 \rightarrow UO_2 + UF_4 + O_2 \qquad (10)$$

$$UO_2F_2 \rightarrow UO_2 + F_2 \qquad (11)$$

$$4UO_2F_2 \rightarrow U_3O_8 + UF_4 + 2F_2 \qquad (12)$$

The evidence is still too meager to permit drawing any definite conclusions as to the predominant reaction. A primary reaction which

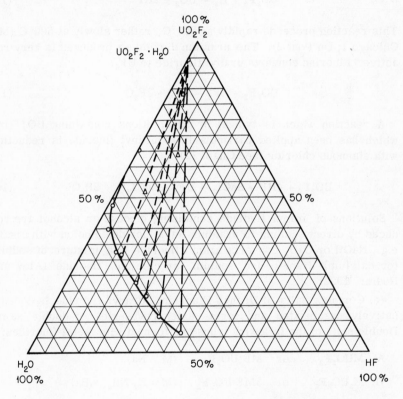

Fig. 16.1 — The uranyl fluoride–hydrogen fluoride–water system at 25°C.

could perhaps account for the observed result (at least in part) is

$$3UO_2F_2 \rightarrow UF_6 + \tfrac{2}{3}U_3O_8 + \tfrac{1}{3}O_2 \qquad (13)$$

Fried and Davidson have estimated the ΔF of this reaction; from their computations it appears that a temperature of at least $1300°K$ would be necessary for it to proceed (MP Chicago 2). However, a number of arbitrary assumptions were involved in computing the free energy of reaction 13, and the possibility that UF_6 actually is formed according to this equation need not be excluded. More work is desirable since formation of UF_6 by decomposition of uranyl fluoride would have some interest.

(d) <u>Some Chemical Reactions of Uranyl Fluoride</u>. Uranyl fluoride undergoes reduction with hydrogen.

$$UO_2F_2 + H_2 \rightarrow UO_2 + 2HF \tag{14}$$

This reaction proceeds rapidly at $600°C$, rather slowly at $500°C$ (MP Chicago 1; Du Pont 2). The uranium dioxide so produced is very reactive. Fluorine converts uranyl fluoride to UF_6.

$$UO_2F_2 + 4F_2 \rightarrow UF_6 + 2F_2O \tag{15}$$

A reaction which is typical of all solutions containing UO_2^{++} but which has been applied particularly to uranyl fluoride is reduction with stannous chloride (see Chap. 12).

$$UO_2F_2 + 4HF + SnCl_2 \rightarrow UF_4 + SnCl_2F_2 + 2H_2O \tag{16}$$

Solutions of uranyl fluoride containing glucose or alcohol are reduced by direct sunlight to uranium tetrafluoride. Fusion with alkali, e.g., NaOH or CaO, converts uranyl fluoride to a mixture of sodium (or calcium) uranate and sodium (or calcium) fluoride (Aloy and Rodier, 1922)

(e) <u>Complex Salts of Uranyl Fluoride</u>. Uranyl fluoride is coordinatively unsaturated and forms extensive series of double salts. Double salts of the following types are formed with metal fluorides:

A. MUO_2F_3 or $MF \cdot UO_2F_2$ (M = Na)

B. $M_3UO_2F_5$ or $3MF \cdot UO_2F_2$ (M = K, NH_4, $\frac{1}{2}Ba$)

C. $M_3(UO_2)_2F_7$ or $3MF \cdot 2UO_2F_2$ (M = K)

D. $M_5(UO_2)_2F_9$ or $5MF \cdot 2UO_2F_2$ (M = K)

Ditte (1884) described compounds of the series $M_4UO_2F_6$ (M = Li, Na, K, Rb, Tl), which he obtained by fusion of U_3O_8 and potassium fluoride with a small amount of carbonate. He also described compounds of the type M$_4$UOF$_8$·nH$_2$O (M = K, Rb). The latter have been shown by Smithells to be in all probability of type B, whereas the former were simply uranates (Smithells, 1883). The preparation of the two compounds Cs$_4$UO$_2$F$_6$ and K$_4$UO$_2$F$_6$ has been reported (SAM Columbia 4), but no details of preparation or analysis were given. The cesium compound forms monoclinic prismatic crystals. Since there does not appear to be any obvious reason that complex compounds of the type 4MF·UO$_2$F$_2$ should not exist, they may well be obtainable although probably not by the methods of Ditte.

Although the UO$_2$F$_2$ complexes were discovered and studied by Bolton (1866b), it was Baker (1879) who first elucidated their relations. When potassium fluoride is added to a solution of uranyl nitrate or uranyl fluoride, K$_3$UO$_2$F$_5$ precipitates. If this salt is recrystallized from pure water (or an aqueous solution containing less than 13 per cent KHF$_2$), K$_5$(UO$_2$)$_2$F$_9$ is formed. Further, if either K$_3$UO$_2$F$_5$ or K$_5$(UO$_2$)$_2$F$_9$ is recrystallized from uranyl nitrate solution, K$_3$(UO$_2$)$_2$F$_7$ is formed.

Of these salts, K$_3$UO$_2$F$_5$ has been studied in some detail. It melts at red heat with decomposition. On heating in air, fluorine is lost, and the compound is converted to uranate. The salt is completely decomposed by warm concentrated sulfuric acid. Ammonia and sodium hydroxide precipitate diuranates, but ammonium and sodium carbonates yield water-soluble complexes. This reaction indicates that the uranium is not held in a very tight complex. A solution of K$_3$UO$_2$F$_5$ gives no precipitate with copper, silver, zinc, mercury, iron, or platinum salts; but barium, calcium, and lead ions precipitate insoluble complex salts of the type Ba$_3$(UO$_2$)$_2$F$_{10}$·2H$_2$O. Photoreduction of an oxalic or formic acid solution of K$_3$UO$_2$F$_5$ results in precipitation of K$_2$UF$_6$. All the complex salts undergo conversion to uranates on heating in air or on fusion with sodium carbonate; the ammonium salt is, of course, converted to U$_3$O$_8$ on ignition, with loss of ammonium fluoride.

Hydrogen peroxide oxidizes aqueous solutions of NaUO$_2$F$_3$ or of K$_3$UO$_2$F$_5$ to insoluble per-compounds which have been formulated as NaUO$_4$F·5H$_2$O and K$_4$U$_4$O$_{15}$F$_6$·4H$_2$O. These formulas cannot be considered as firmly established. The per-compounds decompose above 100°C with evolution of oxygen (Lordkipanidze, 1900). The conductivity of aqueous solutions of K$_3$UO$_2$F$_5$ and (NH$_4$)$_3$UO$_2$F$_5$ has been studied (Miolati and Alvisi, 1897), and the results purport to show the existence of stable (UO$_2$F$_5$)$^{-3}$ ions (compare with crystallographic evidence below). Table 16.5 summarizes some pertinent data on the preparation and crystallographic properties of these complex salts.

Table 16.5 — Preparation and Properties of Complex Compounds of Uranyl Fluoride and Metal Fluorides

Compound	Ratio of metal fluoride to UO$_2$F$_2$	Crystal habit*	Crystallographic* data	Density* at 20°C, g/cc	Solubility	Preparation
NaUO$_2$F$_3$·4H$_2$O	1/1	Monoclinic	a:b:c = 1.0270:1:0.5222 β = 94° 51′			Crystallizes by slow evaporation of a solution of uranyl nitrate and NaF or a solution of sodium uranate in HF; exact conditions are unknown; the dihydrate is obtained on recrystallization from H$_2$O†
K$_3$UO$_2$F$_5$	3/1	Tetragonal; no appreciable fluorescence	a:c = 0.992	4.263	12.5 g/100 g H$_2$O at 21°C;† precipitates by addition of alcohol	Ppts. as a yellow crystalline solid on addition of a slight excess of KF to UO$_2$(NO$_3$)$_2$ solution; this is the primary reaction product of KF and UO$_2$F$_2$ or UO$_2$(NO$_3$)$_2$ solutions*†‡
(NH$_4$)$_3$UO$_2$F$_5$	3/1	Tetragonal; strongly fluorescent in x-ray or ultra-violet light	Refractive index = 1.495§	3.186	10.11 g/100 g solution at 27°C; 20.70 g/100 g solution at 81.3°C;¶ insol. in C$_2$H$_5$OH	Addition of NH$_4$F to a solution of uranyl nitrate*†
Ba$_3$(UO$_2$)$_2$F$_{10}$·2H$_2$O	3/1				Very sparingly sol. in hot water	BaCl$_2$ + K$_3$UO$_2$F$_5$ solution; the dihydrate is obtained by drying at 100°C; Ca and Pb compounds are obtained in the same way†
K$_3$(UO$_2$)$_2$F$_7$·2H$_2$O	3/2	Monoclinic; distinct green fluorescence	a:b:c = 0.918:1:0.978 β = 114° 0′	4.108	Sol. in warm water	Prepared by adding to a solution of UO$_2$F$_2$ an amount of KF insufficient to cause a permanent precipitate and then evaporating or by crystallizing K$_3$UO$_2$F$_5$ or K$_5$(UO$_2$)$_2$F$_9$ from UO$_2$(NO$_3$)$_2$ or UO$_2$F$_2$ solution*
K$_5$(UO$_2$)$_2$F$_9$	5/2	Triclinic; large crystals show distinct fluorescence	a:b:c = 0.6222:1:0.568 α = 72° 38′ β = 116° 23′ γ = 111° 57′	4.379		Prepared by recrystallization of K$_3$UO$_2$F$_5$ from water or from a less than 13% KHF$_2$ solution*

*Baker, 1879.
†Bolton, 1866a.
‡Smithells, 1883.
§Bolland, 1910.
¶Bürger, 1904.

The crystal structure of $K_3UO_2F_5$ has been studied recently (MP Chicago 4). Material prepared by addition of potassium fluoride to a

Table 16.6 — Complex Compounds of Uranyl Fluoride with Organic Bases

Base	Compound	Solubility in H$_2$O at 20°C, g/100 ml
Pyridonium	$C_5H_6NHUO_2F_3 \cdot H_2O$	1.289
Pyridonium	$C_5H_6NH(UO_2)_2F_5 \cdot 3H_2O$	1.952
Quinolonium	$C_9H_8NHUO_2F_3 \cdot H_2O$	0.126
Quinolonium	$C_9H_8NHUO_2F_5 \cdot 2H_2O$	0.979
Tetramethylammonium	$(CH_3)_4N(UO_2)_2F_5 \cdot 2H_2O$	0.143
Tetraethylammonium	$(C_2H_5)_4NUO_2F_3$	0.716
Tetraethylammonium	$(C_2H_5)_4NUO_2F_3 \cdot 2H_2O$	0.771
Trimethyl-p-tolyl ammonium	$(CH_3C_6H_4)(CH_3)_3NUO_2F_3 \cdot 2H_2O$	1.645
Trimethyl-p-tolyl ammonium	$(CH_3C_6H_4)(CH_3)_3N(UO_2)_2F_5 \cdot H_2O$	3.091
Tetramethyl pyridonium	$(C_5H_2)(CH_3)_4NH(UO_2)_3F_7 \cdot 6H_2O$	0.708
Diethyl anilonium	$C_6H_5N(C_2H_5)_2HUO_2F_3 \cdot 2H_2O$	1.759
Diethyl anilonium	$C_6H_5N(C_2H_5)_2H(UO_2)_2F_5 \cdot 2H_2O$	3.896
Triethyl sulfonium	$(C_2H_5)_3S(UO_2)_2F_5 \cdot 2H_2O$	0.897
Trimethylammonium	$(CH_3)_3NH(UO_2)_2F_5 \cdot 2H_2O$	
Propyl ammonium	$C_3H_7NH_3UO_2F_3 \cdot 2H_2O$	
Propyl ammonium	$C_3H_7NH_3(UO_2)_3F_7 \cdot 6H_2O$	
Tetrapropyl ammonium	$(C_3H_7)_4N(UO_2)_3F_7 \cdot 2H_2O$	
Methyl ethyl propyl phenyl ammonium	$(CH_3)(C_2H_5)(C_3H_7)(C_6H_5)N \cdot (UO_2)_3F_7 \cdot 6H_2O$	
Anilonium	$C_6H_5NH_3UO_2F_3 \cdot 3H_2O$	
Dimethyl anilonium	$C_6H_5N(CH_3)_2H(UO_2)_2F_5 \cdot H_2O$	

concentrated uranyl nitrate solution was found to consist of a single-phase, anhydrous $K_3UO_2F_5$. The compound has a tetragonal structure with a body-centered translation group. The unit cell with the dimensions

$$a_1 = 9.05 \pm 0.05 \text{ A} \qquad a_3 = 18.10 \pm 0.10 \text{ A}$$

contains eight molecules. The calculated density is 4.29 g/cc, and the probable space group is I4/amd. An approximate structure has been deduced; an interesting feature is the existence of $(UO_2F_4)^{--}$ units in the lattice with the fifth fluorine bound only to potassium.

Uranyl fluoride forms a series of addition compounds with ammonia (von Unruh, 1909). Liquid ammonia reacts with uranyl fluoride (or lower ammoniated complexes) to give $UO_2F_2 \cdot 4NH_3$, a deep orange-red substance which is more stable than the corresponding derivatives of uranyl bromide and uranyl chloride. Gaseous ammonia reacts with uranyl fluoride to give $UO_2F_2 \cdot 3NH_3$ (orange-yellow), which on warming is converted to $UO_2F_2 \cdot 2NH_3$ (yellow).

Complex compounds with organic bases have also been prepared (Olsson, 1930). A solution of the appropriate base (in dilute hydrofluoric acid) is added to a solution of uranyl nitrate (containing F^-). The precipitate is removed, washed first with dilute hydrofluoric acid, then with water, and finally dried on filter paper. Most of the compounds are stable in air for long periods; heat converts them to U_3O_8. They are somewhat soluble in water, and the solubility usually increases as the temperature is raised. The complexes dissolve in acids; bases such as sodium hydroxide or ammonia precipitate uranates and carbonates form soluble complexes. No solubility is observed in organic solvents, such as glacial acetic acid, alcohol, ether, or acetone. Table 16.6 lists a number of these compounds and their solubilities in water.

The compounds are of three types:

A. $MUO_2F_3 \cdot nH_2O$ or $MF \cdot UO_2F_2$

B. $M(UO_2)_2F_5 \cdot nH_2O$ or $MF \cdot 2UO_2F_2$

C. $M(UO_2)_3F_7 \cdot nH_2O$ or $MF \cdot 3UO_2F_2$

where M is a univalent organic base.

2. URANIUM(IV) OXYFLUORIDE, UOF_2

Giolitti and Agamennone (1905) have claimed preparation of the compound $UOF_2 \cdot 2H_2O$. Giolitti (1904) had developed an analytical method for uranium based on the precipitation of uranium tetrafluoride. The results he obtained by this procedure were very erratic, and in seeking the reason for this he came to the rather surprising conclusion that uranium tetrafluoride does not exist at all and that uranium(IV) oxyfluoride is the compound that precipitates when hydrofluoric acid is added to a uranium(IV) solution.

There can be little doubt that Giolitti's conclusions were based on very inadequate analytical methods. His fluorine determinations could easily have been in error by several hundred per cent.

In the course of this work, Giolitti and Agamennone made several observations of the reaction of hydrofluoric acid and U$_3$O$_8$, which have been confirmed by more recent investigators (MP Ames 3; MP Chicago 5). When U$_3$O$_8$ is treated with hydrofluoric acid, two solid phases appear which can be readily separated by flotation because of the difference in density. The less dense material is blue-green; the more dense material is green. The nature of these two phases is still in doubt.

The Brown University group attempted to prepare uranium(IV) oxyfluoride by treating uranium(IV) oxychloride with liquid anhydrous hydrogen fluoride at room temperature, but uranium tetrafluoride was the only product obtained (Brown 1). An attempt to prepare UOF$_2$ by fusing UO$_2$ with UF$_4$ also failed (MP Ames 13).

3. URANYL CHLORIDE, UO$_2$Cl$_2$

Anhydrous uranyl chloride can be prepared only by high-temperature vapor-phase reactions. Although aqueous solutions of uranyl chloride can be readily obtained, they have not as yet been dehydrated without formation of basic salts.

3.1 Preparation of Anhydrous Uranyl Chloride. The reaction

$$UO_2 + Cl_2 \rightarrow UO_2Cl_2 \tag{17}$$

was employed by Péligot (1842a,b) to prepare anhydrous uranyl chloride. The reaction is carried out at red heat with dry chlorine but is usually incomplete. Regelsberger (1885) employed ether extraction to separate the uranyl chloride from the unreacted oxide. Unfortunately, the ether cannot then be completely removed. The Brown University group found that the reaction between UO$_2$ and commercial dry chlorine proceeds at 500°C, to give a product containing some U$_3$O$_8$ (Brown 2). With UO$_3$, chlorine does not react at 400°C; above 400°C U$_3$O$_8$ is formed (MP Ames 4). Active UO$_2$, prepared by reduction of UO$_3$ with methane (cf. Chap. 11), was reported to yield uranyl chloride on treatment with chlorine (UCRL 1). With U$_3$O$_2$S$_2$(UO$_2$·2US) and chlorine, reaction occurs at 60°C. After it is over, the UCl$_4$ formed can be sublimed away in a stream of chlorine at 600°C, leaving a residue of yellow anhydrous uranyl chloride (MP Ames 5).

A number of other reactions are known in which anhydrous uranyl chloride is formed. When carbon tetrachloride reacts with various uranium oxides, some uranyl chloride is formed which presumably could be separated from concurrently produced uranium tetrachloride either by vacuum sublimation at elevated temperatures or by sublimation in a stream of chlorine (see Chap. 14). Thus when UO_3 is treated with carbon tetrachloride vapor at 290°C, a product containing 23 per cent UO_2Cl_2 and 77 per cent UCl_4 is obtained (UCRL 2). If carbon monoxide or, better, chloroform is introduced with the carbon tetrachloride, the yield of uranyl chloride is increased (UCRL 3). However, these are not particularly convenient preparative reactions.

Hydrogen chloride reacts readily with uranium trioxide to form uranyl chloride. The reaction is exothermic and proceeds spontaneously at room temperature. Moisture must be present, for if anhydrous UO_3 is used, the reaction is extremely slow. A closed system, containing partially hydrated UO_3, is evacuated without heating in order to remove air but leave as much water as possible. Hydrogen chloride is then introduced at a rate sufficient to maintain a pressure of 1 atm in the reaction system. Approximately 18 to 24 hr at room temperature is required for completion of the reaction. Material so prepared has the formula $UO_2Cl_2 \cdot H_2O$. The hydrated uranyl chloride can be dried without decomposition in a current of dry hydrogen chloride gas at 300°C (Brown 3). It has been reported (without detail) that UO_3 suspended in ethanol or carbon tetrachloride gives uranyl chloride on treatment with hydrogen chloride (Purdue 1). The preparative value of this reaction is unknown.

Probably the best method for the preparation of anhydrous uranyl chloride is the reaction of uranium tetrachloride with oxygen at 300 to 350°C.

$$UCl_4 + O_2 \rightarrow UO_2Cl_2 + Cl_2 \tag{18}$$

There is a tendency for the reaction mass to sinter, preventing complete conversion; it can be counteracted by agitating the reaction tube. Mixtures of UCl_5, UCl_6, and UCl_3 also react with oxygen to give uranyl chloride, but UCl_4 appears to be the best starting material (MP Ames 4).

Acetyl chloride, CH_3COCl, reacts with various oxides to give halides or oxyhalides (Chrétien and Oechsel, 1938). With UO_3 and liquid CH_3COCl at room temperature, $UO_2Cl_2 \cdot (CH_3CO)_2O$, a pale-yellow crystalline powder, is obtained. Whether pure uranyl chloride could be obtained from this complex is not stated. Further work would be interesting.

3.2 Preparation of Uranyl Chloride Hydrates. Aqueous solutions of uranyl chloride have been prepared by careful oxidation of a solution of uranium tetrachloride with nitric acid (Arfvedson, 1824), by solution of $UO_3 \cdot H_2O$ in an equivalent amount of hydrochloric acid (Mylius and Dietz, 1901), and also by the addition of the stoichiometric amount of barium chloride to a solution of uranyl sulfate (de Coninck, 1909a). Solid hydrates can be prepared by evaporation from such solutions. Usually, amorphous products are obtained, but slow evaporation in a desiccator yields yellowish-green, doubly refracting, fluorescent, easily decomposed crystals of $UO_2Cl_2 \cdot 3H_2O$ (Mylius and Dietz, 1901). Crystallization can sometimes be induced by introducing into a concentrated solution of uranyl chloride a few seed crystals, prepared by treating some of the semiamorphous material with concentrated hydrochloric acid and allowing it to crystallize spontaneously by slow evaporation in a desiccator.

The monohydrate, $UO_2Cl_2 \cdot H_2O$, has been reported (de Coninck, 1909a) to be formed by slow evaporation of an aqueous solution in dry air. Evaporation of a uranyl chloride solution at 120°C to dryness has also been stated to yield $UO_2Cl_2 \cdot H_2O$ (SAM Columbia 5). Saturation with hydrogen chloride of a saturated aqueous solution of uranyl chloride at −10°C gives $UO_2Cl_2 \cdot HCl \cdot 2H_2O$ as yellow, very unstable crystals (Aloy, 1901a).

All the above-mentioned workers were of the opinion that hydrated uranyl chloride could not be dehydrated without serious decomposition. Mylius and Dietz reported that hydrogen chloride is evolved even at room temperature when evaporation is attempted. The tendency to form basic salts is great. A compound, $UO_2(OH)Cl \cdot 2H_2O$, has been isolated in the form of small yellow needles (Mylius and Dietz, 1901) from the syrupy evaporation product of a uranyl chloride solution. This substance is probably identical with that described much earlier by Lecanu (1825). It is more stable than uranyl chloride hydrate; the water of crystallization is lost at 150°C without further decomposition.

3.3 Physical Properties of Uranyl Chloride. (a) Crystal Structure.

(MP Ames 6.) X-ray diagrams have been obtained for anhydrous uranyl chloride powder and for the needles obtained by condensing uranyl chloride from the vapor (500°C). The two forms are different, as the needles show many maxima that are absent from the powder diagram. The crystal is orthorhombic with four molecules per unit cell. The lattice constants are

Needles: $a_0 = 8.71 \pm 0.01$ A $b_0 = 8.39 \pm 0.01$ A $c_0 = 5.72 \pm 0.01$ A

Powder: $a_0 = 8.69 \pm 0.01$ A $b_0 = 8.39 \pm 0.01$ A $c_0 = 5.70 \pm 0.01$ A

The density calculated from the x-ray data is 5.426 g/cc. A direct measurement of the density by benzene displacement gave 5.28 g/cc (MP Ames 5).

(b) <u>Volatility</u>. Uranyl chloride was described as somewhat volatile in a stream of chlorine or oxygen above 500°C. It has been found, however, that uranyl chloride will not markedly volatilize at 630°C in a chlorine atmosphere in 7 hr (Purdue 2). The volatility has been studied by tracer techniques with results that indicated no substantial volatility below 775°C. These results require confirmation (MP Berkeley 1). Decomposition renders interpretation of the volatility data difficult. Anhydrous uranyl chloride has been reported to melt at a relatively low temperature (red heat). The vapor was reported to be orange-yellow in color (Péligot, 1842a,b).

(c) <u>Miscellaneous</u>. Crystals of anhydrous uranyl chloride show no triboluminescence (Trautz, 1905). Molten uranyl chloride conducts an electric current; chlorine is evolved and uranium dioxide separates out of the melt (Hampe, 1888).

3.4 <u>Chemical Properties of Uranyl Chloride</u>. Anhydrous uranyl chloride is a bright-yellow crystalline substance. The hydrates have a greenish cast and appear to be fluorescent. Both anhydrous and hydrated uranyl chloride are very hygroscopic and rapidly form viscous solutions on exposure to air. In dry air the compounds are stable indefinitely.

(a) <u>Solubility in Water and the Stability of Aqueous Solutions</u>. Uranyl chloride, its hydrates, and basic salts are very soluble in water. The trihydrate is soluble to the extent of 746 parts per 100 parts of water at 18°C and even more at higher temperatures (Mylius and Dietz, 1901), but this seems strange. The solubility is reported to be higher in hydrochloric acid solution (Aloy, 1901b). The density of the saturated aqueous solution at 18°C is 2.740 g/cc, which is sufficiently high to float glass or quartz. The density of solutions containing between 1 and 10 per cent uranyl chloride at 13 to 16°C lies between 1.0056 and 1.0517 g/cc (de Coninck, 1904). The molar heat of solution of $UO_2Cl_2 \cdot H_2O$ (in 2,500 moles of H_2O) has been determined as 6.05 kcal (at 18 to 20°C) (Aloy, 1896). This value is uncertain because of the indefinite composition of the salt used.

Aqueous solutions of uranyl chloride are thermally and photochemically unstable (Mylius and Dietz, 1901). Solutions are usually acid to litmus, indicating appreciable hydrolysis. Although de Coninck (1909b) was unable to verify the existence of $UO_2(OH)Cl \cdot 2H_2O$ (which has been found by other workers), the formation of such basic salts as a result of hydrolysis seems fairly well established.

(b) <u>Nonaqueous Solutions of Uranyl Chloride</u>. Uranyl chloride is said to dissolve in methyl acetate, ethyl acetate, acetone, and pyridine, but whether reaction occurs is not stated (Naumann, 1904, 1909).

Anhydrous uranyl chloride is insoluble in carbon tetrachloride, xylene, and benzene. It does not dissolve in, but reacts with, ethers and chloroform. It is soluble in alcohols, acetophenone, pyridine, and dioxane, but reaction occurs with all these solvents (MP Ames 4). Uranyl chloride trihydrate is soluble in alcohol and ether.

Von Unruh (1909) prepared anhydrous uranyl chloride solutions in amyl alcohol by repeatedly evaporating uranyl acetate, first with hydrochloric acid and then with water, dissolving the hydrated uranyl chloride in amyl alcohol, and distilling the water off with some of the alcohol.

(c) <u>Thermal Stability</u>. Uranyl chloride and its hydrates readily undergo decomposition at elevated temperatures. Uranyl chloride is converted to U$_3$O$_8$ by ignition in air. The Brown University group reports that uranyl chloride decomposes in vacuum with evolution of chlorine at temperatures above 450°C, leaving a mixture of UO$_2$ and U$_3$O$_8$ (Brown 4). The thermal stability has been further studied at Ames (MP Ames 5). In a stream of nitrogen, decomposition to UO$_2$ and chlorine occurs above 400°C. In vacuum it begins at 300°C. In a stream of chlorine, uranyl chloride melts to a reddish-brown liquid at about 500°C. As the temperature is increased to 900°C, decomposition to UO$_2$ and chlorine occurs, with some sublimation of uranyl chloride and of UCl$_5$ formed by reaction with chlorine. The bulk of the uranyl chloride is converted to black crystalline UO$_2$. X-ray studies have shown that the UO$_2$ obtained in this way is slightly different from ordinary brown UO$_2$.

(d) <u>Reducing Agents</u>. Anhydrous uranyl chloride is reduced to UO$_2$ and potassium chloride by metallic potassium (Péligot, 1842a,b). Magnesium at red heat partially reduces uranyl chloride to uranium metal (Seubert and Schmidt, 1892). Hydrogen, zinc, or copper turnings fail to do so (de Coninck, 1904, 1909c). Hydrogen sulfide reduces uranyl chloride to UO$_2$, sulfur, and hydrogen chloride.

(e) <u>Alkalis</u>. (De Coninck, 1904, 1909a,c.) Fusion with potassium hydroxide or sodium hydroxide in air gives a mixture of diuranate with a small amount of uranate. With calcium hydroxide, U$_3$O$_8$ and some CaUO$_4$ form; if air is excluded, UO$_2$ is formed instead of U$_3$O$_8$. Barium hydroxide acts similarly. Calcium oxide and barium oxide in air give some U$_3$O$_8$ and, in the case of CaO, some CaUO$_4$ and CaU$_2$O$_7$ also, whereas BaO gives principally BaU$_2$O$_7$. Strontium hydroxide and strontium oxide behave similarly to the barium compounds.

(f) <u>Acids</u>. (De Coninck, 1903, 1904.) Uranyl chloride evolves chlorine and nitrogen oxides on warming with nitric acid. Concentrated sulfuric acid converts uranyl chloride to uranyl sulfate with evolution of hydrogen chloride. Selenic acid dissolves uranyl chloride; on warming, chlorine is evolved and a solution of uranyl selenite is formed.

(g) Reactions in Aqueous Solutions. Aqueous solutions of uranyl chloride exhibit all the reactions characteristic of the uranyl ion.

(h) Complex Compounds of Uranyl Chloride with Ammonia and Organic Bases. An extensive series of uranyl chloride ammoniates is known. Vacuum-dried uranyl chloride absorbs two molecules of ammonia to form an orange compound, $UO_2Cl \cdot 2NH_3$; of the two NH_3 molecules, one is bound so strongly as to be retained in vacuum (Peters, 1909, 1912). An ether solution of uranyl chloride, when treated with ammonia, forms a precipitate, $UO_2Cl_2 \cdot 2NH_3 \cdot (C_2H_5)_2O$, from which the ether can be removed in vacuum (Regelsberger, 1885). The same etherate can also be obtained from an amyl alcohol solution of uranyl chloride by treatment with ammonia and treatment of the dried precipitate with ether (von Unruh, 1909). The diammoniate is decomposed by warming to 100°C with the formation of ammonium chloride; on ignition in air it is converted to U_3O_8. It is reduced at elevated temperatures by hydrogen or ammonia to UO_2 (Regelsberger, 1885). A triammoniate is also known but only in the form of an etherate, $UO_2Cl_2 \cdot 3NH_3 \cdot (C_2H_5)_2O$ (Regelsberger, 1885). This compound is obtained by treatment of the diammoniate etherate with gaseous ammonia; it is an orange material, stable in air at room temperature, but loses ammonia on heating. An unstable tetraammoniate is formed by treating uranyl chloride with liquid ammonia at 5°C. The orange-red amorphous solid begins to decompose at 10°C (von Unruh, 1909). Rosenheim and Jacobsohn (1906) observed the formation of a gray-green precipitate when uranyl chloride was treated with liquid ammonia, but its composition was not established.

The most precise work on the uranyl chloride—ammonia system is that of Spacu (1936), who studied the reaction of liquid ammonia and uranyl chloride at −78°C. Under these conditions uranyl chloride is found to form decaammoniate, $UO_2Cl_2 \cdot 10NH_3$. As the temperature is raised ammonia is evolved, and vapor pressure measurements have indicated the existence of the complexes $UO_2Cl_2 \cdot 5NH_3$, $UO_2Cl_2 \cdot 4NH_3$, $UO_2Cl_2 \cdot 3NH_3$, $UO_2Cl_2 \cdot 2NH_3$, and $UO_2Cl_2 \cdot NH_3$. Figure 16.2 illustrates the results from which these conclusions were drawn. The temperatures of the isothermal lines in Fig. 16.2 indicate the thermal stability ranges of the various compounds. The pentaammoniate is particularly unstable, even at −44°C, whereas the monoammoniate is quite stable up to 130°C.

The heat of formation of the ammoniates varies from −9.63 kcal per mole for $UO_2Cl_2 \cdot 5NH_3$ to −18.03 kcal per mole for $UO_2Cl_2 \cdot NH_3$.

A number of organic bases also form complex compounds with uranyl chloride. It appears that, although the coordination sphere of uranium in uranyl chloride can accommodate a maximum of four ammonia molecules, steric considerations usually limit the number of organic groups to two. The preparation and properties of a number of such complexes are given in Table 16.7. The complexing groups

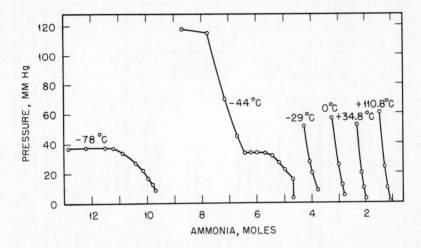

Fig. 16.2 — The uranyl chloride — ammonia system [from P. Spacu, Z. anorg. u. allgem. Chem., 230: 183 (1936)].

also include, in addition to basic nitrogen compounds, compounds whose basic properties are due to oxygen or sulfur.

(i) Complex Compounds of Uranyl Chloride and Metal Halides. Uranyl chloride forms a series of double salts of the type M$_2$UO$_2$Cl$_4$, where M = a univalent metal or an equivalent ion. Aloy (1901b) prepared anhydrous K$_2$UO$_2$Cl$_4$ and Na$_2$UO$_2$Cl$_4$ by treating the appropriate alkali halide with uranyl chloride vapors at red heat. The double salts are described as golden-yellow water-soluble solids that melt at red heat without evolving any vapors.

Compounds of this type can also be prepared from aqueous solution, in which case the dihydrate is usually obtained. The compounds K$_2$UO$_2$Cl$_4$·2H$_2$O and (NH$_4$)$_2$UO$_2$Cl$_4$·2H$_2$O were first prepared by dissolving K$_2$U$_2$O$_7$ or (NH$_4$)$_2$U$_2$O$_7$ in concentrated hydrochloric acid, followed by evaporation until crystallization occurred (Peligot, 1842b). Crystallization of a uranyl chloride solution containing the desired alkali halide also is effective. With M = potassium, it is necessary either

Table 16.7— Complex Compounds of Uranyl Chloride with Organic Bases

Organic base	Complex	Method of preparation	Properties	Solubility							
				Water	Ethanol	Amyl alcohol	Ether	Ethyl acetate	Acetone	Chloroform	Benzene
Ethyl ether[a]	UO₂Cl₂·2(C₂H₅)₂O	Evaporation of a solution of UO₂Cl₄ in ether	Yellow needles; decomposes in moist air; ether cannot be removed								
2,3-Dimethyl chromone[b]	UO₂Cl₂·2C₁₁H₁₀O₂	From conc. HCl solution of the components; recrystallized from conc. HCl	Long glistening bright-yellow prisms; decomposes without melting		Sl. sol.				Sl. sol.	In- sol.	In- sol.
2,3-Dimethyl thiochromone[c]	UO₂Cl₂·2C₁₁H₁₀OS	From conc. HCl solution of the components by evaporation			Sl. sol.				Sl. sol.	In- sol.	In- sol.
Aniline[d]	UO₂Cl₂·2C₆H₅NH₂	From an alcoholic solution of the components; recrystallized from alcohol	Small yellow needles		Sol.						
p-Toluidine[d]	UO₂Cl₂·2C₇H₇NH₂	Evaporation of an alcoholic solution of the components	Yellow-green rhombic crystals								
Pyridine[e]	UO₂Cl₂·2C₅H₅N	From solution of UO₂Cl₂·2H₂O in amyl alcohol and CHCl₃ solution of pyridine, on cooling	Yellow, weak green fluorescence; very hygroscopic	Sol.	Sol. hot						
p-Nitroso dimethyl aniline[f]	UO₂Cl₂·2(CH₃)₂NC₆H₄NO	From a warm alcoholic solution of the components	Brick red; stable in air	Dif. sol.		In- sol.	Sl. sol.		In- sol.	Sl. sol.	
p-Nitroso diethyl aniline[g]	UO₂Cl₂·2(C₂H₅)₂NC₆H₄NO	From a warm alcoholic solution of the components	Orange colored; amorphous	Dif. sol.		In- sol.	Sl. sol.		In- sol.	Sl. sol.	

Table 16.7—(Continued)

Organic base	Complex	Method of preparation	Properties	Water	Ethanol	Amyl alcohol	Ether	Ethyl acetate	Acetone	Chloroform	Benzene
Diketopiperazine[h]	$UO_2Cl_2 \cdot C_4H_6N_2O_2 \cdot 1.5H_2O$	From a warm alcoholic solution	Yellow crystals; stable in air	Sol.	Sol. hot	Sol. hot	V. sl. sol.		Sol.	V. sl. sol.	
Acet-*p*-phenetidine[i]	$UO_2Cl_2 \cdot 2C_{10}H_{13}O_2N$	From a warm amyl alcohol solution	Yellow crystals with green fluorescence; stable in air								
	$UO_2Cl_2 \cdot 3C_{10}H_{13}O_2N$	From an ethyl or amyl alcohol solution of the components	Shining tabular yellow crystals								
Methyl acetanilide[j]	$UO_2Cl_2 \cdot 2C_6H_5N(CH_3)COCH_3$	From an aqueous solution of the components	Stable in air even on heating	Sol.	Dif. sol.		Sol.			Sol.	
Phenyl dimethyl pyrazolone[k] (antipyrine)	$UO_2Cl_2 \cdot 2C_{11}H_{12}N_2O$	From alcoholic solutions of the components at higher temperatures		Sol.	Sol. hot	Sl. sol.	In- sol.	V. sl. sol.	V. sl. sol.	In- sol.	
Bromantipyrine[l]	$UO_2Cl_2 \cdot 2C_{11}H_{11}BrN_2O$	From alcoholic solutions of the components at higher temperatures		Sol.	Sol.	Sl. sol.	In- sol.	V. sl. sol.	Sl. sol.	Sl. sol.	Sl. sol.
Dimethylamino-antipyrine (pyramidon)[m]	$UO_2Cl_2 \cdot C_{11}H_{11}N(CH_3)_2N_2O$	From alcoholic solutions of the components at higher temperatures		Sol.	Sol.	Sl. sol.			Sl. sol.	Sl. sol.	Sl. sol.
	$UO_2Cl_2 \cdot 2C_{11}H_{11}N(CH_3)_2N_2O$	From alcoholic solutions of the components in the cold	Amorphous; yellow								

a Regelsberger, 1885.
b Simonis and Elias, 1915.
c Simonis and Elias, 1916.
d Leeds, 1881.
e Răscanu, 1930–1931a.
f Răscanu, 1931–1932a.
g Răscanu, 1931–1932b.
h Asahina and Dôno, 1930.
i Răscanu, 1930–1931b.
j Răscanu, 1931–1932c.
k Răscanu, 1930–1931c.
l Răscanu, 1931–1933.
m Răscanu, 1930–1931d.

Note: All references to the work of R. Răscanu are from "Gmelins Handbuch der anorganischen Chemie," System No. 55, pp. 135–136, Verlag Chemie, Berlin.

to use an excess of uranyl chloride or to work in concentrated hydrochloric acid solutions to prevent precipitation of potassium chloride. Thus $K_2UO_2Cl_4 \cdot 2H_2O$ was prepared by Rimbach (1904) by crystallizing an aqueous solution containing equimolecular quantities of the components, together with at least 15 per cent hydrochloric acid. $K_2UO_2Cl_4 \cdot 2H_2O$ forms yellow triclinic crystals with a:b:c = 0.607:1: 0.560, $\alpha = 80° 41'$, $\beta = 77° 42'$, and $\gamma = 91° 18'$. It is very soluble in water. Below 60°C dissolution occurs with decomposition, and the undissolved residue is mostly potassium chloride; above 60°C the solute and the undissolved solid have the same composition. The heat of dissolution is about 2 kcal per mole at 18°C, in infinitely dilute solution (1 mole in 2,500 moles) (Aloy, 1896). The compound can be dehydrated at 100°C but only with some decomposition. At red heat it melts with evolution of chlorine. Hydrogen reduces it. In contradistinction to $K_3UO_2F_5$, the compound $K_2UO_2Cl_4 \cdot 2H_2O$ is not reduced in sunlight by formic or oxalic acid solution (Bolton, 1866b).

$Rb_2UO_2Cl_4 \cdot 2H_2O$ and $Cs_2UO_2Cl_4$ are prepared similarly to $K_2UO_2Cl_4$ (Rimbach, 1904). The rubidium compound is isomorphous with the potassium and ammonium salts. The rubidium and cesium salts dissolve in water without decomposition. They are thus more stable than the complex with potassium chloride (and also that with NH_4Cl; cf. below), both of which dissociate upon dissolution at room temperature. A certain correlation appears to exist between stability of the complex and the size of the cation; the larger the latter, the stronger the complex. The cesium compound crystallizes in anhydrous form; the rubidium compound has been observed to do so occasionally. According to Wells and Boltwood (1895), who first studied these compounds, the cesium compound forms shiny rhombic crystals; but Nichols and Howes (1919) report the substance to be triclinic. The potassium, cesium, and ammonium salts have been grown as large crystals, and their absorption and fluorescence spectra studied; details are discussed elsewhere (SAM Columbia 4; Dieke and Duncan, 1949).

Quaternary ammonium salts form an analogous series of compounds. These differ from the complexes given in Sec. 3.4h in that the ligand is a salt rather than a free base. Rimbach (1904) has described compounds derived from ammonium chloride, trimethylammonium chloride, tetramethylammonium chloride, and tetraethylammonium chloride; efforts to prepare hydroxylamine and hydrazine derivatives were unsuccessful. Ammonium uranyl chloride, $(NH_4)_2UO_2Cl_4 \cdot 2H_2O$, can be prepared from a concentrated hydrochloric acid solution of uranyl chloride and ammonium chloride. It forms very unstable crystals isomorphic with the potassium compound. It dissolves in water, with decomposition, below 70°C. Mono-, di-, and trimethylamine hydrochlorides form unstable compounds of the type $[(CH_3)_3N]_2UO_2Cl_4$.

Tetramethylammonium uranyl chloride, $[(CH_3)_4N]_2UO_2Cl_4$, can be prepared from an aqueous solution of the components as greenish-yellow, strongly fluorescent, tetragonal crystals (a:c = 1:0.9057) which dissolve without decomposition in water. Tetraethylammonium uranyl chloride, $[(C_2H_5)_4N]_2UO_2Cl_4$, is likewise prepared by slow crystallization from an aqueous solution containing equivalent proportions of the components. It forms yellow tetragonal crystals, a:c = 1:0.9094, isomorphic with the corresponding tetramethyl compound; like the latter, it dissolves in water without decomposition. Ethylenediammonium uranyl chloride (Grossman and Schuck, 1906), $C_2H_4(NH_3)_2UO_2Cl_4$, best prepared by addition of excess hydrochloric acid to a water solution of ethylenediamine and then addition of an equimolar amount of uranyl chloride, forms very hygroscopic, yellow, prismatic crystals that melt at 219°C (not sharp). The following compounds have also been prepared: pyridinium uranyl chloride, $(C_5H_6N)_2UO_2Cl_4$, yellow crystalline powder, soluble in water and alcohol (Kalischer, 1902); β-lutidinium uranyl chloride, $(C_7H_{10}N)_2UO_2Cl_4$ (Williams, 1881); and quinolinium uranyl chloride, $(C_9H_8N)_2UO_2Cl_4$ (Williams, 1856). The analogous oxonium salt, xanthylium uranyl chloride $[(C_{13}H_9O)_2UO_2Cl_4]$ (Fosse and Lesage, 1906), is also known.

4. URANIUM(IV) OXYCHLORIDE, $UOCl_2$ (URANOUS OXYCHLORIDE)

Several reports appear in the older literature purporting to describe this substance. Benrath (1917) claimed that anhydrous uranous oxychloride resulted from photochemical reduction of an ether solution of uranyl chloride in direct sunlight. The product was described as light green in color. Reduction in alcohol-ether mixtures gave complex mixtures of basic salts; in aqueous solution, only hydrated uranium dioxide formed. Little reliance can be placed on these results; no experimental details were given, and the description of the products evokes doubt as to their correct identification.

Aloy found that evaporation of a uranium tetrachloride solution at low temperatures in vacuum yielded an amorphous precipitate. Redissolution of this precipitate in ethanol and reprecipitation by ether gives a product which, after thorough washing with ether and drying in a desiccator, has the composition $UOCl_2 \cdot 0.5H_2O$. It is described as a bright-green crystalline substance, rather soluble in water; such solutions decompose on heating with separation of hydrated uranium dioxide. These findings are confirmed by observations at the University of California Radiation Laboratory that an aqueous solution of uranium tetrachloride, when dried at 100°C in air or at 120°C in dry hydrogen chloride, gives solids with the composition $UO_{1.1}Cl_{1.9} \cdot 1.2H_2O$ and $UOCl_{2.02} \cdot 1.25H_2O$ (UCRL 4) (see also Chap. 14).

The most direct and generally satisfactory synthesis of uranous oxychloride consists in dissolving UO_2 in excess molten UCl_4 (600°C) (Brown 1,5). The equilibrium

$$UO_2 + UCl_4 \rightleftharpoons 2UOCl_2 \qquad (19)$$

is established under these conditions. After cooling and grinding, the excess UCl_4 can be removed at 450°C in vacuum, conditions under which disproportionation of the uranous oxychloride is negligible.

Uranous oxychloride has also been prepared at the University of California Radiation Laboratory (cf. UCRL 4) by treating UO_2 with UCl_4 vapors. The uranous oxychloride obtained in this way is not as pure as that prepared by liquid-phase reaction.

At Ames two unsuccessful attempts were made to prepare uranium(IV) oxychloride. In one, chlorine was reacted with $U_3O_2S_4$ (in analogy to a reaction successfully used in the preparation of uranous oxybromide), but the product was only a mixture of uranyl chloride with higher uranium chlorides. In the second attempt, superheated water vapor was conducted (in a stream of inert gas) over UCl_4; this, too, failed to produce $UOCl_2$ of satisfactory purity (MP Ames 5).

Uranous oxychloride was described by the Ames group as yellow feather-shaped crystals, but these were called green by workers at Brown University. The yellow crystals are stable in air and dissolve in water to form a green solution. Preliminary x-ray data at Ames (MP Ames 7) on single crystals have been obtained. They appeared to be tetragonal or hexagonal. The spacing along the needle axis is 3.32 A, as derived from layer line spacings. The layer line spacings normal to the needle axis are about 40 A but are indistinct. Some preliminary x-ray work has also been done elsewhere, but no conclusions as to the structure are as yet available (Johns Hopkins 1). A value of 2.4 has been reported for the dielectric constant of solid uranous oxychloride (CEW-TEC 3).

4.1 Heat of Formation of Uranium(IV) Oxychloride. (UCRL 5.) The heat of solution of uranium(IV) oxychloride has been determined as $\Delta H = -16.7 \pm 0.2$ kcal per mole, from which the heat of formation can be calculated.

$$U \text{ (solid)} + Cl_2 \text{ (gas)} + \tfrac{1}{2}O_2 \text{ (gas)} \rightarrow UOCl_2 \text{ (solid)} \qquad (20)$$

$$\Delta H_{298°K} = -261.7 \text{ kcal per mole}$$

4.2 Equilibrium Pressure of Uranium Tetrachloride above Uranous Oxychloride. (UCRL 6.) As mentioned above, the reaction

$$\text{UO}_2 \text{ (solid)} + \text{UCl}_4 \text{ (gas)} \rightleftarrows 2\text{UOCl}_2 \text{ (solid)} \tag{21}$$

is reversible. The equilibrium pressure of UCl$_4$ above uranium(IV) oxychloride has been measured between 460 and 540°C; the results are summarized in Table 16.8. These values yield for the ΔH of reaction 21 a value of 55.1 kcal. The equilibrium decomposition pres-

Table 16.8—Equilibrium Pressure of Uranium Tetrachloride above Pure Uranous Oxychloride

Temp., °C	$\frac{1}{T} \times 10^3$	Pressure of UCl$_4$, mm Hg
460	1.364	7.1×10^{-4}
470	1.346	1.18×10^{-3}
480	1.328	1.95×10^{-3}
490	1.311	3.13×10^{-3}
500	1.294	5.0×10^{-3}
510	1.277	8.1×10^{-3}
520	1.261	1.28×10^{-2}
530	1.245	1.97×10^{-2}
540	1.230	3.03×10^{-2}

Table 16.9—Equilibrium Constants for Hydrogen Reduction of Uranous Oxychloride[*]

Temp., °C	Temp., °K	$\frac{1}{T} \times 10^3$	K
300	573	1.745	<0.004
350	623	1.605	0.0061
385	658	1.520	0.0095
400	673	1.486	0.011
450	723	1.383	0.020

[*]$K = p_{HCl}/p_{H_2}^{1/2}$ (p in atmospheres).

sure of UCl$_4$ over uranium(IV) oxychloride at 500°C is approximately one hundred times smaller than the vapor pressure of UCl$_4$ at the same temperature.

4.3 Equilibrium of Hydrogen Reduction of Uranous Oxychloride. (UCRL 7.) Although the products of reduction of uranous oxychloride by hydrogen have not been identified with certainty, UOCl appears to be the most likely product. Consequently, all the calculations have been referred to the reaction

$$\text{UOCl}_2 \text{ (solid)} + \tfrac{1}{2}\text{H}_2 \text{ (gas)} \rightleftarrows \text{UOCl (solid)} + \text{HCl (gas)} \tag{22}$$

The experimental values of the equilibrium constant are given in Table 16.9. From these data the value $\Delta H_{673°K} = 10.6$ kcal per mole (± 10 to 15 per cent) is obtained; $\Delta F^°_{673°K}$ is then 6.03 kcal per mole, and $\Delta S^°_{673°K} = 6.79$ e.u.

4.4 Thermodynamic Constants. (UCRL 8.) MacWood has given the following values for the thermodynamic constants of uranium oxychloride:

$$\Delta H_{298°K} = -261.7 \text{ kcal per mole}$$

$$\Delta F_{298°K} = -246.3 \pm 1.5 \text{ kcal per mole}$$

$$S_{298°K} = 38.1 \text{ e.u.}$$

For details of the calculations and the assumptions involved, the reader is referred to MacWood's paper.

4.5 Chemical Properties. Few of the chemical properties of uranous oxychloride have been studied. Uranous oxychloride reacts with carbon tetrachloride at 170°C to form uranium tetrachloride. Liquid hydrogen fluoride at room temperature gives uranium tetrafluoride and not uranous oxyfluoride.

5. URANYL BROMIDE, UO_2Br_2

Uranyl bromide appears to be distinctly less stable than the fluoride and chloride. The preparation of anhydrous uranyl bromide was first described by Hermann (1861) and later by von Unruh (1909). Bromine vapors were passed over a mixture of uranium dioxide and charcoal at elevated temperatures, producing a mixture of uranium tetrabromide and uranyl bromide. Von Unruh separated the uranyl bromide by dissolving the reaction product in a mixture of alcohol and ether. Uranyl bromide etherate, $UO_2Br_2 \cdot 2(C_2H_5)_2O$, can be obtained as hygroscopic, red, fluorescent needles from this solution. Most of the ether is lost in vacuum. Richards and Merigold (1902) treated U_3O_8 with bromine or with hydrogen bromide but observed no reaction. A number of workers have prepared anhydrous uranyl bromide by dehydration of uranyl bromide hydrate. This work is discussed in Sec. 5.2.

5.1 Preparation of Anhydrous Uranyl Bromide. (MP Ames 8.) The Ames group prepared anhydrous uranyl bromide by reaction of oxygen with uranium tetrabromide.

$$UBr_4 + O_2 \rightarrow UO_2Br_2 + Br_2 \tag{23}$$

The temperature regulation is important. Below 140°C the reaction is extremely slow; at 200°C considerable amounts of U_3O_8 are produced. The best range is 150 to 160°C; preparations analyzing 96 per cent uranyl bromide can be readily obtained at this temperature. X-ray photographs showed no UO_2, U_3O_8, UO_3, or UBr_4 to be present in the product.

Uranium tribromide, UBr_3, when treated with oxygen at room temperature burns vigorously. This is therefore not a good preparative method. It had been observed (MP Ames 9) that uranium dioxide does not react with bromine even at 720°C. Anhydrous hydrogen bromide reacts at 100°C with dry ammonium diuranate, $(NH_4)_2U_2O_7$, to give, among other products, a water-soluble substance thought to be a double salt of ammonium bromide and uranyl bromide.

5.2 Preparation of Hydrated Uranyl Bromide. Richards and Merigold (1902) prepared aqueous solutions of uranyl bromide by the old method of Berthemot (1830), wherein uranium dioxide suspended in water is heated with bromine. After excess bromine has been removed by evaporation, a solution of uranyl bromide remains which can then be concentrated to a syrup. The yield of crystals from the syrup is small, and their solubility in water and in alcohol is so great that it is almost impossible to wash them free of mother liquor.

Sendtner (1879) obtained crystals of the hydrate by dissolving hydrated uranium dioxide in aqueous hydrobromic acid. The yellow solution was concentrated to a syrup and then dried in a desiccator. The hygroscopic unstable crystals so obtained had the composition $UO_2Br_2 \cdot 7H_2O$. Repeated evaporation of uranyl acetate, first with hydrobromic acid and then with water, also yields an aqueous solution of uranyl bromide (von Unruh, 1909). Large yellow-green crystals form in a desiccator from this solution. These crystals are soluble in amyl alcohol; water of crystallization forms a separate layer, and the alcoholic solution can be decanted. The solution can be dehydrated even more completely by azeotropic distillation. Von Unrun suggested the use of ether for dehydration, but Richards and Merigold (1902) had stated that uranyl bromide reacts with ether. It is doubtful whether alcohol-free uranyl bromide can be obtained at all from alcoholic solutions.

5.3 Properties of Uranyl Bromide. Uranyl bromide is a bright-red, very hygroscopic solid that turns yellow in the presence of water vapor. It dissolves very readily in water to a yellow solution. Ethanol solutions of uranyl bromide are rather stable; in the presence of moisture and light some reduction to uranium(IV) occurs (MP Berkeley 2). Uranyl bromide, as has already been noted, is also soluble in ether and amyl alcohol.

Uranyl bromide hydrate, $UO_2Br_2 \cdot 7H_2O$, decomposes in moist air with evolution of hydrogen bromide and formation of hydrated uranium oxide. The compound is very soluble in water and exhibits reactions characteristic of UO_2^{++} ions. Aqueous solutions undergo extensive hydrolysis on boiling (de Coninck, 1902).

Uranyl bromide is thermally unstable as decomposition with slow liberation of bromine occurs even at room temperature; bromine is more rapidly evolved in a helium atmosphere at 250°C. However, even at 350°C, 48 hr is necessary for the complete decomposition. Since reverse reaction does not take place even at 720°C, the decomposition is irreversible, and its rate depends only on temperature and not on the partial pressure of bromine (MP Ames 10). Ignition of $UO_2Br_2 \cdot 7H_2O$ in the absence of air leads eventually to the formation of uranium dioxide, accompanied by evolution of bromine and hydrogen bromide. The uranium dioxide is red and appears to present one more of the many different varieties of this oxide (see Chap. 11).

Uranyl bromide, like the fluoride and chloride, readily forms double salts. The ammoniates $UO_2Br_2 \cdot 2NH_3$, $UO_2Br_2 \cdot 3NH_3$, and $UO_2Br_2 \cdot 4NH_3$ can be prepared by treatment of ethereal or ethanolic solutions of uranyl bromide with ammonia (von Unruh, 1909). $UO_2Br_2 \cdot 4NH_3$, a deeply colored orange-red compound made by treating the diammoniate with liquid ammonia, decomposes rapidly at room temperature.

Compounds of the type $M_2UO_2Br_4 \cdot 2H_2O$ have also been prepared (Sendtner, 1879). Ammonium or potassium diuranates dissolved in hydrobromic acid and evaporated to crystallization on a water bath form $(NH_4)_2UO_2Br_4 \cdot 2H_2O$ or $K_2UO_2Br_4 \cdot 2H_2O$; these compounds are large rhombic yellow crystals and are very soluble in water. These salts form only if a large excess of acid is used. If the pure salt is dissolved in water, it cannot be regenerated. Ignition of the potassium complex leads to a mixture of uranium oxide and potassium salts. The uranyl bromide double salts are more unstable than the corresponding chlorine compounds. The pyridine compound, $(C_5H_6N)_2UO_2Br_4$, has been prepared by addition of pyridine to a boiling solution of uranium trioxide in excess alcoholic hydrobromic acid. On cooling, yellow crystals of the complex are formed (Loebel, 1907).

Uranyl bromide forms a series of addition compounds with various basic organic compounds. An etherate, $UO_2Br_2 \cdot 2(C_2H_5)_2O$, has been described by von Unruh (1909). It is a yellow-green, fluorescent, crystalline substance, very hygroscopic, which decomposes rapidly in air with evolution of bromine. Răscanu (1930-1931b,c,e; 1932-1933) has examined a series of addition compounds of uranyl bromide with organic nitrogen bases. These are prepared exactly as the very similar uranyl chloride derivatives (cf. Sec. 3.3). The solubilities of the uranyl bromide complexes differ from those of the uranyl chloride compound. They are given in Table 16.10.

The xanthylium compound, $UO_2Br_2 \cdot 2C_{13}H_9OBr$ (yellow crystals), has also been prepared (Fosse and Lesage, 1906).

Table 16.10 — Complex Compounds of Uranyl Bromide with Organic Bases

Organic base	Complex	Description	Solubility
p-Nitroso dimethyl aniline	$UO_2Br_2 \cdot 2(CH_3)_2NC_6H_4NO$	Stable in air; brick-red powder	Sl. sol. in water; insol. in alcohol, ether, acetone, and chloroform
p-Nitroso diethyl aniline	$UO_2Br_2 \cdot 2(C_2H_5)_2NC_6H_4NO$	Dark brick-red powder	Same as above
Methyl acetanilide	$UO_2Br_2 \cdot 2C_6H_5N(CH_3)(C_2H_3O)$	Stable in air; shiny yellow crystals	Sol. in water and alcohol; insol. in ether
Acet-p-phenetidine	$UO_2Br_2 \cdot 4C_{10}H_{13}O_2N$	Orange-yellow; stable in air	Insol. in water; dif. sol. in alcohol, acetone, and chloroform; insol. in ether and amyl alcohol; completely sol. in amyl alcohol on heating
Antipyrine	$UO_2Br_2 \cdot 2C_{11}H_{12}N_2O$	Stable in air	Sol. in water; sl. sol. in alcohol; insol. in ether and chloroform
Bromantipyrine	$UO_2Br_2 \cdot 2C_{11}H_{11}BrN_2O$	Yellow needles	Sol. in warm H_2O or alcohol; sl. sol. in boiling acetone; insol. in ether or $CHCl_3$; sol. in HCl

6. URANIUM(IV) OXYBROMIDE, UOBr₂ (URANOUS OXYBROMIDE)

The preparation of this compound was prompted by the observation (MP Ames 9) that a yellow residue was frequently observed on sublimation of uranium tetrabromide. Its analysis indicated it to be uranous oxybromide. A method of preparation based on the following reactions has been developed at Ames (MP Ames 11):

$$3UO_2 + 2CS_2 \xrightarrow{900°C} U_3O_2S_4 + 2CO_2 \qquad (24)$$

$$U_3O_2S_4 + 6Br_2 \xrightarrow{600°C} 2UOBr_2 + UBr_4 + 2S_2Br_2 \qquad (25)$$

Bromine vapor, carried in a stream of nitrogen, is passed over $U_3O_2S_4$ at 600°C until distillation of UBr_4 and S_2Br_2 ceases. The nonvolatile uranous oxybromide is a greenish-yellow to yellow powder. Analysis shows it to correspond very closely to the composition $UOBr_2$, and x-ray studies show it to be a pure phase and not a mechanical mixture of uranium dioxide and uranium tetrabromide.

Uranium(IV) oxybromide is not particularly hygroscopic, but it dissolves readily in water to give a green solution. The solution is stable

Table 16.11 — Equilibrium Constants for Hydrogen Reduction of Uranous Oxybromide*

Temp., °C	Temp., °K	$\frac{1}{T} \times 10^3$	K
300	573	1.745	<0.004
350	623	1.605	0.0061
375	648	1.543	0.0078
400	673	1.486	0.0108
425	698	1.433	0.0139
440	713	1.403	0.0162

*$K = p_{HBr}/p_{H_2}^{1/2}$ (p in atmospheres).

fcr several hours, after which a black precipitate, presumably hydrated uranium dioxide, begins to precipitate. This precipitate is similar to that obtained when aqueous solutions of uranium tetrabromide are treated with a base. It has been suggested by the Ames workers (MP Ames 12) that stable UO^{++} ions exist in solutions of uranous oxybromide; this interpretation is supported by the fact that electrometric titration shows that four equivalents of hydroxide are required for precipitation per mole of uranium tetrabromide and only two per mole of $UOBr_2$. This point is discussed at greater length in the second portion of this volume.

Uranium(IV) oxybromide appears to be stable and nonvolatile at 600°C in an inert atmosphere; at 800°C, however, disproportionation to uranium dioxide and uranium tetrabromide occurs. U_3O_8 forms on ignition in air.

6.1 Hydrogen Reduction Equilibrium of Uranous Oxybromide. (UCRL 9.) The equilibrium of uranous oxybromide reduction by hydrogen has been studied between 300 and 400°C. Reduction proceeds with the formation of hydrogen bromide; the analysis of the data is made on the assumption that UOBr is the other product.

$$UOBr_2 \text{ (solid)} + \tfrac{1}{2}H_2 \text{ (gas)} \rightarrow UOBr \text{ (solid)} + HBr \text{ (gas)} \qquad (26)$$

The experimental equilibrium constants are given in Table 16.11.

From the data given in Table 16.11, the following values are calculated for reaction 26:

$$\Delta H_{673°K} = 10.1 \text{ kcal per mole}$$

$$\Delta F_{673°K} = 6.06 \text{ kcal per mole}$$

$$\Delta S_{673°K} = 6.01 \text{ e.u.}$$

6.2 Thermodynamic Constants. (UCRL 8.) The following thermodynamic constants have been given by MacWood for uranous oxybromide:

$$\Delta H_{298°K} = -246.9 \pm 0.7 \text{ kcal per mole}$$

$$\Delta F_{298°K} = -231.3 \pm 2.0 \text{ kcal per mole}$$

$$S_{298°K} = 42.9 \text{ e.u. (estimated)}$$

The above heat of formation is based on a calorimetric determination of the heat of solution of uranous oxybromide, which gave $\Delta H = -16.3 \pm 0.2$ kcal per mole (UCRL 5).

7. URANYL IODIDE, UO₂I₂

It is doubtful whether pure uranyl iodide has ever been prepared in the solid state. From what little is known about this substance, it appears likely that it is considerably less stable than uranyl bromide, which already has a distinct tendency to lose its halogen. Various workers have attempted to prepare uranyl iodide by reaction of iodine vapors, hydrogen iodide vapors, or hydrogen iodide with a mixture of uranyl dioxide and carbon but without success (von Unruh, 1909).

Aqueous solutions of uranyl iodide may be readily prepared, however. Von Unruh (1909) prepared such solutions by treating uranyl acetate with hydriodic acid. Concentration of such a solution by heating results in extensive decomposition. Drying in vacuum yields yellow-green fluorescent needles which decompose rapidly in air with liberation of iodine. The substance was not analyzed. Previous attempts by Sendtner (1879) to concentrate solutions of $UO_3 \cdot H_2O$ in hydriodic acid resulted in every case in the evolution of large amounts of iodine (Sendtner, 1879); and a product heavily contaminated with free iodine was obtained. It appears unlikely that aqueous solutions of uranyl iodide can be dehydrated without extensive decomposition.

Solutions of uranyl iodide in water or organic solvents can be readily prepared by double decomposition reactions. The reactants are so chosen that one of the products is uranyl iodide and the other a salt insoluble in the solvent used. Aloy (1901b) dissolved partially dehydrated uranyl nitrate hexahydrate in ether and added a slight excess of barium iodide. On removing the insoluble barium nitrate and concentrating the red solution in vacuum, an unstable, red, very deliquescent, crystalline material was obtained. A methanolic solution of uranyl nitrate treated with sodium iodide yields a solution of uranyl iodide. Von Unruh prepared nonaqueous solutions of uranyl iodide by dissolving in ether the solid material obtained by evaporation of an aqueous solution and then using calcium chloride and sodium metal to remove water and free iodine. Crystals contaminated with iodine and presumed to be $UO_2I_2 \cdot 2(C_2H_5)_2O$ can be obtained from the ethereal solution by evaporation in vacuum or in a stream of dry air. Double decomposition has also been used for preparing aqueous solutions. Thus a solution of uranyl sulfate may be treated with the equivalent amount of barium or calcium iodide (Truttwin, 1925).

It has been reported that uranyl iodide is soluble in methyl acetate, ethyl acetate, acetone, and pyridine (Naumann, 1904, 1909), as well as in water, ether, and methyl, ethyl, or amyl alcohols.

Uranyl iodide forms a series of addition compounds with ammonia (von Unruh, 1909). Uranyl iodide in ether or amyl alcohol solution gives, with gaseous ammonia, $UO_2I_2 \cdot 2NH_3$; more prolonged treatment with ammonia gives $UO_2I_2 \cdot 3NH_3$ as a golden-yellow amorphous solid. Treatment of the diammoniate with liquid ammonia at $0°C$ gives the very unstable $UO_2I_2 \cdot 4NH_3$, which decomposes rapidly above $5°C$.

Aqueous or ethanolic solutions of uranyl iodide have been alleged to possess the property of dissolving normally insoluble heavy metal iodides, such as bismuth iodide, BiI_3, or mercuric iodide, HgI_2. It has been claimed that compounds such as UO_2BiI_5 can be isolated from these solutions (Truttwin, 1925).

REFERENCES

1824 J. A. Arfvedson, Pogg. Ann., 1: 269.
1824 J. J. Berzelius, Pogg. Ann., 1: 34.
1825 L. R. Lecanu, J. pharm. chim., 11: 285.
1830 J. B. Berthemot, Ann. chim. et phys., (2) 44: 387.
1842a E. Péligot, Ann., 43: 278.
1842b E. Péligot, Ann. chim. et phys., (3) 5: 36.
1856 G. Williams, J. prakt. Chem., 69: 358.
1861 H. Hermann, dissertation, University of Göttingen, p. 31.
1866a H. C. Bolton, Z. Chem., (2) 2: 353.
1866b H. C. Bolton, Z. Chem., (2) 2: 356.
1879 H. Baker, J. Chem. Soc., 35: 760.
1879 R. Sendtner, Ann., 195: 325.
1881 A. R. Leeds, J. Am. Chem. Soc., 3: 145.
1881 C. G. Williams, Chem. News, 44: 308.
1883 A. Smithells, J. Chem. Soc., 43: 130.

1884   A. Ditte, Ann. chim. et phys., (6) 1: 341.

1885   F. F. Regelsberger, Ann., 227: 119.

1888   H. Hampe, Chem. Ztg., 12: 106.

1892   K. Seubert and A. Schmidt, Ann., 267: 239.

1895   H. L. Wells and B. B. Boltwood, Z. anorg. Chem., 10: 183.

1896   J. Aloy, Compt. rend., 122: 1541.

1897   A. Miolati and U. Alvisi, Atti accad. nazl. Lincei, (5) 6: II, 380.

1900   S. Lordkipanidze, J. Russ. Phys. Chem. Soc., 32: 286.

1901a   J. Aloy, Bull. soc. chim. Paris, (3) 25: 154.

1901b   J. Aloy, Ann. chim. et phys., (7) 24: 417.

1901   F. Mylius and R. Dietz, Ber., 34: 2774.

1902   W. O. de Coninck, Bull. classe sci. Acad. roy. Belg., 1025.

1902   B. Kalischer, dissertation, University of Berlin, p. 49.

1902   T. W. Richards and B. S. Merigold, Chem. News, 85: 187; Z. anorg. Chem., 31: 244.

1903   W. O. de Coninck, Bull. classe sci. Acad. roy. Belg., 709.

1904   H. Bürger, dissertation, University of Bonn, p. 49.

1904   W. O. de Coninck, Ann. chim. et phys., (8) 3: 500.

1904   F. Giolitti, Gazz. chim. ital., 34: II, 166.

1904   A. Naumann, Ber., 37: 3601, 4328, 4609.

1904   E. Rimbach, Ber., 37: 461.

1905   F. Giolitti and G. Agamennone, Atti accad. nazl. Lincei, (5) 14: I, 114, 165.

1905   M. Trautz, Z. physik. Chem., 53: 48.

1906   R. Fosse and L. Lesage, Compt. rend., 142: 1544.

1906   H. Grossmann and B. Schuck, Z. anorg. Chem., 50: 26.

1906   A. Rosenheim and F. Jacobsohn, Z. anorg. Chem., 50: 306.

1907   H. Loebel, dissertation, University of Berlin, p. 42.

1909a   W. O. de Coninck, Compt. rend., 148: 1769.

1909b   W. O. de Coninck, Bull. classe sci. Acad. roy. Belg., p. 836.

1909c   W. O. de Coninck, Bull. classe sci. Acad. roy. Belg., p. 173.

1909   A. Naumann, Ber., 42: 3790.

1909   W. Peters, Ber., 42: 4833.

1909   A. von Unruh, dissertation, University of Rostock, p. 37.

1910   A. Bolland, Monatsh., 31: 387, 403.

1912   W. Peters, Z. anorg. Chem., 78: 159.

1915   H. Simonis and A. Elias, Ber., 48: 1513.

1916   H. Simonis and A. Elias, Ber., 49: 777.

1917   A. Benrath, Z. wiss. Phot., 16: 258.

1919   E. L. Nichols and H. L. Howes, Carnegie Inst. Wash. Pub., 298: 218.

1922   J. Aloy and E. Rodier, Bull. soc. chim. France, (4) 31: 247.

1925   H. Truttwin, German Patent D.R.P. No. 420,391, from A. Brauer and J. D'Ans, Fortschritte in der anorganisch-chemischen Industrie, III: 1327 (1924-1927).

1930   T. Asahina and T. Dôno, Z. physiol. Chem., 186: 134.

1930   F. Olsson, Z. anorg. u. allgem. Chem., 187: 112.

1930-

1931a   R. Răscanu, Ann. sci. univ. Jassy, 16: 461.

1931b   R. Răscanu, Ann. sci. univ. Jassy, 16: 480, 481, 484, 486, 489.

1931c   R. Răscanu, Ann. sci. univ. Jassy, 16: 54.

1931d   R. Răscanu, Ann. sci. univ. Jassy, 16: 465, 466, 467, 470, 473.

1931e   R. Răscanu, Ann. sci. univ. Jassy, 16: 54, 496.

1931-

1932a   R. Răscanu, Ann. sci. univ. Jassy, 17: 133.

1932b   R. Răscanu, Ann. sci. univ. Jassy, 17: 143.

1932c   R. Răscanu, Ann. sci. univ. Jassy, 17: 75.

1932-

1933   R. Răscanu, Ann. sci. univ. Jassy, 18: 94.
1936   P. Spacu, Z. anorg. u. allgem. Chem., 230: 181.
1938   A. Chrétien and G. Oechsel, Compt. rend., 206: 254-256.
1947   P. F. Wacker and R. K. Cheney, J. Research Natl. Bur. Standards, 39: 317.
1948   W. H. Zachariasen, Acta Crystallographica, 1: 277.
1949   G. H. Dieke and A. B. F. Duncan, "Spectroscopic Properties of Uranium Compounds," National Nuclear Energy Series, Division III, Volume 2, McGraw-Hill Book Company, Inc., New York, 1949.

Project Literature

British 1:   H. S. Peiser and T. C. Alcock, Report BR-589, Mar. 19, 1945.
British 2:   Report BR-442, date received, July 6, 1944.
British 3:   Report [B]LRG-21, June, 1943.
British 4:   Report [B]LRG-22, July, 1943.

Brown 1:   C. A. Kraus, Report CC-1717, July 29, 1944.
Brown 2:   C. A. Kraus, Report A-256, Aug. 14, 1942.
Brown 3:   C. A. Kraus, Report A-360, Oct. 26, 1942, p. 9.
Brown 4:   C. A. Kraus, Report N-1789, July 1, 1943.
Brown 5:   C. A. Kraus, Report CC-342, Nov. 15, 1942.

CEW-TEC 1:   J. W. Gates, L. J. Andrews, and R. B. Pitt, Report CD-4001, Jan. 30, 1945.
CEW-TEC 2:   J. W. Gates, G. H. Clewett, L. J. Andrews, and H. A. Young, Report CD-457, Aug. 4, 1944.
CEW-TEC 3:   C. Sterett, I. Halpern, J. Van Wazer, and H. A. Young, Report CD-GS-28, Mar. 16, 1945.

Du Pont 1:   C. E. Holbrook, Report N-25, Apr. 17, 1943.
Du Pont 2:   I. W. Dobratz, R. C. Downing, C. W. Maynard, S. B. Smith, and L. Spiegler, Report N-33, May 13, 1943.

Johns Hopkins 1:   W. B. Burford III, Report [A]M-2121, February, 1945.

MP Ames   1:   W. D. Cline, Report CC-1983, Dec. 9, 1944, p. 8.
MP Ames   2:   F. Vaslow and A. S. Newton, Report CK-1498, June 17, 1944, p. 10.
MP Ames   3:   N. C. Baenziger and A. D. Tevebaugh, Report CC-1504, Aug. 10, 1944.
MP Ames   4:   O. Johnson, Report CC-1500, June 17, 1944.
MP Ames   5:   O. Johnson, Report CC-1781, Oct. 18, 1944.
MP Ames   6:   N. C. Baenziger and R. E. Rundle, Reports CC-1504, Aug. 10, 1944, and CC-1778, Aug. 18, 1944.
MP Ames   7:   N. C. Baenziger, Report CC-1778, Aug. 18, 1944.
MP Ames   8:   J. Powell and R. W. Nottorf, Report CC-1500, June 17, 1944; J. Powell, Report CC-1504, Aug. 10, 1944.
MP Ames   9:   J. Powell and R. W. Nottorf, Report CC-1496, May 11, 1944.
MP Ames 10:   R. W. Nottorf, Report CC-1778, Aug. 18, 1944.
MP Ames 11:   J. Powell, Report CC-1778, Aug. 18, 1944.
MP Ames 12:   J. Powell and A. S. Newton, Report CC-1778, Aug. 18, 1944.
MP Ames 13:   F. H. Spedding and H. A. Wilhelm, Report CC-1517, Mar. 10, 1944.

MP Berkeley 1:   C. S. Garner and J. W. Kent, Report CN-343, Nov. 15, 1942.
MP Berkeley 2:   B. J. Fontana, Report CK-810, July 20, 1943.

MP Chicago 1: H. S. Brown, O. F. Hill, and A. H. Jaffey, Report CN-343, Nov. 15, 1942.

MP Chicago 2: S. Fried and N. R. Davidson, Report N-1722, Nov. 14, 1944.

MP Chicago 3: W. H. Zachariasen, Reports CP-1249, Jan. 24, 1944; CK-1367, Feb. 15, 1944; and CC-2768, Mar. 12, 1945.

MP Chicago 4: W. H. Zachariasen, Report CK-2737, Mar. 5, 1945, p. 12.

MP Chicago 5: R. Livingston, personal communication.

Montreal 1: J. Hebert and A. G. Maddock, Report MC-43, Mar. 13, 1944.

Natl. Bur. Standards 1: P. F. Wacker and R. K. Cheney, Report MDDC-1096, July 9, 1947.

Purdue 1: E. T. McBee, Report [A]M-2102, Mar. 21, 1945.

Purdue 2: E. T. McBee, Report [A]M-2103, Apr. 21, 1945.

SAM Carbide and Carbon 1: R. Kunin, Report A-3255, May 8, 1945.

SAM Columbia 1: C. F. Hiskey, Report N-673, Feb. 17, 1944.

SAM Columbia 2: M. S. Katz, Report CC-1744, June 17, 1944.

SAM Columbia 3: G. R. Dean, R. Bradt, and M. S. Katz, Report CC-1382, Mar. 6, 1944.

SAM Columbia 4: S. Freed, Report [A]M-144, Sept. 2, 1943.

SAM Columbia 5: J. Spevack and A. V. Grosse, Report A-38, Aug. 21, 1941.

UCRL 1: S. Kilner and M. J. Polissar, Report RL-4.6.42, Nov. 7, 1942.

UCRL 2: S. H. Babcock, Report RL-4.6.57, Jan. 9, 1943.

UCRL 3: S. Rosenfeld, Report RL-4.6.84, Feb. 20, 1943; H. G. Reiber, Report RL-4.6.70, Mar. 13, 1943.

UCRL 4: P. H. Davidson and M. E. Mueller, Report RL-4.6.259, Feb. 26, 1944.

UCRL 5: C. H. Barkelew, Report RL-4.6.937, Oct. 11, 1945.

UCRL 6: P. H. Davidson and I. Streeter, Report RL-4.6.920, June 5, 1945.

UCRL 7: N. W. Gregory, Report RL-4.6.938, Oct. 13, 1945.

UCRL 8: G. E. MacWood, Report RL-4.7.602, Dec. 18, 1945.

UCRL 9: N. W. Gregory, Report RL-4.6.940, Dec. 13, 1945.

INDEX

ON MATHEMATICS AND MATHEMATICIANS (MEMORABILIA MATHEMATICA)

by Robert Edouard Moritz

To Robert Edouard Moritz this work was a ten-year labor of love, and it is a tribute to his discerning and discriminating eye that this selection of passages should remain one of the most stimulating works about mathematics ever published. It was the first collection of its kind in English, and has rarely been equalled in its ability to convey a sense of the full range of mathematics, its enormous accomplishments, and the living personalities of the great mathematicians. To mathematicians it will be a constant source of pleasure, inspiration, and encouragement. To teachers of mathematics and writers about mathematics, it will remain of inestimable value as a source of quotations and ideas. To the layman, it will be a revolution. It should dispel forever the narrow notion that mathematics is a cut-and-dried affair, isolated from other compartments of life and thought.

The more than eleven-hundred fully annotated selections in this book, gathered from the works of over three-hundred authors, cover a vast range of subjects pertaining to mathematics. Grouped in twenty-one chapters, they deal with such topics as the definitions and objects of mathematics; mathematics as a language or as a fine art; the relationship of mathematics to philosophy, to logic, or to science; the nature of mathematics; and the value of mathematics. Other sections contain passages referring to specific subjects in the field, such as arithmetic, algebra, geometry, calculus, and modern mathematics.

Of special interest to nearly everyone interested in mathematics will be the extensive amount of material on the great mathematicians. Over and above the selections by these great men which appear in virtually every section of the book, three sections (chapters VIII, IX, X) are devoted explicitly to the mathematician as a person, and provide irreplaceable glimpses into the lives and personalities of mathematical giants from Archimedes to Euler to Gauss to Klein to Wierstrass.

Formerly entitled "Memorabilia Mathematica." 1140 selected passages. Extensive cross-index. 410pp. 5⅜ x 8.

Paperbound **$1.95**

HIGHER MATHEMATICS FOR STUDENTS OF CHEMISTRY AND PHYSICS

by J. W. Mellor

This standard text, which has aided generations of physicists and chemists, has not been superseded as a practical introduction to higher mathematics. Practical in approach rather than abstract. it builds its many examples and exercises out of familiar material from actual laboratory situations, and presupposes only a working knowledge of elementary algebra and the meaning of a few trigonometric formulas.

Beginning with differential calculus, the author proceeds through analytical geometry, infinite series, probability, determinnts and, similar topics. Treatment is full and detailed, and presentation is exceptionally clear.

CONTENTS: Differential calculus. Coordinate or analytical geometry. Functions with singular properties. Integral calculus. Infinite series and their uses. How to solve numerical equations. How to solve differential equations. Fourier's theorem. Probability and the theory of errors. Calculus of variations. Determinants. Appendix I: Collection of formulae and tables of reference. Appendix II: Reference tables.

"An excellent reference work. If the reader is not familiar with this book . . . it will repay him to examine it," CHEMICAL AND ENGINEERING NEWS. "Has served generations of physicists, chemists and astronomers . . . a classic," ASTROPHYSICAL JOURNAL. "An eminently readable and thoroughly practical treatise," NATURE (London).

Unabridged republication of 4th edition. Prefatory note by Donald G. Miller. Index. 800 problems. 189 figures. xxi + 641pp. 5⅜ x 8.

S193 Paperbound $2.00

THE DYNAMICAL THEORY OF GASES
By Sir James Jeans

This famous book marks an important turning-point in applied mathematics. Written by one of the great scientific expositors of our time, it introduced the notion of dissipation in kinetic theory, an innovation that has had wide application in mathematical physics and engineering.

The book is divided into mathematical and physical chapters, a convenient arrangement that allows the non-mathematical reader to get the gist of Jeans' account without having to struggle with equations intended only for graduate students in mathematics. The chief difference between this fourth edition (originally published in 1925) and the ones that preceded it is in the emphasis given to the quantum theory. Thus the chapter on Radiation and Quantum Theory that terminated the original edition has been partially rewritten. And to this has been added a new chapter on Quantum Dynamics dealing chiefly with the work of Ehrenfest, Sommerfeld, and Epstein.

The Dover edition of **The Dynamical Theory of Gases** is unabridged. it contains 444 pages and measures 6 1/8 x 9 1/4.

S136 Paperbound $2.95

FADS AND FALLACIES
by Martin Gardner

This witty and engaging book examines the various fads, fallacies, strange cults and curious panaceas which at one time or another have masqueraded as science. Not just a collection of anecdotes but a fair, reasoned appraisal of eccentric theory, FADS AND FALLACIES is unique in recognizing the scientific, philosophic and sociological-psychological implications of the wave of pseudoscientific theories which periodically besets the world.

In this second revised edition of a work formerly titled IN THE NAME OF SCIENCE, Martin Gardner has added new, up-to-date material to an already impressive account of hundreds of systematized vagaries. Here you will find discussions of hollow-earth fanatics like Symmes; Velikovsky and wandering planets; Hoerbiger, Bellamy and the theory of multiple moons; Charles Fort and the Fortean Society; dowsing and the other strange methods for finding water, ores and oil. Also covered are such topics as naturopathy, iridiagnosis, zone therapy, food fads; William Reich and orgone sex energy; L. Ron Hubbard and Dianetics; A. Korzybsky and General Semantics. A brand-new examination of Bridey Murphy is included in this edition, along with a new section of bibliographic reference material.

"Although we are amused, we may also be embarrassed to find our friends or even ourselves among the gullible advocates of plausible-sounding doubletalk," SATURDAY REVIEW. "A very able and even-tempered presentation," NEW YORKER.

Second revised edition. 5⅜ x 8.

T394 Paperbound $1.50

THE PHASE RULE
by Alexander Findlay, A. N. Campbell & N. O. Smith

Chemical phenomena of one, two, three, four, and multiple component systems are covered in this completely revised standard text. Clearly written and well diagrammed, THE PHASE RULE will prove especially useful to graduate students in the field.

Brought up to date by Professors A. N. Campbell and N. O. Smith, this volume contains brand new material on such important topics as the treatment of binary and tertiary liquid vapor equilibria, solid solutions in ternary systems, phase changes in cooling of ternary salts, quinary systems of salts and water, salting out, soap systems, and fuming liquids. The revisers have followed modern trends in changing the emphasis in ternary systems from rectangular to triangular coordinates. They have clarified the graphic representation of quarternary reciprocal salt systems; and placed emphasis on the solid model as a means of coordinating p-t, t-x and p-x sections, and on the distinction between sections and projections.

"Recommended with confidence," SCIENCE PROGRESS. "Of particular interest to students of metallurgy, geology, and ceramics, and to the chemist and chemical engineer concerned with phase equilibria," CHEMICAL AND ENGINEERING NEWS. "Should rank as the standard work in English on the subject," NATURE [London].

9th edition. Author, subject indices. 2 appendices. 236 figures, including 14 solid models. 505 footnotes, most bibliographical. xii + 494pp. 5 3/8 x 8.

S92 Paperbound $2.45

THE STORY OF ATOMIC THEORY AND ATOMIC ENERGY
By J. G. Feinberg

In this book nuclear energy is not treated merely as an isolated phenomenon: an Excalibur suddenly spring entire from the lake of human learning. Instead, it is placed in its proper context, as a science rising from and contributing to the long evolution of man's other sciences and his social and moral fabric. The book gives you a wider range of facts on nuclear energy and its cultural implications than any other similar source. Taking you back 600 years before Christ, the author begins by having you sit in at the very birth of recorded atomic theory. Then in a lively, informal manner, he guides you easily through the most involved aspects of atomic theory and energy, pointing out the industrial and political meaning of this new force. You need absolutely no mathematics to follow the exposition.

You see the concept of the atom evolve through the work of Dalton, Avogadro, Mendelejeff, Rutherford, Bohr, and others. You understand the significance of the Wilson Cloud Chamber, Einstein's statement that the destruction of mass results in energy, and such basic concepts as "breeding," fission and fusion, radioactive decay, and binding energy. The author relates dramatically the exciting, dangerous race toward the atom bomb, showing the hurried but often painfully slow attempts to overcome each obstacle. The first grant of $6000 in 1940. The first chain reaction in 1942. The need for pure uranium-235 in quantity. You watch 1945 approach, and you read an eye-witness report of the first atomic explosion.

This book discusses radiation danger, future benefits of atomic power, the effects of a nuclear explosion on London and New York (with maps), the production of direct and indirect power, and provides a look at the accidental atomic "fire" at Windscale reactor, including excerpts from the hourly bulletins. A final chapter, written especially for this edition, brings the story up to 1959.

"Deserves special mention . . . not only authoritative but thoroughly popular in the best sense of the word," Saturday Review.

Formerly "The Atom Story." Expanded with new chapter. 3 appendices. Index. 34 illustrations. vii + 243pp. 5-3/8 x 8.   T625 Paperbound $1.45

FOUNDATIONS OF NUCLEAR PHYSICS
edited by R. T. Beyer

Here, in a single volume, are thirteen fundamental studies in nuclear physics, reproduced in exact facsimile, in the language of their authors. Selected from rare scientific journals often unobtainable except in the largest libraries, these are the thirteen papers you have seen most often cited in footnotes and bibliographies. These are the articles which have introduced momentous discoveries, hypotheses, and inventions in nuclear physics.

Nine Nobel Laureates are included in these papers:

*C. D. Anderson	*F. Joliot
*J. Chadwick	*E. O. Lawrence
*J. D. Cockcroft	M. S. Livingston
*Irene Curie	E. Rutherford
*Enrico Fermi	O. Stern
R. Frisch	F. Strassman
George Gamow	E. T. S. Walton
*O. Hahn	*Hideki Yukawa

This volume also contains the most comprehensive bibliographic guide ever assembled on nuclear physics. 122 double-columned pages long, is classifies over 4,000 articles and books under 12 subject-headings, giving authors, titles, journals, dates, pagination. It also lists 182 journals that publish material on nuclear physics.

Edited by Robert T. Beyer, Brown University, 57 figures, 288pp. 6 ⅛ x 9 ¼.

S19 Paperbound $1.75

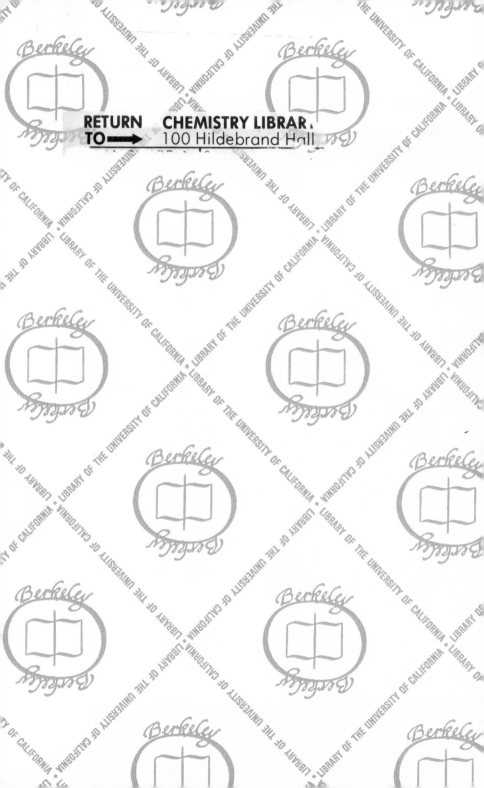